Hydrosystems Engineering and Management

McGraw-Hill Series in Water Resources and Environmental Engineering

Consulting Editors

Paul H. King
Rolf Eliasen, Emeritus

Bailey and Ollis: *Biochemical Engineering Fundamentals*
Bishop: *Marine Pollution and Its Control*
Bouwer: *Groundwater Hydrology*
Canter: *Environmental Impact Assessment*
Chanlett: *Environmental Protection*
Chow, Maidment, and Mays: *Applied Hydrology*
Davis and Cornwell: *Introduction to Environmental Engineering*
Eckenfelder: *Industrial Water Pollution Control*
Linsley and Franzini: *Water Resources Engineering*
Mays and Tung: *Hydrosystems Engineering and Management*
Metcalf & Eddy, Inc.: *Wastewater Engineering: Collection and Pumping of Wastewater*
Metcalf & Eddy, Inc.: *Wastewater Engineering: Treatment, Disposal, Reuse*
McGhee: *Water Supply and Sewerage*
Peavy, Row, and Tchobanoglous: *Environmental Engineering*
Sawyer and McCarty: *Chemistry for Environmental Engineering*
Tchobanoglous, Theisen, and Eliassen: *Solid Wastes: Engineering Principles and Management*

Also Available from McGraw-Hill

Schaum's Outline Series in Engineering

Most outlines include basic theory, definitions, and hundreds of solved problems and supplementary problems with answers.

Titles on the Current List Include:

Acoustics
Advanced Structural Analysis
Basic Equations of Engineering Science
Computer Graphics
Continuum Mechanics
Descriptive Geometry
Dynamic Structural Analysis
Engineering Economics
Engineering Mechanics, 4th edition
Fluid Dynamics, 2d edition
Fluid Mechanics & Hydraulics, 2d edition
Heat Transfer
Introduction to Engineering Calculations
Introductory Surveying
Lagrangian Dynamics
Machine Design
Mathematical Handbook of Formulas & Tables
Mechanical Vibrations
Operations Research
Programming with C
Programming with Fortran
Programming with Pascal
Reinforced Concrete Design, 2d edition
Space Structural Analysis
State Space & Linear Systems
Statistics & Mechanics of Materials
Statics & Strength of Materials
Strength of Materials, 2d edition
Structural Analysis
Structural Steel Design, LFRD Method
Theoretical Mechanics
Thermodynamics, 2d edition
Vector Analysis

HYDROSYSTEMS ENGINEERING AND MANAGEMENT

Larry W. Mays

Professor and Chair
Department of Civil Engineering
Arizona State University

Yeou-Koung Tung

Associate Professor, Department of Statistics
Hydrologist, Wyoming Water Research Center
University of Wyoming

McGraw-Hill, Inc.

New York St. Louis San Francisco Auckland Bogotá
Caracas Lisbon London Madrid Mexico Milan Montreal
New Delhi Paris San Juan Singapore Sydney Tokyo Toronto

This book was set in Times Roman by Electronic Technical Publishing Services.
The editors were B. J. Clark and John M. Morriss;
the production supervisor was Louise Karam.
The cover was designed by Joseph Gillians.
Project supervision was done by Electronic Technical Publishing Services.
R. R. Donnelley & Sons Company was printer and binder.

HYDROSYSTEMS ENGINEERING AND MANAGEMENT

1 2 3 4 5 6 7 8 9 0 DOC DOC 9 0 9 8 7 6 5 4 3 2 1

ISBN 0-07-041146-8

Library of Congress Cataloging-in-Publication Data

Mays, Larry W.
Hydrosystems engineering and management / Larry W. Mays and Yeou-Koung Tung.
p. cm. — (McGraw-Hill series in water resources and environmental engineering)
Includes index.
ISBN 0-07-041146-8
1. Water-supply engineering—Mathematical models. 2. Water-supply—Management—Mathematical models. 3. Operations research. 4. Reliability (Engineering). I. Tung, Yeou-Koung. II. Title. III. Series.
TD353.M39 1992
628.1—dc20 91-24133

ABOUT THE AUTHORS

Larry W. Mays is a Professor and Chair of the Department of Civil Engineering at Arizona State University. Prior to August 1989 he was a professor of Civil Engineering and holder of an Engineering Foundation Endowed Professorship at the University of Texas at Austin where he was on the faculty since 1976. He also served as Director of the Center for Research in Water Resources at Texas. He was a graduate research assistant and then Visiting Research Assistant Professor at the University of Illinois at Urbana-Champaign where he received his Ph.D. in 1976. He received the B.S. (1970) and M.S. (1971) degrees from the University of Missouri at Rolla, after which he served in the U.S. Army stationed at the Lawrence Livermore Laboratory in California. Dr. Mays has been very active in research and teaching throughout his academic career in the areas of hydrology, hydraulics, and water resource systems analysis. He is the co-author of the book *Applied Hydrology*, published by McGraw-Hill and editor and the major contributing author of the book *Reliability Analysis of Water Distribution Systems*, published by the American Society of Civil Engineers. In addition he has served as a consultant to various government agencies and industries. He is a registered engineer in eight states, a registered professional hydrologist and has been active in committees with the American Society of Civil Engineers and other organizations.

Yeou-Koung Tung is an Associate Professor of Statistics and a statistical hydrologist with the Wyoming Water Research Center at the University of Wyoming where he has been since 1985. Prior to that time he was an Assistant Professor of Civil Engineering at the University of Nevada at Reno. He obtained the B.S. (1976) in civil engineering from Tamkang College in the Republic of China, and the M.S. (1978) and Ph.D. (1980) from the University of Texas at Austin. His teaching has been in a wide range of topics including hydrology, water resources, operations research, and probability and statistics. He has published extensively in the areas of application of operations research and probability and statistics in solving hydrologic and water resource problems. In 1987 he received the American Society of Civil Engineers Collingwood Prize. He has been very active with the committees in the American Society of Civil Engineers.

This book is dedicated to the American Indian.

CONTENTS

Part 2 Water Supply Engineering and Management

PREFACE

Hydrosystems is a term originally coined by V. T. Chow to collectively describe the technical areas of hydrology, hydraulics and water resources. Hydrosystems has also been a term used for reference to types of water projects including surface water storage systems, groundwater systems, distribution systems, flood control systems and drainage systems. Hydrosystems, as used in this book, actually applies to both definitions. The intent of this book is to provide a systematic framework for hydrosystems modeling in engineering and management. The book has three major parts: Principles, Water Supply Engineering and Management, and Water Excess Engineering and Management. The part on principles includes chapters on economics, optimization, and probability and reliability analysis to provide the necessary background for the water supply and water excess parts. The premise is that water problems can be subdivided into water supply engineering and management and water excess engineering and management. Water supply management includes chapters on demand forecasting, surface water systems, groundwater systems, and distribution systems, providing a well-rounded treatment of the various topics related to water supply. In a similar manner, the water excess management includes chapters on hydrology and hydraulics, urban drainage design and analysis, flood control systems, and floodplain management. These are topics which typically have not been covered in detail in previous books on water resource systems.

First and foremost, this book is intended to be a textbook for students of water resources engineering and management. This book is not intended to be a review of the literature, but instead is an introduction to methods used in hydrosystems for upper level undergraduate and graduate students. The material can be presented to students with no background in operations research and with only an undergraduate background in hydrology and hydraulics. A major focus is to bring together the use of economics, operations research, probability and statistics with the use of hydrology, hydraulics and water resources for the analysis, design, operation, and management of various types of water projects. This book should also serve as an excellent reference for engineers, water resource planners, water resource systems analysts, and water managers.

Hydrosystems Engineering and Management is a book concerned with the mathematical modeling of problems in water project design, analysis, operation, and management. The quantitative methods include: (a) the simulation of various hydrologic and hydraulic processes; and (b) the use of operations research, probability and statistics, and economics. Typically, these methods have been presented in course work using separate books and have never been integrated in a systematic framework in one single book. An extensive number of example problems are presented for ease in understanding the material. In addition, a large number of end-of-chapter problems are provided for use in homework assignments.

The emphasis in this book is on water quantity management and does not cover water quality management. To cover both water quantity and water quality management in one text would be a voluminous effort. Through the use of a simple hypothetical example of a manufacturing waste treatment example the concepts of linear programming are introduced in Chapter 3. This same example is used in Section 4.6 to illustrate the generalized reduced gradient method of nonlinear programming and once again in Section 5.8 to introduce the concepts of chance-constrained programming.

As expected any book reflects the author's personal perception of the subject evolved over many years of teaching, research, and professional experience. *Hydrosystems Engineering and Management* is aimed at a unified, numerical, rigorous, but practical approach to the subject. The subject matter is analytical in nature introducing the concepts of operations research and reliability analysis into the field of hydrosystems engineering and management. The use of operations research and reliability analysis in hydrosystems engineering has lagged behind the use in other fields of engineering. There are many reasons for this, discussion of which is beyond the purpose of this preface. We have written this book to encourage the use of operations research techniques and the concepts of risk and reliability analysis in the modeling of hydrosystems for engineering and management. The GAMS and GAMS-MINOS software packages for linear and nonlinear programming, respectively, are used as the major vehicles for solving the optimization problems. Hopefully through the study of this book the student will gain an appreciation for the use of optimization techniques and reliability analysis in solving hydrosystems problems.

The manuscript of the book was initially drafted while the first author was at the University of Texas at Austin. This manuscript was used in two graduate courses, a first course in water resources planning and management and a second course in water resource systems analysis. More recent versions of the manuscript were used in two water resource systems analysis courses at Arizona State University. The initial versions of the manuscript were prepared solely for graduate students. Realizing the need to be able to present this material to undergraduates, the later versions included more fundamental material that presented more details of topics such as linear programming, dynamic programming, probability analysis, microeconomics, regression analysis, and econometrics. Also some of the hydrologic and hydraulic fundamentals for groundwater systems, surface water systems, water distribution systems, storm drainage and detention systems, and flood control systems are presented in the final book for the advantage of the undergraduate student.

Hydrosystems Engineering and Management could be used for different types of courses at both the undergraduate and graduate levels. One type of course would be for students with limited background in hydrology and hydraulics and no background in operations research. Such a course could be based upon material in Chapters 1, 2, 3, Sections 5.1–5.7, 6.1–6.3, 7.1–7.6, 8.1–8.3, and 9.1–9.5. For students in a second course, material in Chapter 4, 5.8, 6.4–6.6, 7.7–7.8, 8.4–8.5, 9.5–9.8, Chapters 10, 11, 12, and 13 could be covered. There are many different courses that could be taught from this book at the undergraduate and graduate level

We express our sincere thanks to Jan Hausman and Jody Lester at the University of Texas and Judy Polingyumtewa and Sharyl Hayes at Arizona State University who helped us prepare the manuscript. We also want to acknowledge the assistance provided to us by reviewers of the manuscript including: Nathan Buras, University of Arizona; Paul Chan, New Jersey Institute of Technology; Howard Chang, San Diego State University; Neil Grigg, Colorado State University; Ben Hobbs, Case Western Reserve University; Kevin Lansey, University of Arizona; Ben Chi Yen, University of Illinois at Urbana–Champaign; and our many colleagues and students at the University of Texas at Austin, the University of Wyoming, and Arizona State University.

This book is intended to be a contribution toward the eventual goal of better engineering and management practice in the hydrosystems field. Because of this book there have been many uncaught trout in the streams and lakes in Wyoming and Arizona. We dedicate this book to the American Indian and to humanity and human welfare.

Larry W. Mays
Yeou-Koung Tung

Hydrosystems Engineering and Management

PART I

PRINCIPLES

CHAPTER 1

INTRODUCTION

1.1 BACKGROUND

Hydrosystems is a term originally coined by V. T. Chow to describe collectively the technical areas of hydrology, hydraulics, and water resources including the application of economics, optimization, probability, statistics, and management. Hydrosystems has also been a term used for reference to types of water projects including surface water storage systems, groundwater systems, distribution systems, flood control systems, urban drainage systems, etc. Hydrosystems, as used in this book, actually applies to both definitions. The premise is that hydrosystems engineering and management can be subdivided into (1) water supply engineering and management; and (2) water excess engineering and management. Modern multipurpose water projects are designed for water-supply management and/or water-excess mangement.

Water is a renewable resource that naturally (without human influence) follows the **hydrologic cycle** as shown in Fig. 1.1.1. The hydrologic cycle is basically a continuous process with no beginning or end which can be represented as a system. A **system** is a set of interconnected parts or components that form a whole. The components that form the hydrologic-cycle system are precipitation, evaporation, runoff, and other phases. The **global hydrologic cycle** represented as a system can be divided into three basic subsystems: the **atmospheric water system** containing the precipitation, evaporation, interception, and transpiration processes; the **surface water system** containing the overland flow, surface runoff, subsurface and groundwater outflow, and runoff to streams and the ocean processes; and the **subsurface water system** containing the processes of infiltration, groundwater recharge, subsurface flow, and

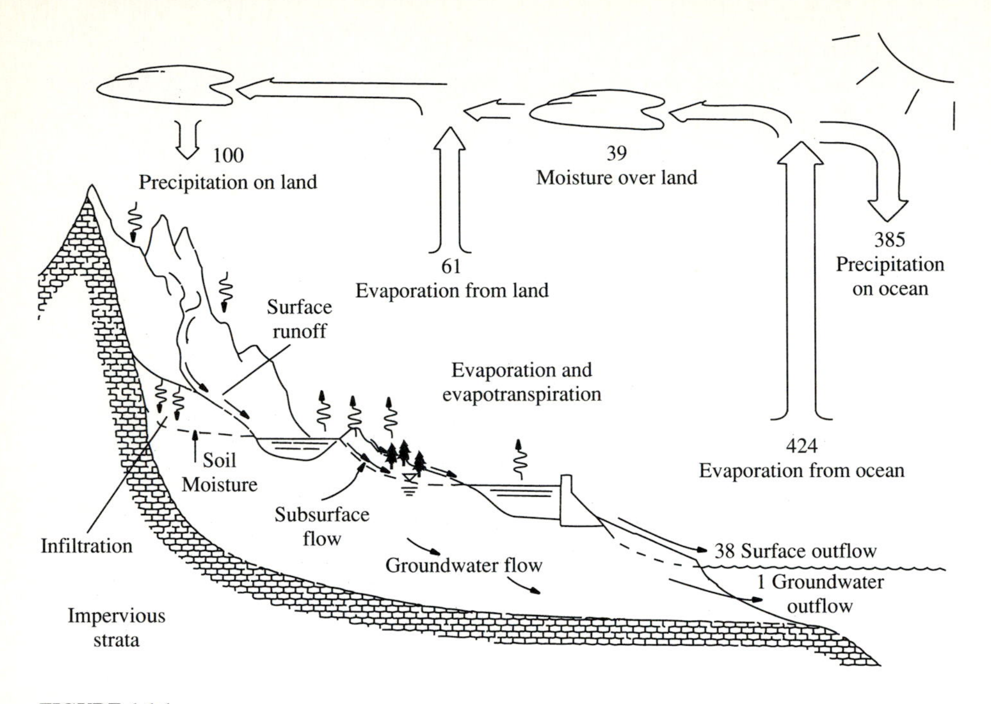

FIGURE 1.1.1
Hydrologic cycle with global annual average water balance given in units relative to a value of 100 for the rate of precipitation on land. (Chow, Maidment, and Mays, 1988)

groundwater flow. This division of subsystems within the global hydrologic system are shown in Fig. 1.1.2.

1.2 DESCRIPTIONS OF HYDROSYSTEMS

Hydrosystems include surface water supply systems, groundwater supply systems, water distribution systems, urban drainage systems, floodplain management systems, and others. This section will present examples of various types of hydrosystems.

SURFACE WATER SUPPLY SYSTEMS. An example of a large comprehensive surface water supply system is the California State Water Project (SWP), which consists of a series of reservoirs linked by rivers, pumping plants, canals, tunnels, and generating plants, as shown in Fig. 1.2.1. In northern California the Feather River flows into Lake Oroville, the SWP principal reservoir. Releases from Lake Oroville flow through the Feather and Sacremento Rivers to the Delta. In the southern part of the Delta, the Harvey Banks pumping plant lifts water into the Bethany Reservoir where water is distributed to SWP's South Bay Aqueduct and to the Governor Edmund G. Brown California Aqueduct, the principal conveyance feature of the SWP. The California Aqueduct is located along the west side of the San Joaquin Valley and flows into the San Luis Reservoir about 100 miles downstream. The aqueduct continues south and

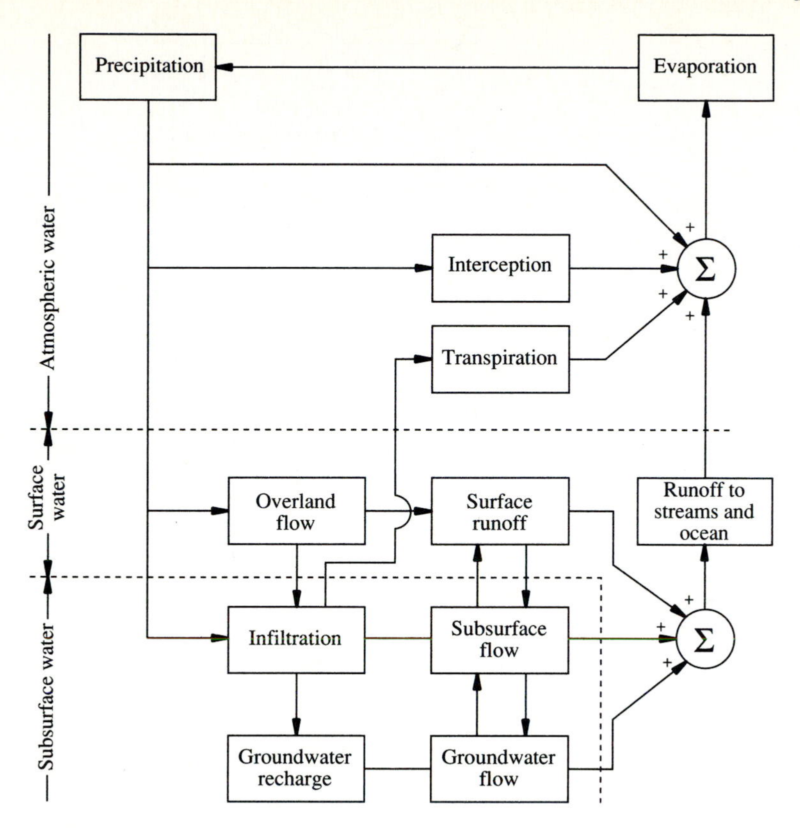

FIGURE 1.1.2
Block-diagram representation of the global hydrologic system. (Chow, Maidment, and Mays, 1988)

water is raised 969 feet by four pumping plants (Dos Amigos, Buena Vista, Wheeler Ridge, and Wind Gap) prior to reaching the Tehachapi Mountains. The Edmonston Pumping Plant lifts the water 1,926 feet to a series of tunnels and siphons. After the Tehachapi Crossing, water enters the East Branch or the West Branch. The East Branch conveys water to Lake Silverwood where it enters the San Bernardino Tunnel and drops 1,418 feet through the Devil Canyon Generating Plant. A pipeline then conveys water to Lake Perris at the southern end of the SWP, 444 miles from the Delta. The West Branch conveys water through the Warne Generating Plant into Pyramid Lake and through Castaic Pumping/Generating Plant into Castaic Lake, at the end of the West Branch. Both simulation and optimization techniques are used by the California Department of Water Resources for the monthly operation of the five main SWP reservoirs (Coe and Rankin, 1989).

URBAN WATER SYSTEMS. On a much smaller scale the urban water system is another example of a hydrosystem. As shown in Fig. 1.2.2 urban water systems include various interfaces with the hydrologic cycle through water supply, discharging wastewater and drainage flows into receiving streams, and conveying of storm and

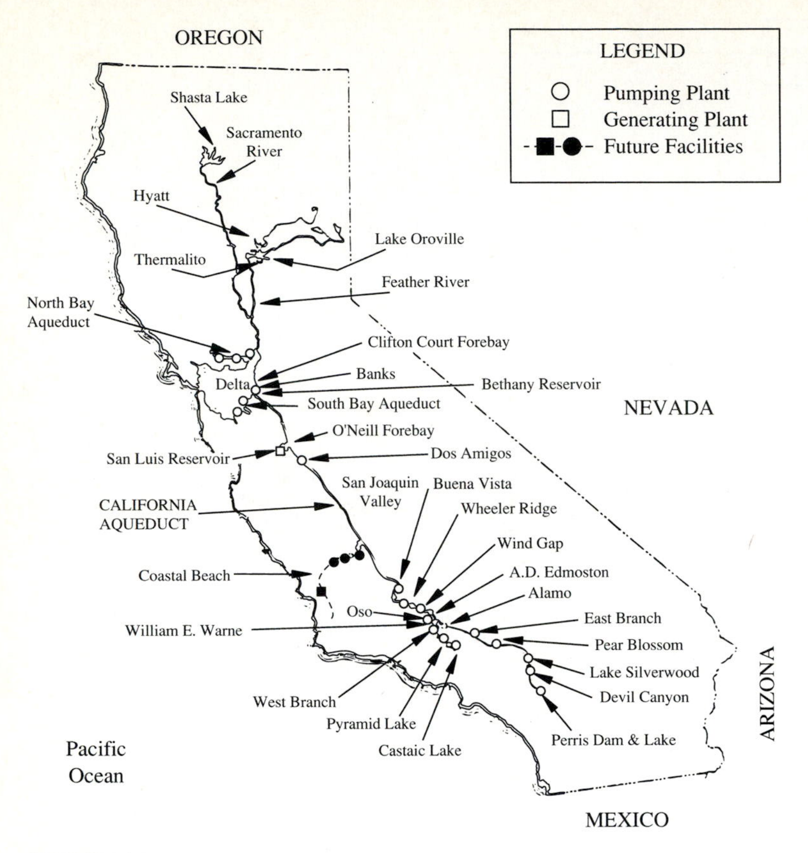

FIGURE 1.2.1
California State Water Project (Sabet and Coe 1986a, 1986b; reprinted by permission in Coe and Rankin, 1989)

flood water through the urban area. The urban water system can actually be divided into other basic systems such as the **water supply system**, the **wastewater system**, the **urban drainage system**, and the **floodplain management system**.

The water supply system can also be subdivided into systems such as the raw water pumping system, the raw water transmission system, the raw water storage, the treatment system, the treated pumping system, and the water distribution system. Many urban wastewater collection systems involve both the urban water runoff component in addition to the combined sewage overflow as shown in Fig. 1.2.3. Such a system includes the runoff, the transport, the storage, and the receiving water.

GROUNDWATER SYSTEMS. Groundwater has been introduced through the hydrologic cycle. Groundwater systems (Fig. 1.2.4) are geologic formations, called **aquifers**, that are capable of storing and transmitting the subsurface waters. These aquifers have varying transmissivities, storages, and water quality properties that affect aquifer pumping and recharge. Groundwater systems can be developed as sources of domestic,

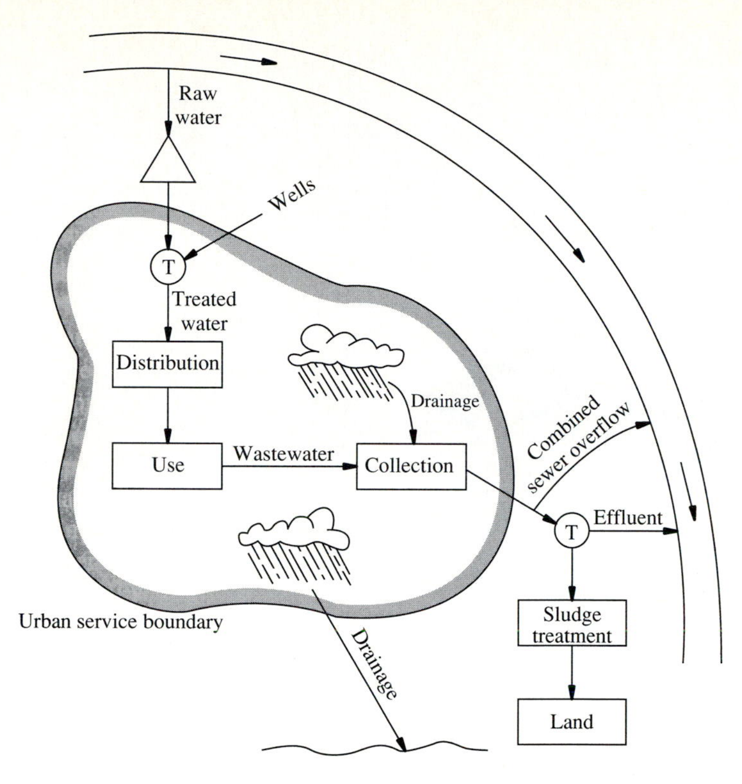

FIGURE 1.2.2
Components of the urban water system. (Grigg, 1986)

industrial, and agricultural water supply. These systems can also provide temporary and/or long term storage and treatment of waste water.

An example of a groundwater system is the Edwards (Balcones fault zone) aquifer shown in Fig. 1.2.5, which extends along the narrow belt from Austin, Texas through San Antonio to Brackettville, Texas. The groundwater is used extensively for public water supply, agriculture, and industry. This aquifer is the water supply source for the city of San Antonio, Texas. In addition it originates several major springs that become recreation centers and also provide the base flow of downstream rivers. These spring fed streams flow into the San Antonio Bay, whose ecosystem is dependent on the freshwater inflows. The present problem facing the Edwards aquifer is the threat of overdrafting of the annual replenishing rate. This aquifer has both unconfined and confined conditions. The confined portion is along the south and is the most productive and the unconfined portion is the recharge zone which yields small to moderate amounts of groundwater. The aquifer thickness ranges in depth from about 400 to 1000 feet deep. Due to high permeability of the aquifer, large volumes of water can move rapidly over wide areas and the response of water levels to high single pumping tends to be regionalized rather than localized.

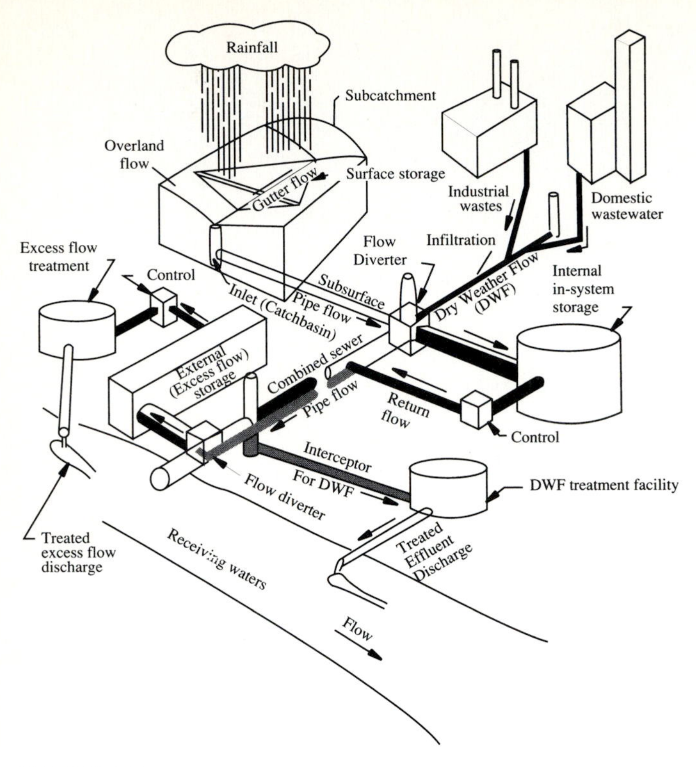

FIGURE 1.2.3
EPA stormwater management model. (Lager et al., 1971)

1.3 THE SYSTEMS CONCEPT

A system, in general, consists of a set of interactive elements that perform independent of each other. A system is characterized by: (1) a **system boundary** which is a rule that determines whether an element is to be considered as a part of the system or of the environment; (2) statement of input and output interactions with the environment; and (3) statements of interrelationships between the system elements, inputs and outputs, called **feedback**.

Depending upon the level of detail used to model the system components, appropriate hydraulic and hydrologic equations are used to describe flow relationships between system components. For example, a continuity equation is always used to describe the flow balance within a water system. The relationship of water quality for a system component would depend upon the nature of that component and its impact on water quality of incoming flows. One of the main tasks of a water resources engineer is to modify the inputs to a hydrosystem so that the desirable outputs are maximized while undesirable outputs are minimized.

Naturally occurring water can be described in terms of the quantitative availability of water as a function of time and location and the quality of water as a function of time and location. The time **t** and the location **x** represent independent variables

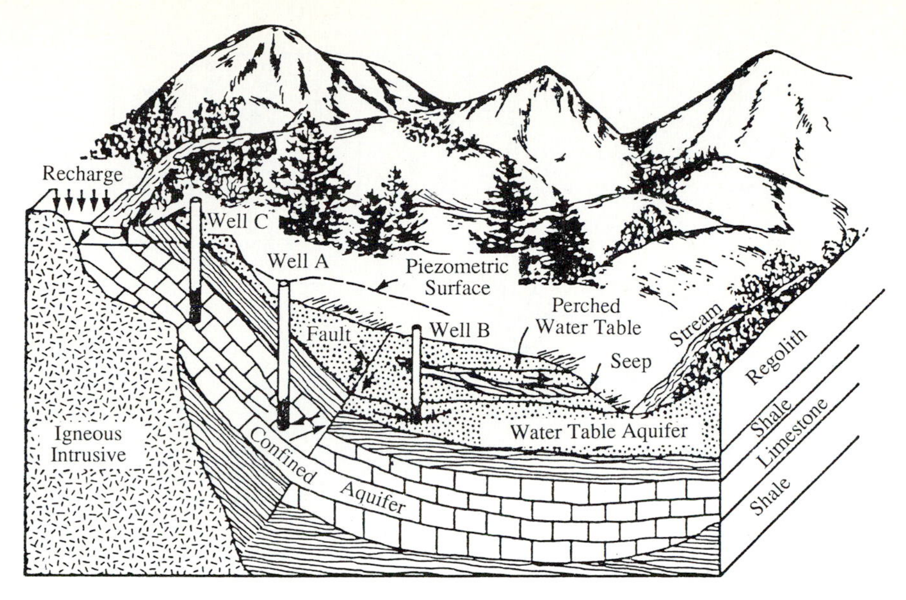

FIGURE 1.2.4
Groundwater system (McWhorter and Sunada, 1977)

describing naturally occurring water. The dependent variables describing the naturally occurring water are the quantity **V** and the quality **Q**. **S** is used to define the **state of the system** which can be expressed as

$$\mathbf{S} = [\mathbf{V}(\mathbf{x},\mathbf{t}),\ \mathbf{Q}(\mathbf{x},\mathbf{t})] \tag{1.3.1}$$

The development of water resources deals with the transformation of the state of the system for naturally occurring water into a desired state $\mathbf{S}^*$, which is a function of the desired quantity $\mathbf{V}^*$ and the desired quality $\mathbf{Q}^*$, which are both a function of the desired time $\mathbf{t}^*$ and desired location $\mathbf{x}^*$. The desired state can then be expressed as

$$\mathbf{S}^* = [\mathbf{V}^*(\mathbf{x}^*,\mathbf{t}^*),\ \mathbf{Q}^*(\mathbf{x}^*,\mathbf{t}^*)] \tag{1.3.2}$$

Evaluation of **S** and $\mathbf{S}^*$ is within the realm of hydrosystems engineering and requires the combined knowledge and use of hydrology and hydraulics. Hydrosystems engineering involves the transformation of **S** into $\mathbf{S}^*$

$$\mathbf{S}^* = \mathbf{WS} + \mathbf{E} \tag{1.3.3}$$

which is the **transformation equation** of the system. The symbol **W** is a **transfer function** between the input **S** and the output $\mathbf{S}^*$ and **E** represents a waste or by-product that is undesirable.

The transfer or transformation function can also be divided into the physical components or hardware, $\mathbf{W}_1$, and the operational aspects or software, $\mathbf{W}_2$, that is,

$$\mathbf{W} = (\mathbf{W}_1, \mathbf{W}_2) \tag{1.3.4}$$

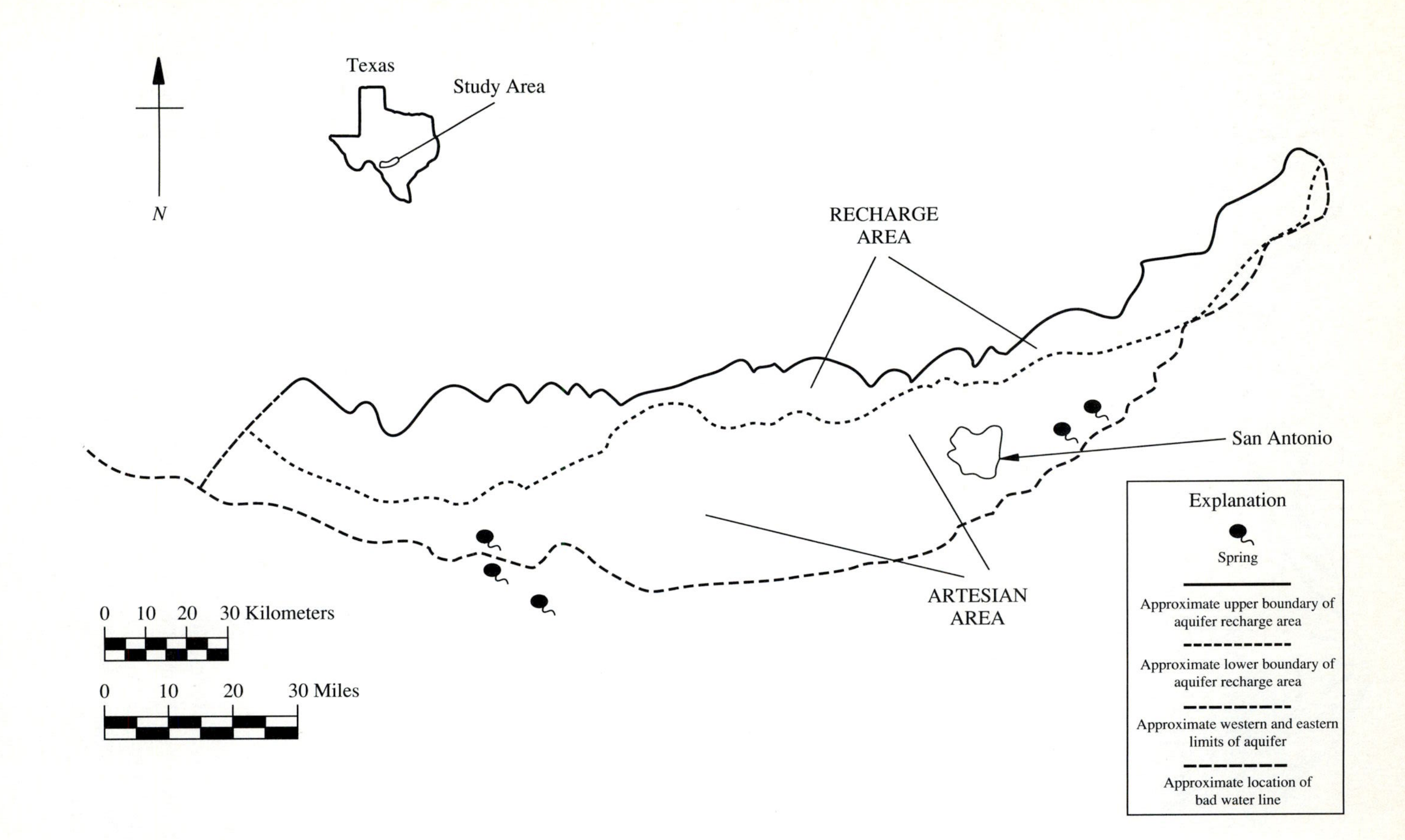

FIGURE 1.2.5
Edwards (Balcones fault zone) aquifer, San Antonio region. (After Guyton and Associates, 1979)

Buras (1972) presented a similar analysis to describe the development problems in water resources engineering. The techniques that can be used to describe the transformation are simulation techniques, optimization techniques, and a combination of these.

1.4 ISSUES IN HYDROSYSTEMS ENGINEERING

The major types of problems (Buras, 1972) that must be solved for various types of hydrosystems are

1. determining the optimal scale of development of the project;
2. determining the optimal dimensions of the various components of the system; and
3. determination of the optimal operation of the system.

If the solutions to these problems are denoted as X_1, X_2, and X_3, then the benefit of these solutions are

$$B = f(X_1, X_2, X_3) \tag{1.4.1}$$

The objective of many hydrosystems projects is to maximize the benefits so that the problem of developing water resources may be stated as

$$\text{Maximize } B = f(X_1, X_2, X_3) \tag{1.4.2}$$

subject to various types of constraints including technological constraints, economic or budgetary constraints, design constraints, operation constraints, demand constraints, and others.

1.4.1 Design Versus Analysis

Hydrosystem problems deal with both design and analysis. Analysis is concerned with determining the behavior of an existing system or a trial system that is being designed. In many cases determination of the behavior of the system is determining operation of the system or the response of a system under specified inputs. The design problem is to determine the sizes of components of the system. As an example, the design of a reservoir system determines the size and location of reservoirs. Analysis of a reservoir system is the process of determining operation policies for the reservoir system. Operation of the reservoir system is required to check the design. In other words, a design is estimated and is then analyzed to see if it performs according to the specifications. If a design satisfies the specifications, then an acceptable design has been found. New designs can be formulated and then analyzed.

1.4.2 Conventional Versus Optimization Procedures

Conventional procedures for design and analysis are basically iterative trial-and-error procedures. The effectiveness of conventional procedures are dependent upon an engineer's intuition, experience, skill, and knowledge of the hydrosystem. Conventional procedures therefore are closely related to the human element, a factor which could lend to inefficient results for the design and analysis of complex systems. Conventional procedures are typically based upon using simulation models in a trial-and-error process. A procedure may be to iteratively use a simulation model to attempt to arrive at an optimal solution. Figure 1.4.1 presents a depiction of the conventional design and analysis procedure. As an example, to determine a least-cost pumping scheme for an aquifer dewatering problem would require a selection of pump sizes and location for the aquifer to be dewatered. Using a trial set of pump sizes and locations, a groundwater simulation model is solved to determine if the water levels are lowered below the desirable elevation. If the pumping scheme (pump size and location) does not satisfy the water levels, then a new pumping scheme is selected and simulated. This iterative process is continued, each time determining the cost of the pumping scheme.

Optimization eliminates the trial-and-error process of changing a design and resimulating with each new design change. Instead, an optimization model automatically changes the design parameters. An optimization procedure has mathematical expressions that describe the system and its response to the system inputs for various design parameters. These mathematical expressions are constraints in the optimization model. In addition, constraints are used to define the limits of the design variables and the performance is evaluated through an objective function, which could be to minimize cost.

An advantage of the conventional process is that the engineer's experience and intuition are used in making conceptual changes in the system or to change or make additional specifications. The conventional procedure can lead to nonoptimal or uneconomical designs and operation policies. Also the conventional procedure can be very time consuming. An optimization procedure requires the engineer to explicitly identify the design variables, the objective function or measure of performance to be optimized, and the constraints for the system. In contrast to the decision-making process in the conventional procedure, the optimization procedure is more organized using a mathematical approach to select the decisions.

1.4.3 Optimization

An optimization problem in water resources may be formulated in a general framework in terms of the decision variables (**x**) with an objective function to

$$\text{Optimize } f(\mathbf{x}) \tag{1.4.3}$$

subject to constraints

$$\mathbf{g}(\mathbf{x}) = \mathbf{0} \tag{1.4.4}$$

and bound constraints on the decision variables

$$\underline{\mathbf{x}} < \mathbf{x} < \bar{\mathbf{x}} \tag{1.4.5}$$

where **x** is a vector of n decision variables $(x_1, x_2, \ldots, x_n)$, $\mathbf{g}(\mathbf{x})$ is a vector of m equations called constraints and $\underline{\mathbf{x}}$ and $\bar{\mathbf{x}}$ represent the lower and upper bounds, respectively, on the decision variables.

Every optimization problem has two essential parts: the objective function and the set of constraints. The **objective function** describes the performance criteria of

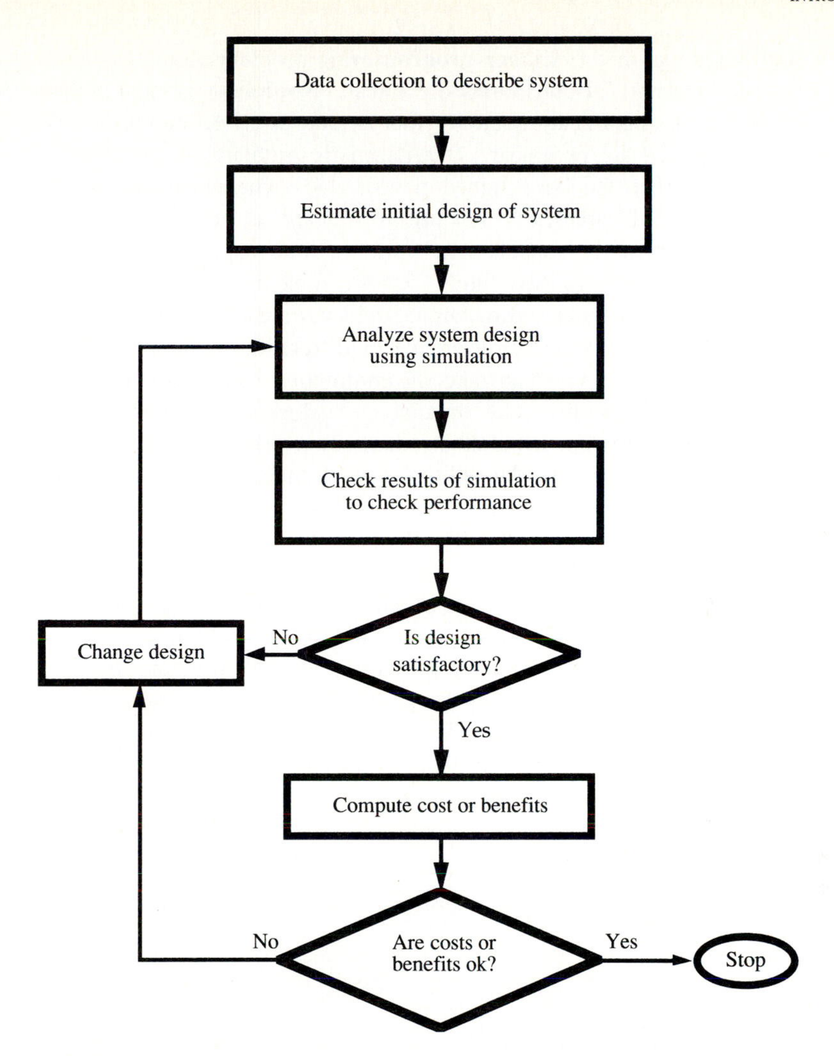

FIGURE 1.4.1
Conventional design and analysis process.

the system. **Constraints** describe the system or process that is being designed or analyzed and can be of two forms: equality constraints and inequality constraints. A **feasible solution** of the optimization problem is a set of values of the decision variables that simultaneously satisfy the constraints. The **feasible region** is the region of feasible solutions defined by the constraints. An **optimal solution** is a set of values of the decision variables that satisfy the constraints and provides an optimal value of the objective function.

Depending upon the nature of the objective function and the constraints, an optimization problem can be classified as: (a) linear vs. nonlinear; (b) deterministic vs. probabilistic; (c) static vs. dynamic; (d) continuous vs. discrete; and (e) lumped parameter vs. distributed parameter. **Linear programming** problems consist of both a linear objective function and all constraints are linear. **Nonlinear programming** problems are represented by nonlinear equations, that is, part or all of the constraints and/or the objective function are nonlinear. **Deterministic** problems consist of coefficients and parameters that can be assigned fixed values whereas **probabilistic** problems consist of uncertain parameters that are considered as random variables. **Static** problems do not explicitly consider the variable time aspect whereas **dynamic** problems do consider the variable time. Static problems are referred to as mathematical programming problems and dynamic problems are often referred to as optimal control problems, which involve difference or differential equations. **Continuous** problems have variables that can take on continuous values whereas with **discrete** problems the variables must take on discrete values. Typically discrete problems are posed as **integer programming** problems in which the variables must be integer values. A **lumped problem** considers the parameters and variables to be homogeneous throughout the system whereas **distributed problems** must take into account detailed variations in the behavior of the system from one location to another.

The method of optimization used depends upon: (1) the type of objective function; (2) the type of constraints; and (3) the number of decision variables. Table 1.4.1 lists some general steps to solve optimization problems. Some problems may not require that the engineer follow the steps in the exact order, but each of the steps should be considered in the process. The overall objective in optimization is to determine a set of values of the decision variables that satisfy the constraints and provide the optimal response to the objective function.

TABLE 1.4.1

The six steps used to solve optimization problems (Edgar and Himmelblau, 1988)

1. Analyze the process itself so that the process variables and specific characteristics of interest are defined, i.e., make a list of all of the variables.
2. Determine the criterion for optimization and specify the objective function in terms of the above variables together with coefficients. This step provides the performance model (sometimes called the economic model when appropriate).
3. Develop via mathematical expressions a valid process or equipment model that relates the input-output variables of the process and associated coefficients. Include both equality and inequality constraints. Use well-known physical principles (mass balances, energy balances), empirical relations, implicit concepts, and external restrictions. Identify the independent and dependent variables to get the number of degrees of freedom.
4. If the problem formulation is too large in scope:
   (a) break it up into manageable parts/or
   (b) simplify the objective function and model.
5. Apply a suitable optimization technique to the mathematical statement of the problem.
6. Check the answers, and examine the sensitivity of the result to changes in the coefficients in the problem and the assumptions.

1.4.4 Single-Objective Versus Multiple-Objective Optimization

The solutions to a growing number of water resource problems facing water resource professionals today are becoming more complex. In most water resource problems, the decision-making process is cultivated by the desire to achieve several goals simultaneously and many of them could be **noncommensurate** and conflicting with each other. In such circumstances improvement of some objectives can not be obtained without the sacrifice of other objectives. Therefore, the ideological theme of "optimality" in the single-objective context is no longer appropriate. Instead, the goal of "optimality" of the single-objective framework is replaced by the concept of "noninferiority" in the multiple-objective analysis.

Cohon (1978) defined the noninferiority in the following passage: "A feasible solution to a multiple-objective programming problem is noninferior if there exists no other feasible solution that will yield an improvement in one objective without causing a degradation in at least one other objective." The idea of a "noninferior" solution set in attempting to maximize two conflicting objectives (Z_1 and Z_2) is shown in Fig. 1.4.2. According to Cohon's definition, it is evident that all interior points, not elements on the curve ABCD, must be inferior solutions to multiple-objective problems because for such inferior points there exists at least one other feasible solution in which the objective function can be improved simultaneously. On the other hand, for any solution point lying on curve ABCD, such as point B, cannot be moved to any other points in the feasible region without degrading the measure of effectiveness in at least one of the objectives. The collection of such noninferior solutions defines the noninferior solution set while the slope of the curve defines the **marginal rate of substitution** representing the trade-off between the conflicting objectives.

The solution to a multiple-objective problem cannot be obtained until the decision-maker provides the characterization of his/her preference between the objectives involved. Information concerning the decision-maker's preference is commonly depicted graphically as an **indifference curve** (see Fig. 1.4.2). The utility of a decision-maker will be the same for combinations of solutions that fall on the same indifference curve. The **best-compromised** solution to a multiple-objective problem is a unique set of alternatives which possess the property of maximum combined utility and are elements in both the noninferior solution set and indifference curve. Such an alternative only exists at the point where the indifference curve and noninferior solution set are tangent.

Mathematically, multiple-objective programming problems can be expressed in terms of **vector optimization** as

$$\text{Optimize } \mathbf{f}(\mathbf{x}) = [f_1(\mathbf{x}), f_2(\mathbf{x}), \ldots, f_K(\mathbf{x})] \tag{1.4.6}$$

in which $\mathbf{f}(\mathbf{x})$ is a K-dimensional vector of objective functions and others are defined previously as for equations (1.4.4)–(1.4.5). There are many methods proposed to derive the solution for a multiple-objective problem. Basically, they can be classified into two categories (Cohon, 1978): generating techniques and techniques incorporating prior knowledge of preference. Descriptions of various techniques for solv-

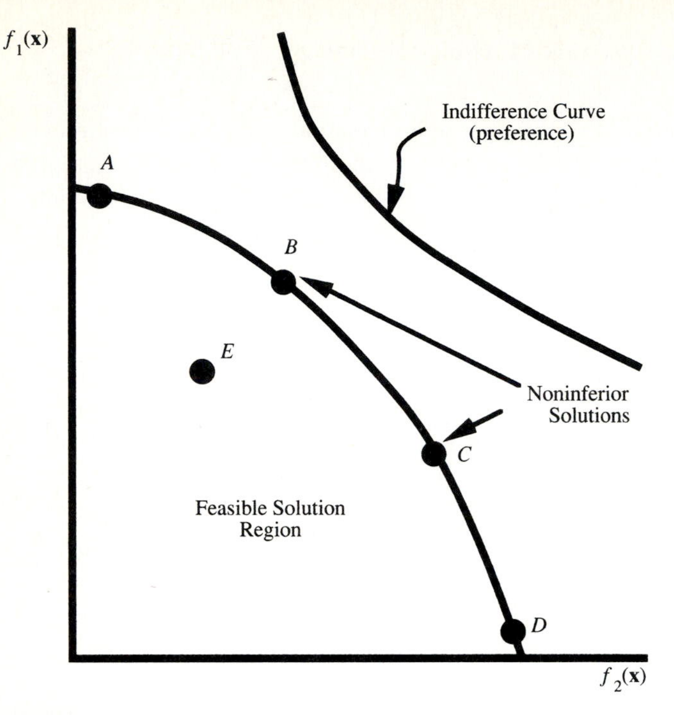

FIGURE 1.4.2
Illustration of the tradeoffs between objectives in a two-dimensional problem setting.

ing multiple-objective problems can be found elsewhere (Cohon, 1978; Chankong and Haimes, 1983). Regardless of the method to be employed to solve a multiple-objective problem, an optimization technique for solving a single-objective problem is still required.

1.4.5 Uncertainties in Hydrosystem Design and Analysis

Risk and reliability analysis in hydrosystem design requires the determination of uncertainty and definitions of risk and reliability. **Uncertainty** could simply be defined as the occurrence of events that are beyond our control. The uncertainty of a hydrosystem is an indeterministic characteristic and is beyond our rigid controls. In the design of hydrosystems, decisions must be made under various kinds of uncertainty. In general, uncertainties in water resource engineering projects can be divided into four basic categories: hydrologic, hydraulic, structural, and economic.

Hydrologic uncertainty for any hydrosystem design problem can further be classified into three types: **inherent**, **parameter**, and **model uncertainties**. The occurrence of various hydrological events such streamflow or rainfall events generally can be considered to be stochastic processes because of the observable natural, or inherent, randomness. Because of the lack of perfect hydrological information about these processes or events, for example, infinitely long historical records, there exist informational uncertainties about the processes. These uncertainties are referred to as the parameter uncertainty and model uncertainty. There is seldom enough information available to accurately evaluate the parameters or statistical characteristics of a probability model.

Hydraulic uncertainty is referred to the uncertainty in the design of hydraulic structures and in the analysis of the performance of hydraulic structures. It mainly arises from three basic types: model, construction and material, and operational conditions of the flow. The model uncertainty results from the use of a simplified or an idealized hydraulic model to describe flow conditions, which contributes to the uncertainty in determining the design capacity of hydraulic structures. For example, flows through or over hydraulic structures are unsteady and nonuniform, which can be described in one-dimensional form by the St. Venant equations. However, equations such as Manning's that cannot adequately describe unsteady and nonuniform flow are quite commonly used in practice. This fact results in additional uncertainty that can be referred to as model error. The construction and material uncertainty results partly from the structure size, for example, a sewer diameter or the culvert width and depth. Manufacturers' tolerances or construction tolerances may vary widely resulting in these uncertainties. Another factor is the misalignment of a hydraulic structure as well as settlement resulting in errors such as in the slope. Material variability could cause variations in the size and distribution of the surface roughness resulting in errors of roughness factors. Changes in resistance coefficients and structural size reduction because of deposition could be referred to uncertainty in operational conditions. Also, factors such as not cleaning the structure to eliminate clogging may result in additional uncertainty.

Structural uncertainty refers to the failure from structural weaknesses. Physical failures of hydraulic structures in hydrosystems can be caused by many things such as water saturation and loss of soil stability, erosion or hydraulic soil failures, wave action, hydraulic overloading, structural collapse, and many others. A good example is the failure of a levee system either in the levee or in the adjacent soil. The structural failure of a levee may be caused by water saturation and loss of soil stability. A flood wave can cause increased saturation of the levee through slumping. Levees can also fail because of hydraulic soil failures and wave action. Methods of analysis rarely include the probability of structural failure in an explicit manner. Results may significantly overestimate the protection offered by a hydraulic structure and underestimate expected damages that may occur.

Economic uncertainty can arise from uncertainties in construction costs, damage costs, projected revenue, operation and maintenance costs, inflation, project life, and other intangible cost and benefit items. Construction, damage, and operation/maintenance costs are all subject to uncertainties because of the fluctuation in the rate of increase of construction materials, labor costs, transportation costs, economic losses, regional differences, and many others. The problem of inflation trends also causes uncertainties and should be accounted for in the evaluation of economic merit of water resource projects. There are also many other economic and social uncertainties that are related to inconvenience losses. An example of this is the failure of a highway crossing caused by flooding resulting in traffic-related losses.

1.4.6 Applications of Optimization in Hydrosystems

Optimization can be applied to many types of application to hydrosystems engineering projects and problems including:

1. Determination of operating policies for reservoirs.
2. Design of reservoir capacities and location.
3. Operation of hydropower plants.
4. Operation of irrigation systems.
5. Operation of regional aquifers to determine recharge and pumpage.
6. Design of aquifer dewatering systems.
7. Design of aquifer reclamation systems.
8. Parameter identification for aquifers.
9. Minimum cost design and operation of water distribution systems.
10. Replacement and rehabilitation of water distribution components.
11. Aqueduct route determination.
12. Minimum cost design of storm sewer systems.
13. Design of detention basins.
14. Determination of flood control systems.
15. Determination of freshwater inflows to bays and estuaries.
16. Determination of firm energy and firm yield.

1.4.7 Building a Model

Development of an optimization model can be divided into five major phases:

1. Collection of data to describe system.
2. Problem definition and formulation.
3. Model development.
4. Model verification and evaluation.
5. Model application and interpretation.

These phases are outlined in Fig. 1.4.3. Data collection can be a very time consuming but extremely important phase of the model-building process. The availability and accuracy of data can have a great deal of effect upon the level and detail of the model that is formulated and the ability to evaluate and verify the model.

The problem definition and formulation includes the steps: identification of decision variables; formulation of the model objective(s); and formulation of the model constraints. In performing these steps one must:

1. Identify the important elements that pertain to the problem.
2. Determine the degree of accuracy required in the model.

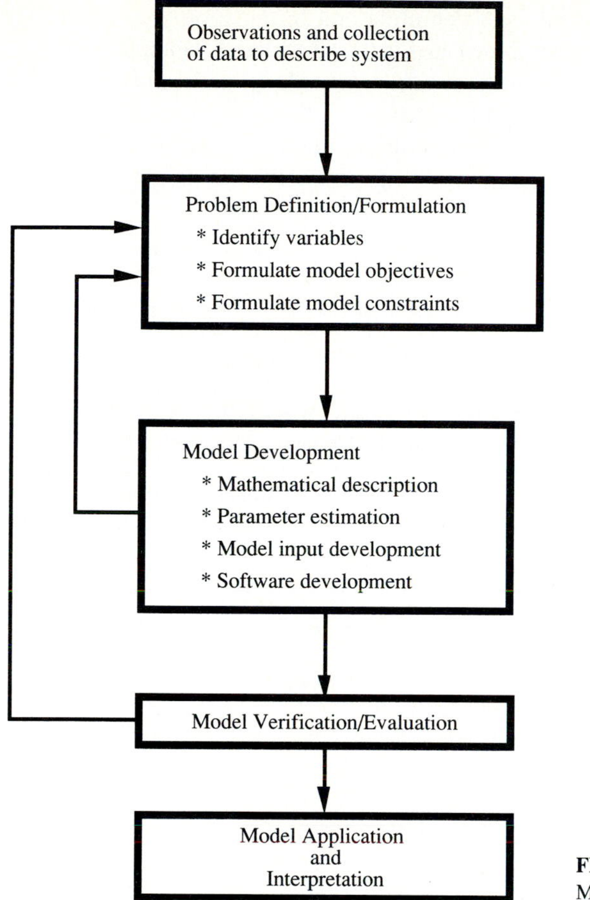

FIGURE 1.4.3
Model building.

3. Determine the potential uses of the model.
4. Evaluate the structure and complexity of the model.
5. Determine the number of independent variables, the number of equations required to describe the system, and the number of unknown parameters.

Model development includes the mathematical description, parameter estimation, model input development, and software development. The model development phase is an iterative process that in many cases will require returning to the model definition and formulation phase.

The model evaluation and verification phase is checking of the model as a whole. This requires first of all testing of the individual model elements which should be carried out early in the model building procedure. In many cases the model builder must determine if the mathematical relationships are describing the actual system properly. A sensitivity analysis should be performed to test the model inputs and parameters. The process of model evaluation and verification is an iterative process and may require returning to the problem definition and formulation phase followed by the model development phase as shown in Fig. 1.4.3. Model validation consists of validation of the logic, validation of the model assumptions, and validation of the model behavior. One important aspect of this process is that in most cases data used in the formulation process should not be used in the validation. Another point to keep in mind is that no single validation procedure is appropriate for all models.

REFERENCES

Buras, N.: *Scientific Allocation of Water Resources*, American Elsevier Publishing Company, New York, 1972.

Chankong, V. and Haimes, Y. Y: *Multiobjective Decision Making: A Theory and Methodology*, Elsevier Science Publishing, New York, 1983.

Chow, V. T., Maidment, D. R., and Mays, L. W.: *Applied Hydrology*, McGraw-Hill, New York, 1988.

Coe, J. Q. and A. W. Rankin: "California's Adaptable Model for Operations Planning for the State Water Project," in *Computerized Decision Support Systems for Water Managers, Proc. 3rd Water Resources Operations Management Workshop* at Colorado State University, Ft. Collins, Colorado, June 27–30, 1988, J. W. Labadie, L. E. Brazil, I. Corbu, and L. E. Johnson, Eds., American Society of Civil Engineers, New York, 1989.

Cohon, J. L.: *Multiobjective Programming and Planning*, Academic Press, New York, 1978.

Edgar, T. F., and D. M. Himmelblau: *Optimization of Chemical Processes*, McGraw-Hill, New York, 1988.

Grigg, N. S.: *Urban Water Infrastructure: Planning, Management, and Operations*, John Wiley & Sons, New York, 1986.

Guyton, W. F., and Associates: "Geohydrology of Comal, San Marcos, and Heuco Springs," Report 234, Texas Department of Water Resources, Austin, Texas, June 1979.

Lager, J. A., R. P. Shubinski, and L. W. Russell: "Development of a simulation model for stormwater management," *J. Water Pollution Control Federation*, Vol. 43, pp. 2424–2435, Dec. 1971.

McWhorter, D. B. and D. K. Sunada: *Groundwater Hydrology and Hydraulics*, Water Resources Publications, Littleton, Colorado, 1977.

BOOKS ON SELECTED TOPICS

Water Resources

Biswas, A. K., (ed.): *Systems Approach to Water Management*, McGraw-Hill, New York, 1976.

Buras, N.: *Scientific Allocation of Water Resources*, American Elsevier Publishing Company, New York, 1972.

Goodman, A. S.: *Principles of Water Resources Planning*, Prentice-Hall, Englewood Cliffs, N.J., 1984.

Grigg, N. S.: *Water Resources Planning*, McGraw-Hill, New York, 1985.

Grigg, N. S.: *Urban Water Infrastructure: Planning, Management, and Operations*, John Wiley & Sons, New York, 1986.

Haimes, Y. Y.: *Hierarchical Analysis of Water Resources Systems*, McGraw-Hill, New York, 1977.

Hall, W. A. and J. A. Dracup: *Water Resources Systems Engineering*, McGraw-Hill, New York, 1970.

Helweg, O. J.: *Water Resources Planning and Management*, John Wiley & Sons, New York, 1985.

Linsley, R. K. and J. B. Franzini: *Water Resources Engineering*, 3d ed., McGraw-Hill, 1979.

Loucks, D. P., J. R. Stedinger, and D. A. Haith: *Water Resource Systems Planning and Analysis*, Prentice-Hall, Englewood Cliffs, N.J., 1981.

Major, D. C.: *Multiobjective Water Resources Planning*, Water Resources Monograph 4, Am. Geophys. Union, Washington, D.C., 1977.

Major, D. C. and R. L. Lenton: *Applied Water Resources Systems Planning*, Prentice-Hall, Englewood Cliffs, N.J., 1979.

Viessman, W. and C. Welty: *Water Management: Technology and Institutions*, Harper and Row, New York, 1985.

Groundwater

Bear, J.: *Hydraulics of Groundwater*, McGraw-Hill, New York, 1979.
Bear, J. and A. Verruijt: *Modeling Groundwater Flow and Pollution*, D. Reidel Publishing Company, Dordrecht, Holland, 1987.
Bouwer, H.: *Groundwater Hydrology*, McGraw-Hill, New York, 1978.
Freeze, R. A. and J. A. Cherry, *Groundwater*, Prentice-Hall, Englewood Cliffs, N.J., 1979.
Kashef, A. A. I.: *Groundwater Engineering*, McGraw-Hill, New York, 1986.
Todd, D. K.: *Groundwater Hydrology*, John Wiley & Sons, New York, 1980.
Willis, R. and W. W-G. Yeh: *Groundwater Systems Planning and Management*, Prentice-Hall, Englewood Cliffs, N.J., 1987.

Hydrology

Bedient, P. B. and W. C. Huber: *Hydrology and Floodplain Analysis*, Addison Wesley, Reading, Mass., 1988.
Bras, R. L.: *Hydrology: An Introduction to Hydrologic Science*, Addison Wesley, Reading, Mass., 1990.
Chow, V. T., D. R. Maidment, and L. W. Mays: *Applied Hydrology*, McGraw-Hill, New York, 1988.
Linsley, Jr., R. K., M. A. Kohler, and J. L. H. Paulaus: *Hydrology for Engineers*, McGraw-Hill, New York, 1982.
Viessman, Jr., W., G. L. Lewis, and J. W. Knapp: *Introduction to Hydrology*, Harper and Row, New York, 1989.

Hydraulics

Chow, V. T.: *Open-Channel Hydraulics*, McGraw-Hill, New York, N.Y., 1959.
French, R. H.: *Open-Channel Hydraulics*, McGraw-Hill, New York, 1985.
Hoggan, D. H.: *Computer-Assisted Floodplain Hydrology and Hydraulics*, McGraw-Hill, New York, 1989.
Hwang, N. H. C., and C. E. Hita: *Fundamentals of Hydraulic Engineering Systems*, Prentice Hall, Englewood Cliffs, N.J., 1987.
Prasuhn, A. L.: *Fundamentals of Hydraulic Engineering*, Holt, Rinehart, and Winston, New York, 1987.
Roberson, J. A., J. J. Cassidy, M. H. Chaudry: *Hydraulic Engineering*, Houghton Mifflin, Boston, Mass., 1988.
Warnick, C. C.: *Hydropower Engineering*, Prentice-Hall, Englewood Cliffs, N.J., 1984.

Economics

Henderson, J. M. and R. E. Quandt: *Microeconomic Theory: A Mathematical Approach*, McGraw-Hill, New York, 1980.

Optimization and Systems Analysis

Arora, J. S.: *Introduction to Optimum Design*, McGraw-Hill, New York, 1989.
Blanchard, B. S.and W. J. Wolter: *Systems Engineering and Analysis*, Prentice-Hall, Englewood Cliffs, N.J., 1981.
Bradley, S. P., A. C. Hax, and T. L. Magnanti: *Applied Mathematical Programming*, Addison-Wesley, Reading, Mass., 1977.
Denardo, E. V.: *Dynamic Programming: Models and Applications*, Prentice-Hall, Englewood Cliffs, N.J., 1982.
Edgar, T. F., and D. M. Himmelblau: *Optimization of Chemical Processes*, McGraw-Hill, New York, 1988.
Gill, P. E., W. Murray, and W. H. Wright: *Practical Optimization*, Academic Press, London, 1981.

Goicoechea, A., D. R. Hansen, and L. Duckstein: *Multiobjective Decision Analysis with Engineering and Business Applications*, John Wiley & Sons, New York, 1982.

Hillier, F. S. and G. J. Lieberman: *Introduction to Operations Research*, McGraw-Hill, New York, 1990.

McCormick, G. P.: *Nonlinear Programming: Theory, Algorithms, and Applications*, John Wiley & Sons, New York, 1983.

Solow, D.: *Linear Programming: An Introduction to Finite Improvement Algorithms*, North-Holland, New York, 1984.

Taha, H. A.: *Operations Research: An Introduction*, Macmillan, New York, 1987.

Reliability Analysis

Duckstein, L. and E. J. Plate (eds.): *Engineering Reliability and Risk in Water Resources*, Martinus Nijhoff Publishers, Dordrecht, Netherlands, 1987.

Harr, M.: *Reliability-Based Design in Civil Engineering*, McGraw-Hill, New York, 1987.

Kapur, K. C. and L. R. Lamberson: *Reliability in Engineering Design*, John Wiley and Sons, New York, 1977.

Mays, L. W. (ed.): *Reliability Analysis of Water Distribution Systems*, American Society of Civil Engineers, New York, 1989.

Yen, B. C. (ed.): *Stochastic and Risk Analysis in Hydraulic Engineering*, Water Resources Publications, Littleton, Colo., 1986.

PROBLEMS

1.3.1 Select a surface water supply system in the state or region of the country that you live in and describe this system in detail. What are the inputs and outputs of this system? What were some of the major engineering accomplishments required in designing, building, and operating this system? What kinds of analysis were used in the hydrologic, hydraulic, and economic studies for this project?

1.3.2 Select an urban drainage system near where you live and describe this system in detail. Describe the inputs and outputs of this system. What kinds of analysis were used in the hydrologic, hydraulic, and economic studies for this project? What design criteria were used?

1.3.3 Select a floodplain management system and describe this system in detail. Describe the inputs and outputs of this system. What kinds of analysis were used in the hydrologic, hydraulic, and economic studies for this project? What design criteria were used?

1.3.4 Describe in detail the water distribution system that delivers water to your place of residence. What kinds of analysis and design criteria were used in the design of this system?

1.3.5 Select a groundwater system in the state or region of the country that you live in and describe this system in detail. Describe the inputs and outputs of the system. Describe any problems that exist with this system?

CHAPTER 2

ECONOMICS FOR HYDROSYSTEMS

2.1 ENGINEERING ECONOMIC ANALYSIS

Engineering economic analysis is an evaluation process that can be used for comparing various water resource project alternatives and selecting the most economical one. This process requires defining feasible alternatives and then applying a discounting technique to select the best alternative. In order to perform this analysis, several basic concepts such as equivalence of kind, equivalence of time, and discounting factors must be understood.

One of the first steps in economic analysis is to find a common value unit such as monetary units. Through the use of common value units, alternatives of rather diverse kinds can be evaluated. The monetary evaluation of alternatives generally occurs over a number of years. Each monetary value must be identified by the amount and the time. The time value of money results from the willingness of people to pay interest for the use of money. Consequently, money at different times cannot be directly combined or compared, but must first be made equivalent through the use of **discount factors**. Discount factors convert a monetary value at one date to an equivalent value at another date.

Discount factors are described using the notation: i is the annual interest rate; n is the number of years; P is the present amount of money; F is the future amount of money; and A is the annual amount of money. Consider an amount of money P

that is to be invested for n years at i-percent interest rate. The future sum F at the end of n years is determined from the following progression:

	Amount at beginning of year	+	Interest	=	Amount at end of year
First year	P	+	iP	=	$(1+i)P$
Second year	$(1+i)P$	+	$iP(1+i)$	=	$(1+i)^2P$
Third year	$(1+i)^2P$	+	$iP(1+i)^2$	=	$(1+i)^3P$
	$\vdots$		$\vdots$		$\vdots$
nth year	$(1+i)^{n-1}P$	+	$iP(1+i)^{n-1}$	=	$(1+i)^nP$

The future sum is then

$$F = P(1+i)^n \tag{2.1.1}$$

and the **single-payment compound amount factor** is

$$\frac{F}{P} = (1+i)^n = \left(\frac{F}{P}, i\%, n\right) \tag{2.1.2}$$

This factor defines the number of dollars which accumulate after n years for each dollar initially invested at an interest rate of i percent. The **single-payment present worth** factor $(P/F, i\%, n)$ is simply the reciprocal of the single-payment compound amount factor. Table 2.1.1 summarizes the various discount factors.

Uniform annual series factors are used for equivalence between present (P) and annual (A) monetary amounts or between future (F) and annual (A) monetary amounts. Consider the amount of money A that must be invested annually (at the end of each year) to accumulate F at the end of n years. The last value of A in the nth year is withdrawn immediately upon deposit so it accumulates no interest. The future value F is

$$F = A + (1+i)A + (1+i)^2A + \cdots + (1+i)^{n-1}A \tag{2.1.3}$$

Equation (2.1.3) is multiplied by $(1 + i)$, and subtract Eq. (2.1.3) from the result to obtain the **uniform annual series sinking fund factor**,

$$\frac{A}{F} = \frac{i}{(1+i)^n - 1} = \left(\frac{A}{F}, i\%, n\right) \tag{2.1.4}$$

The sinking fund factor is the number of dollars A that must be invested at the end of each of n years at i percent interest to accumulate \$1. The **series compound amount factor** (F/A) is simply the reciprocal of the sinking fund factor (Table 2.1.1), which is the number of accumulated dollars if \$1 is invested at the end of each year. The **capital-recovery factor** can be determined by simply multiplying the sinking fund factor (A/F) by the single-payment compound amount factor (Table 2.1.1)

$$\left(\frac{A}{P}, i\%, n\right) = \frac{A}{F}\frac{F}{P} \tag{2.1.5}$$

TABLE 2.1.1
Summary of discounting factors

Type of Discount Factor	Symbol	Given*	Find	Factor	
Single-Payment Factors					
Compound-amount factor	$\left(\frac{F}{P}, i\%, n\right)$	P	F	$(1+i)^n$	P = \$1 F
Present-worth factor	$\left(\frac{P}{F}, i\%, n\right)$	F	P	$\frac{1}{(1+i)^n}$	P F = \$1
Uniform Annual Series Factors					
Sinking-fund factor	$\left(\frac{A}{F}, i\%, n\right)$	F	A	$\frac{i}{(1+i)^n - 1}$	F = \$1 A A A A A
Capital-recovery factor	$\left(\frac{A}{P}, i\%, n\right)$	P	A	$\frac{i(1+i)^n}{(1+i)^n - 1}$	P = \$1 A A A A A
Series compound-amount factor	$\left(\frac{F}{A}, i\%, n\right)$	A	F	$\frac{(1+i)^n - 1}{i}$	A = \$1 F A A A A A
Series present-worth factor	$\left(\frac{P}{A}, i\%, n\right)$	A	P	$\frac{(1+i)^n - 1}{i(1+i)^n}$	P A = \$1 A A A A A
Uniform Gradient Series Factors					
Uniform gradient series present-worth factor	$\left(\frac{P}{G}, i\%, n\right)$	G	P	$\frac{(1+i)^{n+1} - (1+ni+i)}{i^2(1+i)^n}$	P G = \$1 G 2G 3G (n–1)G

*The discount factors represent the amount of dollars for the given amounts of one dollar for P, F, A and G.

This factor is the number of dollars that can be withdrawn at the end of each of n years if \$1 is initially invested. The reciprocal of the capital-recovery factor is the **series present worth factor** (P/A), which is the number of dollars initially invested to withdraw \$1 at the end of each year.

A **uniform gradient series factor** is the number of dollars initially invested in order to withdraw \$1 at the end of the first year, \$2 at the end of the second year, \$3 at the end of the third year, etc.

Example 2.1.1. A water resources project has benefits that equal \$20,000 at the end of the first year and increase on a uniform gradient series to \$100,000 at the end of the fifth year. The benefits remain constant at \$100,000 each year until the end of the year 30, after which they decrease to \$0 on a uniform gradient at the end of year 40. What is the present value of these benefits using a 6-percent interest rate?

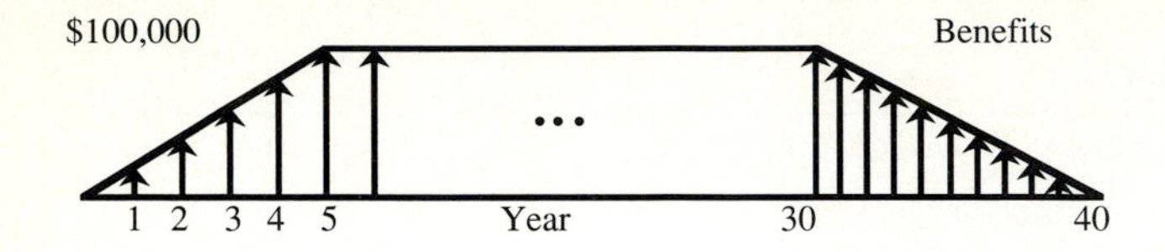

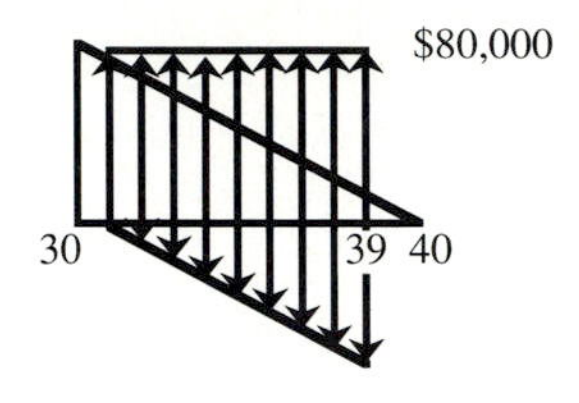

FIGURE 2.1.1
Cash flow diagram.

Solution. The present value of the uniform gradient series for years 1 through 5 is

$$20{,}000\left(\frac{P}{G}, 6\%, 5\right) = 20{,}000(12.1411)$$
$$= \$242{,}822$$

The present value of the annual series for years 6 through 30 is

$$100{,}000\left(\frac{P}{A}, 6\%, 25\right)\left(\frac{P}{F}, 6\%, 5\right) = 100{,}000(12.7834)(0.74726)$$
$$= \$955{,}252$$

Present value of the uniform gradient series for years 31 through 40 is modeled by a series of annual investments of \$80,000 per year for years 31 through 39 and subtracting a uniform gradient series for the same years, as shown in Fig. 2.1.1. The present value is determined by applying the single-payment present-worth factor,

$$80{,}000\left(\frac{P}{A}, 6\%, 9\right)\left(\frac{P}{F}, 6\%, 30\right) - 20{,}000\left(\frac{P}{G}, 6\%, 8\right)\left(\frac{P}{F}, 6\%, 31\right)$$
$$= 80{,}000(6.80170)(0.17411) - 20{,}000(26.05137)(0.16425)$$
$$= \$9{,}159$$

The total present-worth value is

$$\$242{,}822 + \$955{,}252 + \$9{,}159 = \$1{,}207{,}233$$

2.2 BENEFIT-COST ANALYSIS

Water projects extend over time, incur costs throughout the duration of the project, and yield benefits. Typically, costs are large during the initial construction and startup period, followed by only operation and maintenance costs. Benefits typically build up to a maximum over time as depicted in Fig. 2.2.1. The present value of benefits (PVB) and costs (PVC) are, respectively,

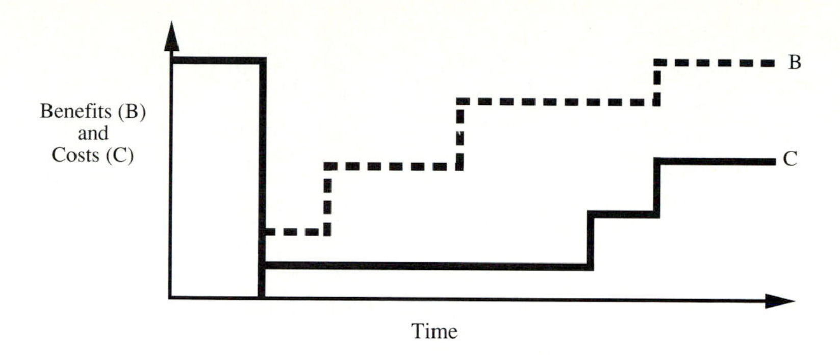

FIGURE 2.2.1
Benefits and costs over time.

(PVB) and costs (PVC) are, respectively,

$$PVB = b_0 + \frac{b_1}{(1+i)} + \frac{b_2}{(1+i)^2} + \cdots + \frac{b_n}{(1+i)^n} \tag{2.2.1}$$

and

$$PVC = c_0 + \frac{c_1}{(1+i)} + \frac{c_2}{(1+i)^2} + \cdots + \frac{c_n}{(1+i)^n} \tag{2.2.2}$$

The present value of net benefits is

$$\begin{aligned} PVNB &= PVB - PVC \\ &= (b_0 - c_0) + \frac{(b_1 - c_1)}{(1+i)} + \frac{(b_2 - c_2)}{(1+i)^2} + \cdots + \frac{(b_n - c_n)}{(1+i)^n} \end{aligned} \tag{2.2.3}$$

In order to carry out benefit-cost analysis, rules for economic optimization of the project design and procedures for ranking projects are needed. Howe (1971) points out that the most important point in project planning is to consider the broadest range of alternatives. The range of alternatives selected are typically restricted by the responsibility of the water resource agency and/or the planners. The nature of the problem to be solved may also condition the range of alternatives. Preliminary investigation of alternatives can help to rule out projects because of technical infeasibility or on the basis of costs.

Consider the selection of an optimal, single-purpose project design such as the construction of a flood-control system or a water-supply project. The optimum size can be determined by selecting the alternative such that the marginal or incremental present value of costs, ΔPVC, is equal to the marginal or incremental present value of the benefits, ΔPVB,

$$\Delta PVB = \Delta PVC$$

The marginal or incremental value of benefits and costs are for a given increase in the size of a project,

$$\Delta PVB = \frac{\Delta b_1}{(1+i)} + \frac{\Delta b_2}{(1+i)^2} + \cdots + \frac{\Delta b_n}{(1+i)^n} \tag{2.2.4}$$

and

$$\Delta \text{PVC} = \frac{\Delta c_1}{(1+i)} + \frac{\Delta c_2}{(1+i)^2} + \cdots + \frac{\Delta c_n}{(1+i)^n} \tag{2.2.5}$$

When selecting a set of projects, one rule for optimal selection is to maximize the present value of net benefits. Another ranking criterion is to use the benefit-cost ratio (B/C), PVB/PVC.

$$\frac{\text{B}}{\text{C}} = \frac{\text{PVB}}{\text{PVC}} \tag{2.2.6}$$

This method has the option of subtracting recurrent costs from the annual benefits or including all costs in the present value of cost. Each of these options will result in a different B/C, with higher B/Cs when netting out annual costs, if the B/C is greater than one. The B/C is frequently used to screen infeasible alternatives whose B/C $<$ 1 from further consideration.

Selection of the optimum alternative is based upon the incremental benefit-cost ratios, ΔB/ΔC, whereas the B/C ratio is used for ranking alternatives. The incremental benefit-cost ratio is

$$\frac{\Delta \text{B}}{\Delta \text{C}} = \frac{\text{PVB}(A_j) - \text{PVB}(A_k)}{\text{PVC}(A_j) - \text{PVC}(A_k)} \tag{2.2.7}$$

where PVB(A_j) is the present value of benefits for alternative A_j. Figure 2.2.2 is a flowchart illustrating the benefit-cost method.

Example 2.2.1. Determine the optimum scale of development for a hydroelectric project using the benefit-cost analysis procedure. The various alternative size projects and corresponding benefits are listed in Table 2.2.1.

Solution. According to Fig. 2.2.2, the benefit-cost analysis procedure first computes the B/Cs of each alternative and ranks the projects with B/C $>$ 1 in order of increasing cost.

TABLE 2.2.1

Determination of optimum scale of development of a hydroelectric plant for Example 2.2.2 (Source: Sewell et al., 1961)

1	2	3	4	5	Incremental		
Scale (kw)	Costs C ($000)	Benefits B ($000)	Net Benefits ($000)	B/C	Costs ΔC ($000)	Benefits ΔB ($000)	ΔB/ΔC
50,000	15,000	18,000	3,000	1.2	—	—	—
60,000	17,400	21,000	3,600	1.2	2,400	3,000	1.3
75,000	21,000	26,700	5,700	1.3	3,600	5,700	1.6
90,000	23,400	29,800	6,400	1.3	2,400	3,100	1.3
*100,000	26,000	32,700	6,700	1.3	2,600	2,900	1.1
125,000	32,500	38,500	6,000	1.2	6,500	5,800	0.9
150,000	37,500	42,500	5,000	1.1	5,000	4,000	0.8
200,000	50,000	50,000	—	1.0	12,500	7,500	0.6

*Optimum Scale of Project

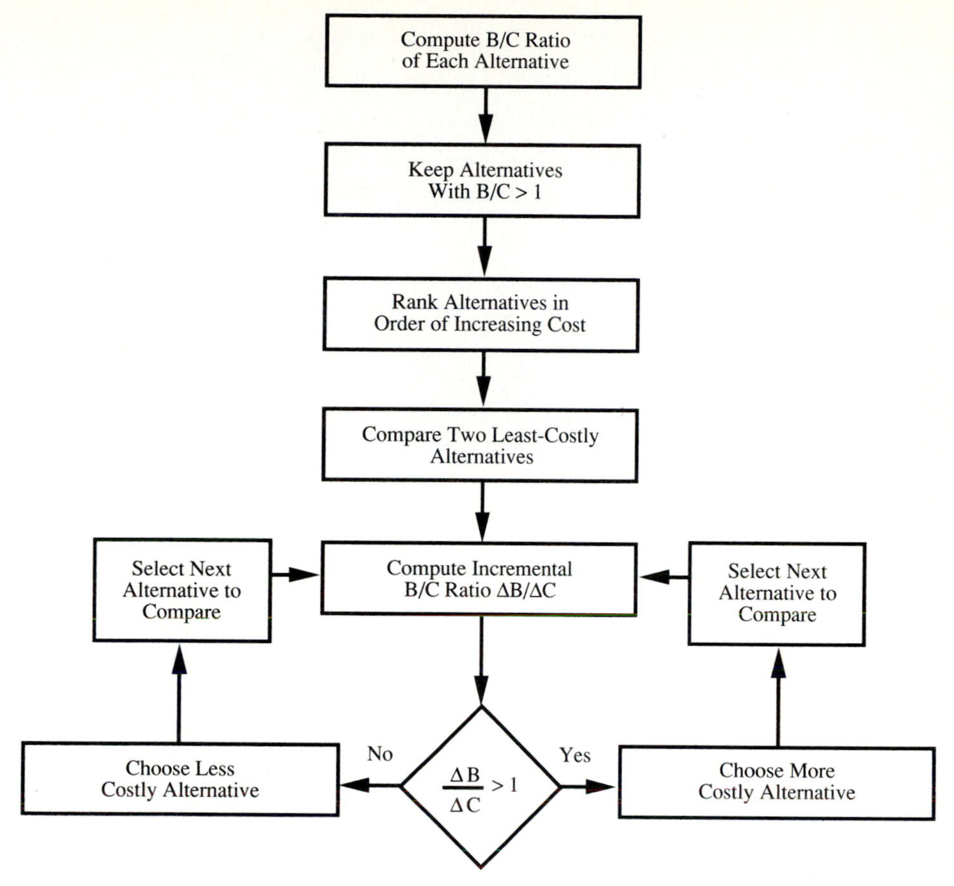

FIGURE 2.2.2
Flowchart for benefit-cost analysis.

Referring to Table 2.2.1, the B/Cs for the alternatives are the incremental benefit-cost ratios, given in column 8. Comparing the 50,000 and 60,000 kW alternatives, the $\Delta B/\Delta C$ is

$$\frac{\Delta B}{\Delta C} = \frac{3000}{2400} = 1.3$$

Note that the incremental benefit-cost ratio is greater than one until the 100,000 and 125,000 kW projects are compared where $\Delta B/\Delta C = 0.9$. This means that the incremental benefits are no longer greater than the incremental costs. The optimum scale of development is the 100,000 kW project, which also has the largest net benefits.

2.3 THEORY OF CONSUMER BEHAVIOR

2.3.1 Utility

A consumer is assumed to choose among alternatives in a manner to gain satisfaction. This assumes that the consumer understands the available alternatives. **Utility func-**

tions contain the information pertaining to the level of satisfaction of each alternative. A utility function with m commodities, $w_1, w_2, \ldots, w_m$, is expressed as

$$u = f(w_1, w_2, \ldots, w_m) \tag{2.3.1}$$

Consider the utility function for a simple case in which a consumer has two commodities to choose from, the utility function is expressed as

$$u = f(w_1, w_2) \tag{2.3.2}$$

where w_1 and w_2 are the quantities of two different commodities. A utility function is assumed continuous with first and second derivatives, and strictly positive first derivatives so that a consumer will always desire more of both commodities. The utility function is defined for consumption during a specified period of time.

A particular level of utility u_0 can be defined as

$$u^0 = f(w_1, w_2) \tag{2.3.3}$$

where u^0 is constant and defines an **indifference curve** which is the locus of combination of commodities for which the consumer has the same level of satisfaction. It is impossible for a single combination of commodities to yield two levels of satisfaction which is to say that indifference curves cannot intersect. Utility function shapes are concave which restricts the shape of indifference curves. For two points, (w_1^0, w_2^0) and (w_1^1, w_2^1) on an indifference curve where $u^0 = f(w_1^0, w_2^0) = f(w_1^1, w_2^1)$, the following is satisfied

$$u = f\left[\theta w_1^0 + (1-\theta)w_1^1, \theta w_2^0 + (1-\theta)w_2^1\right] > u^0 \tag{2.3.4}$$

for all $0 < \theta < 1$. Equation (2.3.4) says that all interior points on a line segment connecting two points on an indifference curve are located on indifference curves of higher levels of satisfaction (see Fig. 2.3.1). An **indifference map** is a system of indifference curves of different levels of utility or satisfaction as shown in Fig. 2.3.2.

Another property of indifference curves is that they tend to approach the axes asymptotically, that is, less and less of one commodity is consumed, the sacrifice of parting with an additional unit becomes greater. Many more units of the second commodity must be substituted to maintain the same level of satisfaction. The total differential of a utility function is

$$du = \frac{\partial f}{\partial w_1} dw_1 + \frac{\partial f}{\partial w_2} dw_2 \tag{2.3.5}$$

where $\partial f / \partial w_1$ and $\partial f / \partial w_2$ are **marginal utilities.**

Moving along an indifference curve substituting one commodity for the other, $du = 0$ so that

$$\frac{\partial f}{\partial w_1} dw_1 + \frac{\partial f}{\partial w_2} dw_2 = 0$$

and by rearranging

$$-\frac{dw_2}{dw_1} = \frac{\dfrac{\partial f}{\partial w_1}}{\dfrac{\partial f}{\partial w_2}} \tag{2.3.6}$$

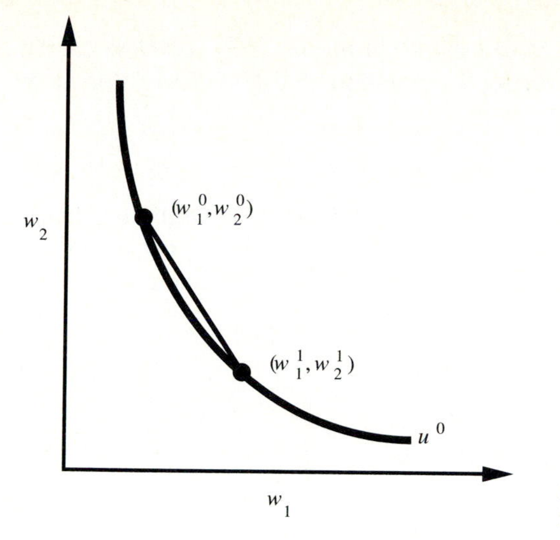

FIGURE 2.3.1
Indifference curve.

which defines the **marginal rate of substitution** or the **rate of commodity substitution** (Henderson and Quandt, 1980). The marginal rate of substitution is the slope of an indifference curve dw_2/dw_1 that defines the rate that a consumer substitutes w_1 for w_2 per unit rate of w_1 to maintain a specified level of utility.

2.3.2 Maximization of Utility

Consider the following consumer budget constraint

$$B^0 = p_1 w_1 + p_2 w_2 \tag{2.3.7}$$

where B^0 represents the consumer income and p_1 and p_2 are the prices of w_1 and w_2, respectively. A consumer wants to maximize the utility function Eq. (2.3.2) subject

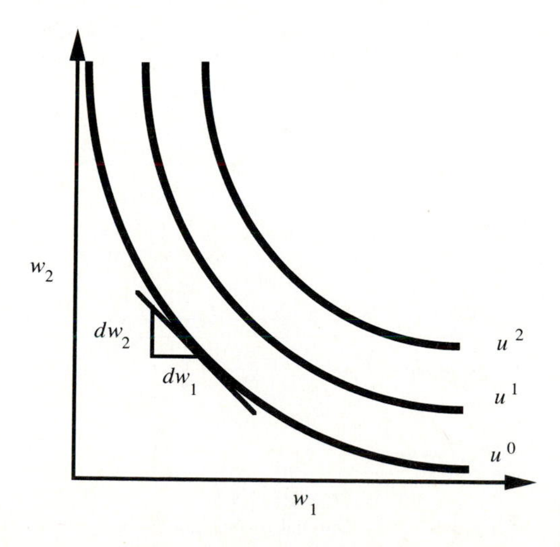

FIGURE 2.3.2
Indifference map.

to the budget constraint Eq. (2.3.7). This constrained maximization problem can be approached through the use of a **Lagrangian function** (see Section 4.5 for more details)

$$L = f(w_1, w_2) + \lambda(B^0 - p_1 w_1 - p_2 w_2) \tag{2.3.8}$$

which combines Eqs. (2.3.2) and (2.3.7) and using λ which is a **Lagrange multiplier**. The unknowns in the Lagrangian function are w_1, w_2 and λ. For optimality (maximization), the following conditions must be satisfied from simple principles in differential calculus:

$$\frac{\partial L}{\partial w_1} = \frac{\partial f}{\partial w_1} - \lambda p_1 = 0 \tag{2.3.9a}$$

$$\frac{\partial L}{\partial w_2} = \frac{\partial f}{\partial w_2} - \lambda p_2 = 0 \tag{2.3.9b}$$

$$\frac{\partial L}{\partial \lambda} = B^0 - p_1 w_1 - p_2 w_2 = 0 \tag{2.3.9c}$$

Combining Eqs. (2.3.9*a*, *b*) results in

$$\frac{\dfrac{\partial f}{\partial w_1}}{\dfrac{\partial f}{\partial w_2}} = \frac{p_1}{p_2} \quad \text{or} \quad \frac{\partial w_2}{\partial w_1} = \frac{p_1}{p_2} \tag{2.3.10}$$

which says that, when the maximum utility is reached, the ratio of the marginal utilities must equal the ratio of prices. The optimal points for income allocation for three budget levels B^0, B^1, and B^2 are illustrated in Fig. 2.3.3.

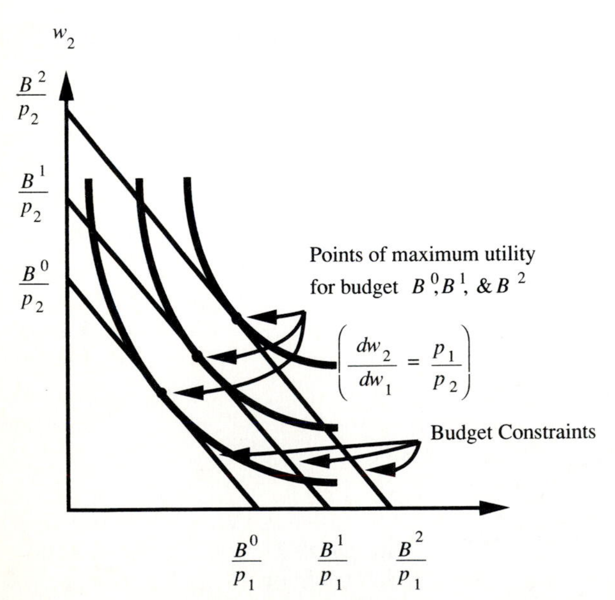

FIGURE 2.3.3
Optimal income allocation.

By Eq. (2.3.6) the left-hand side of Eq. (2.3.10) is the marginal rate of substitution $(-\partial w_2/\partial w_1)$, so that at the maximum utility, the marginal rate of substitution is equal to the price ratio. Equations (2.3.9*a*, *b*) can be written

$$\frac{\dfrac{\partial f}{\partial w_1}}{p_1} = \frac{\dfrac{\partial f}{\partial w_2}}{p_2} = \lambda \tag{2.3.11}$$

which says that the marginal utility divided by the price of the commodity must be the same for all commodities.

Second-order conditions for the maximization of Lagrangian function Eq. (2.3.8) requires that the Hessian determinant is positive (see Section 4.3 for more detail).

$$\begin{vmatrix} \dfrac{\partial^2 f}{\partial w_1^2} & \dfrac{\partial^2 f}{\partial w_1 \partial w_2} & -p_1 \\ \dfrac{\partial^2 f}{\partial w_2 \partial w_1} & \dfrac{\partial^2 f}{\partial w_2^2} & -p_2 \\ -p_1 & p_2 & 0 \end{vmatrix} > 0 \tag{2.3.12}$$

which is

$$2\left(\frac{\partial^2 f}{\partial w_1 \partial w_2}\right) p_1 p_2 - \frac{\partial^2 f}{\partial w_1^2} p_2^2 - \frac{\partial^2 f}{\partial w_2^2} p_1^2 > 0 \tag{2.3.13}$$

From Eqs. (2.3.9*a*, *b*), $p_1 = (\partial f/\partial w_1)\,(1/\lambda)$ and $p_2 = (\partial f/\partial w_2)(1/\lambda)$, which are substituted into Eq. (2.3.13) and multiplying by λ^2 results in

$$2\left(\frac{\partial^2 f}{\partial w_1 \partial w_2}\right)\left(\frac{\partial f}{\partial w_1}\right)\left(\frac{\partial f}{\partial w_2}\right) - \frac{\partial^2 f}{\partial w_1^2}\left(\frac{\partial f}{\partial w_2}\right)^2 - \left(\frac{\partial^2 f}{\partial w_2^2}\right)\left(\frac{\partial f}{\partial w_1}\right)^2 > 0 \tag{2.3.14}$$

which is the strict inequality for a strict quasi-concave function (Section 4.3).

2.3.3 Demand Functions

A consumer's **demand function** describes the quantity of a commodity that the consumer is willing to buy as a function of prices and income. **Demand curves** are generally assumed to be negatively sloped (Fig. 2.3.4), that is, the lower the price the greater the quantity demanded. Demand functions are single-valued functions of prices and income, as shown in the demand curve for water in Fig. 2.3.4. If all prices and income change in the same proportion, then the quantities demanded remain unchanged. This is to say that demand functions are **homogeneous of degree zero in prices and income** (Henderson and Quandt, 1980).

Two forces act on the consumer when the price of a commodity changes: (1) the exchange of commodities or **substitution effect**; and (2) the **income effect** in which income increases if price decreases and decreases if price increases. The substitution effect always occurs such that an increase in price of a commodity results in less of

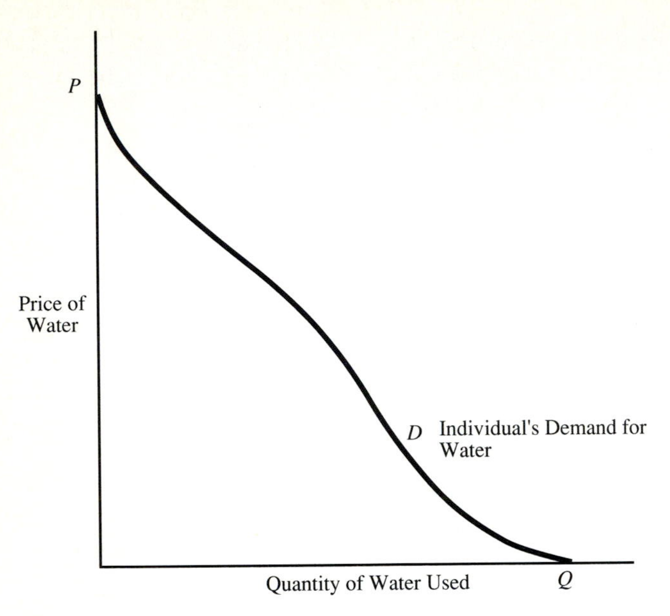

FIGURE 2.3.4
Individual demand curve for water.

the commodity being consumed. A price decrease results in more of the commodity being consumed.

The **elasticity of demand** is the proportionate rate of change in quantity demanded divided by the proportionate rate of change in its own price,

$$\epsilon_{11} = \frac{\partial w_1/w_1}{\partial p_1/p_1} = \frac{p_1}{w_1}\frac{\partial w_1}{\partial p_1} \tag{2.3.15}$$

Commodities with high elasticities ($\epsilon_{11} \geq -1$) are necessities. Large elasticities imply that the quantity demanded is very responsive to price changes.

A consumer's expenditure on a commodity is $p_1 w_1$ so that the change with respect to price is

$$\begin{aligned}\frac{\partial(p_1 w_1)}{\partial p_1} &= w_1 + p_1\frac{\partial w_1}{\partial p_1} = w_1\left(1 + \frac{p_1}{w_1}\frac{\partial w_1}{\partial p_1}\right)\\ &= w_1(1+\epsilon_{11})\end{aligned} \tag{2.3.16}$$

This clearly indicates that if $\epsilon_{11} > -1$, a consumer's expenditures on w_1 will increase with p_1. If $\epsilon_{11} < -1$, a consumer's expenditures will decrease and if $\epsilon_{11} = -1$, expenditures will remain unchanged. Price elasticity of demand is a property of the demand curve so that the elasticity is influenced by those factors which influence demand. Typically, the more elastic a commodity, the more substitutes there are available, the wider the range of uses and the larger the proportion of consumer's income that is spent on the commodity. In summary, demand is relatively inelastic when quantity changes less than proportionately with price and is relatively elastic when quantity changes more than proportionately with price.

The **cross-price elasticity** of demand relates the proportionate change in quantity of a commodity to the proportionate change in price of another commodity,

$$\epsilon_{21} = \frac{p_1}{w_2}\frac{\partial w_2}{\partial p_1} \tag{2.3.17}$$

which may be positive or negative.

Example 2.3.1. Billings and Agthe (1980) developed the following water demand function for Tucson, Arizona

$$\ln(Q) = -7.36 - 0.267\ln(P) + 1.61\ln(I) - 0.123\ln(D) + 0.0897\ln(W)$$

where Q is the monthly water consumption of the average household in 100 ft^3; P is the marginal price facing the average household in cents per 100 ft^3; D is the difference between the actual water and sewer use bill minus what would have been paid if all water was sold at the marginal rate (\$); I is the personal income per household in \$/month; and W is the evapotranspiration minus rainfall (inches). Determine the price elasticity of demand for water.

Solution. This demand function for the commodity, water, can be written as

$$Q = 0.0006362P^{-0.267}I^{1.61}D^{-0.123}W^{0.0897}$$

The price elasticity of demand according to Eq. (2.3.15) is

$$\epsilon = \frac{P}{Q}\frac{dQ}{dP}$$

where

$$\frac{dQ}{dP} = -0.267(0.0006362)P^{-1.267}I^{1.61}D^{0.123}W^{0.0897}$$

$$= -0.267P^{-1}Q$$

The elasticity is then

$$\epsilon = \frac{P}{Q}(-0.267P^{-1}Q)$$

$$= -0.267$$

This price elasticity of -0.267 indicates that for a 1.0 percent increase in price, a 0.267 percent decrease in quantity demanded would be expected or, conversely, a 1.0 percent decrease in price would produce a 0.267 percent increase in quantity demanded.

2.4 THEORY OF THE FIRM

A **firm** is a technical unit in which commodities are produced (Henderson and Quandt, 1980). The **theory of the firm** focuses on explaining how a firm should: (a) allocate its inputs or resources in production of the output or product; (b) decide on the level of production; and (c) respond to a change in price of inputs and outputs. The transformation of inputs into outputs are described through the use of a production function. Hydrosystems can be analyzed using concepts of the theory of the firm.

2.4.1 Basic Concepts

A **production function** expresses the quantity of outputs as a function of the quantities of the variable inputs. Inputs can be thought of as fixed inputs where fixed refers to the fact that the amount of the input does not vary and as variable inputs. The production function could be stated mathematically for one output, q, with m variable inputs x as

$$q = f(x_1, x_2, \ldots, x_m) \tag{2.4.1}$$

A production function presupposes technical efficiency and states the maximum output obtainable from every possible input combination.

Consider the simple production process in Table 2.4.1 which has two variable inputs, irrigation water x_1 and nitrogen fertilizer x_2, with the output being the yield of corn, q.

$$q = f(x_1, x_2) \tag{2.4.2}$$

This production process also has several fixed inputs which include the seed, labor, service of the machinery, and service of the land.

Input and output levels are rates of use or production per unit of time. In the production example in Table 2.4.1 the unit of time is a growing season. In the long run, levels of all inputs are variables whereas in the short run, a fixed input is constant for which the level of availability cannot be altered.

The **total product** of input x_2 in production of q is the quantity of output from input x_2 if x_1 is fixed as $\underline{x}_1$

$$q = f(\underline{x}_1, x_2)$$

so that q is a function of x_2 alone. The relation between q and x_2 is altered by changing $\underline{x}_1$. For each value of $\underline{x}_1$, a **total product curve** can be developed to show the curve of the total product q as a function of the amount of the variable input $\underline{x}_2$.

The **average product** (AP) of x_2 is the total product divided by the quantity of variable input x_2 for fixed input $\underline{x}_1$

$$\text{AP} = \frac{q}{x_2} = \frac{f(\underline{x}_1, x_2)}{x_2} \tag{2.4.3}$$

The **marginal product** (MP_{x_2}) of x_2 is the rate of change of the total product with respect to the quantity of variable input x_2

$$\text{MP}_{x_2} = \frac{\partial q}{\partial x_2} = \frac{\partial f(\underline{x}_1, x_2)}{\partial x_2} = \frac{\Delta q}{\Delta x_2} \tag{2.4.4}$$

The marginal product is the slope of the total product curve. Referring to Fig. 2.4.1, the marginal product increases from the origin to the point of inflection of the total product curve where the slope is a maximum. The marginal product and the average product are equal at the maximum of the average product.

Example 2.4.1. Use the production function presented in Table 2.4.1 to determine the average products and marginal product for irrigation water held constant at 7 inches/acre and considering 40 and 50 lbs of fertilizer per acre.

TABLE 2.4.1

Production schedule of the relationship between irrigation water, nitrogen fertilizer, and yield (in bushels/acre) of corn (Schefter et al., 1978)

Pounds of Nitrogren/Acre	x_1: Inches of Irrigation—Water/Acre													
	0	**1**	**2**	**3**	**4**	**5**	**6**	**7**	**8**	**9**	**10**	**11**	**12**	**13**
0	0.0	0.0	0.0	0.0	0.0	0.0	0.0	0.0	0.0	0.0	0.0	0.0	0.0	0.0
10	0.0	1.8	5.0	9.0	13.2	17.0	19.8	21.0	20.0	16.2	9.0	0.0	0.0	0.0
20	0.0	5.0	12.8	22.2	32.0	41.0	48.0	51.8	51.2	45.0	32.0	11.0	0.0	0.0
30	0.0	9.0	22.2	37.8	54.0	69.0	81.0	88.2	88.0	81.0	63.0	33.0	0.0	0.0
40	0.0	13.2	32.0	54.0	76.8	98.0	115.2	126.0	128.0	118.8	96.0	57.2	0.0	0.0
50	0.0	17.0	41.0	69.0	98.0	125.0	147.0	161.0	164.0	153.0	125.0	77.0	6.0	0.0
60	0.0	19.8	48.0	81.0	115.2	147.0	172.8	189.0	192.0	178.2	144.0	85.8	0.0	0.0
70	0.0	21.0	51.8	88.2	126.0	161.0	189.0	205.8	207.2	189.0	147.0	77.0	0.0	0.0
80	0.0	20.0	51.2	88.0	128.0	164.0	192.0	207.2	204.8	180.0	128.0	44.0	0.0	0.0
90	0.0	16.2	45.0	81.0	118.8	153.0	178.2	189.0	180.0	145.8	81.0	0.0	0.0	0.0
100	0.0	9.0	32.0	63.0	96.0	125.0	144.0	147.0	128.0	81.0	0.0	0.0	0.0	0.0
110	0.0	0.0	11.0	33.0	57.2	77.0	85.8	77.0	44.0	0.0	0.0	0.0	0.0	0.0
120	0.0	0.0	0.0	0.0	0.0	6.0	0.0	0.0	0.0	0.0	0.0	0.0	0.0	0.0
130	0.0	0.0	0.0	0.0	0.0	0.0	0.0	0.0	0.0	0.0	0.0	0.0	0.0	0.0

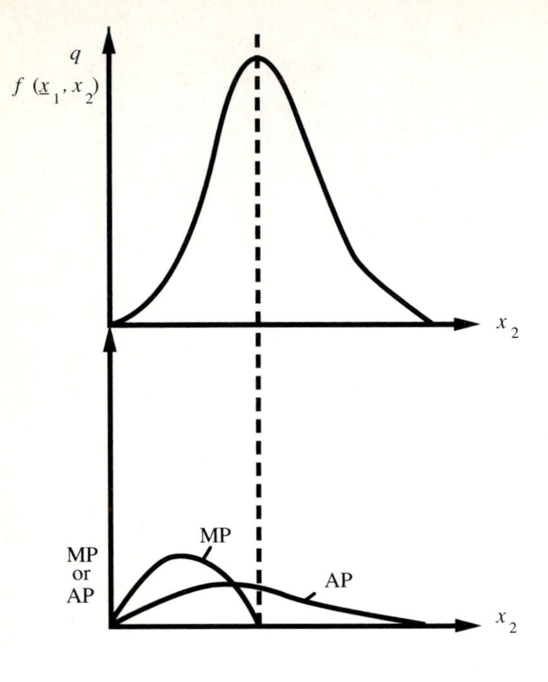

FIGURE 2.4.1
Total product, average product and marginal product.

Solution. Total products for 40 and 50 lbs of fertilizer per acre with water at 7 inches/acre are 126 and 161 bushels, respectively. The average product by Eq. (2.4.3) for 40 lbs of fertilizer is

$$\mathrm{AP} = \frac{q}{x_2} = \frac{126}{40} = 3.15$$

and for 50 lbs is 3.22. The marginal product by Eq. (2.4.4) is

$$\mathrm{MP} = \frac{\Delta q}{\Delta x_2} = \frac{161 - 126}{50 - 40} = 3.5$$

which is the slope of the total product curve.

An **isoquant** defines the locus of all combinations of the variable inputs which yield the same output level, q^0, for example

$$q^0 = f(x_1, x_2) \tag{2.4.5}$$

The isoquants for a two-variable input process are illustrated in Fig. 2.4.2 where the output levels are $q^3 > q^2 > q^1$.

The **rate of technical substitution** (RTS) is the negative of the slope of the isoquant,

$$\mathrm{RTS} = -\frac{dx_2}{dx_1} \tag{2.4.6}$$

which is the rate at which one input must be substituted for another to maintain the

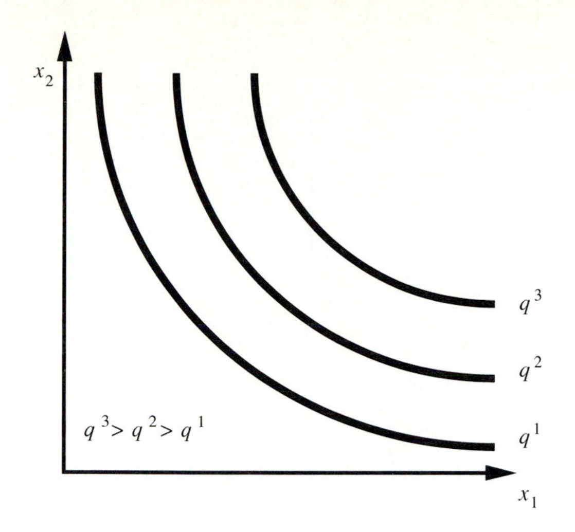

FIGURE 2.4.2
Family of isoquants for two-variable input process.

same output level. The total differential of a production function is

$$dq = \frac{\partial f(x_1, x_2)}{\partial x_1} dx_1 + \frac{\partial f(x_1, x_2)}{\partial x_2} dx_2 \tag{2.4.7a}$$

$$= \text{MP}_{x_1} dx_1 + \text{MP}_{x_2} dx_2 \tag{2.4.7b}$$

To remain on an isoquant for changes in dx_1 and dx_2, then $dq = 0$

$$0 = \text{MP}_{x_1} dx_1 + \text{MP}_{x_2} dx_2 \tag{2.4.8}$$

so that the rate of technical substitution is the ratio of the marginal products

$$\text{RTS} = -\frac{dx_2}{dx_1} = \frac{\text{MP}_{x_1}}{\text{MP}_{x_2}} \tag{2.4.9}$$

The isoquant map of the production process in Table 2.4.1 is shown in Fig. 2.4.3. If MP_{x_1} or MP_{x_2} becomes negative, then too much of the input x_1 or x_2, respectively, is used in the production process and the RTS would be negative. The production process in Fig. 2.4.3 illustrates the area of rational operation where both MP_{x_1} and MP_{x_2} are positive and are enclosed by the **ridge lines** R_1 and R_2. Point B is preferable to point A because much less of x_2 is used at B to produce the same output. Similarly, point B is preferable to point C because less of x_1 at point B is required to produce the same output level. At point D too much of both x_1 and x_2 are required.

In more general form for n variable inputs along an isoquant the following relation holds,

$$\sum_{j=1}^{m} \frac{\partial f(\mathbf{x})}{\partial x_j} dx_j = 0 \tag{2.4.10}$$

where $\mathbf{x} = (x_1, x_2, \ldots, x_m)$.

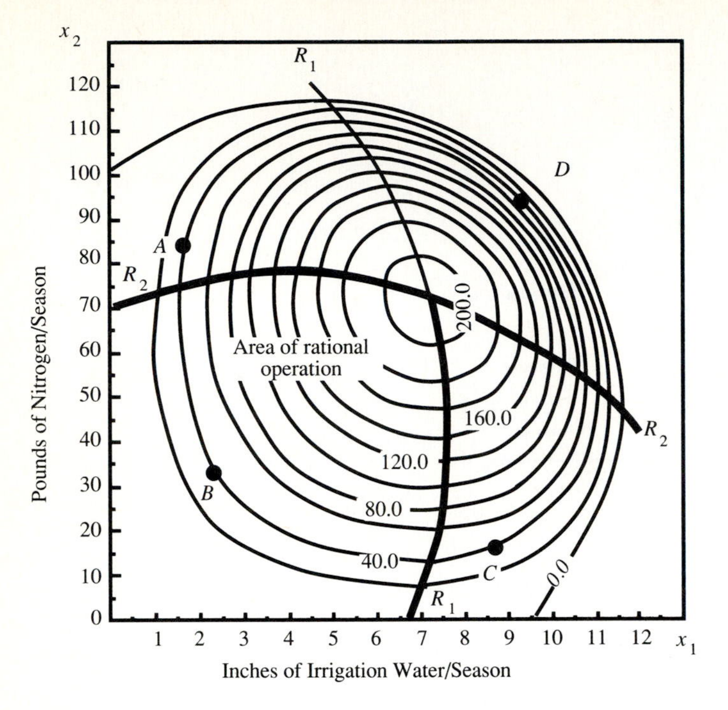

FIGURE 2.4.3
An isoquant map of the production response surface depicted in Table 2.4.1 (Schefter, et al., 1978)

2.4.2 Optimal Input Combinations

Consider the following linear cost function for the two-variable input production process

$$C = r_1 x_1 + r_2 x_2 \tag{2.4.11}$$

where r_1 and r_2 are the prices of inputs x_1 and x_2, respectively. For C fixed at C^0 then an **isocost line** is the locus of input combinations which can be purchased for the fixed cost. Rearranging, an expression for x_2 is

$$x_2 = \frac{C}{r_2} - \frac{r_1}{r_2} x_1 \tag{2.4.12}$$

Fig. 2.4.4 illustrates a family of isocost lines superimposed on a family of isoquants. The optimal input combinations are at input combinations for a particular isoquant where the cost is a minimum. These are at points A, B, and C for output levels q^1, q^2 and q^3, respectively. Differentiating Eq. (2.4.12) with respect to x_1 is

$$\frac{dx_2}{dx_1} = -\frac{r_1}{r_2} \tag{2.4.13}$$

and combining with Eq. (2.4.9) defines the rate of technical substitution as

$$\text{RTS} = -\frac{dx_2}{dx_1} = \frac{\text{MP}_{x_1}}{\text{MP}_{x_2}} = \frac{r_1}{r_2} \tag{2.4.14}$$

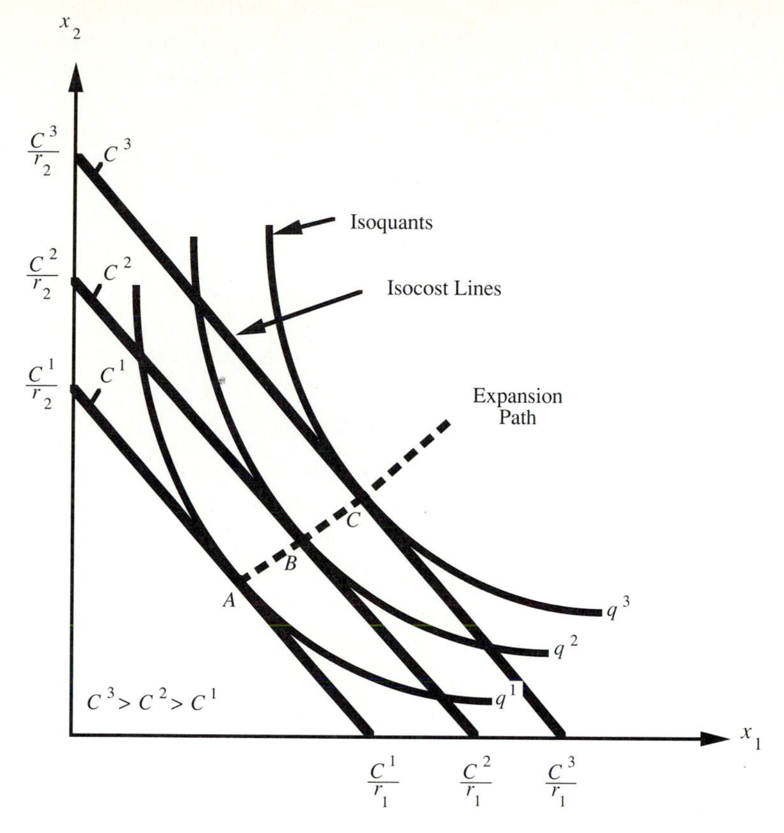

FIGURE 2.4.4
Family of isocost lines superimposed on a family of isoquants.

If a producer is either to maximize output subject to a specified cost or to minimize cost subject to a specified output, then the producer will always use inputs so as to equate the RTS and the input price ratio. Equation (2.4.14) can be rearranged to

$$\frac{\text{MP}_{x_1}}{r_1} = \frac{\text{MP}_{x_2}}{r_2} \tag{2.4.15}$$

which states that the optimal input combination is found where the marginal product of each input per dollar of its cost is equal for all inputs.

The locus of tangency points defines the **expansion path** of the firm as shown in Fig. 2.4.4. Rational operation in a production process only selects input combinations that are on the expansion path. The expansion path can be stated mathematically as

$$g(x_1, x_2) = 0 \tag{2.4.16}$$

for which the first- and second-order conditions for constrained maxima and minima are satisfied (Henderson and Quandt, 1980).

Example 2.4.2. Using the production function in Table 2.4.1, a producer has a cost outlay of \$100/acre for inputs, water and fertilizer. The costs of these inputs are \$10/acre-inch for water and \$2.50/lb/acre for fertilizer. The producer is operating so that the rate of technical substitution is 19 lbs of fertilizer per inch of water per acre and the level of production is 20 bushels/acre. Is the producer operating at optimal?

Solution. The producer is operating such that the rate of technical substitution is 19; however, the optimal RTS is by Eq. (2.4.14)

$$\text{RTS} = \frac{r_1}{r_2} = \frac{10}{2.5} = 4$$

Basically this says that too much fertilizer is being used and that water can be substituted for fertilizer. A producer must use 4 lbs per inch to be at optimum. By substituting one inch of water for 19 lbs of fertilizer, the producer could save the following amount per acre

$$19 \text{ lbs } \times \$2.5/\text{lb } - 1 \text{ in } \times \$10/\text{in } = \$37.50$$

and still produce 20 bushels per acre. This savings could be applied to the purchase of more water in order to increase output beyond the 20 bushels/acre.

This analogy for maximizing output subject to a cost constraint can be derived from the Lagrangian function expressed as

$$L_1 = f(x_1, x_2) - \lambda_1(C^0 - r_1x_1 - r_2x_2) \tag{2.4.17}$$

where λ_1 is the Lagrange multiplier. The partial derivatives of (2.4.17) are

$$\frac{\partial L_1}{\partial x_1} = \frac{\partial f}{\partial x_1} - \lambda_1 r_1 = 0 \tag{2.4.18}$$

$$\frac{\partial L_1}{\partial x_2} = \frac{\partial f}{\partial x_2} - \lambda_1 r_2 = 0 \tag{2.4.19}$$

$$\frac{\partial L_1}{\partial \lambda_1} = C^0 - r_1x_1 - r_2x_2 = 0 \tag{2.4.20}$$

Equations (2.4.18) and (2.4.19) can be used to derive (2.4.15). Also, Eqs. (2.4.18) and (2.4.19) can be used to solve for the Lagrange multiplier

$$\lambda_1 = \frac{\partial f}{\partial x_1}\frac{1}{r_1} = \frac{\partial f}{\partial x_2}\frac{1}{r_2} \tag{2.4.21}$$

Another approach may be to minimize the cost of producing a specified level of output, in which case Eq. (2.4.11) is minimized subject to Eq. (2.4.5). The Lagrangian function is

$$L_2 = r_1x_1 + r_2x_2 + \lambda_2[q^0 - f(x_1, x_2)] \tag{2.4.22}$$

where the partial derivatives of L_2 with respect to x_1, x_2, and λ_2 are

$$\frac{\partial L_2}{\partial x_1} = r_1 - \lambda_2\frac{\partial f}{\partial x_1} = 0 \tag{2.4.23}$$

$$\frac{\partial L_2}{\partial x_2} = r_2 - \lambda_2\frac{\partial f}{\partial x_2} = 0 \tag{2.4.24}$$

$$\frac{\partial L_2}{\partial \lambda_2} = q^0 - f(x_1, x_2) = 0 \tag{2.4.25}$$

Equations (2.4.23) and (2.4.24) can be used to derive Eq. (2.4.15).

Profit maximization is the ultimate goal of the firm in which the levels of both cost and output can vary. **Profit**, P_f, is the difference between total revenue (pq) and total cost (C),

$$\begin{aligned} P_f &= pq - C \\ &= pf(x_1, x_2) - r_1 x_1 - r_2 x_2 \end{aligned} \tag{2.4.26}$$

In order to maximize, set the partial derivatives $\partial P_f/\partial x_1$ and $\partial P_f/\partial x_2$ equal to zero

$$\frac{\partial P_f}{\partial x_1} = p\frac{\partial f}{\partial x_1} - r_1 = 0 \tag{2.4.27}$$

$$\frac{\partial P_f}{\partial x_2} = p\frac{\partial f}{\partial x_1} - r_2 = 0 \tag{2.4.28}$$

which can be used to derive Eq. (2.4.15).

Example 2.4.3. A methodology for estimating the cost of controlling pollution from urban stormwater was developed by Heaney et al. (1978) to determine the optimal combination of storage and treatment. The inputs are the storage volume and the treatment rate, and the output for the production function is the maximum level of control. The isoquants are expressed as

$$T = T_1 + (T_2 - T_1)e^{-KS}$$

where T is the wet-weather treatment rate in inches per hour; T_1 is the treatment rate at which the isoquant is asymptotic to the ordinate, in inches per hour; T_2 is the treatment rate at which isoquants intersect the abscissa in inches per hour; S is the storage volume in inches; and K is a constant per inch. The unit cost of treatment is r_T and the unit cost of storage is r_S. Determine an equation for the optimal amount of storage in inches and the optimal treatment rate in inches per hour to minimize the cost.

Solution. The wet-weather optimization problem is to minimize the cost,

$$\text{Minimize } z = r_S S + r_T T \tag{a}$$

subject to the constraint (isoquants)

$$T = T_1 + (T_2 - T_1)e^{-KS} \tag{b}$$

with T, $S \geq 0$. The Lagrangian function is

$$L = r_S S + r_T T + \lambda[T - T_1 - (T_2 - T_1)e^{-KS}] \tag{c}$$

then the first-order conditions for optimality are

$$\frac{\partial L}{\partial S} = 0 = r_S - \lambda(T_2 - T_1)e^{-KS}(-K) \tag{d}$$

$$\frac{\partial L}{\partial T} = 0 = r_T + \lambda \tag{e}$$

$$\frac{\partial L}{\partial \lambda} = 0 = T - T_1 - (T_2 - T_1)e^{-KS} \tag{f}$$

From (d) then

$$r_S = -K\lambda(T_2 - T_1)e^{-KS} \tag{g}$$

and rearranging

$$-KS = \ln\left(\frac{r_S}{-K\lambda(T_2 - T_1)}\right) \tag{h}$$

and

$$S = -\frac{1}{K}\ln\left(\frac{r_S}{-K\lambda(T_2 - T_1)}\right)$$

$$= \frac{1}{K}\ln\left(\frac{-K\lambda(T_2 - T_1)}{r_S}\right) \tag{i}$$

From (e) substitute $r_T = -\lambda$ so

$$S = \frac{1}{K}\ln\left[\frac{r_T}{r_S}K(T_2 - T_1)\right]$$

Because S cannot be negative, the optimum storage, S^*, would be

$$S^* = \max\left\{\frac{1}{K}\ln\left[\frac{r_T}{r_S}K(T_2 - T_1)\right], 0\right\}$$

The optimum amount of treatment is

$$T^* = T_1 + (T_2 - T_1)e^{-KS^*}$$

2.4.3 Cost in the Short Run

The production function Eq. (2.4.2), the cost Eq. (2.4.11), and the expansion path Eq. (2.4.16) form a system of equations. Now consideration is given to a **short run analysis**. This type of analysis is based on a short enough period of time such that one or more variable input quantities are fixed. Consider a short run so that input x_1 is fixed and x_2 is variable, then Eqs. (2.4.2), (2.4.11) and (2.4.16) become

$$q = f(\underline{x}_1, x_2)$$

$$C = r_1\underline{x}_1 + r_2 x_2$$

$$g(\underline{x}_1, x_2) = 0$$

This system of equations is combined into a single equation

$$C = \phi(q, r_1, r_2) + c \tag{2.4.29}$$

where c is the fixed cost ($c = r_1\underline{x}_1$). Over the short run, the costs of fixed inputs is the **fixed cost**. Assuming that input prices are invariant, then

$$C = \phi(q) + c \tag{2.4.30}$$

which is associated with points along the vertical line in Fig. 2.4.5 with $\underline{x}_1$ fixed.

Total variable cost (TVC) is the money spent on the variable inputs. **Total fixed cost** is the money spent on the fixed explicit cost (for fixed inputs) and the implicit costs of production. **Total cost** (TC) is the sum of the total fixed cost and the total variable costs. The **average total cost** (ATC), the **average variable cost** (AVC), and the **average fixed costs** (AFC) are defined, respectively, as

$$\text{ATC} = \frac{\phi(q) + c}{q} \tag{2.4.31}$$

$$\text{AVC} = \frac{\phi(q)}{q} = \frac{r_2 x_2}{q} \tag{2.4.32}$$

$$\text{AFC} = \frac{c}{q} = \frac{r_1 \underline{x}_1}{q} \tag{2.4.33}$$

where $c = r_1 \underline{x}_1$, as before. The **marginal cost** (MC) is the change in total cost attributable to one-unit change in output

$$\text{MC} = \frac{dC}{dq} = \frac{\Delta \text{TC}}{\Delta q} \tag{2.4.34}$$

Figures 2.4.6*a* and 2.4.6*b* depict a production process in the short run in which the total cost is a cubic function of output. It can be shown that the MC curve passes through the minimum points of the AVC and ATC curves.

The average variable cost Eq. (2.4.32) is

$$\text{AVC} = \frac{\text{TVC}}{q} = \frac{r_2 x_2}{q} = r_2 \frac{1}{\text{AP}} \tag{2.4.35}$$

where AP is the **average product** (q/x_2). The marginal cost from (2.4.34) can be defined further

$$\text{MC} = \frac{\Delta \text{TVC}}{\Delta q} = \frac{r_2 \Delta x_2}{\Delta q} = \frac{r_2}{\text{MP}} \tag{2.4.36}$$

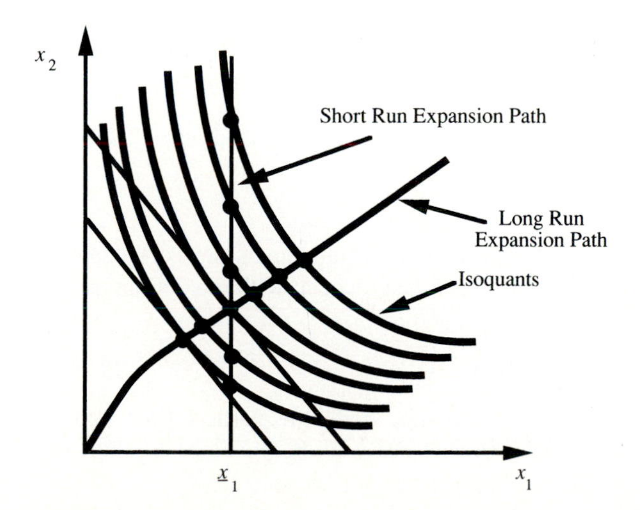

FIGURE 2.4.5
Comparison of short run and long run analysis.

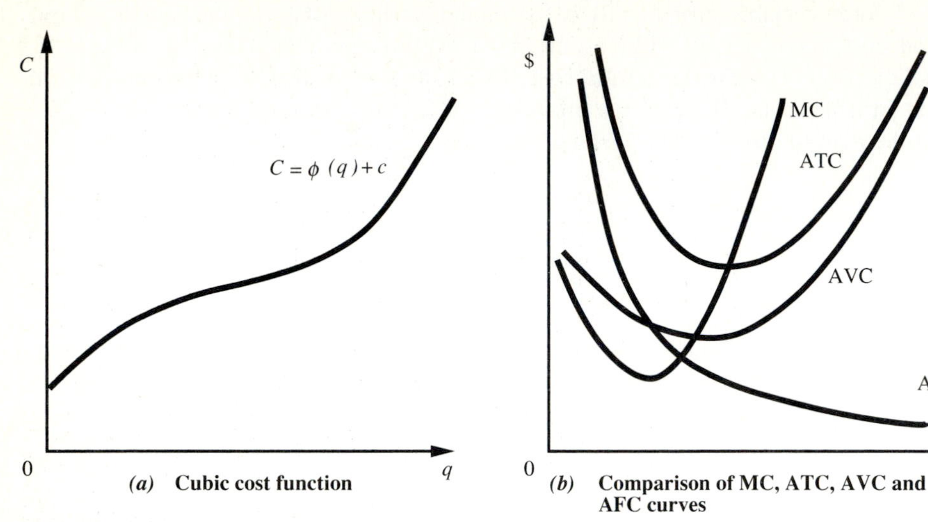

FIGURE 2.4.6
Production process in short run.

where MP is the **marginal product** (2.4.4). This shows that MC will decrease with an increase in MP and increases with a decrease in MP.

The **total revenue** (TR) from sale of a product is TR = pq so that the profit is P_f = TR − TC. To maximize profit

$$\frac{dP_f}{dq} = p - \frac{d\phi}{dq} = 0$$

so that to maximize

$$p = \frac{d\phi}{dq} \tag{2.4.37}$$

The unit price, p, is also the marginal revenue in this case and $d\phi/dq$ is the marginal cost, so that **profit is maximized when the marginal revenue is equal to the marginal cost**.

2.4.4 Cost in the Long Run

Long run refers to a period of time of such length that all inputs are considered as variable. Referring to Fig. 2.4.5, long-run costs are associated with the optimal points along the expansion path

$$g(x_1, x_2) = 0 \tag{2.4.38}$$

whereas short-run costs are associated with points along the vertical line with x_1 fixed. Mathematically, the long-run average-cost curve is the envelope of short-run average-cost curves.

2.4.5 Elasticities of Output and Substitution

The **elasticity of output** with respect to a change in the jth input is

$$\epsilon_j(\mathbf{x}) = \frac{x_j}{f(\mathbf{x})} \frac{\partial f(\mathbf{x})}{\partial x_j} \qquad j = 1, \ldots, n \tag{2.4.39}$$

in which $\mathbf{x} = (x_1, x_2, \ldots, x_n)$, an n-dimensional vector. The elasticity of production is the sum of all elasticities of output with respect to the various inputs,

$$\epsilon(\mathbf{x}) = \sum_{j=1}^{n} \epsilon_j(\mathbf{x}) \tag{2.4.40}$$

Elasticity of substitution measures the rate at which substitution of inputs take place and is the proportionate rate of change of the input ratio divided by the proportionate rate of change of the RTS

$$\sigma = \frac{F}{x_2/x_1} \frac{d(x_2/x_1)}{d(F)} \tag{2.4.41}$$

where

$$F = \frac{\dfrac{\partial f}{\partial x_1}}{\dfrac{\partial f}{\partial x_2}} \tag{2.4.42}$$

which can be reduced to (Henderson and Quandt, 1980)

$$\sigma = \frac{\dfrac{\partial f}{\partial x_1}\left(\dfrac{\partial f}{\partial x_1} x_1 + \dfrac{\partial f}{\partial x_2} x_2\right)}{x_1 x_2 \dfrac{\partial f}{\partial x_2}\left[\dfrac{\partial f}{\partial x_1}\dfrac{\partial F}{\partial x_2} - \dfrac{\partial f}{\partial x_2}\dfrac{\partial F}{\partial x_1}\right]} \tag{2.4.43}$$

using the definition of the total derivative of $\partial(x_2/x_1)/\partial F$. The elasticity of substitution is positive if the production function has convex isoquants. Some production functions have constant elasticities of substitution; however, for most production functions, σ varies from point to point.

2.5 DEMAND, SUPPLY, AND MARKET EQUILIBRIUM

The concepts of demand and supply have been introduced in the previous sections on the theory of the consumer and the theory of the firm. This section brings together the concepts of supply and demand in order to explain the determination of the market price of a commodity or a service and the total quantity traded in a market.

A consumer's demand for commodity w_j depends upon its price of p_j, the prices of all other commodities, and his income B^0. The demand function for consumer i is

$$D_i = D_i(p_1, p_2, \ldots, p_m, B^o) \tag{2.5.1}$$

For all prices remaining constant except p_j and a constant income, the demand function reduces to

$$D_i = D_i(p_j) \tag{2.5.2}$$

The market demand function for a commodity is the sum of the demand functions of individual consumers,

$$D = \sum_i D_i(p) = D(p) \tag{2.5.3}$$

where D is the aggregate demand. This aggregate demand assumes that all prices and the consumers' incomes are constant.

A firm total revenue is TR = pq. The **marginal revenue** is the rate at which total revenue increases as a result of small increases in sales, expressed as

$$\text{MR} = \frac{d(\text{TR})}{dq} = p \tag{2.5.4}$$

The **marginal revenue curve for a firm is the demand curve**.

A **supply function** defines the quantity that a firm will produce as a function of market price. Supply functions for individual firms can be defined for: (a) very short periods of time when output level cannot vary; (b) short run during a time that output level can vary and some inputs are fixed; and (c) long run during which all inputs are considered variable. Supply functions can be derived from the first-order conditions for profit maximization. A firm's short-run MC is a function of the output, MC = $f(q)$. A firm's short-run supply curve is the short-run MC curve which lies above the average variable cost curve (see Fig. 2.5.1). An aggregate supply function is the sum of the individual supply functions,

$$S = \sum_i S_i(p) = S(p) \tag{2.5.5}$$

and an aggregate supply curve is the horizontal sum of individual supply curves. The second-order condition for profit maximization requires the MC curve to be rising.

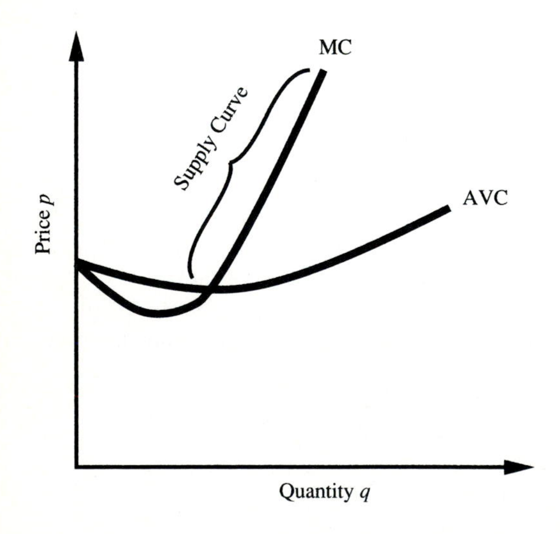

FIGURE 2.5.1
MC and AVC curves for short run. A firm's short-run supply curve is the short-run MC curve that lies above the AVC curve.

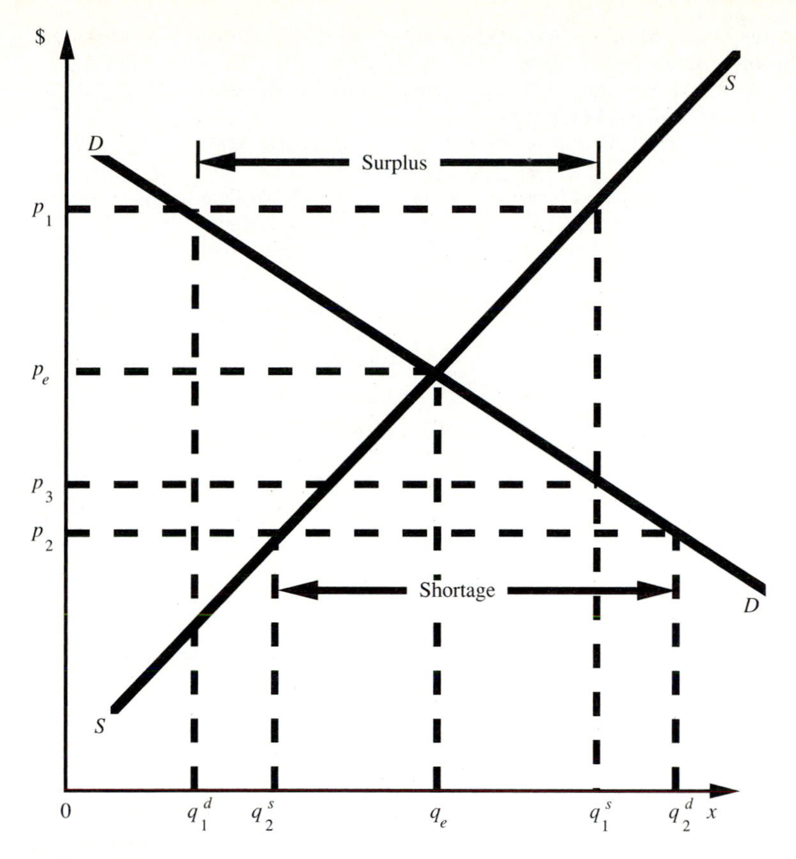

FIGURE 2.5.2
Demand, supply and market equilibrium for x. (Schefter et al., 1978) These curves for supply (S) and demand (D) intersect at the equilibrium point (q_e, p_e). For an increase in price to p_1 the demand would fall to q_d, and the supply would increase to q_s, so that $q_1^s > q_1^d$ resulting in a **surplus**. Competition to sell the surplus would result in a continual lowering of price to p_e. If the price falls to p_2 the quantity demanded would increase to q_2^d and the supply would decrease to q_2^s, so that $q_2^s < q_2^d$ resulting in a **shortage**. Competitive bidding amongst consumers along with increased output by sellers will increase price to p_e.

A firm's long-run optimal output results when the long-run MC is equal to the price. Fig. 2.5.2 shows the supply and demand curves for a commodity. These curves intersect at the **equilibrium price**, p_e, and the **equilibrium quantity** where the quantity demanded and the quantity supplied are equal.

REFERENCES

Billings, R. B. and D. E. Agthe: "Price Elasticities for Water: A Case of Increasing Block Rates," *Land Economics*, vol. 56, no. 1, pp. 73–84, 1980.

Foster, Jr., H. S. and B. R. Beattie: "Urban Residential Demand for Water in the United States," *Land Economics*, vol. 55, no. 1, pp. 43–58, February 1979.

Hanke, S. H. and L. de Maré: "Municipal Water Demands," *Modeling Water Demands*, J. Kindler and C. S. Russell (eds.), Academic Press, 1984.

Heaney, J. P., S. J. Nix, and M. P. Murphy: "Storage-Treatment Mixes of Stormwater Control," *J. Env. Eng. Div.*, ASCE, vol. 104, no. EE4, pp. 581–592, August 1978.

Henderson, J. M. and R. E. Quandt, *Microeconomic Theory: A Mathematical Approach*, McGraw-Hill, New York, 1980.

Howe, C. E. and F. P. Linaweaver, Jr.: "The Impact of Price on Residential Water Demand and Its Relation to System Design and Price Structure," *Water Resources Research*, vol. 3, no. 1, pp. 12–32, 1967.

Howe, C. W.: *Benefit-Cost Analysis for Water System Planning*, Water Resources Monograph 2, American Geophysical Union, Washington, D.C., 1971.

Schefter, J. E., R. M. Hirsh and I. C. James, II: *Natural Resource Economics Course Notes*, Water Resources Division, U.S. Geological Survey, June 1978.

Sewell, W. R. D., J. Davis, A. D. Scott, and D. W. Ross: *Guide to Benefit-Cost Analysis*, Report, Resources for Tomorrow, Information Canada, Ottawa, 1961.

PROBLEMS

2.1.1 Solve Example 2.1.1 using an interest rate of 8 percent.

2.1.2 A water resources project has produced benefits of $40,000 at the end of each of 20 years. Determine the present value of these benefits using a 6 percent interest rate. Determine the value of the benefits at the end of the 20th year.

2.1.3 A water resources project has produced benefits of $20,000 at the end of the first year and increase on a uniform gradient series to $100,000 at the end of the fifth year. The benefits remain constant at $100,000 each year until the end of year 25. Determine the present value of these benefits using a 6 percent interest rate.

2.1.4 A water resources project has produced benefits of $10,000 at the end of the first year and increase on a uniform gradient series to $50,000 at the end of the fifth year. The benefits then decrease to zero on a uniform gradient series to $0 at the end of year 10. Determine the value of these benefits in year 0 and year 25 using an interest rate of 6 percent.

2.2.1 Use the benefit-cost analysis procedure to determine the optimum scale of development for the following alternatives for a small hydroelectric facility.

Alternative (Scale)	Costs $\$ \times 10^4$	Benefits $\$ \times 10^4$
1	650.00	650.00
2	1800.00	2200.00
3	3600.00	4800.00
4	6900.00	9400.00
5	9900.00	14,000.00
6	12,700.00	18,000.00
7	15,400.00	19,700.00
8	17,400.00	20,900.00

2.2.2 Four alternative projects presented here can be used for developing a water supply for a community for the next 40 years. Use the benefit-cost ratio method to compare and select an alternative. Use a 6 percent interest rate.

Year(s)	Project I	Project II	Project III	Project IV
		Construction Cost $\times 10^3$		
0	40,000.00	30,000.00	20,000.00	10,000.00
10	0.00	0.00	0.00	10,000.00
20	0.00	10,000.00	20,000.00	10,000.00
30	0.00	0.00	0.00	10,000.00
		Operation and Maintenance Cost		
0–10	100,000.00	110,000.00	120,000.00	120,000.00
10–20	120,000.00	110,000.00	130,000.00	120,000.00
20–30	140,000.00	120,000.00	140,000.00	130,000.00
30–40	160,000.00	140,000.00	150,000.00	130,000.00

2.3.1 Show that the rate of change of slope of an indifference curve is

$$-\left(\frac{\partial f}{\partial w_2}\right)^{-3}\left[\frac{\partial^2 f}{\partial w_1^2}\left(\frac{\partial f}{\partial w_2}\right)^2 - 2\left(\frac{\partial^2 f}{\partial w_1 \partial w_2}\right)\left(\frac{\partial f}{\partial w_1}\right)\left(\frac{\partial f}{\partial w_2}\right) + \frac{\partial^2 f}{\partial w_2^2}\left(\frac{\partial f}{\partial w_1}\right)^2\right]$$

2.3.2 Find the optimal combination of commodities for a consumer with a utility function, $u = w_1^{1.5} w_2$, and budget constraint, $3w_1 + 4w_2 = 100$ (from Henderson and Quandt, 1980).

2.3.3 The linear demand model derived by Hanke and de Maré (1984) for Malmö, Sweden is

$$Q = 64.7 + 0.00017\ (\text{Inc}) + 4.76\ (\text{Ad}) + 3.92\ (\text{Ch})$$

$$-\ 0.406\ (R) + 29.03\ (\text{Age}) - 6.42\ (P)$$

where

Q = quantity of metered water used per house, per semi-annual period (in m^3).
Inc = real gross income per house per annum (in Swedish crowns; actual values reported per annum and interpolated values used for mid-year periods).
Ad = number of adults per house, per semi-annual period.
Ch = number of children per house, per semi-annual period.
R = rainfall per semi-annual period (in mm).
Age = a dummy variable with a value of 1 for those houses built in 1968 and 1969, and a value of 0 for those houses built between 1936 and 1946.
P = real price in Swedish crowns per m^3 of water, per semi-annual period (includes all water and sewer commodity charges that are a function of water use).

Using the average values of $P = 1.7241$ and $Q = 75.2106$ for the Malmö data, determine the elasticity of demand. Explain the meaning of the result.

2.3.4 Determine the elasticities of demand for the water demand model in Problem 2.3.3 using $P = 1.5$ and $Q = 75.2106$; $P = 2.0$ and $Q = 75.2106$; $P = 1.7241$ and $Q = 50$; and $P = 1.7241$ and $Q = 100$.

2.3.5 Determine the price elasticity of water demand using the linear water demand function developed by Howe and Linaweaver (1967), for nonseasonal use by single-family households, metered with public sewer

$$Q = 206 - 1.3P_w + 3.47V$$

where Q is the average annual quantity demanded for domestic purposes in gallons per dwelling unit per day; P_w is the sum of water and sewer charges that vary with water use (evaluated using a block rate applicable to the average domestic use in each study area); and V is the market value of the dwelling unit in thousands of dollars. Consider the mean water use of 206 gallons/day and the mean water price of 40.1 cents/1000 gallons.

2.3.6 Foster and Beattie (1979) developed a demand function for urban residential water demand for application in the U.S. The demand function is

$$\ln Q = -1.3895 - 0.1278 P_{av} + 0.4619 \ln (I) - 0.1699 \ln (F) + 0.4345 \ln (H)$$

where Q is the quantity of water demanded per meter (1,000 cubic feet per year) P_{av} is the average water price (dollars per 1,000 cubic feet); I is the median household income (dollars per year); F is the precipitation in inches during the growing season; and H is the average number of residents per meter. Determine the price elasticity of demand calculated at the mean price, \$3.67/1,000 gallons.

2.4.1 For the production process in Table 2.4.1, determine and plot the total, average, and marginal product curves for nitrogen fertilizer given that irrigation water is fixed at $x_1 = 7$ in/acre.

2.4.2 Determine and plot the optimal input combination (expansion path) for the production process in Table 2.4.1. The input prices are \$2.50/lb for nitrogen fertilizer and \$10.00/acre-inch for irrigation water.

2.4.3 Determine and plot the long-run total cost, average cost, and marginal cost curve for the production process in Table 2.4.1 using the optimal input combination for the expansion path determined in Problem 2.4.2.

2.4.4 Using the production process in Table 2.4.1, determine and plot the short-run total fixed cost, total variable cost, total cost, average fixed cost (AFC), average variable cost (AVC), average total cost (ATC), and marginal cost (MC). Assume that the irrigation water is fixed at 7 in/acre in the short run. Use input prices of \$2.50/lb for nitrogen fertilizer and \$10.00/acre-inch for irrigation water. Plot the curves for AFC, AVC, ATC, and MC on one plot. Discuss the fact that the MC curve intersects the other three curves.

2.4.5 Using the production process in Table 2.4.1, determine the short-run profit for various levels of the total product. Assuming that corn sells for \$1.49/bushel, input prices are \$2.50/lb for nitrogen fertilizer and \$10.00 acre/inch for irrigation water. The amount of irrigation water is fixed at 7 acre/inches. Use results of Problem 2.4.4 to work this problem. What is the total product that maximizes profit in the short run?

2.4.6 For a linear production function of the form $q = a_1 x_1 + a_2 x_2$ determine the elasticity of production, elasticity of substitution, and marginal product of each input.

2.4.7 Determine the elasticity of output and the elasticity of substitution for the following production function

$$q = b_0 x_1^{b_1} x_2^{b_2}$$

2.5.1 Consider the following demand functions for domestic demand of water (Q_d) and sprinkling demand of water (Q_s), respectively

$$\log Q_d = 2.75 + 0.214 \log p$$

and

$$\log Q_s = 5.131 + 1.57 \log p$$

where Q_d and Q_s are in gallons per household day and p is the price of water in cents per thousand gallons. Domestic demand occurs 12 months of the year and sprinkling demand occurs four months of the year. First, determine the demand curve for domestic and sprinkling, then the aggregate demand curve for a population of one million households. If water prices are set at $0.30/1000 gallons, can a supply of 300 million gallons per day (MGD) meet demands? At what rate would a city have to charge so that the average annual quantity demanded would equal 300 MGD?

CHAPTER 3

LINEAR PROGRAMMING WITH APPLICATIONS TO HYDROSYSTEMS

3.1 LINEAR PROGRAMMING

Linear-programming (LP) models have been applied extensively to optimal resource allocation problems. As the name implies, LP models have two basic characteristics, that is, both the objective function and constraints are linear functions of the decision variables. The general form of an LP model can be expressed as

$$\text{Max (or Min) } x_0 = \sum_{j=1}^{n} c_j x_j \tag{3.1.1a}$$

subject to

$$\sum_{j=1}^{n} a_{ij} x_j = b_i, \qquad \text{for } i = 1,\ 2, \ldots, m \tag{3.1.1b}$$

$$x_j \geq 0, \qquad \text{for } j = 1,\ 2, \ldots, n \tag{3.1.1c}$$

where c_j is the **objective function coefficient**, a_{ij} is the **technological coefficient**, and b_i is the right-hand side (RHS) coefficient.

In algebraic form, this LP model can be expanded as

$$\text{Max (or Min) } x_0 = c_1 x_1 + c_2 x_2 + \cdots + c_n x_n \tag{3.1.2a}$$

subject to

$$\begin{aligned}
a_{11}x_1 + a_{12}x_2 + \cdots + c_{1n}x_n &\le b_1 \\
a_{21}x_1 + a_{22}x_2 + \cdots + a_{2n}x_n &\le b_2 \\
\vdots \quad\quad \vdots \quad\quad\quad \vdots \quad\quad \vdots \quad & \quad \vdots \\
a_{m1}x_1 + a_{m2}x_2 + \cdots + a_{mn}x_n &\le b_m \\
x_1 \ge 0,\ x_2 \ge 0, \ldots, x_n &\ge 0
\end{aligned} \tag{3.1.2b}$$

Alternatively, in matrix form, an LP model can be concisely expressed as

$$\text{Max (or Min) } x_0 = \mathbf{c}^T\mathbf{x} \tag{3.1.3a}$$

subject to

$$\mathbf{Ax} \le \mathbf{b} \tag{3.1.3b}$$

$$\mathbf{x} \ge 0 \tag{3.1.3c}$$

where $\mathbf{c}$ is an n by 1 column vector of objective function coefficients, $\mathbf{x}$ is an n by 1 column vector of decision variables, $\mathbf{A}$ is an m by n matrix of technological coefficients, $\mathbf{b}$ is an m by 1 column vector of the RHS coefficients, and the superscript T represents the transpose of a matrix or a vector. Excellent textbooks on LP include Gass (1985), Taha (1987), Winston (1987), and Hillier and Lieberman (1990).

Example 3.1.1. Consider a system composed of a manufacturing factory and a waste treatment plant owned by the manufacturer (Fiering et al., 1971). The manufacturing plant produces finished goods that sell for a unit price of \$10 K. However, the finished goods cost \$3 K per unit to produce. In the manufacturing process two units of waste are generated for each unit of finished goods produced. In addition to deciding how many units of goods to produce, the plant manager must also decide how much waste will be discharged without treatment so that the total net benefit to the company can be maximized and the water quality requirement of the watercourse is met. The treatment plant has a maximum capacity of treating 10 units of waste with 80 percent waste removal efficiency at a treatment cost of \$0.6 K per unit of waste. There is also an effluent tax imposed on the waste discharged to the receiving water body (\$2 K for each unit of waste discharged). The water pollution control authority has set an upper limit of four units on the amount of waste any manufacturer may discharge. Formulate an LP model for this problem.

Solution. The first step in model development is to identify the system components involved and their interconnections. In this example, the system components are the manufacturing factory, the waste treatment plant, and the watercourse. From the problem statement, the two decision variables to the problem: that is, (1) number of units of finished goods to be produced, x_1; and (2) quantity of waste to be discharged directly to the watercourse without treatment, x_2 can be immediately identified. From the description of the interrelationship of finished goods, waste generated, and treatment plant efficiency a schematic configuration of the system under study can be constructed as shown in Fig. 3.1.1. The amount of waste in each branch is determined by the mass-balance principle.

Furthermore, before the model construction it is essential to identify the problem objective and constraints. In this example, the objective of the problem is profit maximization. The constraints arise primarily from the limitations on treatment capacity and

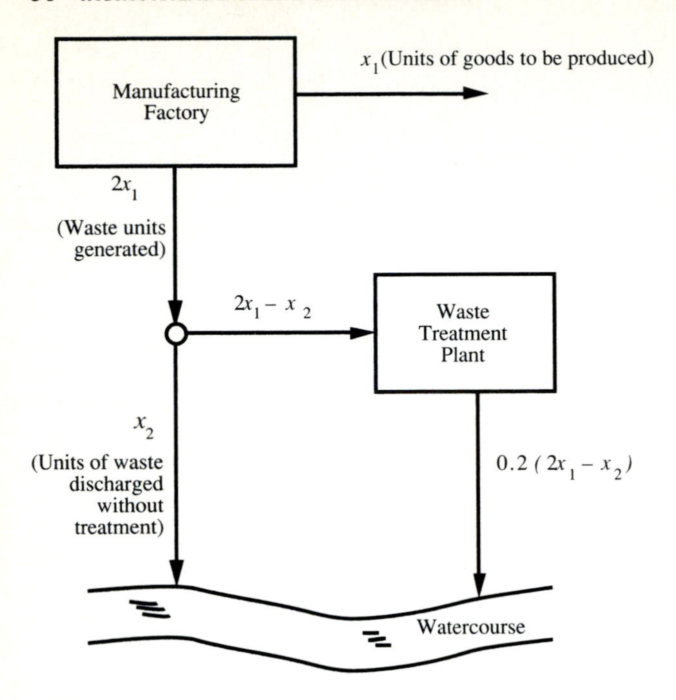

FIGURE 3.1.1
Schematic diagram of manufacturing-waste treatment system.

on the allowable waste discharges set by the water pollution control authority. Recognizing the problem objective and the constraints imposed, the model construction step basically involves the translation of verbal descriptions of the objective and constraints into mathematical expressions in terms of decision variables and parameters. The net revenue of the manufacturer is determined by four items: (a) sales of finished goods (in \$ K), $10x_1$; (b) cost of producing goods (in \$ K), $3x_1$; (c) cost of treating the waste (in \$ K) generated from the production process, $0.6(2x_1 - x_2)$; and (d) effluent tax (in \$ K) charged on untreated waste, $2[x_2 + 0.2(2x_1 - x_2)]$. The profit to the manufacturer is equal to the total sales minus the total costs. The objective function to the problem is to maximize the profit which is $10x_1 - \{3x_1 + 0.6(2x_1 - x_2) + 2[x_2 + 0.2(2x_1 - x_2)]\}$ $= 5x_1 - x_2$. The objective is then expressed as

$$\text{Max } x_0 = 5x_1 - x_2$$

Referring to Fig. 3.1.1, the constraint on the limitation of treatment plant capacity can be expressed mathematically as

$$2x_1 - x_2 \leq 10$$

This constraint states that the amount of waste to be treated, $2x_1 - x_2$, cannot exceed the plant capacity of treating 10 units of waste. Similarly, the constraint associated with the amount of total waste that can be discharged to the watercourse can be expressed as

$$x_2 + 0.2(2x_1 - x_2) \leq 4$$

where the left-hand side (LHS) of the constraint is the total waste to be discharged (see Fig. 3.1.1) and the RHS of 4 is the total allowable waste units set by the water pollution control authority.

In addition to the two obvious constraints indicated here, there exists a rather subtle constraint that must be incorporated in the model. A constraint is needed to ensure that a positive amount of waste is treated. In other words, the model should include a constraint imposing the nonnegativity on the quantity, $2x_1 - x_2$, which can be expressed mathematically as

$$2x_1 - x_2 \geq 0$$

Finally, the two decision variables physically cannot be less than zero. Therefore, non-negativity constraints of the two decision variables, $x_1 \geq 0$ and $x_2 \geq 0$, must be included in the constraint. The final version of the mathematical programming model for the problem, after some algebraic manipulations, can be summarized as

$$\text{Max } x_0 = 5x_1 - x_2$$

subject to

$$2x_1 - x_2 \leq 10$$

$$.4x_1 + .8x_2 \leq 4$$

$$2x_1 - x_2 \geq 0$$

$$x_1 \geq 0 \text{ and } x_2 \geq 0$$

Examining the formulation of the developed optimization model, one recognizes that the resulting model is an LP model. However, strictly speaking, the model presented may not be considered as an LP model if the decision variables, particularly x_1, can only be integer values. If that is the case, the model is a mixed integer programming model that requires a special algorithm to solve it. In fact, there are several assumptions implicitly built into an LP model formulation. These assumptions are discussed in more detail in the following sections.

3.1.1 Assumptions in Linear Programming Models

There are four basic assumptions implicitly built into LP models.

1. Proportionality assumption. This implies that the contribution of the jth decision variable to the effectiveness measure, $c_j x_j$, and its usage of the various resources, $a_{ij} x_j$, are directly proportional to the value of the decision variable.

2. Additivity assumption. This assumption means that, at a given level of activity $(x_1, x_2, \ldots, x_n)$, the total usage of resources and contribution to the overall measure of effectiveness are equal to the sum of the corresponding quantities generated by each activity conducted by itself.

3. Divisibility assumption. Activity units can be divided into any fractional level, so that noninteger values for the decision variables are permissible.

4. Deterministic assumption. All parameters in the model are known constants without uncertainty. The effect of the uncertainty of parameters on the result can be investigated by conducting sensitivity analysis.

3.1.2 Forms of Linear Programming

Because LP models can be presented in a variety of forms (maximization, minimization, $\geq$, $=$, $\leq$), it is necessary to modify these forms to fit a particular solution procedure. Basically, two types of linear programming model formulations are used: **standard form** and **canonical form**.

The standard form is used for solving the LP model algebraically. Its basic characteristics involve the following: (1) all constraints are equality except for the nonnegativity constraints associated with the decision variables which remain inequality of the $\geq$ type; (2) all the RHS coefficients of the constraint equations are nonnegative, that is, $b_i \geq 0$; (3) all decision variables are nonnegative; and (4) the objective function can be either maximized or minimized. An LP model having a standard form can be expressed as

$$\text{Max (or Min) } x_0 = \sum_{j=1}^{n} c_j x_j \tag{3.1.4a}$$

$$\begin{aligned} &\sum_{j=1}^{n} a_{ij} x_j = b_i, && \text{for } i = 1, 2, \ldots, m \\ &x_j \geq 0, && \text{for } j = 1,\ 2, \ldots, n \end{aligned} \tag{3.1.4b}$$

Example 3.1.2. Convert the LP model for the manufacturing-waste treatment example formulated in the previous subsection into the standard form.

Solution. Since the objective function in standard form can be either maximization or minimization, there is no need to modify the objective function. However, the standard form of an LP model requires that all constraints are in the form of equations, which is not the case for all three constraints in the model. Therefore, modifications are required. Note that since the first constraint is of the $\leq$ type, a nonnegative slack variable s_1 can be added to the LHS of the constraint which then becomes

$$2x_1 - x_2 + s_1 = 10$$

Similarly, the second inequality constraint can be transformed to

$$0.4x_1 + 0.8x_2 + s_2 = 4$$

where s_2 is a nonnegative slack variable associated with the second constraint. For the third constraint, since it is of the $\geq$ type, a slack variable $s_3 \geq 0$ can be subtracted from the LHS and the result is

$$2x_1 - x_2 - s_3 = 0$$

The original decision variables x_1, x_2, and the three slack variables s_1, s_2, and s_3 are all nonnegative, then condition 3 is satisfied. Furthermore, the RHS coefficient of the constraints are nonnegative. The resulting standard form of the LP model for the manufacturing-waste treatment example is

$$\text{Max } x_0 = 5x_1 - x_2 + 0s_1 + 0s_2 + 0s_3$$

subject to

$$2x_1 - x_2 + s_1 = 10$$

$$.4x_1 + .8x_2 + s_2 = 4$$

$$2x_1 - x_2 - s_3 = 0$$

All x and s are nonnegative

The **canonical form**, on the other hand, is useful in presenting the duality theory of the LP model. It possesses the following three features in model formulation: (1) all decision variables are nonnegative; (2) all constraints are of the $\leq$ type; and (3) the objective function is of the maximization type. An LP model having a canonical form is

$$\text{Max } x_0 = \sum_{j=1}^{n} c_j x_j \tag{3.1.5a}$$

subject to

$$\sum_{j=1}^{n} a_{ij} x_j \leq b_i, \qquad i = 1, 2, \ldots, m$$
$$x_j \geq 0, \qquad j = 1, 2, \ldots, n \tag{3.1.5b}$$

Note that a negative RHS coefficient, b_i, in the canonical form is permissible.

Example 3.1.3. Convert the original LP model for the manufacturing-waste treatment example into its canonical form.

Solution. Since the original objective function is of the maximization type as required by the canonical form, no modification is needed. With regard to the constraints, the first two constraints are of the $\leq$ type which satisfies the requirement of a canonical form. However, the third constraint is of the $\geq$ type which can be converted to $\leq$ type by multiplying -1 on both sides of the constraint resulting in

$$-2x_1 + x_2 \leq 0$$

Nonnegativity requirement of the decision variables is also satisfied. The resulting LP model in the canonical form can be stated as

$$\text{Max } x_0 = 5x_1 - x_2$$

subject to

$$2x_1 - x_2 \leq 10$$

$$0.4x_1 + .8x_2 \leq 4$$

$$-2x_1 + x_2 \leq 0$$

$$x_1 \geq 0 \text{ and } x_2 \geq 0$$

More often than not, the LP model originally constructed does not satisfy the characteristics of a standard form or a canonical form. The following elementary operations enable one to transform an LP model into any desirable form.

1. Maximization of a function $f(x)$ is equal to the minimization of its negative counterpart, that is, Max $f(x)$ = Min$[-f(x)]$.
2. Constraints of the $\geq$ type can be converted to the $\leq$ type by multiplying by -1 on both sides of the inequality.
3. An equation can be replaced by *two* inequalities of the opposite sign. For example, an equation $g(x) = b$ can be substituted by $g(x) \leq b$ and $g(x) \geq b$.
4. An inequality involving an absolute expression can be replaced by *two* inequalities without an absolute sign. For example, $|g(x)| \leq b$ can be replaced by $g(x) \leq b$ *and* $g(x) \geq -b$.
5. If a decision variable x is unrestricted-in-sign (i.e., it can be positive, zero, or negative), then it can be replaced by the difference of *two* nonnegative decision variables; $x = x^+ - x^-$, where $x^+ \geq 0$ and $x^- \geq 0$.
6. To transform an inequality into an equation, a nonnegative variable can be added or subtracted.

3.2 SOLUTION ALGORITHMS FOR LINEAR PROGRAMMING

3.2.1 Graphical Method

One simple way to solve an LP problem is by using the graphical method. However, this method is limited to LP problems involving at most two decision variables (excluding any of the slack variables). To lay the foundation for geometric interpretation of the algebraic algorithm described later, the graphical method will be used to solve the manufacturing-waste treatment problem in the following example.

Example 3.2.1. Solve the manufacturing-waste treatment problem to find the optimum number of units of finished goods to be produced x_1 and the amount of waste generated to be discharged without treatment x_2 so the net benefit to the manufacturer is maximized.

Solution. Referring to the LP model constructed for the problem (in Example 3.1.1), it involves two decision variables and three constraints (excluding the nonnegativity requirement on the decision variables). The **feasible space** can be defined by all the constraints in the model, including the nonnegativity of decision variables. Because the two decision variables cannot be negative, the feasible space is contained in the northeast quadrant. The feasible space (the cross hatched area) for the example problem is shown in Fig. 3.2.1. Each solid line on Fig. 3.2.1 is defined by the corresponding constraint taken as an equation and the two principal axes represent the nonnegativity of x_1 and x_2. The arrow by each line indicates the **half-plane** in which all points satisfy the associated constraint. The feasible space is therefore the intersection of all feasible half-planes and every point in the feasible space satisfies all constraints simultaneously, which is the feasible solution to the optimization model.

Since the value of the maximum net benefit is not yet known, a trial-and-error procedure must be used. First, a line is drawn for $x_0 = 0$ so that the corresponding objective function equation passes through the origin. Mathematically speaking, all points (x_1, x_2) falling on the line $5x_1 - x_2 = 0$ will yield a value of zero for the total net benefit. Of course, one is only interested in those points that fall within the cross-hatched area which represents the feasible region.

To investigate whether the total net benefit can be further improved, the objective function line can be moved backward or forward parallel to the original line to see how the objective function value varies. The objective function line is moved to the left and values of x_1 and x_2, arbitrarily chosen from the line, are used to compute the value of the objective function. One would find that the x_0 value associated with any line to the left of line $5x_1 - x_2 = 0$ and parallel to it is negative. As the line is moved further to

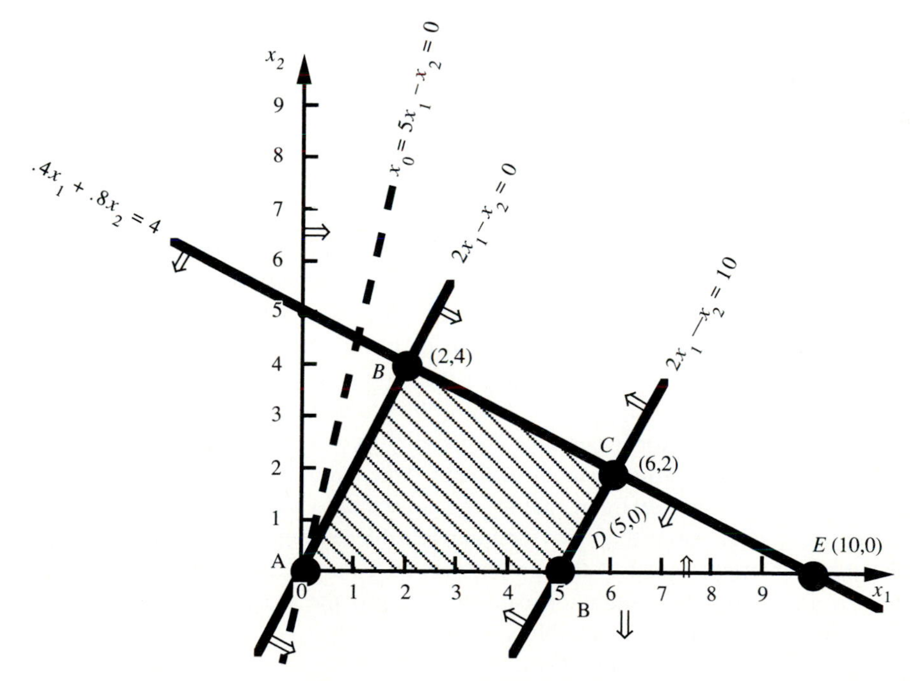

FIGURE 3.2.1
Sketch of feasible space of manufacturing-waste treatment example.

the left, the value of x_0 becomes more and more negative. This indicates that the search is made in the wrong direction since the problem is a maximization type; the higher the x_0 value the better. Accordingly, the search direction should be altered by sliding the objective function line to the right of $5x_1 - x_2 = 0$. Immediately, the value of the net benefit becomes positive and is steadily increasing as the line is moved further to the right. However, the line cannot move to the right indefinitely even if it continuously increases the net benefit. As can be seen, after passing point C the objective function line contains no feasible point. In this example, it can be concluded that the combination of x_1 and x_2 represented by point C, that is, (6, 2), is the optimum solution to the problem. The maximum achievable net benefit to the manufacturer is 5(6) – 1(2) = $28 K.

The graphical method is applicable only to problems involving two decision variables. For problems with more than two decision variables, the shapes of the objective function and constraint equations of constant value are called **hyperplanes** in the n-dimensional space.

3.2.2 Feasible Extreme (or Corner) Points

In the preceding illustration, the resulting optimum solution to the LP problem using the graphical approach happens to be at one corner point in the feasible space (called the **feasible extreme point**). It should be emphasized that there is nothing special or coincidental about the example that leads to such a result. In fact, for all LP problems the optimum solution will always fall on the boundary of the feasible space.

There are three important properties of feasible extreme points in an LP problem. Their implications to the algebraic solution technique will be described later. The following properties will only be stated without providing any mathematical proof that can be found in LP or mathematical programming books (Dantzig, 1963; Bradley, Hax, and Magnanti, 1977; and Taha, 1987).

Property 1a. If there is only one optimal solution to a linear programming model, then it must be a feasible extreme point. 1b. If there are multiple optimal solutions, then at least two must be adjacent feasible extreme points.

As demonstrated in the graphical approach, for any problem having only one optimum solution it is always possible to raise or lower the objective function line (or hyperplane) until it just touches that point, the optimum point, at a corner of the feasible space. As can be imagined, multiple optimum solutions would occur when the objective function line (or hyperplane) is parallel to one of the boundaries of the feasible space. For a two-dimensional problem, when multiple optimum solutions exist, two optimal feasible extreme points are adjacent to each other. For higher dimensional problems, more than two optimal feasible extreme points are adjacent to each other. The implication of this property is that, in searching for the optimum solutions to an LP problem, attention can be given to the feasible extreme points rather than to the interior of the feasible space.

Property 2. There are only a finite number of feasible extreme points.

Consider an LP model having a standard form with m equations and n unknown decision variables where $n > m$. For such indeterminate systems of simultaneous equations, the number of possible solutions is ${}_nC_m = n!/m!(n-m)!$ which is finite. However, this number provides the upper bound on the number of feasible extreme points because many of these solutions are infeasible or nonexistent. This property might suggest that the optimum solution can be obtained by exhaustive enumeration of all feasible extreme points. However, it is generally impractical to do so because the number of feasible extreme points may still be too large to be enumerated efficiently. Furthermore, the optimum solution cannot be identified before all possible feasible extreme points are enumerated.

Property 3. If a feasible extreme point is better (measured with respect to x_0) than all its adjacent feasible points, then it is better than all other feasible extreme points (i.e., it is a global optimum).

From this property, one does not have to exhaustively enumerate all feasible extreme points to obtain the optimum solution of the problem. Instead, the status of one feasible extreme point under investigation can be ascertained by simply comparing its adjacent feasible extreme points. If Property 3 is satisfied, the feasible extreme point under investigation is the **global optimum** to the LP problem.

It should be emphasized that the basic requirement for this property to hold is that the feasible space is convex. Otherwise, the optimum solution obtained is not guaranteed to be the global optimum and is called the **local optimum**. This phenomenon is especially prevalent in nonlinear programming problems. Fortunately, the feasible space is almost always convex for LP problems.

3.2.3 Algorithm for Solving Linear Programming Problems

In this section, the three important properties of feasible extreme points previously discussed are applied to an LP problem and a solution algorithm is devised to solve it. Again, refer to the manufacturing-waste treatment example given previously. As illustrated in Fig. 3.2.1, the LP model has four feasible extreme points. First, one has to determine where the search for the optimum solution begins. In fact any of the feasible extreme points can be used as a starting point. Naturally, if the starting solution is close to the optimum point, one would expect to reach the solution faster. However, knowing a better starting point a priori is generally difficult, especially for multiple dimensional problems. Therefore, it is reasonable to start with a point at the origin $(x_1, x_2) = (0, 0)$ since the origin is a feasible extreme point. Note that, in the course of searching for the optimum solution, it is always desirable to start with a feasible solution.

Having determined the starting feasible extreme point, the corresponding value of the objective function x_0 is computed as the basis for further comparison. In this case, at (0,0), the corresponding x_0 is equal to zero. The next step is to find an improved solution by comparing the objective function values associated with the adjacent feasible extreme points. The two feasible extreme points adjacent to point (0,0) are B at (2,4) and D at (5,0) with the corresponding objective function values 6 and 25, respectively. This result indicates that movement from point A to D yields a better improvement than from A to B. Therefore, D becomes the new base feasible extreme point.

At point D, the comparison process is repeated by identifying the adjacent feasible extreme points of D. In this case, points A and C are the two adjacent points. However, the value of the objective function at A, from the previous comparison, is no better than the current point. Therefore, no comparison is needed for point A. This leaves point C as the feasible extreme point yet to be compared against D. At point C, (6,2), the value of x_0 is $5(6) - 1(2) = 28$. Since this value is higher than 25 at point D, the base feasible extreme point will be updated to point C. At the base feasible extreme point C, there are no new adjacent feasible extreme points to be considered (point B has been compared and outcast by point D in the previous step). The value of x_0 at point C is better than all its adjacent feasible extreme points, that is, points B and D. From the third property of the feasible extreme points discussed above, one can conclude that the optimum solution to the manufacturing-waste treatment example is to produce 6 units of finished goods and to discharge 2 units of waste to the watercourse without treatment, that is to treat $2(6) - 2 = 10$ units of waste. The corresponding optimal net benefit is \$28 K.

The steps described above (utilizing the three properties of a feasible extreme point to solve an LP model) form the underlying concept of the well-known algorithm called the **simplex method**. The method is a general procedure for solving LP problems. It is a very efficient method which has been employed to solve large problems involving thousands of decision variables and constraints on today's computers. Computer codes based on the simplex method are widely available for use.

Example 3.2.2. Algebraically solve the manufacturing waste-treatment problem of Example 3.1.2.

Solution. The objective function and constraints can be written on

$$\begin{array}{lrl}
\text{Objective:} & x_0 - 5x_1 + x_2 + 0s_1 + 0s_2 + 0s_3 &= 0 \\
\text{Constraint 1:} & 2x_1 - x_2 + s_1 + 0s_2 + 0s_3 &= 10 \\
\text{Constraint 2:} & 0.4x_1 + 0.8x_2 + 0s_1 + s_2 + 0s_3 &= 4 \\
\text{Constraint 3:} & -2x_1 + x_2 + 0s_1 + 0s_2 + s_3 &= 0
\end{array}$$

The initial step will be to start at the feasible extreme point of ($x_1 = 0$ and $x_2 = 0$).

Iterative Step 1

The iterative step begins by selecting which variable x_1 or x_2 to increase from zero. Because x_1 has the largest negative coefficient of -5, it has the greatest effect upon increasing the objective function. The next decision is how much x_1 should be increased. Remember the slack variables must never go negative. First check Constraint 1 by solving for x_1 with $s_1 = 0$ and $x_2 = 0$,

$$x_1 = \frac{10}{2} + \frac{x_2}{2} - \frac{s_1}{2} = 5$$

Then substitute $x_1 = 5$ into the other constraints and solve for other slack variables

$$\text{Constraint 2:} \qquad 0.4(5) + 0.8(0) + 0s_1 + s_2 + 0s_3 = 4$$

$$s_2 = 2$$

$$\text{Constraint 3:} \qquad -2(5) + 0 + 0s_1 + 0s_2 + s_3 = 0$$

$$s_3 = 10$$

Because the slack variables remain positive $x_1 = 5$. Now consider Constraint 2 and solve for x_1, with $x_2 = 0$ and $s_2 = 0$

$$\text{Constraint 2:} \qquad 0.4x_1 + 0.8(0) + 0s_1 + 0 + 0s_3 = 4$$

$$x_1 = 10$$

Now check Constraints 1 and 3

$$\text{Constraint 1:} \qquad 2(10) - 0 + s_1 + 0s_2 + 0s_3 = 10$$

$$s_1 = -10$$

$$\text{Constraint 3:} \qquad -2(10) + 0 + 0s_1 + 0s_2 + s_3 = 0$$

$$s_3 = 20$$

According to Constraint 1, $x_1 = 10$ is too large because the slack variable s_1 would be negative ($s_1 = -10$).

Next consider Constraint 3 and solve for x_1 with $x_2 = 0$ and $s_3 = 0$

$$\text{Constraint 3:} \qquad -2x_1 + 0s_1 + 0s_2 + 0 = 0$$

$$x_1 = 0$$

This basically does not move from the starting point.

As a result of the above analysis the largest value that x_1 can be increased to is $x_1 = 5$, where $x_2 = 0$, $s_1 = 0$, $s_2 = 2$, and $s_3 = 10$. Referring to Fig. 3.2.2 this is at point D. The objective function value x_0 corresponding to this feasible solution at point D is

$$x_0 - 5(5) + 0 + 0s_1 + 0s_2 + 0s_3 = 0$$

$$x_0 = 25.$$

Next,

$$x_1 = 5 + \frac{x_2}{2} - \frac{s_1}{2}$$

is substituted into the objective function and constraints.

$$\text{Objective function:} \qquad x_0 - 5\left(5 + \frac{x_2}{2} - \frac{s_1}{2}\right) + x_2 + 0s_1 + 0s_2 + 0s_3 = 0$$

$$x_0 - \frac{3}{2}x_2 + \frac{5}{2}s_1 + 0s_2 + 0s_3 = 25$$

$$\text{Constraint 1:} \qquad x_1 - \frac{1}{2}x_2 + \frac{1}{2}s_1 + 0s_2 + 0s_3 = 5$$

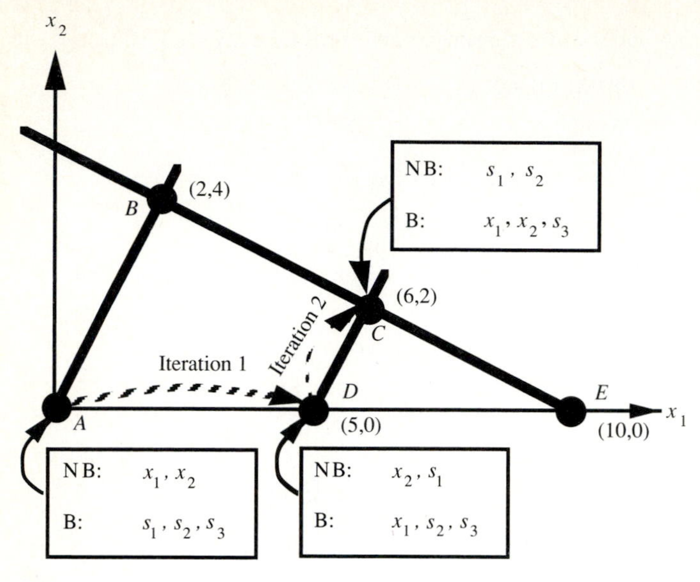

FIGURE 3.2.2
Illustration of algebra of linear programming algorithm.

Constraint 2:
$$0.4\left(5+\frac{x_2}{2}-\frac{s_1}{2}\right)+0.8x_2+0s_1+s_2+0s_3=4$$
$$0x_1+x_2-0.2s_1+s_2+0s_3=2$$

Constraint 3:
$$-2\left(5+\frac{x_2}{2}-\frac{s_1}{2}\right)+x_2+0s_1+0s_2+s_3=0$$
$$0x_1+0x_2+s_1+0s_2+s_3=10$$

Iterative Step 2

The second iterative step begins by defining which variable should be increased from 0. The objective function from above is

$$x_0-\frac{3}{2}x_2+\frac{5}{2}s_1+0s_2+0s_3=25$$

The largest negative coefficient is $-3/2$ for x_2 so it should be increased from 0 to increase the objective function. The next question is how much x_2 can be increased.

Constraint 1: $x_2=2x_1-10$ If both x_1 and $x_2>0$ then $s_1=0$

Constraint 2: $x_2=2+0.2s_1-s_2$ If x_1 and $x_2>0$ then $s_2=0$

$$x_2=2$$

Constraint 3: $0x_1+0x_2+0+0s_2+s_3=10$, so

$$s_3=10$$

Substitute $x_2 = 2$ from Constraint 2 into Constraint 1, $x_2 = 2x_1 - 10$, then $x_1 = 6$. For the objective function substitute $x_2 = 2$

$$x_0 - \frac{3}{2}(2 + 0.2s_1 - s_2) + \frac{5}{2}s_1 + 0s_2 + 0s_3 = 25$$

$$x_0 - 3 - .3s_1 + \frac{3}{2}s_2 + \frac{5}{2}s_1 + 0s_2 + 0s_3 = 25$$

$$x_0 + 2.2s_1 + \frac{3}{2}s_2 + 0s_3 = 28$$

$$x_0 + 0x_1 + 0x_2 + 2.2s_1 + 1.5s_2 + 0s_3 = 28$$

Both x_1 and x_2 have 0 coefficients, therefore no further changes in x_1 and x_2 can be made to increase the objective function.

3.2.4 Basic Algorithm of Solving Linear Programming Problems

As illustrated in the previous example, three basic steps are involved in solving an LP model:

1. *Initialization Step*—Start at a feasible extreme point.
2. *Iterative Step*—Move to a better adjacent feasible extreme point.
3. *Stopping Rule*—Stop iterations when the current feasible extreme point is better than all its adjacent extreme points.

3.3 SIMPLEX METHOD

3.3.1 Basic Algebraic Concepts and Setup

As previously demonstrated, an LP problem using the graphical method as well as the heuristic approach can be easily performed when only two decision variables are involved. The purpose of such an illustration is to demonstrate the geometric concept of the simplex method and its basic algorithm. To solve a problem of a larger scale, in terms of numbers of decision variables and constraints, the methods previously employed cannot be implemented practically or feasibly. Furthermore, to implement a solution algorithm on computers, the problem must be expressed algebraically.

Using algebraic procedures for problem solving, it is generally more convenient to deal with equations than inequality relationships. Hence, to solve an LP model using the simplex method algebraically, the model is first transformed into the standard form which is given in Example 3.1.2. The LP problem involves five decision variables (x_1, x_2, s_1, s_2, and s_3) and three equations.

In linear algebra, the system of equations is called **indeterminant** if there are more unknowns than equations. For problems with n unknowns and m equations, where $n > m$, there is no unique solution to the problem. The possible solutions to an indeterminant system of equations can be obtained by letting $(n - m)$ unknowns equal zero and solving for the remaining m unknowns. The solutions so obtained are called the **basic solutions**. In theory, there will be, at its maximum, a total of ${}_nC_m$ basic solutions to such problems if all exist. For the example under investigation, the problem has at most ${}_5C_3 = 5!/(2!3!) = 10$ basic solutions. Geometrically, each basic solution represents the intersection point of each pair of constraint equations, an **extreme point**, including positivity of x_1 and x_2 axis. The example shown in Fig. 3.2.1 has six basic solutions. Furthermore, among the six existing basic solutions, only four are feasible. These are represented by the four feasible extreme points. Now, let us examine the basic solutions associated with the four feasible extreme points as shown in Table 3.3.1. Every feasible extreme point has exactly two variables (out of 5) which are set equal to zero (indicated by 0). The additional zero associated with feasible extreme points *A* and *B* arises from the third constraint in which its RHS coefficient is zero.

In the preceding exercise, the $(n - m)$ decision variables set to zero are called **nonbasic variables** while the remaining m decision variables, whose values are obtained by solving a system of m equations with m unknowns, are called the **basic variables**. The solution of a basic variable that is nonnegative is called the **basic feasible solution** and identifies the corresponding feasible extreme point in the feasible space. Therefore, in solving an LP model algebraically, one only needs to examine the performance of each basic feasible solution to the model with the standard form.

3.3.2 Algebra of Simplex Method

The simplex method solves an LP model by exploiting the three properties of feasible extreme points previously discussed. The algorithm searches for the optimum of an LP model by abiding by two fundamental conditions: (1) the **optimality condition**; and (2) the **feasibility condition**.

The **optimality condition** ensures that no inferior solutions (relative to the current solution point) are ever encountered. The **feasibility condition** guarantees that, starting with a basic feasible solution, only basic feasible solutions are enumerated during the course of computation.

To solve an LP model algebraically the standard form of the model can be placed into a tabular form as in Table 3.3.2 in which the objective function is expressed as

$$x_0 - 5x_1 + x_2 - 0s_1 - 0s_2 - 0s_3 = 0$$

TABLE 3.3.1
Feasible extreme points for manufacturing-waste treatment example

	$(x_1,$	$x_2,$	$s_1,$	$s_2,$	$s_3)$
A	(0,	0,	10,	4,	0)
B	(2,	4,	10,	0,	0)
C	(6,	2,	0,	0,	10)
D	(5,	0,	0,	2,	10)

TABLE 3.3.2
Simplex tableau of manufacturing-waste treatment example

Basic	x_0	x_1	x_2	s_1	s_2	s_3	Sol'n
x_0	1	−5	1	0	0	0	0
s_1	0	2	−1	1	0	0	10
s_2	0	.4	.8	0	1	0	4
s_3	0	−2	1	0	0	1	0

The LP problem now can be solved following the three steps of the algorithm outlined in Section 3.2.4.

Initialization Step—The simplex method starts from any basic feasible solution. Obviously, slack variables yield a starting feasible solution because, referring to Table 3.3.2: (a) their constraint coefficients form an identity matrix; and (b) all RHS coefficients are nonnegative (property of standard form).

Iterative Step—This step involves two operations according to the optimality condition and feasibility condition. The first operation determines a new basic feasible solution with an improved value of the objective function. The simplex method does this by selecting one of the current nonbasic variables to be increased above zero providing its coefficient in the objective function has the potential to improve the current value of x_0. Since a feasible extreme point in an LP model must have $(n-m)$ nonbasic variables at zero level, one current basic variable must be made nonbasic, providing the solution is feasible. The current nonbasic variable to be made basic is called the **entering variable** while the current basic variable to be made nonbasic is called the **leaving variable**.

For a maximization problem, an entering variable is selected, based on the optimality condition, as the nonbasic variable having the most negative coefficient in the x_0 equation of the simplex tableau. This is equivalent to selecting a variable with the most positive coefficient in the original objective function because the magnitude of the objective function coefficient represents the rate of change of the objective function due to a unit change in the decision variable.

The one with the largest negative value is chosen because it has the greatest potential to improve the objective function value. On the other hand, the rule of selecting the entering variable for a minimization problem is reversed, that is, choose the nonbasic variable with the largest positive coefficient in the objective function row of the simplex tableau as the entering variable. Once the entering variable is determined, one of the current basic variables must be chosen to become nonbasic. The selection of the leaving variable is governed by the feasibility condition to ensure that only feasible solutions are enumerated during the course of the iterations. The leaving variable is selected using the following criterion,

$$\theta_i = \frac{b_i}{a_{ik}}, \qquad \text{for all } a_{ik} > 0$$

with a_{ik} being the constraint coefficients associated with the entering variable x_k.

The current basic variable associated with the row having the minimum $\overline{\theta} = \min(\theta_i)$ is selected as the leaving variable.

Example 3.3.1. Based on Table 3.3.1 select the entering and leaving variables for the first iteration.

Solution. Referring to the simplex tableau in Table 3.3.2, the decision variable x_1 can be chosen as the entering variable. The direction associated with the entering variable indicates the direction along which the improved solution is sought. Referring to Fig. 3.2.1, this represents a move from the current basic feasible solution at point A along the positive direction of the x_1 axis. Moving along the positive x_1 axis, two extreme points D at (5,0) and E at (10,0) can be encountered. In fact, 5 and 10 are the intercepts of the first and second constraint equations on the positive x_1 axis. Referring to Fig. 3.2.1, it is known that point E at (10,0) is infeasible to the problem. Therefore, this indicates that the move along the positive x_1 axis from point A can only be made as far as point D without violating feasibility conditions. The intercepts of constraint equations on the axis indicating the search direction can be obtained from the simplex tableau by computing the ratio of the elements in the solution column to the constraint elements under the column corresponding to the entering variable. Referring to the example illustrated in Table 3.3.2, the two columns used to compute the intercepts are shown below in Table 3.3.3 and following

$$\theta_1 = \frac{b_1}{a_{11}} = \frac{10}{2} = 5$$

$$\theta_2 = \frac{b_2}{a_{21}} = \frac{4}{0.4} = 10$$

The minimum ratio is $\overline{\theta} = \min(\theta_1, \theta_2) = \min(5, 10) = 5$.

Note that the ratio for the last constraint is not determined because the coefficient -2 for the x_1 column indicates a search along the negative x_1 axis, which is infeasible since all decision variables are required to be nonnegative. Compare the values of the positive intercept, the basic variable associated with the smallest intercept, that is, min (5, 10, –) = 5, is s_1. This s_1 will be chosen as the leaving variable and become nonbasic so that the feasibility condition is satisfied. For the sake of discussion, if s_2 is chosen as the leaving variable, the decision would lead us to point E which lies outside of the feasible space and the solution would become infeasible.

Once the entering variable is selected based on the optimality condition and the leaving variable is chosen according to the feasibility condition, the status of

TABLE 3.3.3
Computation of θ for iteration 1

Constraint #	x_1	Sol'n	θ
1	2	10	5
2	.4	4	10
3	−2	0	—

the variables in the basic and nonbasic variable lists must be updated. Using the new variable lists, computations are made to yield a new simplex tableau. In the computation, one should keep in mind that, referring to Table 3.3.2, the elements in the column under each current basic variable have a unity right at the intersection of the leaving variable row and all other elements are zero. The following algebraic operations are made to satisfy such a requirement.

The values of elements in the simplex tableau associated with the new basic and nonbasic variables can be computed by row operations (or **Gauss-Jordan elimination method**). The constraint row associated with the leaving variable is called the **pivoting equation** and serves as the basis for the row operation. The element located at the intersection of the entering column and pivoting row is called the **pivot element**. The pivot equation and pivot element play a central role in the computation. In the row operation, the objective is to transform the tableau into one which has unity at the pivot element and zeros elsewhere in the column associated with the new basic variable.

Example 3.3.2. Referring to Table 3.3.2, perform the pivot operation to update the simplex tableau after x_1 and s_1 are selected as the entering and leaving variables, respectively.

Solution. Knowing that x_1 is the entering variable and s_1 is the leaving variable, the pivot element in Table 3.3.4 is 2, which is enclosed by a circle. The row corresponding to the pivot element is the pivot row. The pivot operation by the Gauss-Jordan elimination method, referring to Table 3.3.4, involves the following two steps:

1. Divide all elements in the pivot equation associated with s_2 by the value of the pivot element.
2. Apply appropriate multiplier to the modified pivoting row of step (1) to other rows in the tableau (see Table 3.3.4) so that all elements other than the pivot element in the entering column have zero values.

The resulting updated simplex tableau is shown in Table 3.3.5 in which the membership list is updated where s_1 is replaced by x_1. The values in the solution column associated with the three constraints are the values corresponding to the current basic variables. The value in the solution column in the x_0 row is the value of objective

TABLE 3.3.4
Illustration of row operations

x_0	x_1	x_2	s_1	s_2	s_3	Sol'n	Row Operation
1	−5	1	0	0	0	0	← (+)(×)(5)
0	(2)	−1	1	0	0	10	(÷)(2) →
0	.4	.8	0	1	0	4	← (+)(×)(−.4)
0	−2	1	0	0	1	0	← (+)(×)(2)

TABLE 3.3.5
Results of iteration 1

Basic	x_0	x_1	x_2	s_1	s_2	s_3	Sol'n
x_0	1	0	-1.5	2.5	0	0	25
x_1	0	1	$-.5$	.5	0	0	5
s_2	0	0	1	$-.2$	1	0	2
s_3	0	0	0	1	0	1	10

function at the current solution point. Given by the current simplex tableau, the new solution is $x_1 = 5$ and $x_2 = 0$ (because of its nonbasic status) with the associated objective function value $x_0 = 25$. Values of slack variables are unimportant in the present context because they do not affect the value of the objective function.

In actual computation, elements in the columns of basic variables do not need to be calculated. In fact, not every single element in the tableau has to be calculated during every iteration. The row operation described above can be modified to increase its flexibility of being able to calculate the values of any desired elements in the tableau. Suppose that, in any iteration of the simplex method, element a_{ij} in the ith row and jth column is the pivot element. The value of the element located at the intersection of the kth row and the lth column, that is, a_{kl} can be calculated by

$$a'_{kl} = (a_{kl}a_{ij} - a_{kj}a_{il})/a_{ij} \tag{3.3.1}$$

where a'_{kl} is the new value replacing the old a_{kl} in the previous simplex tableau. The required information used in Eq. (3.3.1) is shown schematically in Fig. 3.3.1.

Once the new simplex tableau is generated, it is examined to see if the optimum is reached. This is done by checking the values of the current nonbasic variables in the objective function row of the simplex tableau. Using the same argument as employed in the first iteration, one wants to see if any of the remaining nonbasic variables have the potential to further improve the current x_0 value. For maximization problems, any nonbasic variable associated with a negative coefficient in the x_0 row of the simplex tableau is a candidate for the entering variable in the next iteration. If such a situation is identified, the current simplex tableau is reoptimized using the same procedure as described above. If all the objective coefficients in the x_0 row of the simplex tableau are nonnegative, the optimum solution is reached because no variable has any potential to further increase the value of x_0.

Examining the current simplex tableau, one observes that the objective coefficient associated with x_2 (a nonbasic variable) in the x_0 row is -1.5. This indicates that an increase in the value of x_2 from the zero level would further improve the current value of $x_0 = 25$. Therefore, x_2 is selected as the entering variable. To choose the leaving variable, ratios of the solutions of current basic variables (x_1, s_2, s_3) to the elements in the tableau under the entering column are calculated. The current basic variable associated with the smallest positive ratio is chosen as the leaving variable. That

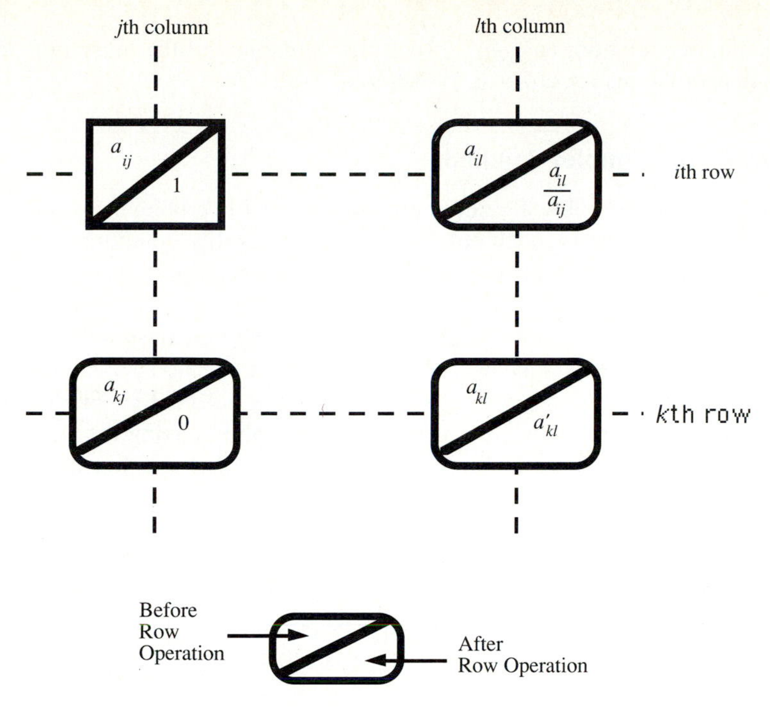

FIGURE 3.3.1
Row operations by elements.

is, the current basic variable, s_2, corresponding to min (–, 2/1, 10/0) = min(–, 2, ∞) = 2 is the leaving variable in the second iteration of the simplex method. The pivot row is the s_2 row and the pivot element is 1. After the row operations, the new simplex tableau is updated as Table 3.3.6. The feasible extreme point associated with this tableau is $(x_1, x_2) = (6, 2)$ and the corresponding objective function value, x_0, is 28 which is higher than the previous value of 25. Examining the objective coefficients in the x_0 row of the tableau, one observes that all coefficients are nonnegative. For a maximization problem, which is the case in this manufacturing-waste treatment example, this indicates that no nonbasic variables (s_1 and s_2) exist which have the potential to further improve the current value of the objective function. With this, one

TABLE 3.3.6
Results of iteration 2

Basic	x_0	x_1	x_2	s_1	s_2	s_3	Sol'n
x_0	1	0	0	2.2	1.5	0	28
x_1	0	1	0	.4	.5	0	6
x_2	0	0	1	−.2	1	0	2
s_3	0	0	0	1	0	1	10

concludes that the current solution, $(x_1^*,\ x_2^*) = (6, 2)$, is optimum and the maximum achievable net benefit to the manufacturer is x_0^* = \$28 K.

3.3.3 Summary of the Simplex Method

From the descriptions of the simplex algorithm for solving an LP problem, the solution procedure follows two basic conditions, that is, the optimality condition and the feasibility condition. More specifically on the algebraic operations, these two conditions can be phrased as follows.

The optimality condition dictates the selection of an entering variable which has the potential to further improve that value of the current objective function. Given the x_0 row reexpressed in terms of the nonbasic variables only, one selects the entering variable in maximization (minimization) from the nonbasic variables having the most negative (most positive) coefficient in the x_0 row. When all the LHS coefficients of the x_0 row in the simplex tableau are nonnegative (nonpositive), the optimum solution to the problem is reached.

The feasibility condition dictates the selection of a leaving variable so that solutions obtained during simplex iterations always remain feasible. The leaving variable is the basic variable corresponding to the smallest positive ratio of the current value of the basic variables to the positive constraint coefficients of the entering variable regardless of whether the problem is a maximization or minimization type.

The following steps summarize the simplex method for maximization:

Step 0: Express the problem in standard form with a starting basic feasible solution and then develop the tableau format. The initial tableau must always contain a basic feasible solution (check for an identity matrix).

Step 1: Scan the x_0 row; if all elements are nonnegative, stop; the optimal solution has been found; otherwise, go to step 2.

Step 2: Select the entering variable as the one corresponding to the most negative x_0 coefficient. This identifies the pivot column or the key column.

Step 3: Scan the pivot (key) column coefficient; if all are nonpositive, stop; the solution is unbounded (see Section 3.6). If at least one element is positive, go to step 4.

Step 4: Calculate

$$\theta_i = b_i / a_{ik} \qquad \text{for all } a_{ik} > 0$$

where a_{ik} is the ith element of the pivot column. Then find $\bar{\theta} = \min(\theta_i)$. The variable defined in step 2 replaces the variable of the pivot row in the next solution.

Step 5: To get the next tableau divide the pivot row by the pivot element. Now use this row to perform row operations (addition of multiples of this row) on the other rows to get all zeros in the rest of the pivot column (including the x_0 row). Return to step 1.

Pivot equation:

$$\text{New pivot equation} = \text{old pivot equation} \div \text{pivot element}$$

Other equations:

$$\text{New equation} = \text{old equation} - \begin{pmatrix}\text{its entering}\\ \text{column}\\ \text{coefficient}\end{pmatrix} \times \begin{pmatrix}\text{new pivot}\\ \text{equation}\end{pmatrix}$$

3.4 ARTIFICIAL VARIABLE METHODS

The simplex algorithm described in the previous section is directly applicable to problems whose initial basic variables are immediately available after the initialization step. This would be the case if all constraints in the model are $\leq$ type where slack variables can be added to make them equalities. The matrix of technological coefficients of slack variables at the initialization step form an identity matrix. However, some modifications to the model formulation will be needed for problems where the starting basic feasible solution cannot be readily found. This frequently occurs for models involving constraints of $\geq$ or = types. The third constraint in the manufacturing-waste treatment example, originally, is of the $\geq$ type. By simply subtracting the slack variable, s_3, from the LHS of the standard form of the constraint results in

$$2x_1 - x_2 - s_3 = 0$$

The resulting matrix associated with the slack variables (s_1, s_2, and s_3) is

s_1	s_2	s_3
1	0	0
0	1	0
0	0	−1

which is not an identity matrix. Thus, s_3 cannot be used as one of the starting basic variables. In the earlier illustration of the simplex method, both sides of the third constraint are multiplied by -1 making it a $\leq$ constraint so that the technological coefficient associated with s_3 is +1. This is permissible because the RHS of the third constraint is zero and manipulation of the sign does not violate the requirement of a nonnegative RHS of the standard form. However, if the RHS of the third constraint is positive, as are the other two constraints, the change of direction of inequality would violate the nonnegative RHS requirement of the standard form.

The **artificial variable method** is simply a mathematical trick that, through addition of so-called **artificial variables**, enables one to solve an LP model using the regular simplex algorithm previously described. Basically, artificial variables are used in two situations: (a) for constraints of the $\geq$ type, the constraints are written as

$$\sum_{j=1}^{n} a_{ij}x_j - s_i + r_i = b_i; \tag{3.4.1}$$

while (b) for constraints of = type, the constraints are written as

$$\sum_{j=1}^{n} a_{ij}x_j + r_i = b_i \tag{3.4.2}$$

where r_i is the nonnegative artificial variable.

Note that the coefficient associated with the artificial variables in the constraints is always +1. The main purpose of introducing artificial variables is to use them as the starting basic feasible variables for applying the regular simplex algorithm. Another reason for introducing artificial variables is that the addition of these might cause a violation of the corresponding constraint, unless they are zero. Artificial variables can be used as indicators to tell whether or not the model formulated has a feasible solution. If all the artificial variables in the problem are driven to the nonbasic at the zero level when the simplex algorithm stops, the problem has at least one feasible solution. On the other hand, if one or more artificial variables remain in the basics when optimum is reached, the feasible space of the problem does not exist and the problem is infeasible.

Using artificial variables as the starting basic solution in the simplex iteration, one should realize that the starting solution is not feasible to the original constraints of the problem. To find the optimum feasible solution to the original problem, all artificial variables must be driven to zero, if possible. There are many ways to do so. The following subsections describe two commonly used methods of solving an LP model when artificial variables are present. The same manufacturing-waste treatment example will be used to illustrate these methodologies.

3.4.1 Big-M Method

Without manipulating the sign of the third constraint of the original manufacturing-waste treatment example, the constraint set can be written in the following standard form

$$\begin{aligned} 2x_1 - x_2 \quad + s_1 \quad\quad &= 10 \\ .4x_1 + .8x_2 \quad\quad + s_2 &= 4 \\ 2x_1 - x_2 - s_3 \quad\quad + r_3 &= 0 \end{aligned}$$

(All x_1, x_2, s_1, s_2, s_3 and r_3 are nonnegative)

where r_3 is an artificial variable which will be used, along with s_1 and s_2, as the starting basic variables.

As mentioned above, one should try to eliminate all artificial variables from the basic variable set. For a maximization problem, this can be done by assigning a large negative coefficient, say $-M$, to the artificial variables in the objective function.

$$\text{Max } x_0 = \sum_{j=1}^{n} c_j x_j - \sum_{i=1}^{m} M r_i$$

Since the problem is a maximization type, the association of a large negative coefficient with artificial variables imposes a stiff 'penalty' to the existence of any positive artificial variables. Therefore, in the course of maximization, the algorithm would automatically attempt to avoid using any artificial variable as the basic variable. By the same token, if the problems are of the minimization type, a large positive coefficient (+M) would be assigned to each of the artificial variables in the objective function.

$$\text{Min } x_0 = \sum_{j=1}^{n} c_j x_j + \sum_{i=1}^{m} M r_i$$

This big-M can be considered as the *unit penalty* for any violation of model constraints. The objective function for the example can be formulated as

$$\text{Max } x_0 = 5x_1 - x_2 + 0s_1 + 0s_2 + 0s_3 - Mr_3$$

There are two basic shortcomings of the big-M method. They are:

1. The final step has to be reached before discovering that the problem has no feasible solution, that is, some artificial variables are positive.
2. In computer implementation of the big-M method, one has to assume a numerical value of M. In doing this, computational errors and instability could occur during the course of iteration due primarily to a scaling problem.

Therefore, the second method, called the two-phase method, which circumvents the two drawbacks of the big-M method is used in LP computer software packages.

3.4.2 Two-Phase Method

The method gets its name from the fact that computations proceed in two phases. The first phase of the computations simply attempts to drive the artificial variables out of the solution using the simplex method thus forming a starting solution, without the artificial variables, for the second phase. The second phase merely moves to the optimum solution using the simplex algorithm.

In the Phase I computation, the model to be solved has an objective function described as

$$\text{Min } r_0 = \sum_{k=1}^{K} r_k \tag{3.4.3}$$

where K is the total number of artificial variables in the model, subject to the constraints in the standard form after the necessary artificial variables are added. The objective function of this phase is the same for both maximization and minimization problems. Since all artificial variables are nonnegative, it is obvious that the minimum achievable r_o is zero which occurs only when all artificial variables are zero. In a case where the value of $r_o \neq 0$ when optimality is reached, the problem does not have a feasible space.

Upon the completion of the Phase I, Phase II optimization follows, provided that all artificial variables become nonbasic variables at the zero level. In Phase II computations, the final tableau of the Phase I computation is modified by dropping all columns associated with the artificial variables. Further, the r_0 row of the tableau is replaced by the x_0 row of the original objective function. Before checking the optimality condition, one must check that the tableau satisfies the requirement that the x_0 row can only be a function of nonbasic variables. If the condition is not satisfied, row operations must be performed to satisfy the requirement. Once this condition is satisfied, the simplex algorithm is applied to optimize the problem.

Example 3.4.1. Solve the manufacturing-waste treatment problem using the two-phase method.

Solution. The Phase I formulation of the manufacturing-waste treatment problem can be expressed as

$$\text{Min } r_0 = r_3$$

subject to

$$2x_1 - x_2 + s_1 = 10$$

$$.4x_1 + .8x_2 + s_2 = 4$$

$$2x_1 - x_2 - s_3 + r_3 = 0$$

All x_1, x_2, s_1, s_2, s_3 and r_3 are nonnegative.

The simplex tableau corresponding to this model formulation is shown in Table 3.4.1(*a*). Again, before proceeding with the simplex iteration, the objective function row in the simplex tableau must be expressed only as a function of the nonbasic variables x_1, x_2, and s_3. This can be accomplished by adding the r_3 row to the r_0 row. The resulting tableau is shown in Table 3.4.1(*b*).

The Phase I model can now be optimized by selecting x_1 as the entering variable according to the optimality condition for minimization problems. The selection of the leaving variable is the same as before according to the feasibility condition. After one iteration, the optimum solution to Phase I of this example is reached based on Table 3.4.1(*c*). Based on the final tableau of the Phase I procedure, the starting tableau of Phase II is shown in Table 3.4.1(*d*). Note that the original maximization problem is considered in the Phase II computations.

Notice that Table 3.4.1(*d*) does not satisfy the requirement that the x_0 row can only be a function of nonbasic variables. The basic variable x_1 has a nonzero coefficient in the x_0 row. Again, row operations are applied to eliminate the -5 associated with x_1 in the x_0 row. After the row operations, the resulting simplex tableau can be rewritten as Table 3.4.1(*e*). Obviously the optimality condition is not satisfied because the objective function coefficients associated with nonbasic variables x_2 and s_3 are negative. After one more iteration, the optimum solution is reached with the final tableau shown in Table 3.4.1(*g*).

3.5 INTERPRETING THE SIMPLEX TABLEAU

From the final simplex tableau one is able to obtain information regarding the optimal solution and the corresponding objective function value. Is there any other information provided by the simplex tableau? The answer is yes. It is important to be able to

TABLE 3.4.1
Manufacturing-waste treatment problem solved by two-phase method

	Basic	r_0	x_1	x_2	s_3	s_1	s_2	r_3	Sol'n
				Phase I					
	r_0	1	0	0	0	0	0	−1	0
(*a*)	s_1	0	2	−1	0	1	0	0	10
	s_2	0	.4	.8	0	0	1	0	4
	r_3	0	2	−1	−1	0	0	1	0
	r_0	1	2	−1	−1	0	0	0	0
(*b*)	s_1	0	2	−1	0	1	0	0	10
	s_2	0	.4	.8	0	0	1	0	4
	r_3	0	(2)	−1	−1	0	0	1	0
	r_0	1	0	0	0	0	0	−1	0
(*c*)	s_1	0	0	0	1	1	0	−1	10
	s_2	0	0	1	.2	0	1	−.2	4
	x_1	0	1	−.5	−.5	0	0	.5	0
				Phase II					
	x_0	1	−5	1	0	0	0		0
(*d*)	s_1	0	0	0	1	1	0		10
	s_2	0	0	1	.2	0	1		4
	x_1	0	1	−.5	−.5	0	0		0
	x_0	1	0	−1.5	−2.5	0	0		0
(*e*)	s_1	0	0	0	(1)	1	0		10
	s_2	0	0	1	.2	0	1		4
	x_1	0	1	−.5	−.5	0	0		0
	x_0	1	0	−1.5	0	2.5	0		25
(*f*)	s_3	0	0	0	1	1	1		10
	s_2	0	0	(1)	0	−.2	1		2
	x_1	0	1	−.5	0	.5	0		5
	x_0	1	0	0	0	2.2	1.5		28
(*g*)	s_3	0	0	0	1	1	0		10
	x_2	0	0	1	0	−.2	1		2
	x_1	0	1	0	0	.4	.5		6

interpret the information contained in the simplex tableau because, in practice, the computation aspect of a simplex algorithm is taken over by computers. One should know how to interpret the final solution and possibly how to perform sensitivity analysis. Consider the final tableau, that is, Table 3.4.1(g), of the manufacturing-waste treatment problem solved by the two-phase method in Example 3.4.1. Much useful information can be obtained from the tableau as described in the following subsections.

3.5.1 Optimal Solution

The optimal basic feasible variables are listed in the first column of the final tableau and the corresponding variable values are given in the last column of the tableau. The optimum objective function value is given in the solution column associated with the objective function row. For example, the optimum solution from the tableau shown in Table 3.4.1(g) is

$$(s_3^*, x_2^*, x_1^*) = (10, 2, 6) \qquad \text{with} \qquad x_0^* = 28$$

All other decision variables are zero because they are nonbasic variables. In interpreting the solution, values of the optimum feasible slack variables are ignored because they do not yield any information regarding the optimum course of action to be determined by the analysis.

3.5.2 Status of Resources

In many LP models, the RHS coefficients represent the limitation of resources or can be viewed as such, especially for $\leq$ type constraints. Although the values of slack variables bear no direct significance regarding the optimal level of activities to the problem, their status does reveal the status of resources or constraints at the current optimal solution.

If a slack variable is in the basic feasible solution set at optimality, the corresponding constraint is **nonbinding** and the corresponding resource is **abundant**. Otherwise, the constraint is **binding** and the resource is **scarce**. Referring to Table 3.4.1(g) in Example 3.4.1, notice that s_1 and s_2 are nonbasic while s_3 is basic. This implies that constraints 1 and 2 are binding and the corresponding available "resources" are used up. To put this interpretation in the problem context, the total capacity of the treatment plant and the limitation of allowable untreated waste are both consumed at the optimum decision level. Since the LHS of the third constraint represents the amount of waste to be treated, the slack variable $s_3^* = 10$ is actually the optimum amount of waste to be treated.

3.5.3 Per-Unit Worth of Resources (Shadow Prices)

Knowledge concerning the per-unit worth of a resource enables one to prioritize future funding allocations to various resources. Mathematically, the per-unit worth of resources (i.e., the RHS values) can be expressed as $\partial x_0/\partial b_i$ for $i = 1, 2, \ldots, m$. For nonbinding constraints where the corresponding resources are abundant, an increase or decrease of one unit will not have any effect on the current optimum allocation decision. For binding constraints whose corresponding resources are scarce, the per-unit worth of resource would no longer be zero. Refer again to Fig. 3.2.1. By increasing the capacity of the treatment plant b_1 or the allowable limit of untreated waste in the water course b_2, the domain of feasible space will be expanded to the right. The corresponding optimum objective function value would also increase accordingly. Similarly, if b_1 or b_2 are decreased, the optimum x_0 value would also decrease. Hence, any adjustment of the RHS coefficients of binding constraints would result in changes in the optimum objective function value, for better or worse. This implies that the per-unit worth of resource for a binding constraint is nonzero. The values of per-unit worth of a resource are given in the final tableau under the starting basic feasible variables. In the example considered, the starting basic variables are s_1, s_2, and r_3 and the corresponding coefficients in the x_0 row are 2.2, 1.5, and 0 (since r_3 in Phase II can be treated as a nonbasic variable). Therefore, the per-unit worth of resource for the constraints 1, 2, and 3 in the example are

$$\left(\frac{\partial x_0}{\partial b_1}, \frac{\partial x_0}{\partial b_2}, \frac{\partial x_0}{\partial b_3}\right) = (2.2, 1.5, 0)$$

The values mean that an increase of one unit of treatment capacity b_1 would increase the optimal x_0 value by \$2.2 K while an increase of one unit of allowable limitation on untreated waste would increase the value of x_0 \$1.5 K. With this understanding, information on the per-unit worth of a resource can be used to prioritize the allocation of additional funds used for expanding the operation.

3.6 CASES OF SIMPLEX METHOD APPLICATION

There are several unusual situations that could happen during the course of a simplex iteration. Each has different implications to the solution of an LP model. These situations are discussed in the following sections.

3.6.1 Degeneracy

Degeneracy occurs when, in any iteration, the value of one or more basic variables becomes zero. The solution for this case is said to be degenerate. When the simplex iteration enters the degenerate situation, there is no assurance that the value of the objective function will improve. The simplex iteration might enter an endless loop which will repeat the same sequence of iterations without ever reaching the optimum solution which is called the **cycling problem**. Sometimes, such degeneracy is temporary in that the solution is trapped without any improvement for only a finite number of iterations. In fact, the manufacturing-waste treatment example is an illustration of a temporary degeneracy case because one of the starting basic variables is zero, that is, $r_3 = 0$ in Table 3.4.1(*b*). The value of the objective function stays the same for the first iteration. After that, the objective function value starts to improve.

Aside from the cycling problem associated with degeneracy, it raises another question regarding the termination of the simplex iteration. Should one stop the iterations when a degenerate solution first appears? If not, how many iterations should we try? In general, a degenerate solution occurs when there are more constraints than needed to determine the feasible extreme point. Referring to Fig. 3.2.1, notice that only two constraints are needed to define the feasible extreme point A at (0,0) while, in fact, three constraint equations (x_1 = 0, x_2 = 0, and $2x_1 - x_2 \geq 0$) intersect at point A.

3.6.2 Unbounded Solution

An **unbounded solution** occurs when the solution space is unbounded so that the value of the objective function can be increased or decreased indefinitely. For example, without the second constraint in the manufacturing-waste treatment example associated with the limitation of allowable waste amount in the stream, the resulting feasible space would appear as Fig. 3.6.1. The feasible space is made of the strip bounded by constraints $2x_1 - x_2 \geq 0$ and $2x_1 - x_2 \leq 10$ but unbounded on the top. Using the argument presented in the graphical method, one can push the objective function line indefinitely to the right and the objective function value would increase continuously

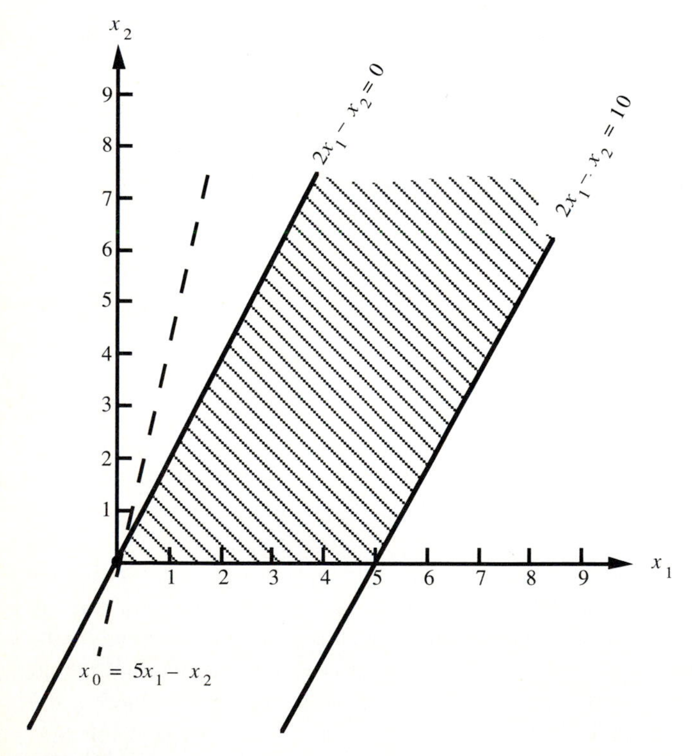

FIGURE 3.6.1
Unbounded feasible space with unbounded solution.

without limitation. However, it is not necessary that an unbounded solution space would yield an unbounded value for the objective function.

Applying the simplex algorithm to the modified model using the big-M method, the simplex tableau is shown as Table 3.6.1(c). One can immediately observe that results in Table 3.6.1 have not reached the optimality of the problem. The choice of x_2 as the entering variable leads to the smallest ratio being equal to $10/0 = \infty$. In general, an unbounded solution can be detected if, at any iteration, any of the candidates for the entering variable has all negative or zero coefficients in the constraints.

3.6.3 Alternative Optimal Solutions

Alternative optimal solutions occur when the objective function is parallel to a binding constraint (i.e., a constraint which is satisfied in equality at the optimum solution). In this case, the objective function may assume the same optimum value at more than one basic solution. The case of alternative optimal solutions can be detected when the optimal tableau is reached and the nonbasic variable has a zero coefficient in the x_0 row.

3.6.4 Nonexisting Feasible Solution

The **nonexisting feasible solution** case occurs when the feasible solution space does not exist, that is, there is no single point that can be satisfied by all the constraints.

TABLE 3.6.1
Solution of manufacturing-waste treatment problem without water quality constraints

$$\begin{aligned} \textbf{Max} \quad & x_0 = 5x_1 - x_2 \\ \textbf{S.t.} \quad & 2x_1 - x_2 \leq 10 \\ & 2x_1 - x_2 \geq 0 \end{aligned}$$

	Basic	x_0	x_1	x_2	s_3	s_1	r_3	Sol'n
	x_0	1	−5	2	0	0	M	0
(a)	s_1	0	2	−1	0	1	0	10
	r_3	0	2	−1	−1	0	1	0
	x_0	1	−5 −2M	2 +M	M	0	0	0
(b)	s_1	0	2	−1	0	1	0	10
	r_3	0	2	−1	−1	0	1	0
	x_0	1	0	−.5	−2.5 +.5M	0	2.5 +.5M	0
(c)	s_1	0	0	0	1	1	−1	10
	x_1	0	1	−.5	−.5	0	.5	0

We discussed this concept when the artificial variable method was demonstrated. To re-iterate, a nonexisting feasible solution can be detected when the calculation satisfies the optimality condition and there still exists one or more artificial variables in the basic variable set at a positive level.

3.7 DUALITY OF THE SIMPLEX METHOD

Every LP problem has two problems associated with it. One problem is called the **primal** and the other is called the **dual**. The two problems are very closely related so that the optimum solution for one problem yields complete information about the optimum solution to the other. In certain cases, this relationship can reduce the computational effort needed to solve an LP problem. Further, it is more important to use these relations to perform sensitivity analysis regarding the variation in the optimal solution due to changes in the coefficients and formulation of the problem. In summary, the main purposes for exploring the primal-dual relationship are: (a) the computational effort for solving LP problems can be reduced in certain cases; and (b) to perform sensitivity analysis.

3.7.1 Definition of Dual Problem

This subsection describes the mechanism of translating a primal LP model to its dual form.

DUAL PROBLEM WHEN PRIMAL IS IN CANONICAL FORM. The procedures involved are:

1. Each constraint in one problem corresponds to a variable in the other problem.
2. The elements of the RHS of constraints in one problem are equal to the respective coefficients of the objective function in the other problem.
3. One problem seeks maximization while the other seeks minimization.
4. Maximization problems have constraints of $\leq$ type and the minimization problems are of the $\geq$ type.
5. The decision variables in both problems are nonnegative.

More specifically, consider the **primal problem** having the canonical form

$$\text{Max } x_0 = \sum_{j=1}^{n} c_j x_j \tag{3.7.1a}$$

subject to

$$\sum_{j=1}^{n} a_{ij} x_j \leq b_i, \qquad i = 1, 2, \ldots, m \tag{3.7.1b}$$

$$x_j \geq 0, \qquad j = 1, 2, \ldots, n \tag{3.7.1c}$$

The corresponding **dual problem** can be formulated as

$$\text{Min } y_0 = \sum_{i=1}^{m} b_i y_i \tag{3.7.2a}$$

subject to

$$\sum_{i=1}^{m} a_{ij} y_i \geq c_j, \qquad j = 1, 2, \ldots, n \tag{3.7.2b}$$

$$y_i \geq 0, \qquad i = 1, 2, \ldots, m \tag{3.7.2c}$$

In the dual formulation, the nonnegative decision variables y are called the **dual variables.** As can be seen, the number of decision variables and the number of constraints in the primal and dual problems are interchanged.

DUAL PROBLEM WHEN PRIMAL IS IN STANDARD FORM. Formulation of the dual problem with the primal problem in standard form can also be made using modifications of the procedures outlined in the previous subsection. A primal problem in standard form is written as

$$\text{Max } x_0 = \sum_{j=1}^{n} c_j x_j \tag{3.7.3a}$$

subject to

$$\sum_{j=1}^{n} a_{ij} x_j = b_i, \qquad i = 1, 2, \ldots, m \tag{3.7.3b}$$

$$x_j \geq 0, \qquad j = 1, 2, \ldots, n \tag{3.7.3c}$$

The corresponding dual formulation when the primal has the standard form, can be expressed as

$$\text{Min } y_0 = \sum_{i=1}^{m} b_i y_i \tag{3.7.4a}$$

subject to

$$\sum_{i=1}^{m} a_{ij} y_i \leq c_j, \qquad j = 1, 2, \ldots, n \tag{3.7.4b}$$

$$\text{All } y_i\text{'s are unrestricted-in-sign.} \tag{3.7.4c}$$

In summary, the form of the primal-dual problem can be presented by Table 3.7.1. Computational effort for solving an LP model is largely dictated by the number of constraints. Therefore, for a linear programming problem having n decision variables, m constraints and $m > n$, it is more computationally efficient to solve its dual counterpart. Since there exists a close relationship between primal and dual problems, the solution of the dual can be used to find the solution of the original problem.

3.7.2 Primal-Dual Relationship

This subsection discusses a few important primal-dual relationships that enable us to find the optimal solution of one problem given the solution to the other. The primal problem of a maximization type will be assumed.

TABLE 3.7.1
Primal-dual table

Dual Variable	Primal Problem x_1	x_2	$\cdots$	$\cdots$	$\cdots$	x_n		RHS	
y_1	a_{11}	a_{12}	$\cdots$	$\cdots$	$\cdots$	a_{1n}	$\leq$	b_1	
y_2	a_{21}	a_{22}	$\cdots$	$\cdots$	$\cdots$	a_{2n}	$\leq$	b_2	Coefficients
$\cdots$	$\cdots$	$\cdots$		$\cdots$		$\cdots$	$\cdots$	$\cdots$	of Objective
$\cdots$	$\cdots$	$\cdots$		$\cdots$		$\cdots$	$\cdots$	$\cdots$	Function
$\cdots$	$\cdots$	$\cdots$		$\cdots$		$\cdots$	$\cdots$	$\cdots$	(Min)
y_m	a_{m1}	a_{m2}	$\cdots$	$\cdots$	$\cdots$	a_{mn}	$\leq$	b_m	Dual
RHS	$\geq c_1$	$\geq c_2$	$\cdots$	$\cdots$	$\cdots$	$\geq c_n$			
	Coefficients of Objective Function (Max) Primal								

1. For any two feasible solutions of the primal and dual problems, the value of the objective function of the primal is always less than or equal to that of the dual, that is, $x_0 \leq y_0$.
2. At optimality, both the primal and dual problems have identical values of the objective function, that is, $x_0^* = y_0^*$.
3. Optimal solution to the primal problem directly provides the optimal solution to the dual problem and vice versa.

Example 3.7.1. Formulate and solve the dual problem of the manufacturing-waste treatment example and cross compare the final simplex tableaus of the two problems.

Solution. Recall that the original problem formulation is

$$\text{Max } x_0 = 5x_1 - x_2$$

subject to

$$2x_1 - \quad x_2 \leq 10$$

$$0.4x_1 + 0.8x_2 \leq 4$$

$$2x_1 - \quad x_2 \geq 0$$

$$x_1 \geq 0 \qquad \text{and} \qquad x_2 \geq 0$$

and the final simplex tableau using the two-phase method is given in Example 3.4.1. The corresponding dual formulation of the original model is

$$\text{Min } y_0 = 10y_1 + 4y_2$$

subject to

$$2y_1 + .4y_2 - 2y_3 \geq 5$$

$$-y_1 + .8y_2 + \quad y_3 \geq -1$$

$$y_1 \geq 0, y_2 \geq 0, y_3 \geq 0$$

To solve the dual formulation, the problem is first converted to its standard form as

$$\text{Min } y_0 = 10y_1 + 4y_2$$

subject to

$$2y_1 + .4y_2 - 2y_3 - u_1 \qquad + v_1 = 5$$

$$y_1 - .8y_2 - \quad y_3 \qquad + u_2 \qquad = 1$$

$$y_1 \geq 0, \quad y_2 \geq 0, \quad y_3 \geq 0, \quad u_1 \geq 0, \quad u_2 \geq 0, \quad v_1 \geq 0$$

where u_1 and u_2 are slack variables and v_1 is an artificial variable. Using the two-phase method and v_1 and u_2 as the starting basic variables, the algebra of solving the above dual problem is given in Table 3.7.2. From the tableau, the optimal solution to the dual problem is $(y_1^*, y_2^*, y_3^*) = (2.2, 1.5, 0)$. Examining the final tableau of the primal problem, one observes that the optimal solution y_1^* and y_2^*, which correspond to the first two constraints, are given in the x_0 row under the starting basic variables (s_1, s_2 of the same constraints). The optimal value for y_3^* is also given in the x_0 row of the final primal tableau under r_3, the starting basic variable for the third constraint. Recall the discussion on the per-unit worth of resources of an LP problem, one recognizes that the per-unit worth of resource of the primal problem is the optimal solution of the dual problem.

On the other hand, the optimal solution to the primal problem can be found from the final tableau of the dual problem, that is, Table 3.7.2(*f*). Again, focus is on the coefficients in the objective function row under the starting basic variables. The starting basic variables of the dual problem are v_1 and u_2 where v_1 is an artificial variable. The optimal value of x_1, corresponding to the first dual constraint, is given in the y_0 row under v_1 while the optimal x_2, corresponding to the second constraint in dual problem, is given in the same row under u_2 by ignoring the sign of the coefficients.

In summary, the following rules can be used to obtain the optimum solution to the dual variable from the optimum primal tableau.

1. If the dual variable corresponds to a slack starting variable in the primal problem, its optimal value is given directly by the coefficient of this slack variable in the optimal x_0 row.
2. If the dual variable corresponds to an artificial starting variable in the primal problem, its optimal value is given by the coefficient of this artificial variable in the optimal x_0 row after ignoring the sign.

3.8 MATRIX VERSION OF A SIMPLEX TABLEAU

This section lays the notational foundation to some of the important algebraic manipulations of nonlinear optimization algorithms described in Chapter 4. Consider the matrix version of an LP formulation in standard form

$$\text{Max (or Min) } x_0 = \mathbf{c}^T\mathbf{x} \tag{3.8.1a}$$

subject to

$$(\mathbf{A}, \mathbf{I})\ \mathbf{x} = \mathbf{b} \tag{3.8.1b}$$

$$\mathbf{x} \geq \mathbf{0} \tag{3.8.1c}$$

TABLE 3.7.2
Solution of dual formulation of manufacturing-waste treatment problem

	Basic	y_0	y_1	y_2	y_3	u_1	v_1	u_2	Sol'n
	Phase I: Min $v_0 = v_1$								
(*a*)	v_0	1	0	0	0	0	−1	0	0
	v_1	0	2	.4	−2	−1	1	0	5
	u_2	0	1	−.8	−1	0	0	1	1
(*b*)	v_0	1	2	.4	−2	−1	0	0	5
	v_1	0	2	.4	−1	−1	1	0	5
	u_2	0	1	−.8	−1	0	0	1	1
(*c*)	v_0	1	0	2	0	−1	0	−2	3
	v_1	0	0	2	0	−1	1	−2	3
	y_1	0	1	−.8	−1	0	0	1	1
(*d*)	v_0	1	0	0	0	0	−1	0	0
	y_2	0	0	1	0	−.5	.5	−1	1.5
	y_1	0	1	0	−1	−.4	.4	.2	2.2
	Phase II: Min $y_0 = 10y_1 + 4y_2$								
(*e*)	y_0	1	−10	−4	0	0	0	0	0
	y_2	0	0	1	0	−.5	.5	−1	1.5
	y_1	0	1	0	−1	−.4	.4	.2	2.2
(*f*)	y_0	1	0	0	−10	−6	6*	−2	28
	y_2	0	0	1	0	−.5	.5	−1	1.5
	y_1	0	1	0	−1	−.4	.4	.2	2.2

* v_1 cannot be the entering variable because it remains as a nonbasic variable in Phase II computation.

where **I** is an identity matrix and

$$\mathbf{c} = \begin{pmatrix} \mathbf{c}_{\mathrm{I}} \\ \mathbf{c}_{\mathrm{II}} \end{pmatrix}; \qquad \mathbf{x} = \begin{pmatrix} \mathbf{x}_{\mathrm{I}} \\ \mathbf{x}_{\mathrm{II}} \end{pmatrix}$$

in which $\mathbf{c}_{\mathrm{I}}$ and $\mathbf{c}_{\mathrm{II}}$ are vectors of objective function coefficients corresponding to the starting nonbasic variable vector $\mathbf{x}_{\mathrm{I}}$ and the starting basic variable vector $\mathbf{x}_{\mathrm{II}}$, respectively. The expansion of Eq. (3.8.1) leads to

$$\text{Max } x_0 - \mathbf{c}_{\mathrm{I}}^T\mathbf{x}_{\mathrm{I}} - \mathbf{c}_{\mathrm{II}}^T\mathbf{x}_{\mathrm{II}} = 0 \qquad (3.8.2a)$$

subject to

$$\mathbf{A}\mathbf{x}_{\mathrm{I}} + \mathbf{I}\mathbf{x}_{\mathrm{II}} = \mathbf{b} \qquad (3.8.2b)$$

At any iteration of the simplex method, the problem can also be expressed in terms of a basic variable set, $\mathbf{x}_B$, and a nonbasic variable set, $\mathbf{x}_N$, as

$$\text{Max } x_0 = \mathbf{c}_B^T \mathbf{x}_B + \mathbf{c}_N^T \mathbf{x}_N \tag{3.8.3a}$$

subject to

$$\mathbf{B}\mathbf{x}_B + \mathbf{N}\mathbf{x}_N = \mathbf{b} \tag{3.8.3b}$$

where $\mathbf{c}_B$ and $\mathbf{c}_N$ are the vectors of objective function values associated with the current basic variable set $\mathbf{x}_B$ and nonbasic variable set $\mathbf{x}_N$, respectively. $\mathbf{B}$ and $\mathbf{N}$ are matrices formed by the elements in constraint equations associated with the current basic variable $\mathbf{x}_B$ and nonbasic variable $\mathbf{x}_N$, respectively. The matrix $\mathbf{B}$ is called the **basis matrix.**

Because at any iteration of the simplex method all nonbasic variables are zero, Eqs. (3.8.3*a* and *b*) can be reduced to

$$\text{Max } x_0 = \mathbf{c}_B^T \mathbf{x}_B \tag{3.8.4a}$$

subject to

$$\mathbf{B}\mathbf{x}_B = \mathbf{b} \tag{3.8.4b}$$

Then, the solution to Eqs. (3.8.4*a* and *b*) for x_0 and $\mathbf{x}_B$ can be obtained as

$$\begin{aligned}\begin{pmatrix} x_0 \\ \mathbf{x}_B \end{pmatrix} &= \begin{pmatrix} 1 & -\mathbf{c}_B^T \\ 0 & \mathbf{B} \end{pmatrix}^{-1} \begin{pmatrix} 0 \\ \mathbf{b} \end{pmatrix} \\ &= \begin{pmatrix} 1 & -\mathbf{c}_B^T\mathbf{B}^{-1} \\ 0 & \mathbf{B}^{-1} \end{pmatrix} \begin{pmatrix} 0 \\ \mathbf{b} \end{pmatrix} = \begin{pmatrix} \mathbf{c}_B^T\mathbf{B}^{-1}\mathbf{b} \\ \mathbf{B}^{-1}\mathbf{b} \end{pmatrix}\end{aligned} \tag{3.8.5}$$

The simplex tableau corresponding to the current solution can be obtained by combining Eqs. (3.8.2) and (3.8.5) which yields

$$\begin{pmatrix} 1 & \mathbf{c}_B^T\mathbf{B}^{-1}\mathbf{A} & \mathbf{c}_B^T B^{-1} - \mathbf{c}_{\text{II}}^T \\ 0 & B^{-1}A & B^{-1} \end{pmatrix} \begin{pmatrix} x_0 \\ \mathbf{x}_{\text{I}} \\ \mathbf{x}_{\text{II}} \end{pmatrix} = \begin{pmatrix} \mathbf{c}_B^T\mathbf{B}^{-1}\mathbf{b} \\ \mathbf{B}^{-1}\mathbf{b} \end{pmatrix} \tag{3.8.6}$$

Hence, at any iteration of the simplex method, the tableau can be expressed as in Table 3.8.1.

One important item that can be observed in Table 3.8.1 is that at any iteration, the simplex tableau can be completely generated from the original LP model given that the basic variables are identified. This implies that there is no need to store all the elements of the simplex tableau during the course of computation. Further,

TABLE 3.8.1
Matrix version of simplex tableau

Basic	x_0	x_{I}^T	x_{II}^T	Solution
x_0	1	$\mathbf{c}_B^T\mathbf{B}^{-1}\mathbf{A} - \mathbf{c}_{\text{I}}^T$	$\mathbf{c}_B^T\mathbf{B}^{-1} - \mathbf{c}_{\text{II}}^T$	$\mathbf{c}_B^T\mathbf{B}^{-1}\mathbf{b}$
$\mathbf{x}_B$	0	$\mathbf{B}^{-1}\mathbf{A}$	$\mathbf{B}^{-1}$	$\mathbf{B}^{-1}\mathbf{b}$

the inverse of the basis matrix **B** appears everywhere in the tableau. In the simplex computation, it is worthwhile to employ the most efficient numerical algorithm to obtain the inverse of **B** which, in most cases, represents the most time consuming part of the computation. The term $\mathbf{c}_B^T\mathbf{B}^{-1}$, called the **simplex multiplier**, also plays an important role in determining the per-unit worth of the resources.

3.9 NEW METHODS FOR SOLVING LINEAR PROGRAMMING PROBLEMS

The simplex algorithm searches for the optimality of an LP problem by examining the feasible extreme points. The path of optimum seeking is made along the boundary of the feasible region. The algorithm, since its conception, has been widely used and considered as the most efficient approach for solving LP problems (Dantzig, 1963). Only recently have two new methods for solving LP problems been developed which have an entirely different algorithmic philosophy (Khatchian, 1979; Karmarkar, 1984).

In contrast to the simplex algorithm, Khatchian's ellipsoid method and Karmarkar's projective scaling method seek the optimum solution to an LP problem by moving through the interior of the feasible region. A schematic diagram illustrating the algorithmic differences between the simplex and the two new algorithms is shown in Fig. 3.9.1. Khatchian's ellipsoid method approximates the optimum solution of an LP problem by creating a sequence of ellipsoids (an ellipsoid is the multidimensional analog of an ellipse) that approach the optimal solution. Both Khatchian's method and Karmarkar's method have been shown to be polynomial time algorithms. This means that the time required to solve an LP problem of size n by the two new methods would take at most an^b where a and b are two positive numbers.

On the other hand, the simplex algorithm is an exponential time algorithm in solving LP problems. This implies that, in solving an LP problem of size n by a simplex algorithm, there exists a positive number c such that for any n the simplex algorithm would find its solution in a time of at most $c2^n$. For a large enough n (with positive a, b, and c), $c2^n > an^b$. This means that, in theory, the polynomial time algorithms are computationally superior to exponential time algorithms for large LP problems.

REFERENCES

Bradley, S., A. Hax, and T. Magnanti: *Applied Mathematical Programming*, Addison-Wesley, Reading, Mass., 1977.

Brooke, A., D. Kendrick, and A. Meerhaus, *GAMS: A User's Guide*, The Scientific Press, Redwood City, Calif., 1988.

Dantzig, G. B.: *Linear Programming and Extensions*, Princeton University Press, Princeton, N.J., 1963.

Fiering, M. B., J. J. Harrington, and R. J. deLucia: "Water Resource System Analysis," Research Paper No. 3, Policy and Research Coordination Branch, Department of Energy, Mines and Resources, Ottawa, Canada, 1971.

Gass, S. I.: *Linear Programming: Methods and Application*, McGraw-Hill Inc., New York, 1985.

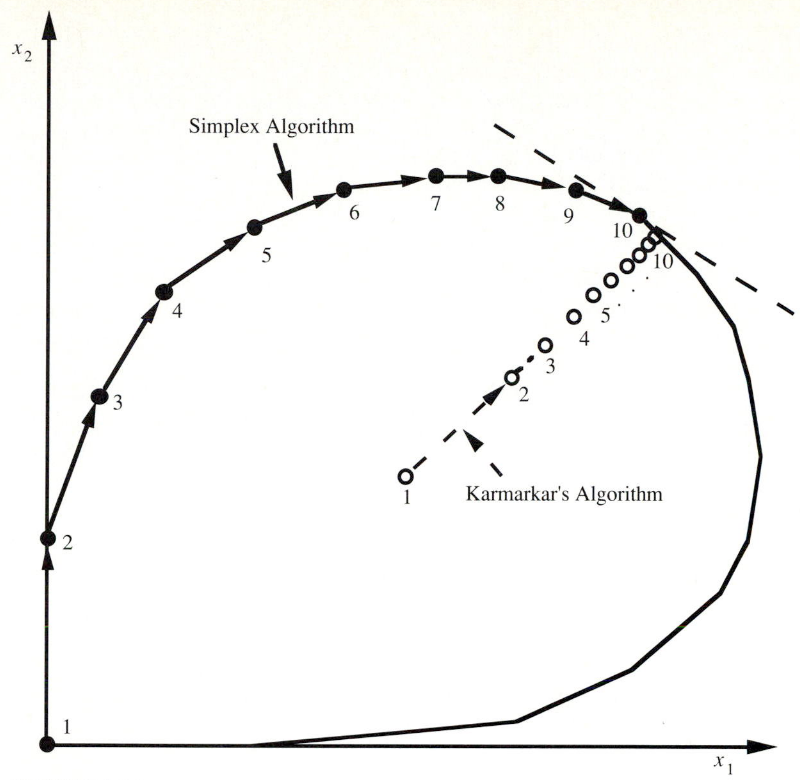

FIGURE 3.9.1
Difference in optimum search path between simplex algorithm and Karmarkar's algorithm.

Hillier, F. S. and G. J. Lieberman: *Introduction to Operations Research*, 5th ed., McGraw-Hill Inc., New York, 1990.

Karmarkar, N.: "A New Polynomial-Time Algorithm for Linear Programming," *Combinatorica*, **4**(4): 373–395, 1984.

Khatchian, L.: "A Polynomial Algorithm in Linear Programming," *Soviet Mathematics*, Doklady 20, pp. 191–194, 1979.

Murtagh, B. A. and M. A. Saunders: MINOS 5.1 User's Guide, Report SOL 83-20R, Stanford University, Dec. 1983, revised 1987.

Singal, J., R. E. Marston, and T. Morin: "Fixed Order Branch and Bound Methods for Mixed-Integer Programming: The Zoom System," Working Paper, Mangement Information Science Department, The University of Arizona, Tucson, Dec. 1987.

Taha, A. T.: *Operations Research: An Introduction*, Macmillan, New York, 1987.

Winston, W. L.: *Operations Research: Applications and Algorithms*, PWS-Kent Publisher, Boston, 1987.

PROBLEMS

3.1.1 Consider the following LP model

$$\text{Minimize } x_0 = 2x_1 + 3x_2 + 5x_3$$

$$\text{subject to} \quad \begin{aligned} x_1 + x_2 - x_3 &\geq -5 \\ -6x_1 + 7x_2 - 9x_3 &= 15 \\ |19x_1 - 7x_2 + 5x_3| &\leq 13 \end{aligned}$$

$x_1 \geq 0, \quad x_2 \geq 0, \quad x_3$ is unrestricted-in-sign

(*a*) Transform the model to its standard form.

(*b*) Transform the model to its canonical form.

3.1.2 As described in Section 3.1.2, an unrestricted-in-sign variable, say x_j, in an LP model can be replaced as $x_j = x_j^+ - x_j^-$ with $x_j^+ \geq 0$ and $x_j^- \geq 0$. In doing so, the corresponding number of nonnegative decision variables in the LP model will be doubled. Show that it is possible to replace k unrestricted variables with exactly $k+1$ nonnegative variable, $x_j = x_j' - w, j = 1, 2, \ldots, k, x_j' \geq 0$, $w \geq 0$, and $w \geq \max\{x_j^-\}$. After Taha (1987).

3.1.3 Show how to linearize the following objective function corresponding to the min–max principle? After Taha (1987).

$$\text{Minimize } x_0 = \max\left\{\left|\sum_{j=1}^{n} c_{1j}x_j\right|, \left|\sum_{j=1}^{n} c_{2j}x_j\right|, \ldots, \left|\sum_{j=1}^{n} c_{Kj}x_j\right|\right\}$$

3.1.4 Show how to linearize a max–min objective function

$$\text{Maximize } x_0 = \min\left\{\left|\sum_{j=1}^{n} c_{ij}x_j\right|, \left|\sum_{j=1}^{n} c_{2j}x_j\right|, \ldots, \left|\sum_{j=1}^{n} c_{Kj}x_j\right|\right\}$$

3.2.1 Consider the following LP problem

$$\text{Maximize } x_0 = 3x_1 + 5x_2$$

$$\text{subject to} \quad \begin{aligned} x_1 \qquad &\leq 4 \\ x_2 &\leq 6 \\ 3x_1 + 2x_2 &\leq 18 \end{aligned}$$

(*a*) Identify the feasible extreme points for the problem.

(*b*) Solve the problem graphically.

3.2.2 Referring to the graphical solutions obtained in Problem 3.2.1, determine the amounts of reduction of nonbinding constraints that will not affect the feasibility of the current optimal solution.

3.2.3 Referring to the graphical solutions obtained in Problem 3.2.1, determine the range of objective function coefficient of x_2 so that the current optimal solutions remain feasible.

3.2.4 Is it possible to change the values of RHS coefficients of binding constraints without affecting the feasibility of a current optimal solution?

3.2.5 Solve the following LP problem graphically

$$\text{Maximize } x_0 = 3x_1 + 2x_2$$

subject to

$$x_1 + 2x_2 \le 6$$
$$2x_1 + x_2 \le 8$$
$$-x_1 + x_2 \le 1$$
$$x_2 \le 2$$

3.2.6 Referring to the graphical solutions obtained in Problem 3.2.3, determine the amounts of reduction of nonbinding constraints without affecting the feasibility of the current optimal solutions.

3.2.7 Determine the range of change of the objective function coefficient of x_1 so that the current optimal solution remains feasible for Problem 3.2.3.

3.3.1 Solve Problem 3.2.1 algebraically.

3.3.2 Solve Problem 3.2.1 using the simplex method.

3.3.3 Solve Problem 3.2.5 algebraically.

3.3.4 Solve Problem 3.2.5 using the simplex method.

3.4.1 Solve the following LP problem using the big-*M* method. (Taha, 1987)

$$\text{Maximize } x_0 = 2x_1 + 3x_2 - 5x_3$$

subject to

$$x_1 + x_2 + x_3 = 7$$
$$2x_1 + 5x_2 + x_3 \ge 10$$

3.4.2 Solve the LP problem in Problem 3.4.1 using the two-phase method.

3.4.3 Solve the following LP problem by the two-phase method (Taha, 1987).

$$\text{Maximize } x_0 = 2x_1 - x_2 + x_3$$

subject to

$$2x_1 + 3x_2 - 5x_3 \ge 4$$
$$-x_1 + 9x_2 - x_3 \ge 3$$
$$4x_1 + 6x_2 + 3x_3 \le 8$$
$$x_1, x_2, x_3 \ge 0$$

3.4.4 Solve LP in Problem 3.4.3 using GAMS.

3.4.5 Referring to Fig. 3.P.1, water is available at supply points 1, 2, and 3 in quantities of 4, 8, and 12 thousand barrels, respectively. All of this water must be shipped to destinations A, B, C, D, and E, which have requirements of 1, 2, 3, 8, and 10 thousand barrels, respectively. The following table gives the cost (in dollars) of shipping one barrel of water from the given supply point to the given destination.

		Destination				
		A	B	C	D	E
Sources	1	7	10	5	4	12
	2	3	2	0	9	1
	3	8	13	11	6	14

Find the shipping schedule which minimizes the total cost of transportation and the corresponding cost of transportation using GAMS.

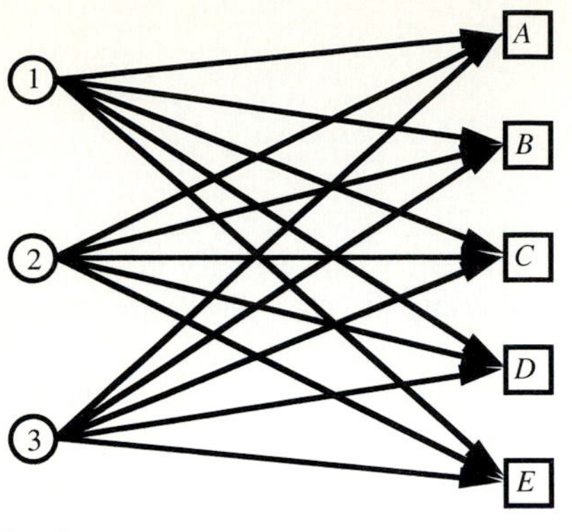

FIGURE 3.P.1
Water transportation problem.

3.4.6 Solve the manufacturing-waste treatment problem (Examples 3.1.1 and 3.4.1) using the two-phase method with a new selling unit price of \$15 K.

3.4.7 Solve Problem 3.4.6 using GAMS (or any other available LP code).

3.4.8 Solve the manufacturing-waste treatment problem (Examples 3.1.1 and 3.4.1) using the two-phase method with a different effluent tax of \$3 K for each unit of waste discharged.

3.4.9 Solve Problem 3.4.8 using GAMS (or any other available LP code).

3.4.10 Solve the manufacturing-waste treatment problem (Examples 3.1.1 and 3.4.1) using the two-phase method with a new upper limit of five units on the amount of waste any manufacturer may discharge.

3.4.11 Solve Problem 3.4.10 using GAMS (or any other available LP code.)

3.4.12 Resolve Problem 3.4.5 when demand for each destination reduces by 500 barrels.

3.4.13 For equitable reasons, the water received at a given destination has to be more or less uniformly distributed from the three sources. For this equity consideration, a restriction is imposed that, for all destinations, the difference in water received from two sources cannot be greater than 10 percent of the total received for that destination. Formulate the transportation model by incorporating equity considerations.

3.4.14 Solve Problem 3.4.13 numerically for an optimal transportation scheme of water when the demand for each destination is reduced by 500 barrels.

3.4.15 Compare the results of Problems 3.4.12 and 3.4.14 and discuss the differences.

3.4.16 Solve the manufacturing-waste treatment problem (Example 3.1.1) using the big-M method.

3.5.1 Based on the results obtained in Problem 3.3.2,
(*a*) determine the status of the constraints.
(*b*) what are the per-unit worth of RHS coefficients?
(*c*) assume that the RHS coefficients represent the resources available and each costs 1, 2, and 3 units, respectively. If future expansion for resources is possible, how do you prioritize these resources?

3.5.2 Based on the results obtained in Problem 3.3.4,
(*a*) determine the status of the four constraints.
(*b*) what are the per-unit worths of the RHS coefficients?
(*c*) How do you prioritize the four resources for future expansion if the unit costs of each RHS resources are 1/2, 1/2, 1/2, and 1/2, respectively?

3.6.1 Consider the following LP problem (after Taha, 1987)

$$\text{Maximize } x_0 = 3x_1 + x_2$$

subject to

$$x_1 + 2x_2 \leq 5$$

$$x_1 + x_2 - x_3 \leq 2$$

$$7x_1 + 3x_2 - 5x_3 \leq 20$$

$$x_1, x_2, x_3 \geq 0$$

Solve the problem by the simplex algorithm and point out any unusual condition such as degeneracy, alternative solutions, unboundedness, etc. during the course of computations.

3.7.1 Based on the LP model developed for the water transportation problem, that is, Problem 3.4.5, formulate its dual model.

3.7.2 Based on the LP model developed for the water transportation problem with equity consideration, that is, Problem 3.4.13, formulate its dual model.

3.8.1 Refer to the manufacturing-waste treatment in the standard form, that is, Example 3.1.2. Suppose that the current basic variables are identified to be $\mathbf{x_B} = (x_1, x_2)^T$. Use the matrix version of a simplex tableau to generate the entire tableau. Check your results with the final simplex tableau of the manufacturing-waste treatment problem given in Table 3.3.6.

3.9.1 Solve Problem 3.4.3 using the dual simplex method and compare it with the two primal algorithms used in Problems 3.4.3 and 3.4.4. Discuss which algorithm is preferable.

3.9.2 Solve the manufacturing-waste treatment example by the revised simplex algorithm.

3.10.1 Solve the manufacturing-waste treatment example by Karmarkar's algorithm.

3.10.2 Describe in detail Karmarkar's algorithm for solving LP problems.

APPENDIX 3A
GAMS (GENERAL ALGEBRAIC MODELING SYSTEM)

Throughout this book many linear and nonlinear programming examples and end of chapter problems are presented that are far too difficult to solve by hand. Probably the best software, from the viewpoint of teaching through this book, and to solve real-world problems will be the GAMS family of software (Brooke et al., 1988) which consists of three software modules. These are: GAMS, which can solve LP problems (Brooke et al., 1988); GAMS/MINOS, which is an adaption of MINOS (Modular In-Core Nonlinear Optimization System) for both linear and nonlinear programming problems (Murtagh and Saunders, 1987); and GAMS/ZOOM that is an adaptation of ZOOM (Zero/One Optimization Method) for mixed integer programming problems (Singhal, Marston, and Morin, 1987). All three of these software modules are available in PC, workstation, and mainframe versions. GAMS was developed by an economic modeling group at the World Bank in an effort to provide a system structure and programming language in which the conciseness of expression, generality, and portability could be maintained and to use the computer to keep track of as many programming details as possible.

3A.1 GAMS INPUT

GAMS (General Algebraic Modeling System) is a programming **language** that is used for the compact representation of large and complex models (Brooke, Kendrick, and Meerhaus, 1988). A **GAMS model** is a collection of statements in the GAMS language that is input. GAMS models consist of statements that define data structures, initial values, data modifications, and symbolic relationships (equations). The input is free format and the ordering of statements is done such that an entity of the model cannot be referenced before it is declared to exist. Basic components of a GAMS model are shown in Table 3.A.1.

Example 3.A.1. Develop the GAMS model (input) for the manufacturing waste treatment Example 3.4.1. The manufacturing-waste treatment problem is used to illustrate the development of a GAMS model.

Solution. The input can begin with comment statements such as the following for this example.

```
$ TITLE WASTE TREATMENT PLANT PROBLEM
*
* The Problem: find the quantity of goods to be produced and
*              the quantity of waste to be discharged
```

TABLE 3.A.1
Structure of a GAMS model

Inputs
SETS
Declaration
Assignment of members
Data (PARAMETERS, TABLES, SCALARS)
Declaration
Assignment of values
VARIABLES
Declaration
Assignment of type
(Optional) Assignment of bounds and/or initial values
EQUATIONS
Declaration
Definition
MODEL AND SOLVE STATEMENTS
(Optional) DISPLAY statements
Outputs
Echo Print
Reference maps
Equation Listings
Status Reports
Results

```
*              without treatment which maximizes the
*              total net benefit of the company
```

The $ and * are in column 1 of the input.

Sets are the basic building blocks and correspond to the indices in the algebraic representation. In the manufacturing-waste treatment problem there is only one SET statement:

```
SETS C constraint      /PLANTCAP, DISCLIMIT, MAXWASTE/
     A amount          /GOODS, WASTE/                 ;
```

Two sets are declared with the names C and A. The lower case words, constraint and amount, are text, which is optional. PLANTCAP is used to refer to plant capacity, DISCLIMIT refers to discharge limits, and MAXWASTE refers to the maximum amount of waste. GOODS refers to goods manufactured and WASTE refers to waste generated. Alternatively the above could be expressed in separate sets:

```
SET C constraint/PLANTCAP, DISCLIMIT, MAXWASTE/;

SET A amount/GOODS, WASTE/;
```

Data can be entered by one of three formats: lists, tables, or direct assignments. Using data entry by lists the PARAMETER statements are:

```
PARAMETER       CC (A)   Coefficients of objective function
                /GOODS   5
                WASTE   -1/                                ;

PARAMETER       RHS (C)  right-hand-sides of constraints
                /PLANTCAP 10
                DISCLIMIT  4
                MAXWASTE   0/                          ;
```

The above parameter statements declare the existence of two parameters, CC(A) with a domain A and RHS (C) with a domain C. The **domain** is the set over which a parameter, variable or equation is defined. The statement also provides documentary text for each parameter and assigns values of CC (A) and RHS (C). The following table defines the remaining data entry which are the coefficients in the constraints.

```
TABLE COEFF(C,A)  Coefficients in constraints
                  GOODS        WASTE
PLANTCAP          2            -1
DISCLIMIT         0.4           0.8
MAXWASTE          2            -1
```

The decision variables of a GAMS model must be declared with a VARIABLES statement. Each variable is given a name, a domain if needed and optional text. The VARIABLES statement for the treatment example is

```
VARIABLES  X(A)                decision variables
           MAXBENEFIT
POSITIVE VARIABLE X;
```

GAMS is an algebraic modeling language that allows a group of equations or inequalities of a model to be created simultaneously not individually. The equation declarations for the manufacturing waste treatment example is

```
EQUATIONS  C1       plant capacity constraint
           C2       waste discharge limit constraint
           C3       no more waste constraint
           PROFIT   profit of company;
```

The constraints are as defined in Example 3.1.1 and the objective function is to minimize profit.

Equation summations are based on the idea that a summation is an operator with two arguments, SUM (index of summation, summand), which are the index of the summation and the summand. As an example, the plant capacity constraint of the general form $\sum_j c_{i,j}x_j$ is $\sum_A$ COEFF (C,A)*X(A) where C is i, A is j, COEFF (C,A) is $c_{i,j}$, X(A) is x_j and * is a multiplication sign.

The components of equation definition are in order: 1) name of equation being defined; 2) the domain; 3) an optimal domain restriction condition; 4) the symbol; 5) LHS expression; 6) relational operator (less than or equal to =L=, greater than or equal to =G=, and equal to =E=); and 7) the LHS expression.

A complete set of constraints can now be defined. The capacity constraint $\left(\sum_i c_{i,j}x_j \leq b_i\right)$ is:

```
C1..SUM (A, COEFF ('PLANTCAP', A) * X(A)) =L=RHS ('PLANTCAP')
```

As defined above, RHS ( ) defines the right-hand side (b_i) of the constraint and the constraint is a less than or equal to constraint. The complex set of equations are:

```
C1..SUM (A, COEFF ('PLANTCAP', A) * X (A)) =L=RHS ('PLANTCAP');
C2..SUM (A, COEFF ('DISCLIMIT', A) * X (A)) =L=RHS ('DISCLIMIT');
C3..SUM (A, COEFF ('MAXWASTE', A) * X (A)) =G=RHS ('MAXWASTE');
PROFIT..SUM (A, CC (A) * X (A)) =E=MAXBENEFIT;
```

GAMS does not use an explicit entity called the objective function. A variable is used to specify the function to be optimized; this variable is unconstrained in sign and scalar-valued (no domain) in the equation definitions. In the above equation definitions the word MAXBENEFIT is used to define the objective.

The model, solve, and display statements are now explained. In GAMS the word MODEL has a precise meaning, a collection of EQUATIONS. The MODEL is given a name in a declaration statement by first using the keyword MODEL followed by the model's name, followed by a list of equation names enclosed in slashes / /. If all previously defined equations are included then the word /ALL/ is used, instead of the complete list. For the manufacturing waste-treatment example the declaration statement is:

```
MODEL WSTPLNT /ALL/;
```

If the list is used rather than the shortcut, the model statement is:

```
MODEL WSTPLNT /C1, C2, C3, PROFIT/;
```

This statement is very useful when a user wants to create several models in one GAMS run. In other words the list option would be used only when a subset of the existing equations comprise a specific model.

After a model has been declared and assigned equations, the solver is called using a SOLVE statement. For the manufacturing-waste treatment example, the SOLVE statement is:

```
SOLVE WSTPLNT USING LP MAXIMIZING MAXBENEFIT;
```

The format of the SOLVE statement is: 1) keyword SOLVE; 2) name of the model to be solved; 3) the keyword USING; 4) the available solution procedure such as LP for LP, NLP for nonlinear programing, or MIP for mixed integer programming; 5) the keyword MINIMIZING or MAXIMIZING; and 6) the name of the variable to be optimized, such as MAXBENEFIT in the manufacturing waste treatment example.

The DISPLAY statement is used to define the level of output. For the manufacturing waste treatment example the following display statement is used

```
DISPLAY X.L, X.M;
```

in which X.L calls for a printout of the final values of the decision variables and X.M calls for a printout of the marginal values of the variables.

The complete GAMS model (input) is shown in Figure 3.A.1.

3A.2 GAMS OUTPUT

Table 3.A.1 illustrates that the output includes the echo print, reference maps, equation listings, status reports and results. The echo print is a printout of the input file such as shown in Fig. 3.A.1, for the manufacturing waste treatment example. If errors are found by the GAMS coded errors are inserted in the echo print on the line immediately following the error. The messages start with "****" and contain a "$" directly below the point where the error occurred.

```
1$
2 *
3 * THE PROBLEM:  FIND THE QUANTITY OF GOODS TO BE PRODUCED
4 *               AND THE QUANTITY OF WASTE TO BE DISCHARGED
5 *               WITHOUT TREATMENT WHICH MAXIMIZE THE TOTAL
6 *               NET BENEFIT OF THE COMPANY
7 *
8 *------------------------------------------------------------
9
10 SETS  C  CONSTRAINT  /PLANTCAP, DISCLIMIT, MAXWASTE/
11       A  AMOUNT      /GOODS, WASTE/     ;
12
13 *------------------------------------------------------------
14
15 PARAMETER    CC(A)  COEFFICIENTS OF OBJECTIVE FUNCTION
16                /GOODS            5
17                 WASTE           -1/      ;
18
19 PARAMETER  RHS(C)  RIGHT-HAND-SIDES OF CONSTRAINTS
20                    /PLANTCAP   10
21                     DISCLIMIT   4
22                     MAXWASTE    0/      ;
23
24 TABLE  COEFF(C,A)  COEFFICIENTS IN CONSTRAINTS
25                    GOODS        WASTE
26        PLANTCAP      2           -1
27        DISCLIMIT     0.4          0.8
28        MAXWASTE      2           -1      ;
29
30 *------------------------------------------------------------
31
32 VARIABLES  X(A)      DECISION VARIABLES
33            MAXBENEFIT
34
35 POSITIVE VARIABLE X;
36
37 *------------------------------------------------------------
38
39 EQUATIONS   C1       PLANT CAPACITY CONSTRAINT
40             C2       WASTE DISCHARGE LIMIT CONSTRAINT
41             C3       NO MORE WASTE CONSTRAINT
42             PROFIT   PROFIT OF COMPANY
43
44 C1..SUM (A, COEFF ('PLANTCAP', A) *X (A)) =L=RHS ('PLANTCAP');
45
46 C2..SUM (A, COEFF ('DISCLIMIT', A) *X (A)) =L=RHS ('DISCLIMIT');
```

FIGURE 3.A.1

GAMS model for the manufacturing-waste treatment example

```
47
48 C3.. SUM (A, COEFF ('MAXWASTE', A) *X (A)) =G=RHS ('MAXWASTE');
49
50 PROFIT..SUM (A, CC (A) *X (A)) =E=MAXBENEFIT;
51
52 *----------------------------------------------------------------
53 MODEL WSTPLNT /ALL/  ;
54 *----------------------------------------------------------------
55 SOLVE WSTPLNT USING LP MAXIMIZING MAXBENEFIT;
56 *----------------------------------------------------------------
57 DISPLAY X.L, X.M;
```

FIGURE 3.A.1 (continued)

The next section of output are the reference maps. A cross-reference map is the first map which is an alphabetical, cross-referenced list of all entities (sets, parameters, variables, and equations) of the model. The cross-reference map for the manufacturing-waste treatment example is as follows

```
Symbol        Type References
A              SET  DECLARED  11  DEFINED   11        REF   15
                          24  32      2*44 2*46       2*48 2*50
                     CONTROL  44        46   48         50
C              SET  DECLARED  10  DEFINED   10        REF   19
                          24
C1             EQU  DECLARED  39  DEFINED   44   IMPL-ASN   55
                         REF  53
C2             EQU  DECLARED  40  DEFINED   46   IMPL-ASN   55
                         REF  53
C3             EQU  DECLARED  41  DEFINED   48   IMPL-ASN   55
                         REF  53
CC           PARAM  DECLARED  15  DEFINED   16        REF   50
COEFF        PARAM  DECLARED  24  DEFINED   24        REF   44
                          46  48
MAXBENEFIT     VAR  DECLARED  33 IMPL-ASN   55        REF   50
                          55
PROFIT         EQU  DECLARED  42  DEFINED   50   IMPL-ASN   55
                         REF  53
RHS          PARAM  DECLARED  19  DEFINED   20        REF   44
                          46  48
WSTPLNT      MODEL  DECLARED  53  DEFINED   53        REF   55
X              VAR  DECLARED  32 IMPL-ASN   55        REF   35
                          44  46        48   50       2*57
```

As an example the cross-reference list says that `A` is a set that is declared in line `11`, defined in line `11` and referenced once in lines `15`, `24`, and `32` and referenced twice (e.g. `2 * 44`) in lines `44`, `46`, `48`, and `50`.

The second part of the reference map lists the model entities by type, as shown below for the manufacturing-waste treatment example.

```
Sets
A             AMOUNT
C             CONSTRAINT

PARAMETERS

CC            COEFFICIENTS OF OBJECTIVE FUNCTION
COEFF         COEFFICIENTS IN CONSTRAINTS
RHS           RIGHT-HAND-SIDES OF CONSTRAINTS

VARIABLES

MAXBENEFIT
X             DECISION VARIABLES

EQUATIONS

C1            PLANT CAPACITY CONSTRAINT
C2            WASTE DISCHARGE LIMIT CONSTRAINT
C3            NO MORE WASTE CONSTRAINT
PROFIT        PROFIT OF COMPANY

MODELS

WSTPLNT
```

This list also provides the associated documentary text.

If no compilation errors occur then the equation listing is the next output. This is a product of the SOLVE statement that shows the model with current values of the sets and parameters. The equation listing for the manufacturing-waste treatment example is

```
----    C1         =L=  PLANT CAPACITY CONSTRAINT

C1..    2*X(GOODS)-X(WASTE)=L=10  ;

----    C2         =L=  WASTE DISCHARGE LIMIT CONSTRAINT

C2..    0.4*X(GOODS)+0.8*X(WASTE)=L=4 ;

----    C3         =G=  NO MORE WASTE CONSTRAINT

C3..   2*X(GOODS)-X(WASTE)=G=0  ;

----   PROFIT      =E=  PROFIT OF COMPANY

PROFIT..5*X(GOODS)-X(WASTE)-MAXBENEFIT=E=0  ;
```

Default output also presents a section called column listing which is analogous to the equation listing. The column listing for the manufacturing-waste treatment example is

```
----    X          DECISION VARIABLES

X (GOODS)
                   (.LO,   .L,   .UP = 0,   0,  +     INF)
   2               C1
   0.4             C2
   2               C3
   5               PROFIT

X (WASTE)
                   (.LO,   .L,   .UP = 0,   0,  +     INF)
  -1               C1
   0.8             C2
  -1               C3
  -1               PROFIT

---- MAXBENEFIT

MAXBENEFIT
                   (.LO,   .L,   .UP = -INF,  0,      +INF)
  -1               PROFIT
```

The symbols `.LO`, `.L`, and `.UP` refer to lower bound, level or primal value, and upper bound, respectively. For `X(GOODS)` the lower bound, primal, and upper bound are, `0`, `0`, and `+INF` (positive infinity). Bounds on the variables were set automatically as `0` and `+INF`; however, these can be overwritten in the input. For example if the lower and upper bounds on `X (GOODS)` were `0.5` and `10.5` this would be specified as

```
X.LO (GOODS) = 0.5

X.UP (GOODS) = 10.5
```

These statements must appear in the input after the variable declaration and before the solve statement.

Model statistics is the last section of output that GAMS produces before the solver begins solving the model which is shown below for the manufacturing-waste treatment example,

```
MODEL STATISTICS

BLOCKS OF EQUATIONS   4        SINGLE EQUATIONS    4
BLOCKS OF VARIABLES   2        SINGLE VARIABLES    3
NON ZERO ELEMENTS     9
```

A `SOLVE SUMMARY` is printed once the solver has executed. This has two important parts, the `SOLVER STATUS` and the `MODEL STATUS`. These are shown below for the manufacturing-waste treatment example,

```
SOLVE SUMMARY

      MODEL     WSTPLNT         OBJECTIVE  MAXBENEFIT
      TYPE  LP                  DIRECTION  MAXIMIZE
      SOLVER    BDMLP           FROM LINE  55

****  SOLVER STATUS        1 NORMAL COMPLETION
****  MODEL STATUS         1 OPTIMAL
****  OBJECTIVE VALUE            28.0000

RESOURCE USAGE, LIMIT      0.033         1000.000
ITERATION COUNT, LIMIT     1             1000
```

Note that the `SOLVER STATUS` is a normal completion, however there are five other possibilities. The `MODEL STATUS` was `1 OPTIMAL`. There are a total of 11 possible statuses including `3 UNBOUNDED` and `4 INFEASIBLE` for LP.

The final information in the output is the solution reports, as shown below for the manufacturing-waste treatment example,

```
                         LOWER    LEVEL     UPPER    MARGINAL

---- EQU C1              -INF    10.000    10.000       2.200
---- EQU C2              -INF     4.000     4.000       1.500
---- EQU C3D                .    10.000     +INF       -1.000
---- EQU PROFIT             .        .         .        -1000

       C1       PLANT CAPACITY CONSTRAINT
       C2       WASTE DISCHARGE LIMIT CONSTRAINT
       C3       NO MORE WASTE CONSTRAINT
       PROFIT   PROFIT OF COMPANY

---- VAR X                     DECISION VARIABLES

                LOWER    LEVEL      UPPER     MARGINAL
GOODS             .      6.000      +INF        .
                  .      2.000      +INF        .

                         LOWER      LEVEL     UPPER     MARGINAL

---- VAR MAXBENEFIT                 -INF      28.000    +INF
```

Single dots shown above represent zeros. The next printout is a report summary which defines the number of nonoptimal, infeasible, and unbounded rows and columns.

```
****  REPORT SUMMARY                      0       NONOPT
                                          0  INFEASIBLE
                                          0   UNBOUNDED
```

Once the solution reports are written the DISPLAY statement output is given

```
----  57    VARIABLE X.L          DECISION VARIABLES
GOODS 6.000,  WASTE 2.000

----  57    VARIABLE X.M          DECISION VARIABLES
                 ALL      0.000
```

The GAMS PC software and user guide (Brooke, Kendrick, and Meerhaus, 1988) can be purchased from The Scientific Press, 507 Seaport Court, Redwood City, CA 94063-2731.

CHAPTER

4

DYNAMIC AND NONLINEAR PROGRAMMING WITH APPLICATIONS TO HYDROSYSTEMS

Earlier hydrosystems applications of operations research techniques relied mainly upon the use of linear and dynamic programming techniques. The use of these techniques applied to solving hydrosystem problems has been rather widespread in the literature. Linear programming codes are widely available whereas dynamic programming requires a specific code for each application. The use of nonlinear programming in solving hydrosystems problems has not been as widespread even though most of the problems requiring solutions are nonlinear problems. The recent development of new nonlinear programming techniques and the availability of nonlinear programming codes have attracted new applications of nonlinear programming in hydrosystems. The first two sections of this chapter describe the basics of dynamic programming. Then, unconstrained nonlinear optimization procedures are described followed by descriptions of constrained nonlinear optimization procedures.

4.1 DYNAMIC PROGRAMMING

Dynamic programming (DP) transforms a sequential or multistage decision problem that may contain many interrelated decision variables into a series of single-stage problems, each containing only one or a few variables. In other words, the dynamic programming technique decomposes an N-decision problem into a sequence of N separate, but interrelated, single-decision subproblems. Decomposition is very useful in solving large, complex problems by decomposing a problem into a series of smaller subproblems and then combining the solutions of the smaller problems to obtain the solution of the entire model composition. The reason for using decomposition is to solve a problem more efficiently which can lead to significant computational savings. As a rule of thumb, computations increase exponentially with the number of variables, but only linearly with the number of subproblems. This chapter only briefly describes dynamic programming. Books that deal with dynamic programming are Dreyfus and Law (1977), Cooper and Cooper (1981), and Denardo (1982).

To describe the general philosophy of the dynamic programming technique, consider the following resource allocation problem. Suppose that funds are allocated to three water development projects, A, B, and C, to maximize the total expected revenue. Each development project consists of different alternative configurations that require different funding levels and yield different revenues. Due to budget limitations, the total available funds for the entire development is fixed. If the number of alternatives for each project is not too large, exhaustive enumeration of all possible combinations of the project alternatives is perhaps practical to identify the optimal combination of alternatives for the entire project development. Naturally, this **exhaustive enumeration** approach possesses three main shortcomings: (1) it would become impractical if the number of alternative combinations is large; (2) the optimal course of action cannot be verified, even if it is obtained in the early computation, until all the combinations are examined; and (3) infeasible combinations (or solutions) cannot be eliminated in advance.

In dynamic programming the alternative for each project is considered individually without ignoring the interdependence among the projects through the total available budget. Since the total funds are limited, the amount available to each project depends on the allocations to the other projects. Whatever funds are given to projects A and B, the allocation to the remaining project, C, must be made to optimize its return with respect to the remaining capital available. In other words, the optimal allocation to project C is conditioned upon the available funds for C after allocation to projects A and B are made. Since the optimal allocations to projects A and B are not known, the optimal allocation and revenue from project C must be determined for all possible remaining funds, after allocations to projects A and B have been made. Furthermore, whatever amount is allocated to project A, the allocations to projects B and C must be made optimally with respect to the remaining funds after the allocation is made to project A. To find the optimal allocation to project B, the allocation maximizing the revenue from project B together with the optimal revenue from C, as a function of the remaining funds from the allocation to project B, is sought. Finally, the optimal allocation to project A is determined, to maximize the revenue from project A plus the combined optimal revenue from both B and C, as a function of the funds remaining after the allocation to project A.

In reality, funds are allocated to the three projects simultaneously. The sequential allocation is a mathematical artifact allowing one to make the decisions sequentially. Dynamic programming can overcome the shortcomings of an exhaustive enumeration procedure using the following concepts.

1. The problem is decomposed into subproblems and the optimal alternative is selected for each subproblem sequentially so that it is never necessary to enumerate all combinations of the problem in advance.
2. Because optimization is applied to each subproblem, nonoptimal combinations are automatically eliminated.
3. The subproblems should be linked together in a special way so that it is never possible to optimize over infeasible combinations.

4.1.1 Elements of Dynamic Programming Model

The water project funding allocation example described previously can be modeled mathematically as

$$\text{Max} \sum_{i=1}^{N} \sum_{j=1}^{M_i} r_{ij} x_{ij} \tag{4.1.1a}$$

subject to

$$\sum_{i=1}^{N} \sum_{j=1}^{M_i} c_{ij} x_{ij} \leq F \tag{4.1.1b}$$

$$\sum_{j=1}^{M_i} x_{ij} \leq 1, \qquad \text{for } i = 1, 2, 3, \ldots, N \tag{4.1.1c}$$

$$\text{All } x_{ij} = 0 \text{ or } 1 \tag{4.1.1d}$$

where r_{ij} is the revenue that can be generated from alternative j of project i, c_{ij} is the funding requirement for alternative j of project i, x_{ij} is a decision variable which can take either 0 or 1 with 0 indicating alternative j of project i not selected and 1 otherwise, F is the total funds available for the development, N is the total number of projects under consideration, and M_i is the number of alternatives of project i. The second set of constraints represents that not all projects under consideration have to be funded. Furthermore, alternatives within each project are mutually exclusive, that is, only one alternative per project can be chosen. According to the general discussion of the dynamic programming solution approach, the above mathematical programming model can be depicted as Fig. 4.1.1.

Referring to Fig. 4.1.1, the basic elements and terminologies in a dynamic programming formulation are introduced as follows.

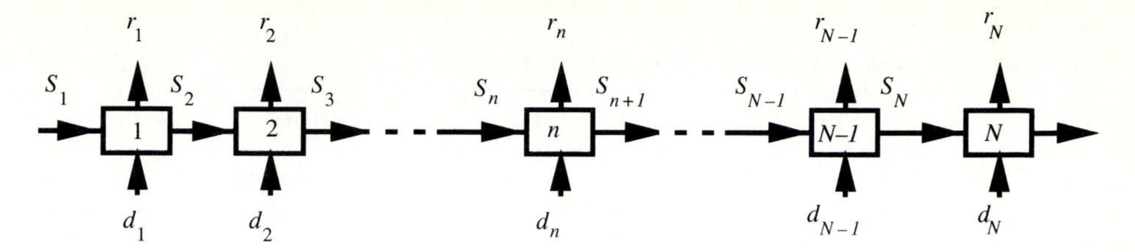

FIGURE 4.1.1
Sequential representation of serial dynamic programming problems.

1. **Stages (n)** are the points of the problem where decisions are to be made. In the funds allocation example, each project represents a stage in the dynamic programming model. If a decision making problem can be decomposed into N subproblems, there will be N stages in the dynamic programming formulation.
2. **Decision Variables (d_n)** are courses of action to be taken for each stage. The decision in the project funding example is the alternative within the project to be selected. The number of decision variables, d_n, in each stage is not necessarily equal to one.
3. **State Variables (S_n)** are variables describing the state of a system at any stage n. A state variable can be discrete or continuous, finite or infinite. Referring to Fig. 4.1.1, at any stage n, there are input states, S_n, and output states, S_{n+1}. The state variables of the system in a dynamic programming model have the function of linking succeeding stages so that, when each stage is optimized separately, the resulting decision is automatically feasible for the entire problem. Furthermore, it allows one to make optimal decisions for the remaining stages without having to check the effect of future decisions for decisions previously made.
4. **Stage Return (r_n)** is a scalar measure of the effectiveness of decision making in each stage. It is a function of the input state, the output state, and the decision variables of a particular stage. That is, $r_n = r(S_n, S_{n+1}, d_n)$.
5. **Stage Transformation** or **State Transition (t_n)** is a single-valued transformation which expresses the relationships between the input state, the output state, and the decision. In general, through the stage transformation, the output state at any stage n can be expressed as the function of the input state and the decision as

$$S_{n+1} = t_n(S_n, d_n) \tag{4.1.2}$$

To demonstrate the algebraic algorithm of the dynamic programming approach in optimizing a problem, the project funding allocation problem is solved in the following example.

Example 4.1.1. Table 4.1.1 contains the required funding (in millions of dollars) for each alternative and r_i indicates the revenue (in millions) that can be generated for each alternative. The total budget available for the development is $7 million. Assume that all projects under consideration must be implemented. Determine the optimal combination of alternatives that maximizes the total revenue.

TABLE 4.1.1
Costs and revenues for alternatives in Example 4.1.1

	Project A		Project B		Project C	
Alternative	Cost c_A (\$1M)	Revenue r_A (\$1M)	Cost c_B (\$1M)	Revenue r_B (\$1M)	Cost c_C (\$1M)	Revenue r_C (\$1M)
1	1	5	2	8	1	3
2	2	6	3	9	3	5
3	3	8	4	12	—	—

Solution. The basic elements for the dynamic programming model are defined.

1. **Stage** (n): each project represents a stage with $n = A, B$, and C.
2. **State variable** (S_n): the state variable at each stage is the set of alternatives considered.
3. **Decision variable** (d_n): the decision variable is the alternative selected for the next stage (project).
4. **Stage return** (r_n): the revenue generated from the selected alternative.
5. **State transition function** (t_n): $S_n = d_n$ for $n = C$ and $S_{n+1} = d_n$ for $n = A$ and B.

Schematically, the problem can be depicted as Fig. 4.1.2*a* in the form of a sequential decision problem. More explicitly, the system can be expanded to a network representation as shown in Fig. 4.1.2*b* when feasible states (i.e., alternatives) in each stage (i.e., project) are shown. Assume that the sequence of decisions follows the project order as shown in Fig. 4.1.2*a*; the analysis can proceed backward by starting with the last project (i.e., project *C*). The order of the projects is immaterial in this example.

The backward algorithm starts with stage $n = C$ which has two feasible states (alternatives) as shown in Fig. 4.1.2*b*. Since this is the terminal stage, no optimization is involved. In other words, the optimal decision for project C is $d_c^* = S_c^*$ because the corresponding optimization can be stated as

$$\text{Max } f_c(S_c) = r_c(d_c = S_c)$$

In tabular form, the computation for this stage is shown in Table 4.1.2(*a*).

Move one stage backward to consider project B. For each feasible state in stage $n = B$, the objective is to identify the best connection (in terms of accumulated revenue

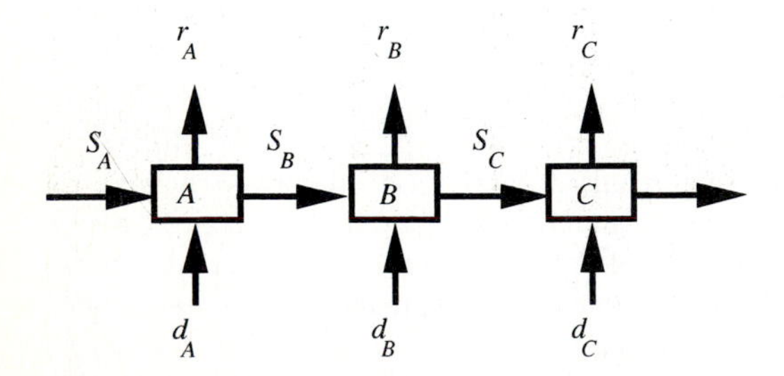

FIGURE 4.1.2*a*
Sequential representation of the funding allocation example problem.

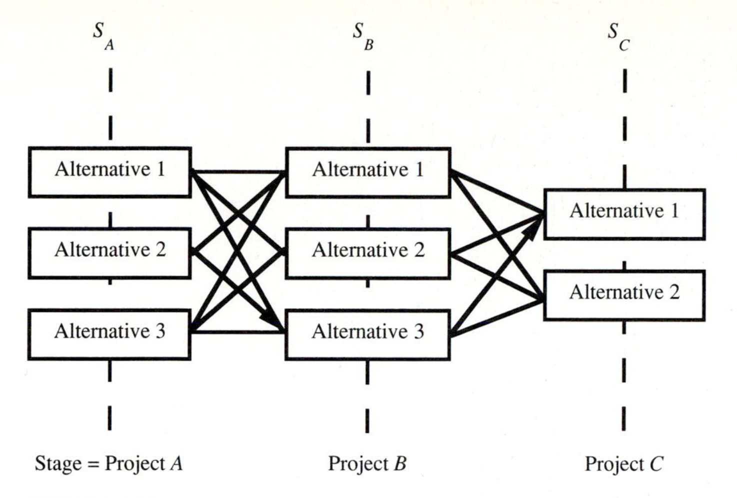

FIGURE 4.1.2*b*
Network representation of the funding allocation example using alternatives as state variables.

up to project B) to all the feasible states in the immediate downstream stages (only stage C at the present time). The optimization problem is

$$\text{Max } f_B(S_B) = r_B(S_B) + f_c^*(d_B)$$

Computations to identify the optimal connections for each feasible state in project B are given in Table 4.1.2(*b*). Note that the problem is a sequential selection of alternatives subject to the budget constraint. It is necessary to keep track of the accumulated expenditures as the optimization procedure moves from one stage to the next.

In Table 4.1.2(*b*) for each of the feasible states in project B, that is, S_B = (Alt-1, Alt-2, Alt-3), there corresponds two possible decisions connecting to the states in the next stage (project C), $d_C = S_C$ = (Alt-1, Alt-2). To explain the computations involved in Table 4.1.2(*b*) consider the last row corresponding to S_B = Alt-3. The accumulated return associated with d_C = Alt-1 is \$12 M + \$3 M = \$15 M in which the first number (\$12 M) is the revenue generated from the selection of Alt-3 for project B and the second number (\$3 M) results from the selection of Alt-1 for project C as shown in Table 4.1.2(*a*). The accumulated cost corresponding to this connection of states between projects B and C (i.e., S_B = Alt-3 and S_C = Alt-1) is \$4 M + \$1 M = \$5 M, which is feasible within the budget limit of \$7 M. With regard to S_B = Alt-3 and $d_B = S_C$ = Alt-2, the accumulated revenue of \$17 M is greater than the former one of \$15 M; however, the accumulated cost is \$7 M which uses up all the budget leaving no funding for project A. This violates the constraint requiring that all projects must be implemented. Hence, decision $d_B = S_C$ = Alt-2 is infeasible and should be discarded. The last two columns of Table 4.1.2(*b*) register the optimal accumulated revenue and optimal decision associated with each of the feasible states for project B. The optimal connections for each feasible state in each stage to the immediate downstream stage are highlighted by the boxes in Table 4.1.2*b*.

Upon the completion of project B allocations, the analysis proceeds backward to consider project A. The optimal solution for project A is Alt-3 which has a total revenue of \$21 M for the three projects. Using the same procedure employed for project B, the computational table for project A is shown in Table 4.1.2(*c*). Examining the last column

TABLE 4.1.2(*a*)
DP computations for project *C*

S_C	Optimal return f_C^*($1M)	Optimal decision d_c^*	
Alt-1	3 (1)†	Alt-1	(Cost = $1M)
Alt-2	5 (3)	Alt-2	(Cost = $3M)

†Quantities in ( ) are accumulated costs for project *C*.

TABLE 4.1.2(*b*)
DP computations for project *B*

S_B	$f_B(S_B, d_B) = r_B(S_B) + f_C^*(d_B)$		$f_B^*(\ )$	d_B^*
	$d_B = S_C$			
	Alt-1	Alt-2		
Alt-1	8 + 3 = 11 (2 + 1 = 3)†	8 + 5 = [13] (2 + 3 = 5)	13	Alt-2 ($5)
Alt-2	9 + 3 = 12 (3 + 1 = 4)	9 + 5 = [14] (3 + 3 = 6)	14	Alt-2 ($6)
Alt-3	12 + 3 = [15] (4 + 1 = 5)	12 + 5 = 17 (4 + 3 = 7, inf)#	15	(Alt-1) ($5)

†Quantities in ( ) are accumulated costs for projects *B* and *C*.
#Infeasible (inf), this decision uses up all the $7M budget leaving no funding for project *A*.

TABLE 4.1.2(*c*)
DP computations for project *A*

S_A	$f_A(S_A, d_A) = r_A(S_A) + f_B^*(d_A)$			$f_A^*(\ )$	d_A^*
	$d_A = S_B$				
	Alt-1	Alt-2	Alt-3		
Alt-1	5 + 13 = 18 (1 + 5 = 6)†	5 + 14 = 19 (1 + 6 = 7)	5 + 15 = [20] (1 + 5 = 6)	20	Alt-3 ($6)
Alt-2	6 + 13 = 19 (2 + 5 = 7)	6 + 14 = 20 (2 + 6 = 8, inf)	6 + 15 = [21] (2 + 5 = 7)	21	(Alt-3) ($7)
Alt-3	8 + 13 = 21 (3 + 5 = 8, inf)	8 + 14 = 22 (3 + 6 = 9, inf)	* + 15 + 23 (3 + 5 = 8, inf)	—	— —

†Quantities in ( ) are accumulated costs for projects *A*, *B* and *C*.

in Table 4.1.2(c) corresponding to the optimal decision d_A^*, it is observed that the total budget of \$7 M is used up when the optimal solution is obtained.

The final step of the dynamic programming technique is to "trace back" through the three stages in reverse order following the sequence of A→B→C utilizing the corresponding computation table. Note that, from Table 4.1.2(c), the maximum achievable total revenue is \$21 M associated with S_A = Alt-2. At S_A = Alt-2, the optimal decision is d_A^* = S_B = Alt-3 as circled. Proceed backward to Table 4.1.2(b), with S_B = Alt-3, the corresponding optimal decision is d_B^* = S_C = Alt-1. This then completes the trace-back procedure which identifies the optimal selection of alternatives as (d_A^*, d_B^*, d_C^*) = (Alt-2, Alt-3, Alt-1) and the optimal (maximum) total revenue is \$21 M. This trace-back procedure is illustrated in Fig. 4.1.2b indicated by heavy lines with arrows. The problem can also be solved by defining the state variable as the amount of budget available (Problem 4.1.2).

4.1.2 Operational Characteristics of Dynamic Programming

From Example 4.1.1, the basic features that characterize all dynamic programming problems are as follows:

1. The problem is divided into stages, with decision variables at each stage.
2. Each stage has a number of states associated with it.
3. The effect of the decision at each stage is to produce return, based on the stage return function, and to transform the current state variable into the state variable for the next stage, through the state transform function.
4. Given the current state, an optimal policy for the remaining stages is independent of the policy adopted in previous stages. This is called **Bellman's principle of optimality**, which serves as the backbone of dynamic programming.
5. The solution begins by finding the optimal decision for each possible state in the last stage (called the **backward recursive**) or in the first stage (called the **forward recursive**). A forward algorithm computationally advances from the first to the last stage whereas a backward algorithm advances from the last stage to the first.
6. A recursive relationship that identifies the optimal policy for each state at any stage n can be developed, given the optimal policy for each state at the next stage, $n + 1$. This backward recursive equation, referring to Fig. 4.1.1, can be written as

$$f_n^*(S_n) = \underset{d_n}{\text{opt.}} \left\{r_n\left(S_n, d_n\right) \circ f_{n+1}^*\left(S_{n+1}\right)\right\}$$

$$= \underset{d_n}{\text{opt.}} \left\{r_n\left(S_n, d_n\right) \circ f_{n+1}^*[t_n(S_n, d_n)]\right\} \tag{4.1.3}$$

where $\circ$ represents an algebraic operator which can be +, $-$, $\times$, or whichever is appropriate to the problem. The recursive equation for a forward algorithm is stated as

$$f_n^*(S_n) = \underset{d_n}{\text{opt.}} \left\{r_n\left(S_n, d_n\right) \circ f_{n-1}^*\left(S_{n-1}\right)\right\} \tag{4.1.4}$$

Note that, in the example, the computations for stage 3 (project C) do not involve the term $f_{n+1}(S_{n+1})$ in the recursive Eq. (4.1.3). This would generally be the case using backward dynamic programming optimization. Therefore, the recursive equation for the backward dynamic programming technique can be written as

$$f_n^*(S_n) = \begin{cases} \underset{d_n}{\text{opt.}} \left[r_n(S_n, d_n) \right], & \text{for } n = N \quad (4.1.5a) \\ \underset{d_n}{\text{opt.}} \left[r_n(S_n, d_n) \circ f_{n+1}^*(S_{n+1}) \right], & \text{for } n = 1 \text{ to } N-1 \quad (4.1.5b) \end{cases}$$

Example 4.1.2. Define the backward recursive equations for the water development funding Example 4.1.1.

Solution. For the third stage ($n = 3$) $f_3^*(S_3) = \underset{d_c}{\text{Max}}(r_c)$; for the second stage $f_2(S_2) = \underset{d_B}{\text{Max}}\left(r_B + f_C^*\right)$; and for stage 1 $f_1(S_1) = \underset{d_A}{\text{Max}}\left(r_A + f_B^*\right)$.

Example 4.1.3. The inflows to a reservoir having a total capacity of four units of water are 2, 1, 2, and 1 unit, respectively, for the four seasons in a year. For convenience, the amount of water is counted only by integer units of water. Thereby, the releases from the reservoir for water supply to a city and farms are sold for $2000 for the first unit of release, $1500 for the second unit, $1000 for the third unit, and $500 for the fourth unit. When the reservoir is full and spills one unit of water, a minor flood will result in a damage of $1500. When the spill reaches 2 units, a major flood will result in a damage of $4000. Determine the operation policy for maximum annual returns by dynamic programming using a backward algorithm, considering any amount of storage at the end of the year.

Solution. The first step is to explain the benefit function. One unit of release has a return (benefit) of $2000; two units of release have a return of $2000 + $1500 = $3500; three units of release have a return of $3500 + $1000 = $4500; four units of release have a return of $4500 + $500 = $5000; five units of release (when the reservoir is only four units in size mean that the reservoir becomes full and one unit is released; the benefit is $5000 (four units for water supply) − $1500 (one unit of spill) = $3500; and for 6 units of release the benefit is $5000 − $4000 (2 units of spills) = $1000.

Each season represents a stage as shown in Fig. 4.1.3. The state variable for this reservoir operation problem is the storage, that is, the initial or beginning of season storage $S_n = ST_n$ and the final or end of season storage $\tilde{S}_n = ST_{n+1}$ for season n. Reservoir release is the decision variable denoted as $d_n = R_n$ for season n. The transformation function is simply the continuity equation relating the initial season storage to the final season storage of a stage n as

$$\tilde{S}_n = S_n + QF_n - R_n$$

where QF_n is the inflow for stage n. The final storage of a particular stage (season) is the initial storage of the next state, that is,

$$\tilde{S}_n = S_{n+1}$$

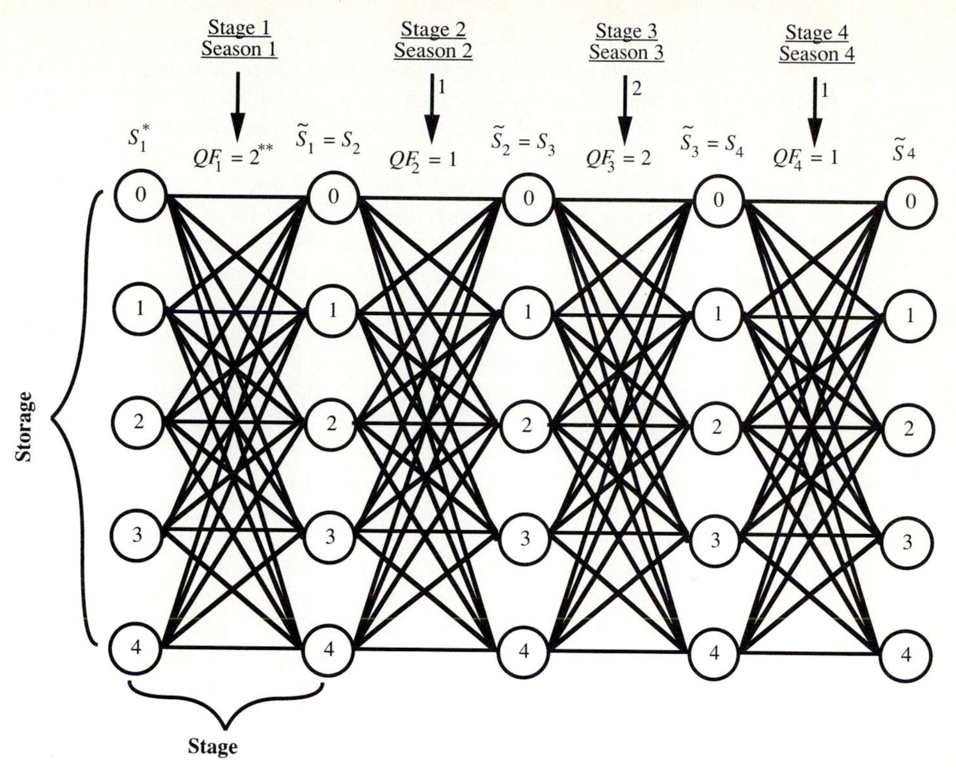

FIGURE 4.1.3
Stages and state space for Example 4.1.3.

The DP backward recursive equation for this example is

$$f_n^*(S_n) = \begin{cases} \underset{d_n}{\text{Max}}\,[r_n(S_n, d_n)], & \text{for } n = 4 \\ \underset{d_n}{\text{Max}}\left[r_n(S_n, d_n) + f_{n+1}^*(S_{n+1})\right], & \text{for } n = 1, \ldots, 3 \end{cases}$$

DP computations are shown in Tables 4.1.3(*a*)–(*d*). Referring to Table 4.1.3(*a*) for the stage 4 computations, the first column represents the initial storage (state) in season 4 (Stage 4). The second column is the initial storage S_4 plus the inflow $QF_4 = 1$. The next five columns are the heart of the DP computations where the DP recursive equation is solved. Each column is for a possible final storage $\tilde{S}_4 = 0,\ 1,\ 2,\ 3$, and 4. For the last stage 4 this is a trivial computation because the recursive equation is $f_4^*(S_4) = r_4(S_4, d_4)$. As an example for the initial storage $S_4 = 0$ and final storage $\tilde{S}_4 = 0$ refers to a release $R_4 = 1$. This is determined by solving the state transformation function for R_4

$$R_4 = S_4 + QF_4 - \tilde{S}_4 = 0 + 1 - 0 = 1$$

For a release of 1 unit the benefit is \$2000 so $f_4(S_4) = r_4(S_4 = 0, d_4 = R_4 = 1) = \$2000$. In a manner similar for $S_4 = 1$ and $\tilde{S}_4 = 0$, $R_4 = 1 + 1 - 0 = 2$ then $f_4(S_4 = 1,\ d_4 = R_4 = 2) = \$3500$. The columns maximum benefit $f_4^*(S_4)$ defines the maximum of the benefits for each initial storage considering the final storage which is the same as considering each possible release (decision) for the initial storage. The optimal release for each initial storage is listed in the next column and the optimal final storage is listed in the last column.

Table 4.1.3(*b*) presents the DP computations for the third stage (third season) which has an inflow of 2 units. To illustrate the computation procedure, consider $S_3 = 0$ and $\tilde{S}_3 = 0$. The release represented by this combination of initial and final storage is $R_3 = S_3 + QF_3 - \tilde{S}_3 = 0 + 2 - 0 = 2$, then the return or benefit is $r_3(S_3 = 0,\ d_3 = R_3 = 2) = \$3500$. The optimum cumulative benefit for stage 4 was $f_4^*(S_4 = 0) = \$2000$ from Table 4.1.3(*a*) for $\tilde{S}_3 = S_4$. The computations are repeated backwards for stages 2 (season 2) and 1 (season 1) which are shown in Tables 4.1.3(*c*) and (*d*), respectively.

Once the DP computations are completed for stage 1, a traceback is performed to identify the optimum set of releases. Referring to the DP computations for stage 1 in Table 4.1.3(*d*), the maximum benefit for each of the beginning storages S_1 are: \$11,000 for $S_1 = 0$, \$12,500 for $S_1 = 1$, \$14,000 for $S_1 = 2$, \$15,000 for $S_1 = 3$, and \$16,000 for $S_1 = 4$. The maximum benefit is therefore \$16,000. A traceback can be performed for each of initial storages. For purposes of illustration a traceback is performed for the maximum benefit of \$16,000 for $S_1 = 4$. The optimal release for this case is either $d_1^* = 2$ or 3 units with respective final storages for stage 1 of $\tilde{S}_1 = 4$ or 3. Referring to Table 4.1.3(*c*) with $S_2 = \tilde{S}_1 = 4$ or 3 the optimal releases for $S_2 = 3$ are $d_2^* = 2$ or 3 for the respective final storages for stage 2 of $\tilde{S}_2 = 2$ or 1. For $S_2 = 4$ the optimal releases are $d_2^* = 2$ or 3 for the respective final storages of $\tilde{S}_2 = 3$ or 2. So $S_3 = \tilde{S}_2 = 1,\ 2$, or 3. Now referring to Table 4.1.3(*b*) for stage 3 the optimal releases are: for $S_3 = 1$, $d_2 = 2$ and $\tilde{S}_3^* = 1$; for $S_3 = 2$, $d_3^* = 2$ or 3 and $\tilde{S}_3^* = 2$ or 3; and for $S_3 = 3$, $d_3^* = 3$ and $\tilde{S}_3^* = 2$. The optimal final storages for stage 3 are $\tilde{S}_3^* = 1$ or 2. The traceback now continues to stage 4 in Table 4.1.3(*a*). For $S_4 = \tilde{S}_4^* = 1$ or 2 then the optimal decisions are $d_4^* = 2$ or 3, respectively so that $S_4^* = 0$ for both cases.

In summary the maximum benefit for this problem would be starting out with the reservoir full at the beginning of the year and to have it empty at the end of the year. This is obvious for this example, but from the viewpoint of operation of an actual reservoir this policy would not be followed. The optimal traceback is further illustrated in Fig. 4.1.4 with the optimal releases shown.

Although dynamic programming possesses several advantages in solving water resources problems, especially for those involving the analysis of multistage processes,

TABLE 4.1.3(*a*)
DP computations for stage 4 of Example 4.1.3

(*a*) Stage 4-Season 4

Initial storage S_4	Total storage $S_4 + QF_4$	Benefits $f_4(S_4) = r_4(S_4, d_4)$ — Final storage $\tilde{S}_4$: 0	1	2	3	4	Maximum benefit $f_4^*(S_4)$	Decision (Release) $d_4 = R_4^*$	Final storage $\tilde{S}_4^*$
0	1	2000	0	—	—	—	2000	1	0
1	2	3500	2000	0	—	—	3500	2	0
2	3	4500	3500	2000	0	—	4500	3	0
3	4	5000	4500	3500	2000	0	5000	4	0
4	5	3500	5000	4500	3500	2000	5000	4	1

TABLE 4.1.3(*b*)

DP computations for stage 3 of Example 4.1.3

(*b*) Stage 3-Season 3

Initial storage S_3	Total storage $S_3 + QF_3$	Benefits $f_3(S_3) = r_3(S_3, d_3) + f_4^*(S_4)$ Final storage $\tilde{S}_3 = S_4$: 0	1	2	3	4	Maximum benefit $f_3^*(S_3)$	Decision (Release) $d_3 = R_3^*$	Final storage $\tilde{S}_3^* = S_4$
0	2	3500 + 2000 = 5500	2000 + 3500 = 5500	0 + 4500 = 4500	—	—	5500	2, 1	0, 1
1	3	4500 + 2000 = 6500	3500 + 3500 = 7000	2000 + 4500 = 6500	0 + 5000 = 5000	—	7000	2	1
2	4	5000 + 2000 = 7000	4500 + 3500 = 8000	3500 + 4500 = 8000	2000 + 5000 = 7000	0 + 5000 = 5000	8000	3, 2	1, 2
3	5	3500 + 2000 = 5500	5000 + 3500 = 8500	4500 + 4500 = 9000	3500 + 5000 = 8500	2000 + 5000 = 7000	9000	3	2
4	6	1000 + 2000 = 3000	3500 + 3500 = 7000	5000 + 4500 = 9500	4500 + 5000 = 9500	3500 + 5000 = 8500	9500	4, 3	2, 3

TABLE 4.1.3(*c*)
DP computations for stage 2 of Example 4.1.3

(*c*) Stage 2-Season 2

Initial storage S_2	Total storage $S_2 + QF_2$	Benefits $f_2(S_2) = r_2(S_2, d_2) + f_3^*(S_3)$					Maximum benefit $f_2^*(S_2)$	Decision (Release) $d_2 = R_2^*$	Final storage $\tilde{S}_2^* = S_3$
		Final storage $\tilde{S}_2 = S_3$							
		0	1	2	3	4			
0	1	2000 + 5500 = 7500	0 + 7000 = 7000	—	—	—	7500	1	0
1	2	3500 + 5500 = 9000	2000 + 7000 = 9000	0 + 8000 = 8000	—	—	9000	2, 1	0, 1
2	3	4500 + 5500 = 10,000	3500 + 7000 = 10,500	2000 + 8000 = 10,000	0 + 9000 = 9000	—	10,500	2	1
3	4	5000 + 5500 = 10,500	4500 + 7000 = 11,500	3500 + 8000 = 11,500	2000 + 9000 = 11,000	0 + 9500 = 9500	11,500	3, 2	1, 2
4	5	3500 + 5500 = 9000	5000 + 7000 = 12,000	4500 + 8000 = 12,500	3500 + 9000 = 12,500	2000 + 9500 = 11,500	12,500	3, 2	2, 3

TABLE 4.1.3(*d*)

DP computations for stage 1 of Example 4.1.3

(*d*) Stage 1-Season 1									
		Benefits $f_1(S_1) = r_1(S_1, d_1) + f_2^*(S_2)$							
		Final storage $\tilde{S}_1 = S_2$							
Initial storage S_1	Total storage $S_1 + QF_1$	0	1	2	3	4	Maximum benefit $f_1^*(S_1)$	Decision (Release) $d_1 = R_1^*$	Final storage $\tilde{S}_1^* = S_2$
0	2	3500 + 7500 = 11,000	2000 + 9000 = 11,000	0 + 10,500 = 10,500	—	—	11,000	2, 1	0, 1
1	3	4500 + 7500 = 12,000	3500 + 9000 = 12,500	2000 + 10,500 = 12,500	0 + 11,500 = 11,500	—	12,500	2, 1	1, 2
2	4	5000 + 7500 = 12,500	4500 + 9000 = 13,500	3500 + 10,500 = 14,000	2000 + 11,500 = 13,500	0 + 12,500 = 12,500	14,000	2	2
3	5	3500 + 7500 = 11,000	5000 + 9000 = 14,000	4500 + 10,500 = 15,000	3500 + 11,500 = 15,000	2000 + 12,500 = 14,500	15,000	3, 2	2, 3
4	6	1000 + 7500 = 8500	3500 + 9000 = 12,500	5000 + 10,500 = 15,500	4500 + 11,500 = 16,000	3500 + 12,500 = 16,000	16,000	3, 2	3, 4

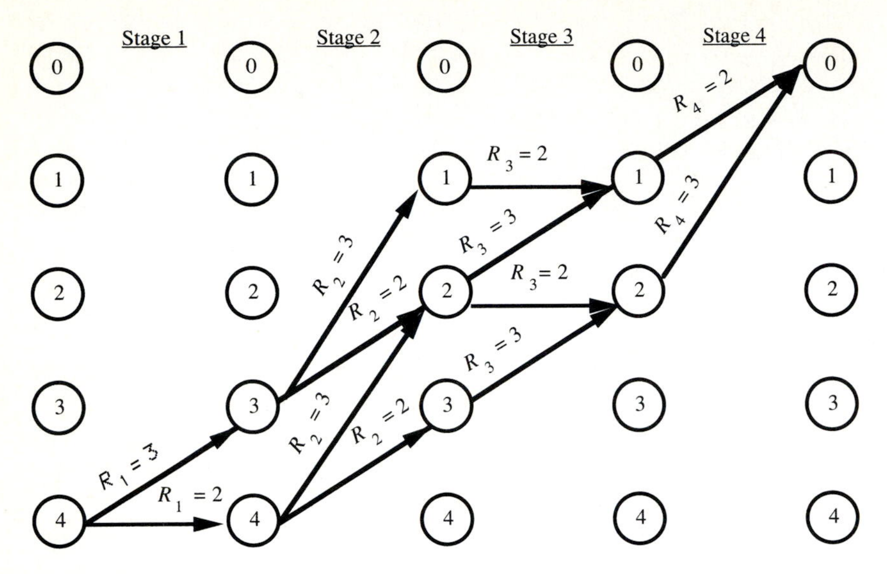

FIGURE 4.1.4
Traceback for Example 4.1.3.

it has two disadvantages, that is, the computer memory and time requirements. These disadvantages could become especially severe under two situations: (1) when the number of state variables is large; and (2) when the dynamic programming is applied in a discrete fashion to a continuous state space. The problem associated with the latter case is that there exist difficulties in obtaining the true optimal solution without a considerable increase in discretization of state space. With the advancement in computer technology those disadvantages are becoming less and less severe.

An increase in the number of discretizations and/or state variables would geometrically increase the number of evaluations of the recursive formula and core-memory requirement per stage. This problem of rapid growth of computer time and core-memory requirement associated with multiple-state variable dynamic programming problems is referred to as **curse of dimensionality**. From the problem-solving viewpoint, the problem of increased computer time is of much less concern than that of the increased computer storage requirement. An increase in required computer core memory might result in exceeding the available storage capacity of a particular computer facility in use and the problem cannot be solved. On the other hand, an increase in computer time requires one to be more patient for the final result. Therefore, the rapid growth in memory requirements associated with multiple-state variable problems can make the difference between solvable and unsolvable problems.

4.2 DISCRETE DIFFERENTIAL DYNAMIC PROGRAMMING

Discrete differential dynamic programming (DDDP) is an iterative DP procedure which is specially designed to overcome some of the difficulties of the DP approach previously mentioned. DDDP uses the same recursive equation as DP to search among the discrete states in the stage-state domain. Instead of searching over the entire stage-state domain for the optimum, as is the case for DP, DDDP examines only a portion of the stage-state domain saving computer time and memory (Chow et al., 1975). This optimization procedure is solved through iterations of trial states and decisions to search for the optimum for a system subject to the constraints that the trial states and decisions should be within the respective **admissible domain**, that is, feasible in the state and decision spaces.

In DDDP, the first step is to assume a trial sequence of admissible decisions called the **trial policy** and the state vectors of each stage are computed accordingly. This sequence of states within the admissible state domain for different stages is called the **trial trajectory**. An alternative to the above procedure is first to assume a trial trajectory and then use it to compute the trial policy. Several states, located in the neighborhood of a trial trajectory, can be introduced to form a band called a **corridor** around the trail trajectory (see Fig. 4.2.1). It is a common practice to discretize the state space into uniform increments, called the **state increments**, where the total number of discretizations, referred to as **grid points** or **lattice points**, for each state variable is the same. The decisions, therefore, have to be made with respect to the method of discretizing the state variables. The interval of the decision variable is dependent on the corresponding interval of the state variable.

The traditional dynamic programming approach is applied in a DDDP problem to the states within the corridor using the recursive relationship for an improved trajectory defining a new policy or set of decisions within the introduced corridor. The improved trial trajectory is then adopted as the new trajectory to form a new corridor. This process of corridor formation, optimization with respect to the states within the corridor and trace-back to obtain an improved trajectory for the entire system is called an **iteration**.

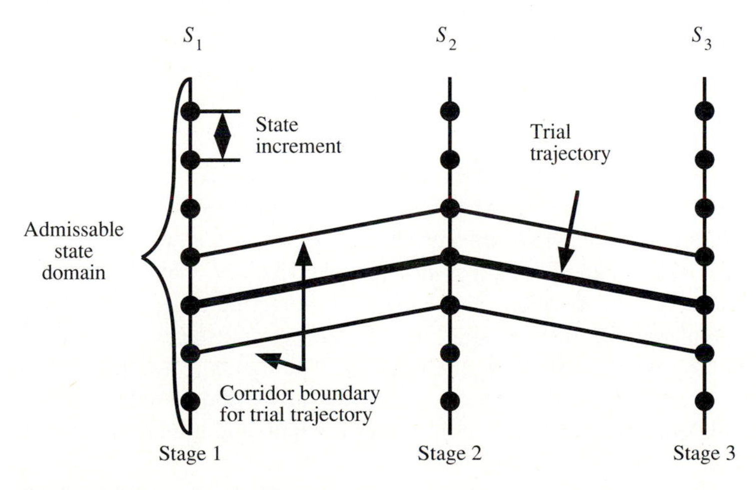

FIGURE 4.2.1
Formation of corridor around trial trajectory for a three-stage problem.

The new trajectory, defining the system through optimization of the trial corridor, is subsequently adopted as the new trial trajectory to establish the improved corridor for the next iteration. This procedure is repeated beyond some iteration k which produces a corridor providing a system return f_k such that further iterations with this size of corridor will produce a difference in system return, $f_k - f_{k-1}$ less than a specified tolerance. At this point in the optimization procedure, the size of the state increment can be reduced to set up a new corridor in which the states or lattice points are closer together. A smaller corridor is formed around the improved trajectory from the last iteration completed. The iterations continue reducing the size of the state increment through the system accordingly until a specified minimum corridor size is reached.

The criterion used to determine when the state increment size should be reduced is based on the relative change of the optimal objective function value of the previous iteration, that is,

$$|f_k - f_{k-1}|/f_{k-1} \leq \epsilon \tag{4.2.1}$$

where ϵ is the specified tolerance level. Fig. 4.2.2 is a flow chart showing the general algorithm of the DDDP procedure.

Although the use of DDDP partially circumvents the problem associated with the curse of dimensionality of regular DP, it introduces other considerations involved with choosing the initial trajectory, changing the lattice point or state spacing, and converging to a local optimum instead of a global optimum. The major factors affecting the performance of DDDP include:

1. the number of lattice points;
2. the initial state increment along with the number of lattice points which determine corridor widths;
3. the initial trial trajectory used to establish the location of the corridors within the state space for the first iteration; and
4. the reduction rate of the state increment size which determines the corridor width for various iterations.

These factors are somewhat dependent upon one another for expedient and efficient use of DDDP for water resources problems. The choice of the initial trial trajectory and the initial corridor width are interdependent. For example, if a small corridor width is chosen in conjunction with an initial trial trajectory, which is far from the optimal region, unnecessary iterations are required to move the trajectory into the optimal or near optimal region. It is possible under some circumstances for DDDP to converge to a solution which is far from the optimal. The number of lattice points and the initial state increment together determine the initial corridor width. Using a combination of a small number of lattice points and a small initial state increment could also result in local optimal solutions far from the optimal.

The effect of choosing a poor initial trajectory can be reduced if a large number of lattice points and/or large initial state increments are used. In essence, the better the

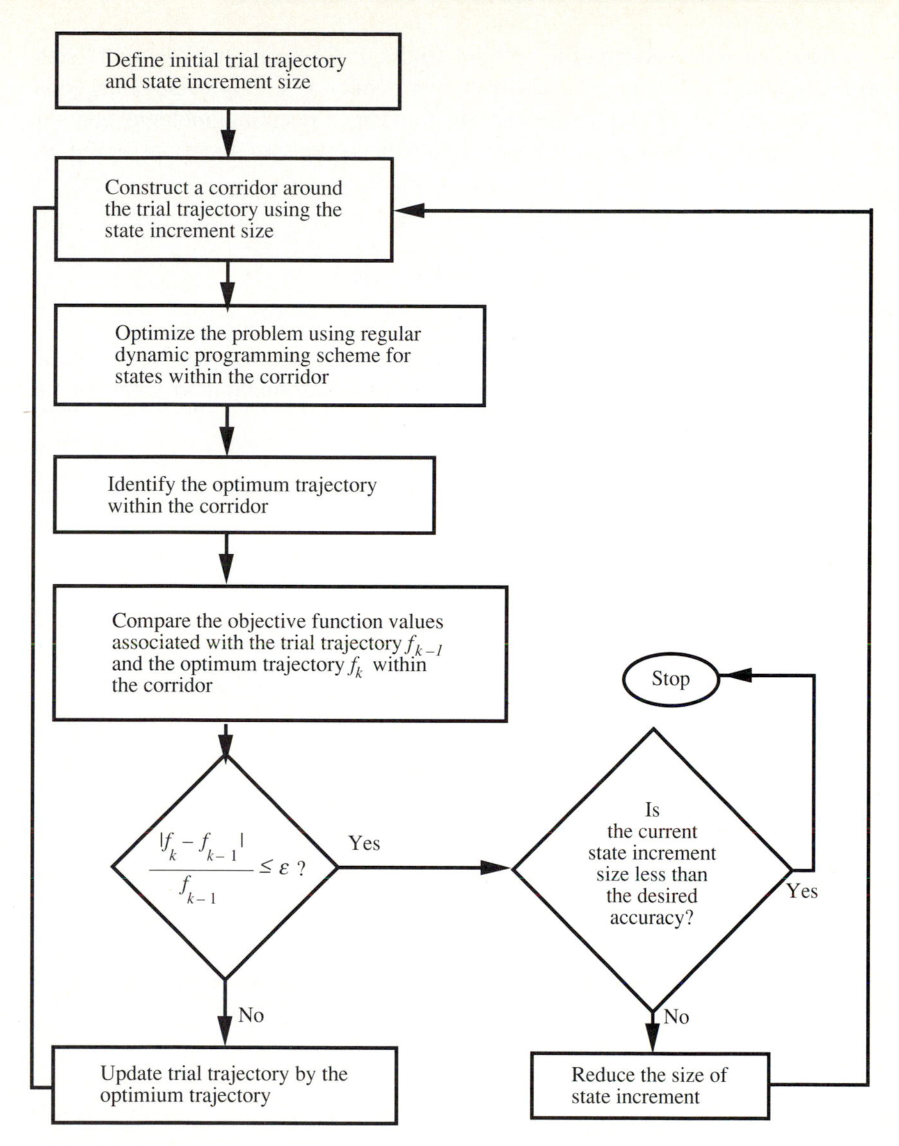

FIGURE 4.2.2
General algorithm for discrete differential dynamic programming.

initial trajectory the smaller the number of lattice points and initial state increments, or simply, the smaller the initial corridor width that can be used. Very small state increments with more lattice points can also be used to establish a small corridor width. This can result in improved convergence; however, increasing the number of lattice points increases the computation time and storage requirement. Computation time can be improved by increasing the reduction rate of the state increment at each iteration; however, too large a reduction rate of the state increment may miss the optimal region thus not providing the optimal solution. Choosing a large initial state increment and a large reduction rate of the state-increment size may be advantageous. However, when the initial state increment is too large, resulting in large corridor widths, unnecessary computations are performed in regions of the state space far from the optimal.

4.3 MATRIX ALGEBRA FOR NONLINEAR PROGRAMMING

To explain the concepts of nonlinear programming, various techniques of matrix algebra and numerical linear algebra are used. A brief introduction to some of the concepts is provided in this section.

A **line** is a set of points

$$\mathbf{x} = \mathbf{x}^0 + \beta \mathbf{d} \tag{4.3.1}$$

in which $\mathbf{x}^0$ is a fixed point on the line, β is a scalar step size, and $\mathbf{d}$ is the direction of the line. Figure 4.3.1 shows a line in two dimensions where $\mathbf{x}^0$ is the point (1,1) and $\mathbf{d}$ is the direction vector $(2, 1)^T$ indicated by the arrow in which superscript T represents the transpose of a vector or matrix.

A **function** of many variables $f(\mathbf{x})$ at point $\mathbf{x}$ is also an important concept. For a function that is continuous and continuously differentiable, there is a vector of first partial derivatives called the **gradient** or **gradient vector**

$$\nabla f(\mathbf{x}) = \left[\frac{\partial f}{\partial \mathbf{x}}\right] = \left(\frac{\partial f}{\partial x_1}, \frac{\partial f}{\partial x_2}, \ldots, \frac{\partial f}{\partial x_n}\right)^T \tag{4.3.2}$$

where ∇ is the vector of **gradient operators** $(\partial/\partial x_1, \ldots, \partial/\partial x_n)^T$. Geometrically, the gradient vector at a given point represents the direction along which the maximum rate of increase in function value would occur. For $f(\mathbf{x})$ twice continuously differentiable there exists a matrix of second partial derivatives called the **Hessian matrix** or **Hessian**.

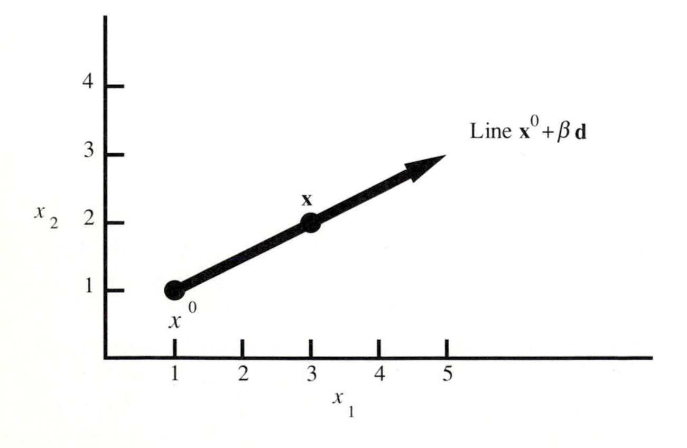

FIGURE 4.3.1
Definition of a line in two dimensions.

$$\mathbf{H}(\mathbf{x}) = \nabla^2 f(\mathbf{x}) = \begin{bmatrix} \dfrac{\partial^2 f}{\partial x_1^2} & \dfrac{\partial^2 f}{\partial x_1 \partial x_2} & \cdots & \dfrac{\partial^2 f}{\partial x_1 \partial x_n} \\ \dfrac{\partial^2 f}{\partial x_2 \partial x_1} & \dfrac{\partial^2 f}{\partial x_2^2} & \cdots & \dfrac{\partial^2 f}{\partial x_2 \partial x_n} \\ \cdots & \cdots & \cdots & \cdots \\ \dfrac{\partial^2 f}{\partial x_n \partial x_1} & \cdots & \cdots & \dfrac{\partial^2 f}{\partial x_n^2} \end{bmatrix} \tag{4.3.3}$$

The Hessian is a square and symmetric matrix.

Example 4.3.1. Consider the following quadratic function (Edgar and Himmelblau, 1988)

$$f(\mathbf{x}) = 4x_1^2 + x_2^2 - 2x_1x_2$$

The contours of this function are shown in Fig. 4.3.2. Determine the gradient and Hessian for this function at point $\mathbf{x} = (1, 1)$.

Solution. The gradient by Eq. (4.3.2) is

$$\nabla f(\mathbf{x}) = \begin{pmatrix} 8x_1 - 2x_2 \\ 2x_2 - 2x_1 \end{pmatrix}$$

and the Hessian by Eq. (4.3.3) is

$$\mathbf{H}(\mathbf{x}) = \nabla^2 f(\mathbf{x}) = \begin{bmatrix} 8 & -2 \\ -2 & 2 \end{bmatrix}$$

At point $\mathbf{x} = (1, 1)$, the gradient is

$$\nabla f(1, 1) = \begin{bmatrix} 6 \\ 0 \end{bmatrix}$$

while the Hessian is invariant to solution points.

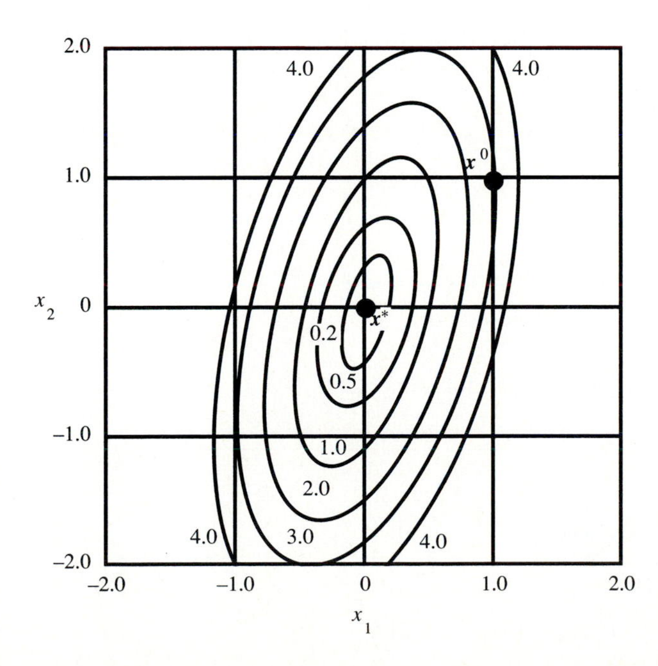

FIGURE 4.3.2
Contours of quadratic function ($f(x) = 4x_1^2 + x_2^2 - 2x_1x_2$) in Example 4.3.1 (Edgar and Himmelblau, 1988).

The concepts of **convexity** and **concavity** are used to establish whether a **local optimum**, **local minimum**, or **local maximum**, is also the **global optimum**, which is the best among all solutions. In the univariate case, a function $f(x)$ is said to be **convex** over a region if for every x_a and x_b, $x_a \neq x_b$, the following holds

$$f[\theta x_a + (1-\theta)x_b] \leq \theta f(x_a) + (1-\theta)f(x_b), 0 \leq \theta \leq 1 \quad (4.3.4)$$

The function is **strictly convex** when the above relation holds with a less than ($<$) sign.

Conversely, a function is **concave** over a region if for every x_a and x_b, $x_a \neq x_b$, the following holds

$$f[\theta x_a + (1-\theta)x_b] \geq \theta f(x_a) + (1-\theta)f(x_b), 0 \leq \theta \leq 1 \quad (4.3.5)$$

The function is **strictly concave** when the above relation holds with a greater than ($>$) sign. Figure 4.3.3 further illustrates the above concepts.

Equations (4.3.4) and (4.3.5) are not convenient to use in testing for convexity or concavity of a univariate function. Instead, it is easier to examine the sign of its second derivative, $d^2f(x)/dx^2$. From fundamental calculus, as shown in Fig. 4.3.3, if

$$\frac{d^2f}{dx^2} < 0$$

then the function is concave and if

$$\frac{d^2f}{dx^2} > 0$$

then the function is convex.

The convexity and concavity of multivariable functions $f(\mathbf{x})$ can also be determined using the Hessian matrix. The definitions of **positive definite**, **negative definite**, and **indefinite** are used to identify the type of Hessian, that is,

Positive definite **H**:

$$\mathbf{x}^T\mathbf{H}\mathbf{x} > 0 \qquad \text{for all } \mathbf{x} \neq 0$$

Negative definite **H**:

$$\mathbf{x}^T\mathbf{H}\mathbf{x} < 0 \qquad \text{for all } \mathbf{x} \neq 0$$

Indefinite **H**:

$$\mathbf{x}^T\mathbf{H}\mathbf{x} < 0 \qquad \text{for some } \mathbf{x};$$
$$> 0 \qquad \text{for other } \mathbf{x}$$

Positive semidefinite **H**:

$$\mathbf{x}^T\mathbf{H}\mathbf{x} \geq 0 \qquad \text{for all } \mathbf{x}$$

Negative semidefinite **H**:

$$\mathbf{x}^T\mathbf{H}\mathbf{x} \leq 0 \qquad \text{for all } \mathbf{x}$$

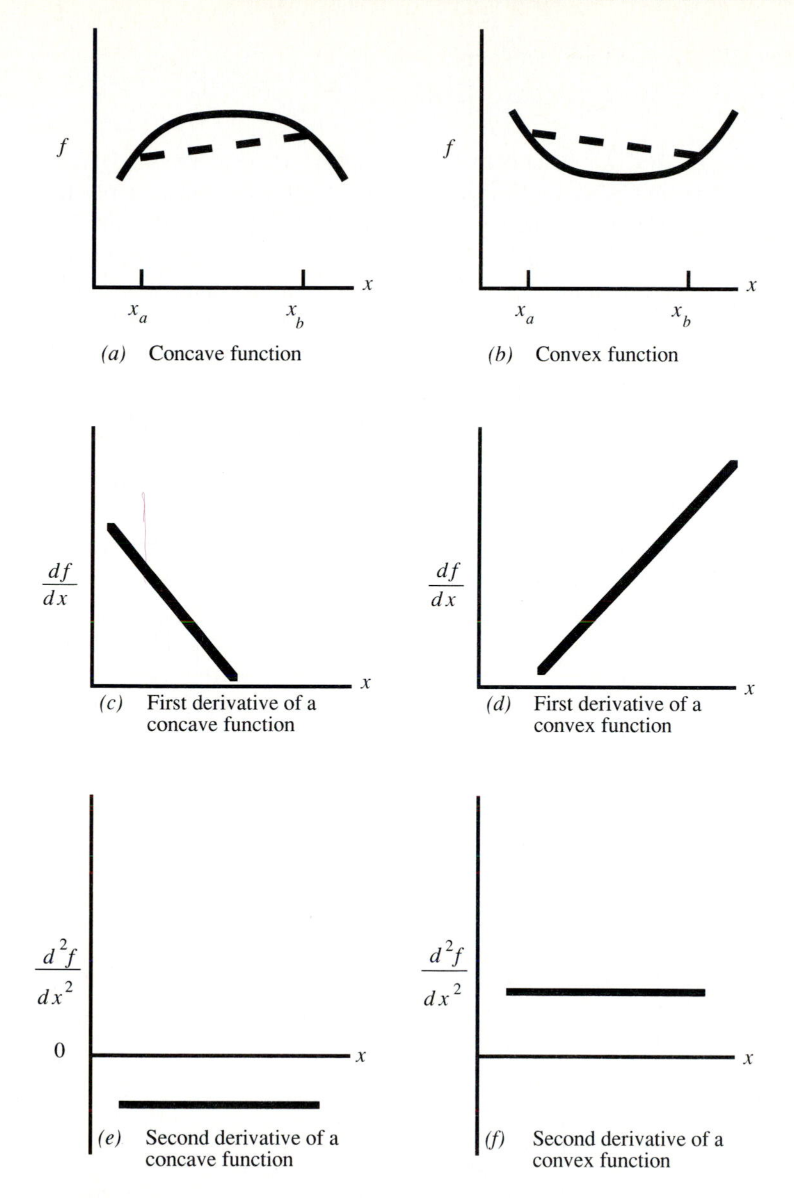

FIGURE 4.3.3
Comparison of concave and convex quadratic functions.

The basic rules for convexity and concavity of a multivariate function $\mathbf{f}(\mathbf{x})$ with continuous second partial derivatives are:

1. $\mathbf{f}(\mathbf{x})$ is concave, $\mathbf{H}(\mathbf{x})$ is negative semidefinite;
2. $\mathbf{f}(\mathbf{x})$ is strictly concave, $\mathbf{H}(\mathbf{x})$ is negative definite;

3. **f(x)** is convex, **H(x)** is positive semidefinite; and
4. **f(x)** is strictly convex, **H(x)** is positive definite.

To test the status of **H(x)** for strict convexity, two tests are available (Edgar and Himmelblau, 1988). The first is that all diagonal elements of **H(x)** must be positive and the determinants of all leading principal minors, det $\{M_i(\mathbf{H})\}$, and also of **H(x)**, det (**H**) are positive (> 0) (see Problem 4.3.2). Another test is that all eigenvalues of **H(x)** are positive (> 0) (see Problem 4.3.2). For strict concavity all diagonal elements must be negative and det (**H**) and det $\{M_i(\mathbf{H})\} > 0$ if i is even ($i = 2, 4, 6, \ldots$); det (**H**) and det $\{M_i(\mathbf{H})\} < 0$ if i is odd ($i = 1, 3, 5, \ldots$). The strict inequalities > or < in these tests are replaced by $\geq$ or $\leq$, respectively, to test for convexity and concavity.

Convex regions or sets are used to classify constraints. A convex region exists if for any two points in the region, $\mathbf{x}_a \neq \mathbf{x}_b$, all points $\mathbf{x} = \theta\mathbf{x}_a + (1 - \theta)\mathbf{x}_b$, where $0 \leq \theta \leq 1$, are on the line connecting $\mathbf{x}_a$ and $\mathbf{x}_b$ are in the set. Fig. 4.3.4 illustrates convex and nonconvex regions.

The convexity of a feasible region and the objective function in nonlinear optimization has an extremely important implication with regard to the type of optimal solution to be obtained. For linear programming problems (in Chapter 3), the objective function and feasible region both are convex therefore the optimal solution is a global one. On the other hand, the convexity of both the objective function and feasible region in a nonlinear programming problem cannot be ensured, the optimal solution achieved, therefore, cannot be guaranteed to be global.

Example 4.3.2. Classify the following function

$$f(x) = 4x_1^2 - 4x_1x_2 + 4x_2^2$$

Solution. Determine the Hessian

$$\mathbf{H} = \begin{pmatrix} \dfrac{\partial^2 f}{\partial x_1^2} & \dfrac{\partial^2 f}{\partial x_1 \partial x_2} \\ \dfrac{\partial^2 f}{\partial x_2 \partial x_1} & \dfrac{\partial^2 f}{\partial x_2^2} \end{pmatrix} = \begin{pmatrix} 8 & -4 \\ -4 & 8 \end{pmatrix}$$

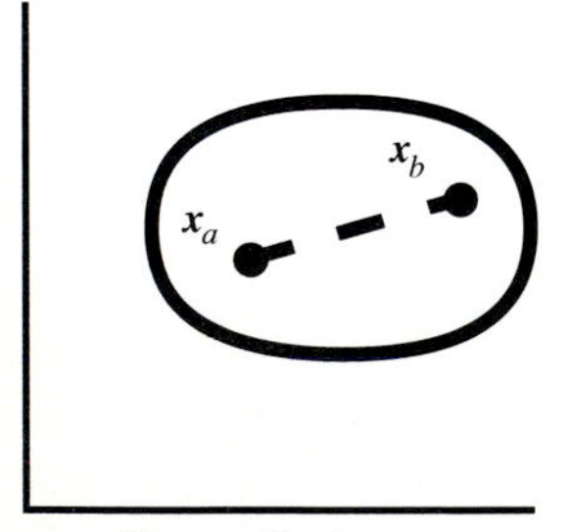

Convex Region

Nonconvex Region

FIGURE 4.3.4
Illustration of convex and nonconvex regions.

Both diagonal elements are positive. Next compute the leading principal minors

$$\mathbf{M}_1 = 8 \qquad \det \mathbf{M}_1 = 8$$

$$\mathbf{M}_2 = \mathbf{H} \qquad \det \mathbf{M}_2 = (8)(8) - (-4)(-4) = 48$$

The Hessian is positive definite because both diagonal elements are positive and both leading principal minors are positive; therefore $f(\mathbf{x})$ is strictly convex.

Example 4.3.3. Classify the region defined by the following set of constraints.

$$-x_1^2 + x_2 \geq 2$$

$$x_1 - x_2 \leq -3$$

Solution. Figure 4.3.5 shows a plot of the two functions,

$$g_1(\mathbf{x}) = -x_1^2 + x_2 - 2 \geq 0$$

$$g_2(\mathbf{x}) = -x_1 + x_2 - 3 \geq 0$$

The solution must describe the convexity of the two functions using the Hessians

$$\mathbf{H}[g_1(\mathbf{x})] = \begin{bmatrix} -2 & 0 \\ 0 & 0 \end{bmatrix}$$

$$\mathbf{H}[g_2(\mathbf{x})] = \begin{bmatrix} 0 & 0 \\ 0 & 0 \end{bmatrix}$$

Both $g_1(\mathbf{x})$ and $g_2(\mathbf{x})$ are concave resulting in a convex region.

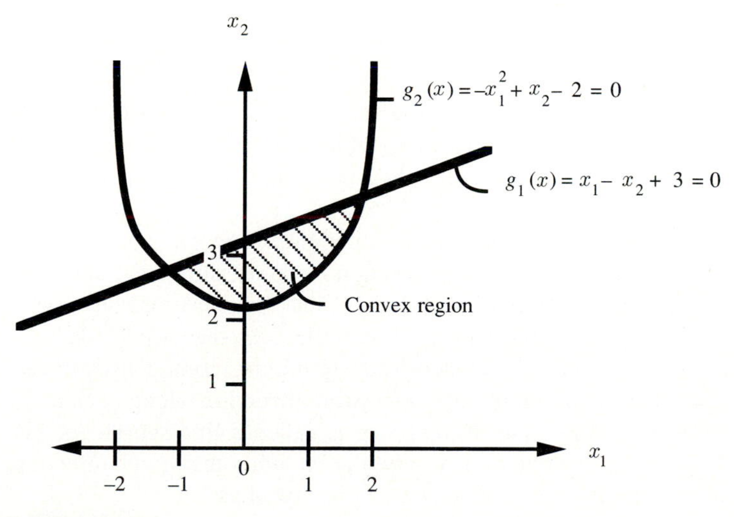

FIGURE 4.3.5
Example of a convex region.

4.4 UNCONSTRAINED NONLINEAR OPTIMIZATION

This section describes the basic concepts of unconstrained nonlinear optimization including the necessary and sufficient conditions of a local optimum. Further, unconstrained optimization techniques for univariate and multivariate problems are described. Understanding unconstrained optimization procedures is important because these techniques are the fundamental building blocks in many of the constrained nonlinear optimization algorithms.

4.4.1 Basic Concepts

The problem of unconstrained minimization can be stated as

$$\text{Minimize } f(\mathbf{x}) \qquad (4.4.1)$$

$$\mathbf{x} \in E^n$$

in which $\mathbf{x}$ is a vector of n decision variables $\mathbf{x} = (x_1, x_2, \ldots, x_n)^T$ defined over the entire Euclidean space E^n. Since the feasible region is infinitely extended without bound, the optimization problem does not contain any constraints.

Assume that $f(\mathbf{x})$ is a nonlinear function and twice differentiable; it could be convex, concave, or a mixture of the two over E^n. In the one-dimensional case, the objective function $f(\mathbf{x})$ could behave as shown in Fig. 4.4.1*a* consisting of peaks, valleys, and inflection points. The necessary conditions for a solution to Eq. (4.4.1) at $\mathbf{x}^*$ are: (1) $\nabla f(\mathbf{x}^*) = \mathbf{0}$; and (2) $\nabla^2 f(\mathbf{x}^*) = \mathbf{H}(\mathbf{x}^*)$ is semi-positive definite. The **sufficient conditions** for an unconstrained minimum are: (1) $\nabla f(\mathbf{x}^*) = \mathbf{0}$; and (2) $\nabla^2 f(\mathbf{x}^*) = \mathbf{H}(\mathbf{x}^*)$ is strictly positive definite.

In theory, the solution to Eq. (4.4.1) can be obtained by solving the following system of n nonlinear equations with n unknowns,

$$\nabla f(\mathbf{x}^*) = \mathbf{0} \qquad (4.4.2)$$

The approach has been viewed as indirect in the sense that it backs away from the original problem of minimizing $f(\mathbf{x})$. Furthermore, an iterative numerical procedure is required to solve the system of nonlinear equations which tends to be computationally inefficient.

By contrast, the preference is given to those solution procedures which directly attack the problem of minimizing $f(\mathbf{x})$. Direct solution methods, during the course of iteration, generate a sequence of solution points in E^n that terminate or converge to a solution to Eq. (4.4.1). Such methods can be characterized as search procedures.

In general, all search algorithms for unconstrained minimization consist of two basic steps. The first step is to determine the **search direction** along which the objective function value decreases. The second step is called a **line search** (or **one dimensional search**) to obtain the optimum solution point along the search direction. Mathematically, minimization for the line search can be stated as

$$\underset{\beta}{\text{Min}}\ f(\mathbf{x}^0 + \beta \mathbf{d}) \qquad (4.4.3)$$

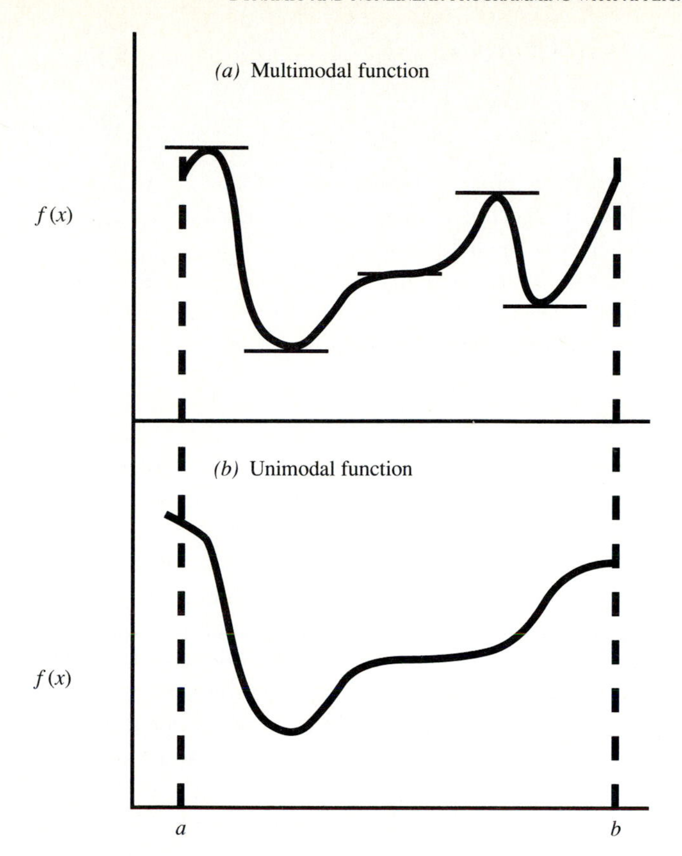

FIGURE 4.4.1
Types of one-dimensional nonlinear functions.

in which $\mathbf{x}^0$ is the current solution point, $\mathbf{d}$ is the vector indicating the search direction, and β is a scalar, $-\infty < \beta < \infty$, representing the step size whose optimal value is to be determined. There are many search algorithms whose differences primarily lie in the way the search direction $\mathbf{d}$ is determined.

Due to the very nature of search algorithms, it is likely that different starting solutions might converge to different local minima. Hence, there is no guarantee of finding the global minimum by any search technique applied to solve Eq. (4.4.1) unless the objective function is a convex function over E^n.

In implementing search techniques, specification of convergence criteria or stopping rules is an important element that affects the performance of the algorithm and the accuracy of the solution. Several commonly used stopping rules in an optimum seeking algorithm are

(*a*)
$$\|\mathbf{x}^k - \mathbf{x}^{k+1}\| < \epsilon_1; \tag{4.4.4a}$$

(*b*)
$$\frac{\|\mathbf{x}^k - \mathbf{x}^{k+1}\|}{\|\mathbf{x}^k\|} < \epsilon_2; \tag{4.4.4b}$$

(*c*) $$|f(\mathbf{x}^k) - f(\mathbf{x}^{k+1})| < \epsilon_3; \tag{4.4.4c}$$

(*d*) $$\left|\frac{f(\mathbf{x}^k) - f(\mathbf{x}^{k+1})}{f(\mathbf{x}^k)}\right| < \epsilon_4; \tag{4.4.4d}$$

in which superscript k is the index for iteration, ϵ represents the tolerance or accuracy requirement, $\|\mathbf{x}\|$ is the length of the vector $\mathbf{x}$, and $|x|$ is the absolute value. The specification of the tolerance depends on the nature of the problem and on the accuracy requirement. Too small a value of ϵ (corresponding to high accuracy requirement) could result in excessive iterations. On the other hand, too large a value of ϵ could make the algorithm terminate prematurely at a nonoptimal solution.

4.4.2 One-Dimensional Search

The line search techniques for solving one-dimensional optimization problems form the backbone of nonlinear programming algorithms. Multidimensional problems are ultimately solved by executing a sequence of successive line searches. One-dimensional search techniques can be classified as curve-fitting (approximation) techniques or as interval elimination techniques. **Interval elimination** techniques for a one-dimensional search essentially eliminate or delete a calculated portion of the range of the variable from consideration in each successive iteration of the search for the optimum of $f(\mathbf{x})$. After a number of iterations when the remaining interval is sufficiently small the search procedure terminates. These methods determine the minimum value of a function over a closed interval $[a, b]$ assuming that a function is **unimodal**, that is, it has only one minimum value in the interval (Fig. 4.4.1*b*). Two interval elimination techniques commonly used are the golden section method and the Fibonacci search method. Because these two methods are similar, only the golden section is described.

GOLDEN SECTION METHOD. The golden section is based upon splitting a line segment into two segments in which the ratio of the whole line to the larger segment is the same as the ratio of the larger segment (Δ_L) to the smaller segment (Δ_S) (Fig. 4.4.2). Let the whole interval length be $\Delta_S + \Delta_L = 1$ where Δ_L is the larger subinterval and Δ_S is the smaller subinterval. Then $1/\Delta_L = \Delta_L/\Delta_S$ or $1/(1-\Delta_S) = (1-\Delta_S)/\Delta_S$ which is $(\Delta_S)^2 - 3\Delta_S + 1 = 0$. This quadratic function has two roots 2.618 and 0.382, of which only 0.382 has meaning. This shows that the two segments are $\Delta_L = F_L = (1 - 0.382) = 0.618$ and $\Delta_S = F_S = 0.382$. The objective of the golden section algorithm is to apply the fractions F_L and F_S for any particular interval to compute the proper distances.

The golden section algorithm for minimizing a function, $f(x)$, can be stated as follows:

Step 0 $k = 0$. Select values of a^0 and b^0 that bracket the minimum of $f(x)$.

Step 1 Determine the interior points $x_1^k = a^k + 0.382(b^k - a^k)$ and $x_2^k = b^k - 0.382(b^k - a^k) = a^k + 0.618(b^k - a^k)$.

Step 2 Determine $f(x_1^k)$ and $f(x_2^k)$.

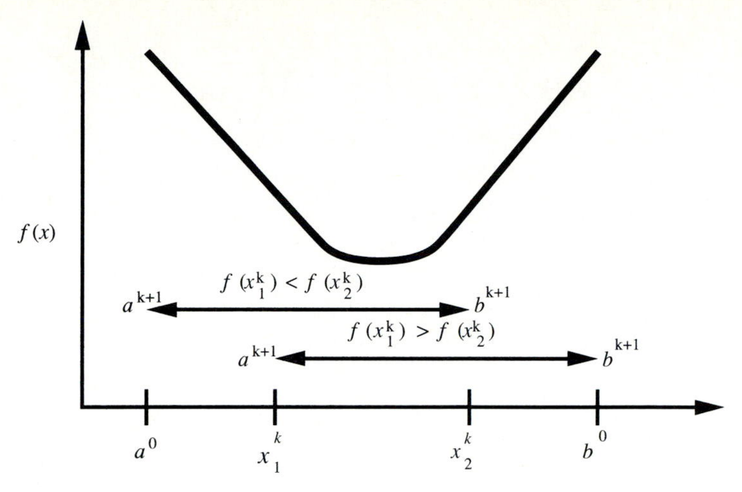

FIGURE 4.4.2
Golden section method.

Step 3

If $f(x_1^k) < f(x_2^k)$: $a^{k+1} = a^k$ and $b^{k+1} = x_2^k$.

If $f(x_1^k) > f(x_2^k)$: $a^{k+1} = x_1^k$ and $b^{k+1} = b^k$.

If $f(x_1^k) = f(x_2^k)$: $a^{k+1} = a^k$ and $b^{k+1} = x_2^k$ or $a^{k+1} = x_1^k$ and $b^{k+1} = b^k$.

Step 4 If the convergence criteria are not satisfied ($k = k + 1$) and return to Step 1. After k iterations, the subinterval has length $(b^k - a^k) = (0.618)^k(b^0 - a^0)$.

Example 4.4.1. Determine the minimum of the function $f(x) = x^3 - 6x$ using the golden section method with a convergence criteria of $b^k - a^k < 0.1$.

Solution. The second derivative is $6x$ so that the function is convex for $x \geq 0$. Following the above algorithm:

Step 0

$$\text{Select } a^0 = 0 \text{ and } b^0 = 4.$$

Iteration $k = 0$:

Step 1

$$x_1^0 = 0 + 0.382(4 - 0) = 1.528$$

$$x_2^0 = 0 + 0.618(4 - 0) = 2.472$$

Step 2

$$f(x_1^0) = 3.567 - 9.168 = -5.601$$

$$f(x_2^0) = 15.106 - 14.832 = 0.274$$

Step 3

$$f(x_1^0) < f(x_2^0), \text{ therefore } a^1 = a^0 = 0 \text{ and } b^1 = x_2^0 = 2.472$$

Step 4

$$\text{Convergence } b^1 - a^1 = 2.472 > 0.1$$

Iteration $k = 1$:

Step 1

$$x_1^1 = 0 + 0.382(2.472 - 0) = 0.944$$

$$x_2^1 = 0 + 0.618(2.472 - 0) = 1.528$$

Step 2

$$f(x_1^1) = 0.841 - 5.664 = -4.823$$

$$f(x_2^1) = 3.668 - 9.168 = -5.600$$

Step 3

$$f(x_1^1) > f(x_2^1), \text{ therefore } a^2 = x_1^1 = 0.944 \text{ and } b^2 = b^1 = 2.472$$

Step 4

$$\text{Convergence } b^2 - a^2 = 2.472 - 0.944 = 1.528 > 0.1$$

Iteration $k = 2$:

Step 1

$$x_1^2 = 0.944 + 0.382(1.528) = 1.527$$

$$x_2^2 = 0.944 + 0.618(1.528) = 1.888$$

Step 2

$$f(x_1^2) = 3.561 - 9.162 = -5.601$$

$$f(x_2^2) = 6.730 - 11.328 = -4.598$$

Step 3

$$f(x_1^2) < f(x_2^2), \text{ therefore } a^3 = a^2 = 0.944 \text{ and } b^3 = x_2^2 = 1.888$$

Step 4

$$\text{Convergence } b^3 - a^3 = 0.944 > 0.1$$

Several more iterations would be required to obtain the optimal solution. Obviously, for this simple problem the minimum can be determined much more simply by setting the first derivative equal to zero and solving

$$\nabla f(x) = 3x^2 - 6 = 0$$

resulting in

$$x^* = \sqrt{2} = 1.414$$

The above simple example is used to illustrate the convergence of the golden section method.

4.4.3 Multivariable Methods

Unconstrained optimization problems can be stated in a general form as

$$\text{Minimize } z = f(\mathbf{x}) = f(x_1, x_2, \ldots, x_n) \tag{4.4.5}$$

For maximization, the problem is to minimize $-f(\mathbf{x})$. The solution of these types of problems can be stated in an algorithm involving the following basic steps or phases:

Step 0 Select an initial starting point $\mathbf{x}^{k=0} = (x_1^0, x_2^0, \ldots, x_n^0)$.

Step 1 Determine a search direction, $\mathbf{d}^k$.

Step 2 Find a new point $\mathbf{x}^{k+1} = \mathbf{x}^k + \beta^k \mathbf{d}^k$ where β^k is the step size, a scalar, which minimizes $f(\mathbf{x}^k + \beta^k \mathbf{d}^k)$.

Step 3 Check the convergence criteria such as Eqs. (4.4.4*a–e*) for termination. If not satisfied, set $k = k + 1$ and return to Step 1.

The various unconstrained multivariate methods differ in the way the search directions are determined. The recursive line search for an unconstrained minimization problem is expressed in Step 2 as

$$\mathbf{x}^{k+1} = \mathbf{x}^k + \beta^k \mathbf{d}^k \tag{4.4.6}$$

Table 4.4.1 lists the equations for determining the search direction for four basic groups of methods: descent methods, conjugate direction methods, quasi-Newton methods and Newton's method. The simplest are the steepest descent methods while the Newton methods are the most computationally intensive.

In the **steepest descent method** the search direction is $-\nabla f(\mathbf{x})$. Since $\nabla f(\mathbf{x})$ points to the direction of the maximum rate of increase in objective function value, therefore, a negative sign is associated with the gradient vector in Eq. (4.4.6) because the problem is a minimization type. The recursive line search equation for the steepest descent method is reduced to

$$\mathbf{x}^{k+1} = \mathbf{x}^k - \beta^k \nabla f(\mathbf{x}^k) \tag{4.4.7}$$

Using **Newton's method**, the recursive equation for line search is

$$\mathbf{x}^{k+1} = \mathbf{x}^k - \mathbf{H}^{-1}(\mathbf{x}^k) \nabla f(\mathbf{x}^k) \tag{4.4.8}$$

Although Newton's method converges faster than most other algorithms, the major

TABLE 4.4.1
Computation of search directions*

Search direction	Definition of Terms
Steepest Descent $\mathbf{d}^{k+1} = -\nabla f(\mathbf{x}^{k+1})$	
Conjugate gradient methods (1) Fletcher-Reeve $\mathbf{d}^{k+1} = -\nabla f(\mathbf{x}^{k+1}) + a_1 \mathbf{d}^k$	$a_1 = \dfrac{\nabla^T f(\mathbf{x}^{k+1}) \nabla f(\mathbf{x}^{k+1})}{\nabla^T f(\mathbf{x}^k) \nabla f(\mathbf{x}^k)}$ $\mathbf{d}^0 = -\nabla f(\mathbf{x}^0)$
(2) Polak-Ribiere $\mathbf{d}^{k+1} = -\nabla f(\mathbf{x}^{k+1}) + a_2 \mathbf{d}^k$	$a_2 = \dfrac{\nabla^T f(\mathbf{x}^{k+1}) \mathbf{Y}^{k+1}}{\nabla^T f(\mathbf{x}^k) \nabla f(\mathbf{x}^k)}$ $\mathbf{Y}^{k+1} = \nabla f(\mathbf{x}^{k+1}) - \nabla f(\mathbf{x}^k)$ $\mathbf{d}^0 = -\nabla f(\mathbf{x}^0)$
(3) 1-Step BFGS $\mathbf{d}^{k+1} = -\nabla f(\mathbf{x}^{k+1}) + a_3(a_4 \mathbf{S}^{k+1} + a\mathbf{Y}^k)$	$a_3 = \dfrac{1}{(\mathbf{S}^{k+1})^T \mathbf{Y}^{k+1}}$ $a_4 = -\left(1 + \dfrac{(\mathbf{Y}^{k+1})^T \mathbf{Y}^{k+1}}{(\mathbf{S}^{k+1})^T \mathbf{Y}^{k+1}}\right)(\mathbf{S}^{k+1})^T \nabla f(\mathbf{x}^{k+1})$ $+ (\mathbf{Y}^{k+1})^T \nabla f(\mathbf{x}^{k+1})$ $a_5 = (\mathbf{S}^{k+1})^T \nabla f(\mathbf{x}^k)$ $\mathbf{S}^{k+1} = \mathbf{x}^{k+1} - \mathbf{x}^k$ $\mathbf{d}^0 = -\nabla f(\mathbf{x}^0)$
Quasi-Newton methods (1) Davidon-Fletcher-Powell DFP method (variable metric method) $\mathbf{d}^{k+1} = \mathbf{G}^{k+1} \nabla f(\mathbf{x}^{k+1})$	$\mathbf{G}^{k+1} = \mathbf{G}^k + \dfrac{\mathbf{S}^k (\mathbf{S}^k)^T}{(\mathbf{S}^k)^T \mathbf{Y}^k} - \dfrac{\mathbf{G}^k \mathbf{Y}^k (\mathbf{G}^k \mathbf{Y}^k)^T}{(\mathbf{Y}^k)^T \mathbf{G}^k \mathbf{Y}^k}$
(2) Broyden-Fletcher-Goldfarb-Shanno (BFGS) method $\mathbf{d}^{k+1} = \mathbf{G}^{k+1} \nabla f(\mathbf{x}^{k+1})$	$\mathbf{G}^{k+1} = \mathbf{G}^k + \left(\dfrac{1 + (\mathbf{Y}^k)^T \mathbf{G}^k \mathbf{Y}^k}{(\mathbf{Y}^k)^T \mathbf{S}^k}\right) \dfrac{\mathbf{S}^k (\mathbf{S}^k)^T}{(\mathbf{S}^k)^T \mathbf{Y}^k}$ $- \dfrac{\mathbf{Y}^k (\mathbf{S}^k)^T \mathbf{G}^k + \mathbf{G}^k \mathbf{S}^k (\mathbf{Y}^k)^T}{(\mathbf{Y}^k)^T \mathbf{S}^k}$
(3) Broyden family $\mathbf{G}^\phi = (1 - \phi)\mathbf{G}^{DFP} + \phi \mathbf{G}^{BFGS}$	
Newton method $\mathbf{d}^{k+1} = -\mathbf{H}^{-1}(\mathbf{x}^k) \nabla f(\mathbf{x}^k)$	

*Formulas for other search directions can be found in Luenberger (1984).

disadvantage is that it requires inverting the Hessian matrix in each iteration which is a computationally cumbersome task.

The **conjugate direction methods** and **quasi-Newton methods** are intermediate between the steepest descent and Newton's method. The conjugate direction methods are motivated by the need to accelerate the typically slow convergence of the steepest descent method. Conjugate direction methods, as can be seen in Table 4.4.1, define the search direction by utilizing the gradient vector of the objective funciton of the current iteration and the information on the gradient and search direction of the previous iteration. The motivation of quasi-Newton methods is to avoid inverting the Hessian matrix as required by Newton's method. These methods use approximations to the inverse Hessian with a different form of approximation for the different quasi-Newton methods. Detailed descriptions and theoretical development can be found in textbooks such as Fletcher (1980), Gill, Murray, and Wright (1981), Dennis and Schnable (1983), Luenberger (1984), and Edgar and Himmelblau (1988).

Example 4.4.2. Apply the steepest descent method to minimize the function

$$f(\mathbf{x}) = 4x_1^2 + x_2^2 - 2x_1x_2$$

considered in Example 4.3.1. Use $(x_1, x_2) = (1, 1)$ as the starting solution point.

Solution. From Example 4.3.1, the gradient of the above objective function is

$$\nabla f(\mathbf{x}) = \begin{pmatrix} 8x_1 - 2x_2 \\ 2x_2 - 2x_1 \end{pmatrix}$$

For iteration 0 ($k = 0$): $\left(x_1^{(0)}, x_2^{(0)}\right) = (1, 1)$

$$\mathbf{d}^0 = -\nabla f(1, 1) = \begin{pmatrix} -6 \\ 0 \end{pmatrix}$$

The line search is, then

$$\mathbf{x}^1 = \mathbf{x}^0 + \beta\mathbf{d}^0 = \begin{pmatrix} 1 \\ 1 \end{pmatrix} + \beta\begin{pmatrix} -6 \\ 0 \end{pmatrix} = \begin{pmatrix} 1 - 6\beta \\ 1 \end{pmatrix}$$

in which β is to be determined to minimize

$$f(\mathbf{x}^1) = f(\beta) = 4(1 - 6\beta)^2 + (1)^2 - 2(1 - 6\beta)(1)$$

$$= 4(1 - 6\beta)^2 - 2(1 - 6\beta) + 1$$

The optimal step size β can be obtained by the golden section method or by solving $df(\beta)/d\beta = 0$ because $f(\beta)$ is convex, $\frac{d^2 f}{d\beta^2} > 0$. For simplicity, the latter is used.

$$\frac{df(\beta)}{d\beta} = 288\beta - 36 = 0$$

$$\therefore\ \beta^0 = 0.125$$

Therefore, the new solution can be determined as

$$\mathbf{x}^1 = \begin{pmatrix} 1 - 6(0.125) \\ 1 \end{pmatrix} = \begin{pmatrix} 0.25 \\ 1 \end{pmatrix}$$

For iteration 1 ($k = 1$) : $\left(x_1^1, x_2^1\right) = (0.25, 1)$

$$\mathbf{d}^1 = -\nabla f(0.25, 1) = \begin{pmatrix} 0 \\ -1.5 \end{pmatrix}$$

The line search is

$$\mathbf{x}^2 = (\mathbf{x}^1 + \beta \mathbf{d}^1) = \begin{pmatrix} 0.25 \\ 1 \end{pmatrix} + \begin{pmatrix} 0 \\ -1.5\beta \end{pmatrix} = \begin{pmatrix} 0.25 \\ 1 - 1.5\beta \end{pmatrix}$$

$$f(\mathbf{x}^2) = f(\beta) = 4(0.25)^2 + (1 - 1.5\beta)^2 - 2(0.25)(1 - 1.5\beta)$$

$$= 0.25 + (1 - 1.5\beta)^2 - 0.5(1 - 1.5\beta)$$

The optimal step size β is determined from solving

$$\frac{\partial f(\beta)}{\partial \beta} = 2(1 - 1.5\beta)(-1.5) + 0.75$$

$$= -4.5\beta + 2.25 = 0$$

$$\therefore \beta^1 = 2.25/4.5 = 0.5$$

$$\text{Hence, } \mathbf{x}^2 = \begin{pmatrix} 0.25 \\ 1 - 1.5(0.5) \end{pmatrix} = \begin{pmatrix} 0.25 \\ 0.25 \end{pmatrix}$$

The iteration continues until the solution approaches the optimal solution $\mathbf{x}^* = (0, 0)$. As shown in Fig. 4.4.3 the path of steepest descent is zigzagging with each move perpendicular to the previous one. The method will be inefficient if the contours of the objective function become more and more *elongated* (stretched) along a particular direction. On the other hand, the steepest descent method would only take one iteration to converge if the contour is a perfect circle.

4.5 CONSTRAINED OPTIMIZATION: OPTIMALITY CONDITIONS

4.5.1 Lagrange Multiplier

Consider the general nonlinear programming problem with the nonlinear objective:

$$\text{Minimize } f(\mathbf{x}) \tag{4.5.1a}$$

subject to

$$g_i(\mathbf{x}) = 0 \qquad i = 1, \ldots, m \tag{4.5.1b}$$

and

$$\underline{x}_j \leq x_j \leq \overline{x}_j \qquad j = 1, 2, \ldots, n \tag{4.5.1c}$$

in which Eq. (4.5.1c) is a bound constraint for the jth decision variable x_j with $\underline{x}_j$ and $\overline{x}_j$ being the lower and upper bounds, respectively.

In a constrained optimization problem, the feasible space is not infinitely extended, unlike an unconstrained problem. As a result, the solution that satisfies the optimality condition of the unconstrained optimization problem does not guarantee to be feasible in constrained problems. In other words, a local optimum for a constrained problem might be located on the boundary or a corner of the feasible space at which the gradient vector is not equal to zero. Therefore, modifications to the optimality conditions for unconstrained problems must be made.

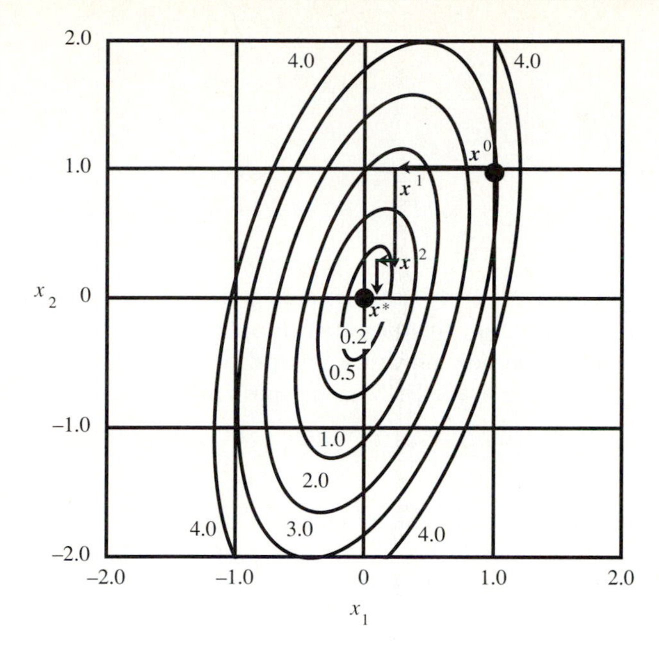

FIGURE 4.4.3
Search pattern of steepest descent method (after Edgar and Himmelblau, 1988).

The most important theoretical results for nonlinear constrained optimization are the **Kuhn-Tucker conditions**. These conditions must be satisfied at any constrained optimum, local or global, of any linear and nonlinear programming problems. They form the basis for the development of many computational algorithms.

Without losing generality, consider a nonlinear constrained problem stated by Eq. (4.5.1) with no bounding constraints. Note that constraint Eq. (4.5.1*b*) are all equality constraints. Under this condition, the **Lagrange multiplier method** converts a constrained nonlinear programming problem into an unconstrained one by developing an augmented objective function, called the **Lagrangian**. For a minimization, the **Lagrangian function** $L(\mathbf{x}, \boldsymbol{\lambda})$ is defined as

$$L(\mathbf{x}, \boldsymbol{\lambda}) = f(\mathbf{x}) + \boldsymbol{\lambda}^T \mathbf{g}(\mathbf{x}) \tag{4.5.2}$$

in which $\boldsymbol{\lambda}$ is the vector of Lagrange multipliers and $\mathbf{g}(\mathbf{x})$ is a vector of constraint equations. Algebraically, Eq. (4.5.2) can be written

$$L(x_1, \ldots, x_n, \lambda_1, \ldots, \lambda_m) = f(x_1, \ldots, x_n) + \sum_{i=1}^{m} \lambda_i g_i(x_1, \ldots, x_n) \tag{4.5.3}$$

$L(\mathbf{x}, \boldsymbol{\lambda})$ is the objective function, with $m + n$ variables, that is to be minimized. The necessary and sufficient conditions for $\mathbf{x}^*$ to be the solution for minimization are:

1. $f(\mathbf{x}^*)$ is convex and $g(\mathbf{x}^*)$ is convex in the vicinty of $\mathbf{x}^*$;
2.

$$\frac{\partial L(\mathbf{x}^*)}{\partial x_j} = \frac{\partial f}{\partial x_j} + \sum_{i=1}^{m} \lambda_i \frac{\partial g_i}{\partial x_j} = 0 \qquad j = 1, \ldots, n; \tag{4.5.4a}$$

3.

$$\frac{\partial L}{\partial \lambda_i} = g_i(\mathbf{x}) = 0 \qquad i = 1, \ldots, m; \tag{4.5.4b}$$

4.

$$\lambda_i \text{ is unrestricted-in-sign} \qquad i = 1, \ldots, m. \tag{4.5.4c}$$

Solving Eqs. (4.5.4*a*) and (4.5.4*b*) simultaneously provides the optimal solution.

Lagrange multipliers have an important interpretation in optimization. For a given constraint, these multipliers indicate how much the optimal objective function value will change for a differential change in the RHS of the constraint. That is,

$$\left.\frac{\partial f}{\partial b_i}\right|_{\mathbf{x}=\mathbf{x}^*} = \lambda_i$$

illustrating that the Lagrange multiplier λ_i is the rate of change of the optimal value of the original objective function with respect to a change in the value of the RHS of the ith constraint. The λ_is are called **dual variables** or **shadow prices** with the economic interpretation given in Section 3.5.

Example 4.5.1. Consider the following problem

$$\text{Minimize } (x_1 - 1)^2 + (x_2 - 2)^2 \tag{a}$$

subject to

$$x_1 - 2x_2 = 0 \tag{b}$$

Use the method of Lagrange multipliers to solve this problem.

Solution. The Lagrangian function is

$$L(x_1, x_2, \lambda) = (x_1 - 1)^2 + (x_2 - 2)^2 + \lambda(x_1 - 2x_2) \tag{c}$$

Applying Eqs. (4.5.4*a*) and (4.5.4*b*), the optimal solution must satisfy the following equations.

$$\frac{\partial L}{\partial x_1} = 2(x_1 - 1) + \lambda = 0 \tag{d}$$

$$\frac{\partial L}{\partial x_2} = 2(x_2 - 2) - 2\lambda = 0 \tag{e}$$

$$\frac{\partial L}{\partial \lambda} = x_1 - 2x_2 = 0 \tag{f}$$

Solving these equations, the optimal solutions to the problem are $x_1^* = 1.6, x_2^* = 0.8$, and $\lambda^* = -1.2$.

4.5.2 Kuhn-Tucker Conditions

Equations (4.5.4*a*)-(4.5.4*c*) form the optimality conditions for an optimization problem involving only equality constraints. The Lagrange multipliers associated with the equality constraints are unrestricted-in-sign. Using the Lagrange multiplier method, the optimality conditions for the following generalized nonlinear programming problem can be derived.

$$\text{Minimize } f(\mathbf{x})$$

subject to

$$g_i(\mathbf{x}) = 0 \qquad i = 1, \ldots, m$$

and

$$\underline{x}_j \leq x_j \leq \overline{x}_j \qquad j = 1, \ldots, n$$

In terms of the Lagrangian method, the above nonlinear minimization problem can be written as

$$\text{Min } L = f(\mathbf{x}) + \boldsymbol{\lambda}^T \mathbf{g}(\mathbf{x}) + \underline{\boldsymbol{\lambda}}^T(\underline{\mathbf{x}} - \mathbf{x}) + \overline{\boldsymbol{\lambda}}^T(\mathbf{x} - \overline{\mathbf{x}}) \tag{4.5.5}$$

in which $\boldsymbol{\lambda}, \underline{\boldsymbol{\lambda}}$, and $\overline{\boldsymbol{\lambda}}$ are vectors of Lagrange multipliers corresponding to constraints $g(\mathbf{x}) = \mathbf{0}, \underline{\mathbf{x}} - \mathbf{x} \leq \mathbf{0}$, and $\mathbf{x} - \overline{\mathbf{x}} \leq \mathbf{0}$, respectively. The Kuhn-Tucker conditions for the optimality of the above problem are

$$\nabla_x L = \nabla_x f + \boldsymbol{\lambda}^T \nabla_x \mathbf{g} - \underline{\boldsymbol{\lambda}} + \overline{\boldsymbol{\lambda}} = \mathbf{0} \tag{4.5.6a}$$

$$g_i(\mathbf{x}) = 0, \qquad i = 1, 2, \ldots, m \tag{4.5.6b}$$

$$\underline{\lambda}_j(\underline{x}_j - x_j) = \overline{\lambda}_j(x_j - \overline{x}_j) = 0, \qquad j = 1, 2, \ldots, n \tag{4.5.6c}$$

$$\lambda \text{ unrestricted-in-sign, } \underline{\boldsymbol{\lambda}} \geq \mathbf{0}, \overline{\boldsymbol{\lambda}} \geq \mathbf{0} \tag{4.5.6d}$$

4.6 CONSTRAINED NONLINEAR OPTIMIZATION: GENERALIZED REDUCED GRADIENT (GRG) METHOD

4.6.1 Basic Concepts

Similar to the linear programming simplex method, the fundamental idea of the generalized reduced gradient method is to express m (number of constraint equations) of the variables, called **basic variables**, in terms of the remaining $n - m$ variables, called **nonbasic variables**. The decision variables can then be partitioned into the basic variables, $\mathbf{x}_B$, and the nonbasic variables, $\mathbf{x}_N$,

$$\mathbf{x} = (\mathbf{x}_B, \mathbf{x}_N)^T \tag{4.6.1}$$

Nonbasic variables not at their bounds are called **superbasic variables**, Murtaugh and Saunders (1978).

The optimization problem can now be restated in terms of the basic and nonbasic variables

$$\text{Minimize } f(\mathbf{x}_B, \mathbf{x}_N) \tag{4.6.2a}$$

subject to

$$\mathbf{g}(\mathbf{x}_B, \mathbf{x}_N) = \mathbf{0} \tag{4.6.2b}$$

and

$$\underline{\mathbf{x}}_B \leq \mathbf{x}_B \leq \overline{\mathbf{x}}_B \tag{4.6.2c}$$

$$\underline{\mathbf{x}}_N \leq \mathbf{x}_N \leq \overline{\mathbf{x}}_N \tag{4.6.2d}$$

The m basic variables in theory can be expressed in terms of the $n-m$ nonbasic variables as $\mathbf{x}_B(\mathbf{x}_N)$. Assume that constraints $\mathbf{g}(\mathbf{x}) = \mathbf{0}$ is differentiable and the m by m **basis matrix B** can be obtained as

$$\mathbf{B} = \left[\frac{\partial \mathbf{g}(\mathbf{x})}{\partial \mathbf{x}_B}\right]$$

which is nonsingular such that there exists a unique solution of $\mathbf{x}_B(\mathbf{x}_N)$. **Nonsingularity** means that det $(\mathbf{B}) \neq 0$.

The objective called a **reduced objective** can be expressed in terms of the nonbasic variables as

$$F(\mathbf{x}_N) = f(\mathbf{x}_B(\mathbf{x}_N), \mathbf{x}_N) \tag{4.6.3}$$

The original nonlinear programming problem is transformed into the following **reduced problem**

$$\text{Minimize } F(\mathbf{x}_N) \tag{4.6.4a}$$

subject to

$$\underline{\mathbf{x}}_N \leq \mathbf{x}_N \leq \overline{\mathbf{x}}_N \tag{4.6.4b}$$

which can be solved by an unconstrained minimization technique with slight modification to account for the bounds on nonbasic variables. Generalized reduced gradient algorithms, therefore, solve the original problem (4.5.1) by solving a sequence of reduced problems (4.6.4*a,b*) using unconstrained minimization algorithms.

4.6.2 General Algorithm and Basis Changes

Consider solving the reduced problem (4.6.4) starting from an initial feasible point $\mathbf{x}^0$. To evaluate $F(\mathbf{x}_N)$ by Eq. (4.6.3), the values of the basic variables $\mathbf{x}_B$ must be known. Except for a very few cases, $\mathbf{x}_B(\mathbf{x}_N)$ cannot be determined in closed form; however, it can be computed for any $\mathbf{x}_N$ by an iterative method which solves a system of m nonlinear equations with the same number of unknowns as equations. A procedure for solving the reduced problem starting from the initial feasible solution $x^{k=0}$ is:

Step 0 Start with an initial feasible solution $\mathbf{x}^{k=0}$ and set $\mathbf{x}_N^k = \mathbf{x}^{k=0}$

Step 1 Substitute $\mathbf{x}_N^k$ into Eq. (4.6.2*b*) and determine the corresponding values of $\mathbf{x}_B$ by an iterative method for solving m nonlinear equations $\mathbf{g}(\mathbf{x}_B(\mathbf{x}_N^k), \mathbf{x}_N^k) = \mathbf{0}$.

Step 2 Determine the search direction $\mathbf{d}^k$ for the nonbasic variables.

Step 3 Choose a step size for the line search scheme, β^k such that

$$\mathbf{x}_N^{k+1} = \mathbf{x}_N^k + \beta^k \mathbf{d}^k \tag{4.6.5}$$

This is done by solving the one-dimensional search problem

$$\text{Minimize } F(\mathbf{x}_N^k + \beta \mathbf{d}^k)$$

with $\mathbf{x}$ restricted so that $\mathbf{x}_N^k + \beta \mathbf{d}^k$ satisfies the bounds on $\mathbf{x}_N$. This one-dimensional search requires repeated applications of Step 1 to evaluate F for the different β values.

Step 4 Test the current point $\mathbf{x}^k = (\mathbf{x}_B^k, \mathbf{x}_N^k)$ for optimality, if not optimal, set $k = k+1$ and return to Step 1.

Refer to Fig. 4.6.1, the optimization problem can be stated as

$$\text{Minimize } f(x_1, x_2)$$

subject to

$$g_1(x_1, x_2) \geq 0$$

$$g_2(x_1, x_2) \geq 0$$

$$x_1, x_2 \geq 0$$

The two inequality constraints can be converted to equality constraints using slack variables x_3 and x_4

$$g_1(x_1, x_2) - x_3 = 0$$

$$g_2(x_1, x_2) - x_4 = 0$$

$$x_i \geq 0 \qquad i = 1, \ldots, 4$$

The initial point A is on the curve $g_2(x_1, x_2, x_3) = 0$ where the only variable that cannot be basic is x_4 which is at its lower bound of zero. The reduced objective is $F(x_2, x_4)$ which is the objective function $f(\mathbf{x})$ evaluated on $g_2(x_1, x_2, x_3) = 0$.

For purposes of illustrating the basis changes, assume that the algorithm moves along the curve $g_2(x_1, x_2, x_3) = 0$ as indicated by the arrow in Fig. 4.6.1 until the curve $g_1(x_1, x_2, x_4) = 0$ is reached. It should be kept in mind that an algorithm could move interior from the initial point A, releasing x_4 from its lower bound of zero, but for the sake of illustration of basis changes, the procedure here will stay on the curves. At the point B where constraints g_1 and g_2 intersect, the slack variable x_3 goes to zero. Because x_3 is originally basic it must leave the basis and be replaced by one of the nonbasics, x_2 or x_4. Because x_4 is zero, x_2 becomes basic and the new reduced objective is $F_2(x_3, x_4)$ with x_3 and x_4 at their lower bounds of zero. Once again, assuming the algorithm moves along the curve $g_1(x_1, x_2, x_4) = 0$ towards the x_2 axis, F_2 is minimized at point C where x_1 becomes zero (nonbasic) and x_4 becomes basic. The procedure would then move along the x_2 axis to point D which is obviously the minimum.

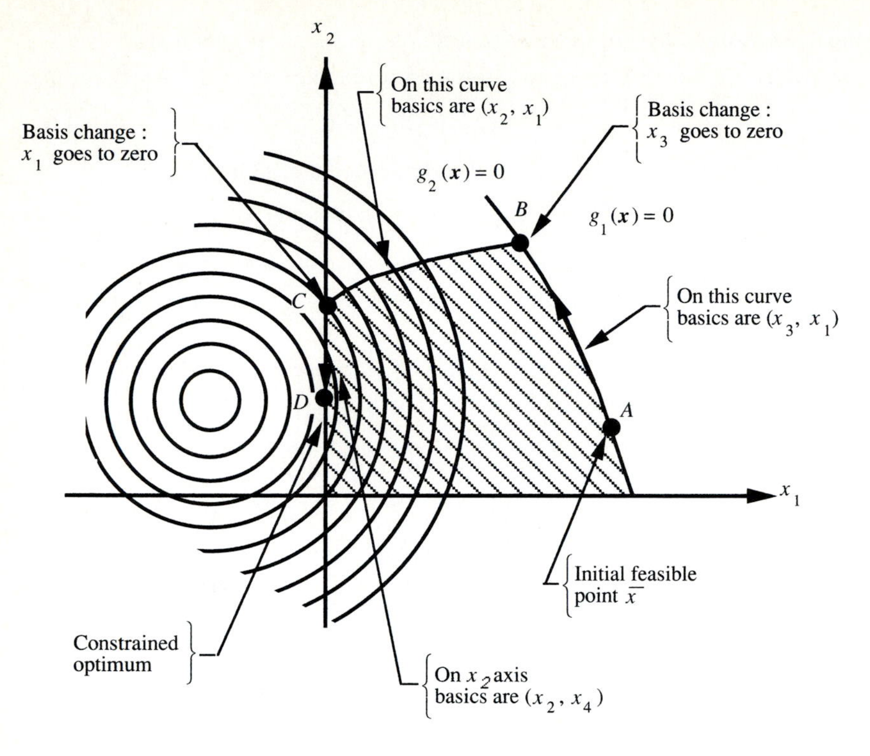

The constraints

$$\left.\begin{aligned} g_1(x_1, x_2) - x_3 &= 0 \\ g_2(x_1, x_2) - x_4 &= 0 \\ x_i \geq 0, i &= 1 \ldots. 4 \end{aligned}\right\} \text{Define the shaded feasible region}$$

FIGURE 4.6.1
Sequence of basis changes (adapted from Lasdon et al., 1974).

4.6.3 The Reduced Gradient

Computation of the reduced gradient is required in the **generalized reduced gradient** (GRG) method in order to define the search direction. Consider the simple problem

$$\text{Minimize } f(x_1, x_2)$$

subject to

$$g(x_1, x_2) = 0$$

The total derivative of the objective function is

$$df(\mathbf{x}) = \frac{\partial f(\mathbf{x})}{\partial x_1} dx_1 + \frac{\partial f(\mathbf{x})}{\partial x_2} dx_2 \tag{4.6.6}$$

and the total derivative of the constraint function is

$$dg(\mathbf{x}) = \frac{\partial g(\mathbf{x})}{\partial x_1} dx_1 + \frac{\partial g(\mathbf{x})}{\partial x_2} dx_2 = 0 \tag{4.6.7}$$

The **reduced gradients** are $\nabla f(\mathbf{x})$ and $\nabla g(\mathbf{x})$ defined by the coefficients in the total derivatives,

$$\nabla f(\mathbf{x}) = \left[\frac{\partial f}{\partial x_1}, \frac{\partial f}{\partial x_2}\right]^T \tag{4.6.8}$$

$$\nabla g(\mathbf{x}) = \left[\frac{\partial g}{\partial x_1}, \frac{\partial g}{\partial x_2}\right]^T \tag{4.6.9}$$

Consider the basic (dependent) variable to be x_1 and the nonbasic (independent) variable to be x_2. Equation (4.6.7) can be used to solve for dx_1

$$dx_1 = -\frac{\partial g(\mathbf{x})/\partial x_2}{\partial g(\mathbf{x})/\partial x_1} dx_2 \tag{4.6.10}$$

which is then substituted into Eq. (4.6.6) in order to eliminate dx_1. The resulting total derivative of the objective function $f(\mathbf{x})$ can be expressed as

$$df(\mathbf{x}) = \left\{ -\left(\frac{\partial f(\mathbf{x})}{\partial x_1}\right)\left(\frac{\partial g(\mathbf{x})}{\partial x_1}\right)^{-1}\left(\frac{\partial g(\mathbf{x})}{\partial x_2}\right) + \left(\frac{\partial f(\mathbf{x})}{\partial x_2}\right)\right\} dx_2 \tag{4.6.11}$$

The reduced gradient is the expression in brackets { } and can be reduced to

$$\frac{df(\mathbf{x})}{dx_2} = \frac{\partial f(\mathbf{x})}{\partial x_2} - \left(\frac{\partial f(\mathbf{x})}{\partial x_1}\right)\left(\frac{\partial x_1}{\partial x_2}\right) \tag{4.6.12}$$

which is scalar because there is only one nonbasic variable x_2.

The reduced gradient can be written in vector form for the multiple variable case as

$$\nabla_N F = \left[\frac{\partial F}{\partial \mathbf{x}_N}\right] = \left[\frac{\partial f(\mathbf{x})}{\partial \mathbf{x}_N}\right] - \left[\frac{\partial f(\mathbf{x})}{\partial \mathbf{x}_B}\right]^T \left[\frac{\partial \mathbf{x}_B}{\partial \mathbf{x}_N}\right] \tag{4.6.13}$$

in which

$$\left[\frac{\partial \mathbf{x}_B}{\partial \mathbf{x}_N}\right] = \left[\frac{\partial \mathbf{g}(\mathbf{x})}{\partial \mathbf{x}_B}\right]^{-1} \left[\frac{\partial \mathbf{g}(\mathbf{x})}{\partial \mathbf{x}_N}\right] = \mathbf{B}^{-1} \left[\frac{\partial \mathbf{g}(\mathbf{x})}{\partial \mathbf{x}_N}\right] \tag{4.6.14}$$

The **Kuhn-Tucker multiplier** vector $\boldsymbol{\pi}$ is defined by

$$\left[\frac{\partial f(\mathbf{x})}{\partial \mathbf{x}_B}\right]^T \left[\frac{\partial \mathbf{g}(\mathbf{x})}{\partial \mathbf{x}_B}\right]^{-1} = \left[\frac{\partial f(\mathbf{x})}{\partial \mathbf{x}_B}\right]^T \mathbf{B}^{-1} = \boldsymbol{\pi}^T \tag{4.6.15}$$

Using these definitions the reduced gradient in Eq. (4.6.13) can be expressed as

$$\nabla_N F = \left[\frac{dF}{d\mathbf{x}_N}\right] = \left[\frac{\partial f(\mathbf{x})}{\partial \mathbf{x}_N}\right] - \boldsymbol{\pi}^T \left[\frac{\partial \mathbf{g}(\mathbf{x})}{\partial \mathbf{x}_N}\right] \tag{4.6.16}$$

4.6.4 Optimality Conditions for GRG Method

Consider the nonlinear programming problem

$$\text{Minimize f}(\mathbf{x})$$

subject to

$$g_i(\mathbf{x}) = 0 \qquad i = 1, \ldots, m$$
$$\underline{x}_j \leq x_j \leq \overline{x}_j \qquad j = 1, \ldots, n$$

In terms of basic and nonbasic variables, the Lagrangian function for the problem can be stated as

$$\begin{aligned} L &= f(\mathbf{x}) + \boldsymbol{\lambda}^T \mathbf{g}(\mathbf{x}) + \underline{\boldsymbol{\lambda}}^T(\underline{\mathbf{x}} - \mathbf{x}) + \overline{\boldsymbol{\lambda}}^T(\mathbf{x} - \overline{\mathbf{x}}) \\ &= f(\mathbf{x}_B, \mathbf{x}_N) + \boldsymbol{\lambda}^T \mathbf{g}(\mathbf{x}_B, \mathbf{x}_N) + \underline{\boldsymbol{\lambda}}_B^T(\underline{\mathbf{x}}_B - \mathbf{x}_B) + \underline{\boldsymbol{\lambda}}_N^T(\underline{\mathbf{x}}_N - \mathbf{x}_N) \\ &\quad + \overline{\boldsymbol{\lambda}}_B^T(\mathbf{x}_B - \overline{\mathbf{x}}_B) + \overline{\boldsymbol{\lambda}}_N^T(\mathbf{x}_N - \overline{\mathbf{x}}_N) \end{aligned} \tag{4.6.17}$$

in which $\boldsymbol{\lambda}_N$ and $\boldsymbol{\lambda}_B$ are vectors of Lagrange multipliers for nonbasic and basic variables, respectively.

Based on Eq. (4.5.6), the Kuhn-Tucker conditions for optimality in terms of the basic and nonbasic variables are

$$\nabla_B L = \nabla_B f + \boldsymbol{\lambda}^T \nabla_B \mathbf{g} - \underline{\boldsymbol{\lambda}}_B + \overline{\boldsymbol{\lambda}}_B = \mathbf{0} \tag{4.6.18a}$$
$$\nabla_N L = \nabla_N f + \boldsymbol{\lambda}^T \nabla_N \mathbf{g} - \underline{\boldsymbol{\lambda}}_N + \overline{\boldsymbol{\lambda}}_N = \mathbf{0} \tag{4.6.18b}$$
$$\underline{\boldsymbol{\lambda}}_B \geq \mathbf{0} \qquad \underline{\boldsymbol{\lambda}}_N \geq \mathbf{0} \tag{4.6.18c}$$
$$\overline{\boldsymbol{\lambda}}_B \geq \mathbf{0} \qquad \overline{\boldsymbol{\lambda}}_N \geq \mathbf{0} \tag{4.6.18d}$$
$$\underline{\boldsymbol{\lambda}}_B^T(\underline{\mathbf{x}}_B - \mathbf{x}_B) = \overline{\boldsymbol{\lambda}}_B^T(\mathbf{x}_B - \overline{\mathbf{x}}_B) = \mathbf{0} \tag{4.6.18e}$$
$$\underline{\boldsymbol{\lambda}}_N^T(\underline{\mathbf{x}}_N - \mathbf{x}_N) = \overline{\boldsymbol{\lambda}}_N^T(\mathbf{x}_N - \overline{\mathbf{x}}_N) = \mathbf{0} \tag{4.6.18f}$$

If $\mathbf{x}_B$ is strictly between its bounds then $\underline{\boldsymbol{\lambda}}_\mathbf{B} = \overline{\boldsymbol{\lambda}}_\mathbf{B} = \mathbf{0}$ by Eq. (4.6.18*e*) so that from Eq. (4.6.18*a*),

$$\boldsymbol{\lambda}^T = \left[-\frac{\partial f}{\partial \mathbf{x}_B}\right]^T \left[\frac{\partial \mathbf{g}}{\partial \mathbf{x}_B}\right]^{-1} = \left[-\frac{\partial f}{\partial \mathbf{x}_B}\right]^T \mathbf{B}^{-1} = -\boldsymbol{\pi}^T \tag{4.6.19}$$

In other words, when $\underline{\mathbf{x}}_B < \mathbf{x}_B < \overline{\mathbf{x}}_B$, the Kuhn-Tucker multiplier vector $\boldsymbol{\pi}$ is the Lagrange multiplier vector for the equality constraints $\mathbf{g}(\mathbf{x}) = \mathbf{0}$.

If $\mathbf{x}_N$ is strictly between its bounds, that is, $\underline{\mathbf{x}}_N < \mathbf{x}_N < \overline{\mathbf{x}}_N$, then $\underline{\boldsymbol{\lambda}}_N = \overline{\boldsymbol{\lambda}}_N = 0$ by Eq. (4.6.18*f*) so that

$$\left[\frac{\partial F}{\partial \mathbf{x}_N}\right] = \mathbf{0} \tag{4.6.20}$$

From Eqs. (4.6.16) and (4.6.18*b*), $\nabla_N L = \nabla_N F = \mathbf{0}$ for $\mathbf{x}_N$ strictly between its bounds. If $\mathbf{x}_N$ is at its lower bound, $\mathbf{x}_N = \underline{\mathbf{x}}_N$, then $\overline{\boldsymbol{\lambda}}_N = \mathbf{0}$ so

$$\left[\frac{\partial F}{\partial \mathbf{x}_N}\right] = \underline{\boldsymbol{\lambda}}_N \geq \mathbf{0} \tag{4.6.21}$$

If $\mathbf{x}_N$ is at its upper bound, $\mathbf{x}_N = \bar{\mathbf{x}}_N$, then $\underline{\boldsymbol{\lambda}}_N = \mathbf{0}$ so that

$$\left[\frac{\partial F}{\partial \mathbf{x}_N}\right] = \bar{\boldsymbol{\lambda}}_N \leq \mathbf{0} \tag{4.6.22}$$

The above three equations, Eqs. (4.6.20)–(4.6.22), define the optimality conditions for the reduced problem defined by Eq. (4.6.4). The Kuhn-Tucker conditions for the original problem may be viewed as the optimality conditions for the reduced problem.

Example 4.6.1. Consider the manufacturing waste-treatment plant problem of Example 3.1.1. The manufacturing plant can produce finished goods (x_1) that sell at a unit price of \$10 K. Finished goods cost \$3 K per unit to produce. The waste generated is $2x_1^{0.8}$ instead of the $2x_1$ considered in Example 3.1.1 (see Fig. 4.6.2). The treatment plant has a maximum capacity of treating 10 units of waste with 80 percent waste removal capability at a treatment cost of \$0.6 K per unit of waste. The effluent tax imposed on the waste discharged to the receiving water body is \$2 K for each unit of waste discharged. The amount of waste discharged without treatment is x_2. This problem is now a nonlinear programming problem with a nonlinear objective function and constraints.

Solution. The treatment cost is $0.6(2x_1^{0.8} - x_2)$ and the effluent tax on untreated waste is $2[x_2 + 0.2(2x_1^{0.8} - x_2)]$. The objective function is to maximize the profit which is

$$\begin{aligned}\text{Max } x_0 &= 10x_1 - [3x_1 + 0.6(2x_1^{0.8} - x_2) + 2[x_2 + 0.2(x_1^{0.8} - x_2)]\} \\ &= -2x_1^{0.8} + 7x_1 - x_2\end{aligned}$$

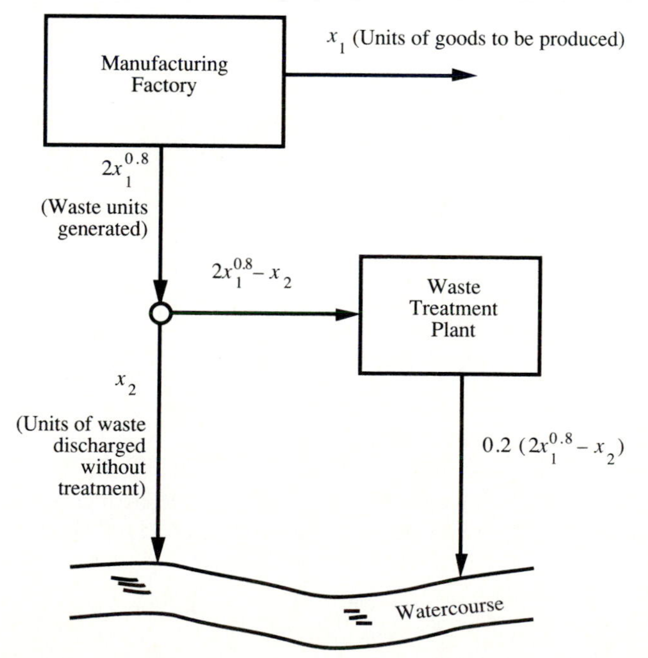

FIGURE 4.6.2
Schematic diagram of manufacturing-waste treatment system.

subject to constraints

a. Maximum treatment plant capacity

$$2x_1^{0.8} - x_2 \leq 10$$

b. Maximum amount of waste that can be discharged

$$0.4x_1^{0.8} + 0.8x_2 \leq 4$$

c. Amount of waste to treatment plant is positive.

$$2x_1^{0.8} - x_2 \geq 0$$

$$x_1 \geq 0 \text{ and } x_2 \geq 0$$

The problem is illustrated graphically in Fig. 4.6.2.

To solve the problem, nonnegative slack variables x_3, x_4, and x_5 are introduced to obtain equality constraints

$$g_1(\mathbf{x}) = 2x_1^{0.8} - x_2 + x_3 - 10 = 0$$

$$g_2(\mathbf{x}) = 0.4x_1^{0.8} + 0.8x_2 + x_4 - 4 = 0$$

$$g_3(\mathbf{x}) = 2x_1^{0.8} - x_2 - x_5 = 0$$

To speed up the convergence of this nonlinear optimization problem, the LP optimal solution to the Example 3.1.1 $(x_1, x_2) = (6, 2)$ is used as the initial solution to the nonlinear problem. To test the feasibility of this initial solution $(6, 2)$, it is substituted into the constraints to see if all are satisfied. In fact, $(6, 2)$ is a feasible solution.

At $(x_1, x_2) = (6, 2)$, the values of slack variables are $x_3 = 3.614$, $x_4 = 0.723$, and $x_5 = 6.386$. Here, one can choose $\mathbf{x}_N = (x_1, x_2)^T$, $\mathbf{x}_B = (x_3, x_4, x_5)^T$ in which x_1 and x_2 are superbasic variables. Now, the basic variables can be expressed in terms of nonbasic variables as

$$x_3 = -2x_1^{0.8} + x_2 + 10$$

$$x_4 = -0.4x_1^{0.8} - 0.8x_2 + 4$$

$$x_5 = 2x_1^{0.8} - x_2$$

Express basic variables in the objective function in terms of nonbasic ones and the reduced objective is

$$F(x_1, x_2) = -2x_1^{0.8} + 7x_1 - x_2$$

To determine the search direction for improving the current solution, the reduced gradient must be computed. The reduced gradient vector can be computed by using Eq. (4.6.16) as

$$\nabla_N F = \left[\frac{\partial f}{\partial \mathbf{x}_N}\right] - \boldsymbol{\pi}^T \left[\frac{\partial \mathbf{g}}{\partial \mathbf{x}_N}\right]$$

in which

$$\left[\frac{\partial f}{\partial \mathbf{x}_N}\right] = \begin{bmatrix} \dfrac{\partial f}{\partial x_1} \\ \dfrac{\partial f}{\partial x_2} \end{bmatrix}_{(x_1,x_2)=(6,\,2)} = \begin{bmatrix} -1.6x_1^{-0.2}+7 \\ -1 \end{bmatrix}_{(6,\,2)} = \begin{bmatrix} 5.88 \\ -1 \end{bmatrix}$$

$$\left[\frac{\partial \mathbf{g}}{\partial \mathbf{x}_N}\right] = \begin{bmatrix} \dfrac{\partial g_1}{\partial x_1} & \dfrac{\partial g_1}{\partial x_2} \\ \dfrac{\partial g_2}{\partial x_1} & \dfrac{\partial g_2}{\partial x_2} \\ \dfrac{\partial g_3}{\partial x_1} & \dfrac{\partial g_3}{\partial x_2} \end{bmatrix}_{(x_1,x_2)=(6,\,2)} = \begin{bmatrix} 1.6x_1^{-0.2} & -1 \\ 0.32x_1^{-0.2} & 0.8 \\ 1.6x_1^{-0.2} & -1 \end{bmatrix} = \begin{bmatrix} 1.12 & -1 \\ 0.22 & 0.8 \\ 1.12 & -1 \end{bmatrix}$$

$$\boldsymbol{\pi}^T = \left[\frac{\partial f}{\partial x_B}\right]^T \mathbf{B}^{-1} = \left(\frac{\partial f}{\partial x_3}, \frac{\partial f}{\partial x_4}, \frac{\partial f}{\partial x_5}\right) \begin{bmatrix} \dfrac{\partial g_1}{\partial x_3} & \dfrac{\partial g_1}{\partial x_4} & \dfrac{\partial g_1}{\partial x_5} \\ \dfrac{\partial g_2}{\partial x_3} & \dfrac{\partial g_2}{\partial x_4} & \dfrac{\partial g_2}{\partial x_5} \\ \dfrac{\partial g_3}{\partial x_3} & \dfrac{\partial g_3}{\partial x_4} & \dfrac{\partial g_3}{\partial x_5} \end{bmatrix}$$

$$= (0,0,0) \begin{bmatrix} 1 & 0 & 0 \\ 0 & 1 & 0 \\ 0 & 0 & 1 \end{bmatrix} = (0,0,0)$$

Therefore, the vector of reduced gradient ($\nabla_N F$) at the current solution point $\mathbf{x}_N = (x_1, x_2)^T = (6, 2)^T$ and $\mathbf{x}_B = (x_3, x_4, x_5)^T = (3.614, 0.723, 6.386)^T$ is

$$\nabla_N F = \begin{pmatrix} 5.88 \\ -1 \end{pmatrix} - (0,0,0) \begin{bmatrix} 1.12 & -1 \\ 0.22 & 0.8 \\ 1.12 & -1 \end{bmatrix} = \begin{pmatrix} 5.88 \\ -1 \end{pmatrix}$$

In this example, since the reduced objective function can be explicitly expressed in terms of current nonbasic variables, the reduced gradient can be evaluated, alternatively, as

$$\nabla_N F = \left[\frac{\partial F}{\partial \mathbf{x}_N}\right] = \begin{bmatrix} \frac{\partial F}{\partial x_1} \\ \frac{\partial F}{\partial x_2} \end{bmatrix}_{(x_1,x_2)=(6,2)} = \begin{bmatrix} -1.6x_1^{-0.2}+7 \\ -1 \end{bmatrix}_{(6,2)} = \begin{pmatrix} 5.88 \\ -1 \end{pmatrix}$$

Adopting the steepest descent method, the search direction at $(x_1, x_2) = (6, 2)$ is determined as

$$\mathbf{d} = \begin{pmatrix} d_1 \\ d_2 \end{pmatrix} = \nabla_N F = \begin{pmatrix} \dfrac{\partial F}{\partial x_1} \\ \dfrac{\partial F}{\partial x_2} \end{pmatrix} = \begin{pmatrix} -1.6x_1^{-0.2}+7 \\ -1 \end{pmatrix}_{(6,2)} = \begin{pmatrix} 5.88 \\ -1 \end{pmatrix}$$

Starting from $(x_1, x_2) = (6, 2)$ along $\mathbf{d} = (5.88, -1)^T$ the line search is performed to determine the optimal feasible step size (β) that

$$\begin{aligned} \text{Maximize } F(\mathbf{x}) &= F(x_1 + \beta d_1, x_2 + \beta d_2) = F(6 + 5.88\beta, 2 - \beta) \\ &= -2(6 + 5.88\beta)^{0.8} + 7(6 + 5.88\beta) - (2 - \beta) \\ &= -2(6 + 5.88\beta)^{0.8} + 42.16\beta + 40 \end{aligned}$$

Because the reduced objective is convex, β cannot be determined from differentiating the objective. Line search for determining the optimal and feasible β can be performed using a simple trial-and-error procedure as follows:

	$\mathbf{x}_N$			$\mathbf{x}_B$		
β	x_1	x_2	$F(\mathbf{x})$	x_3	x_4	x_5
0	6.000	2	31.61	3.614	0.723	6.386
0.1	6.588	1.9	35.18	2.863	0.673	7.137
0.3	7.764	1.7	42.34	1.394	0.579	8.606
0.5	8.940	1.50	49.54	−0.04	0.493	10.037
0.494	8.905	1.506	49.33	0	0.495	9.995

For $\beta = 0.5$ the solution is infeasible because x_3 is negative. With $\beta = 0.494$, the basic variable x_3 drops to zero which now becomes a nonbasic variable in the next iteration. The new solution point is indicated as point 2 on Fig. 4.6.3. The line search could alternatively be performed using a one-dimensional search procedure such as the golden section method.

At the new solution point 2, $(x_1, x_2) = (8.905, 1.506)$, the nonbasic and basic variables are identified as $\mathbf{x}_N = (x_1, x_3)^T$ and $\mathbf{x}_B = (x_2, x_4, x_5)^T$. Either x_1 or x_2 can become a basic variable because neither are on their bounds. For the purposes of this example, x_2 was arbitrarily chosen to become basic. Refer to the constraint equations,

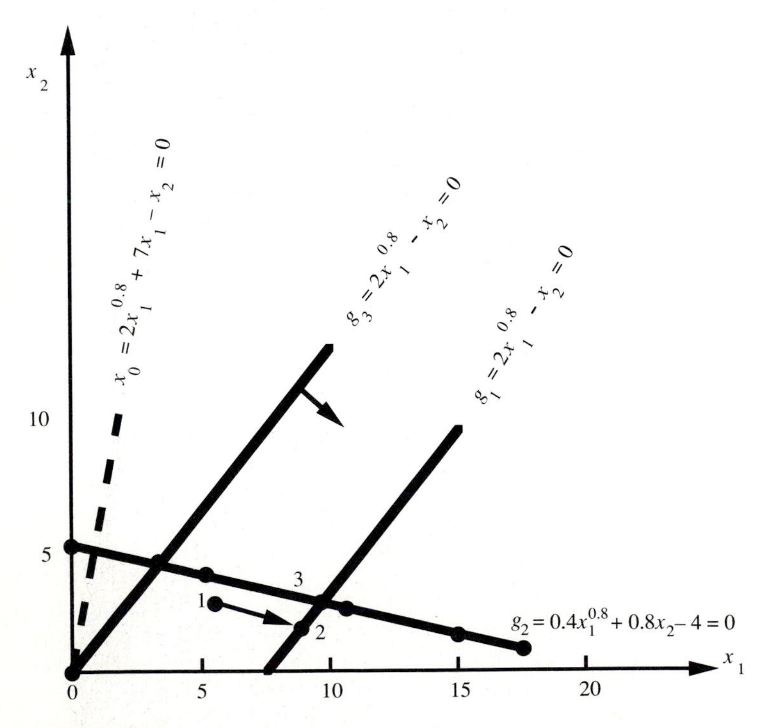

FIGURE 4.6.3
Graphical illustration of nonlinear manufacturing-waste treatment example.

the basic variables can be expressed in nonbasic terms as

$$x_2 = 2x_1^{0.8} + x_3 - 10$$

$$x_4 = -0.4x_1^{0.8} - 0.8x_2 + 4 = -2x_1^{0.8} - 0.8x_3 + 12$$

$$x_5 = 2x_1^{0.8} - x_2 = 10 - x_3$$

Hence, the reduced objective is

$$F(x_1, x_3) = -2x_1^{0.8} + 7x_1 - x_2 = -2x_1^{0.8} + 7x_1 - 2x_1^{0.8} - x_3 + 10$$

$$= -4x_1^{0.8} + 7x_1 - x_3 + 10$$

The reduced gradient at $(x_1, x_3) = (8.905, 0)$ defines the search direction,

$$\mathbf{d} = \begin{pmatrix} d_1 \\ d_3 \end{pmatrix} = \begin{pmatrix} \partial F/\partial x_1 \\ \partial F/\partial x_3 \end{pmatrix} = \begin{pmatrix} -3.2x_1^{-0.2} + 7 \\ -1 \end{pmatrix}_{(8.905,0)} = \begin{pmatrix} 4.934 \\ -1 \end{pmatrix}$$

Again, starting from point 2, $(x_1, x_2) = (8.905, 1.506)$, along the search direction $\mathbf{d} = (4.934, -1)^T$, the optimal step size β that maximizes the reduced objective is sought.

$$\text{Maximize } F(\mathbf{x}) = F(x_1 + \beta d_1, x_3 + \beta d_3) = F(8.905 + 4.934\beta, -\beta)$$

$$= -4(8.905 + 4.934\beta)^{0.8} + 7(8.905 + 4.934\beta) + \beta + 10$$

$$= -4(8.905 + 4.934\beta)^{0.8} + 35.538\beta + 72.335$$

Note that the nonbasic variable x_1 is superbasic (nonbasic variable not at its bounds) while x_3 is at the lower bound of zero. Therefore, determination of the optimal feasible β by the trial-and-error line search can be performed similar to the previous iteration by only considering superbasic variables. This is shown in the following table. Because d_3 is negative and $x_3 = 0$, which is its lower bound, it must remain as a nonbasic variable at its lower bound.

	$\mathbf{x}_N$			$\mathbf{x}_B$		
β	x_1	x_3	$F(x_2, x_3)$	x_2	x_4	x_5
0	8.905	0	49.33	1.506	0.494	10
0.1	9.398	−0.1	51.873	2.008	−0.008(inf)	10
0.098	9.389	−0.098	51.822	1.998	0.002	10
0.0985	9.390	−0.0985	51.734	2.000	0.000	10

Now, at the new solution point $(x_1, x_2) = (9.390, 2.000)$, the basic variable x_4 decreases to zero, and, becomes a nonbasic variable at its lower bound. Along with $x_3 = 0$, this indicates that constraints $g_1(\mathbf{x})$ and $g_2(\mathbf{x})$ are binding. Hence, the two nonbasic variables x_3 and x_4 are zero at their lower bounds. The basic variables (x_1, x_2, x_5), then, in terms of nonbasics (x_3, x_4) are

$$x_1 = (6 - 0.4x_3 - 0.5x_4)^{1.25}$$

$$x_2 = 2 + 0.2x_3 - x_4$$

$$x_5 = 10 - x_3$$

Then, the reduced objective is

$$\begin{aligned} F(x_3, x_4) = & -2(6 - 0.4x_3 - 0.5x_4) + 7(6 - 0.4x_3 - 0.5x_4)^{1.25} \\ & - (2 + 0.2x_3 - x_4) \\ = & -14 + 0.6x_3 + 1.5x_4 + 7(6 - 0.4x_3 - 0.5x_4)^{1.25} \end{aligned}$$

with reduced gradient

$$\begin{aligned} \nabla_N F &= \begin{pmatrix} \partial F/\partial x_3 \\ \partial F/\partial x_4 \end{pmatrix} = \begin{pmatrix} 0.6 - 3.5(6 - 0.4x_3 - 0.5x_4)^{0.25} \\ 1.5 - 4.375(6 - 0.4x_3 - 0.5x_4)^{0.25} \end{pmatrix}_{(0,0)} \\ &= \begin{pmatrix} -4.878 \\ -5.347 \end{pmatrix} < \begin{pmatrix} 0 \\ 0 \end{pmatrix} \end{aligned}$$

which satisfies the optimality condition since the problem is a maximization. The negativity of ∇F_N for the nonbasic variables (x_3, x_4), which are both at their lower bounds, indicates that an increase in any one of them above zero would decrease the current objective function. Therefore, the optimal solution to this nonlinear manufacturing-waste treatment problem is $(x_1^*, x_2^*) = (9.39, 2.00)$ with an objective function value of \$51.834 K.

4.7 CONSTRAINED NONLINEAR OPTIMIZATION: PENALTY FUNCTION METHODS

The essential idea of penalty function methods is to transform constrained nonlinear programming problems into a sequence of unconstrained optimization problems. The basic idea of these methods is to add one or more functions of the constraints to the objective function in order to delete the constraints. Basic reasoning for such approaches is that unconstrained problems are much easier to solve. Using a penalty function a constrained nonlinear programming problem is transformed to an unconstrained problem.

$$\left.\begin{array}{ll} & \text{Minimize } f(\mathbf{x}) \\ \text{subject to} & \mathbf{g}(\mathbf{x}) \end{array}\right\} \Rightarrow \text{Minimize } L[f(\mathbf{x}), \mathbf{g}(\mathbf{x})]$$

where $L[f(\mathbf{x}), \mathbf{g}(\mathbf{x})]$ is a **penalty function**. Various forms of penalty functions have been proposed which can be found elsewhere (Gill, Murray, and Wright, 1981; McCormick, 1983). The penalty function is minimized by stages for a series of values of parameters associated with the penalty. In fact, the Lagrangian function (described in Section 4.5.) is one form of penalty function. For many of the penalty functions, the Hessian of the penalty function becomes increasingly **ill-conditioned** (i.e., the function value is extremely sensitive to a small change in the parameter value) as the solution approaches the optimum. This section briefly describes a penalty function method called the **augmented Lagrangian method**.

The augmented Lagrangian method adds a quadratic penalty function term to the Lagrangian function (Eq. 4.5.2), to obtain

$$L_A(\mathbf{x}, \boldsymbol{\lambda}, \psi) = f(\mathbf{x}) + \sum_{i=1}^{m} \lambda_i g_i(\mathbf{x}) + \frac{\psi}{2} \sum_{i=1}^{m} g_i^2(\mathbf{x})$$

$$= f(\mathbf{x}) + \boldsymbol{\lambda}^T \mathbf{g}(\mathbf{x}) + \frac{\psi}{2} \mathbf{g}(\mathbf{x})^T \mathbf{g}(\mathbf{x}) \tag{4.7.1}$$

where ψ is a positive penalty parameter. Some desirable properties of Eq. (4.7.1) are discussed by Gill, Murray and Wright (1981).

Example 4.7.1. Gill, Murray, and Wright (1981) and Edgar and Himmelblau (1988) used the following minimization problem to illustrate the augmented Lagrangian.

$$\text{Minimize } f(x) = x^3$$

subject to

$$x + 1 = 0$$

Solution. The augmented Lagrangian by Eq. (4.7.1) is

$$L_A(x, \lambda, \psi) = x^3 + \lambda(x+1) + \frac{\psi}{2}(x+1)^2$$

For $\lambda^* = -3$ and $\psi = 9$, the augmented Lagrangian function is

$$L_A(x, -3, 9) = x^3 - 3(x+1) + \frac{9}{2}(x+1)^2$$

At $x = -1$, the gradient of the augmented Lagrangian is

$$\begin{aligned} \nabla_x L_A(x, -3, 9) &= 3x^2 - 3 + 9(x+1) \\ &= 3(-1)^2 - 3 + 9(-1+1) \\ &= 3 - 3 + 0 \\ &= 0 \end{aligned}$$

and the second derivative is

$$\begin{aligned} \nabla_x^2 L_A(x, -3, 9) &= 6x + \psi \\ &= 6(-1) + 9 \\ &= 3 \end{aligned}$$

The second derivative is positive definite for $\psi > 6$. The plot of $L_A(x, -3, 9)$ in Fig. 4.7.1 illustrates that at $x = -1$, $L_A(x, -3, 9)$ is a local minimum. It should be noted that this augmented Lagrangian function is unbounded below for any value of ψ.

For ideal circumstances, $\mathbf{x}^*$ can be computed by a single unconstrained minimization of the differentiable function (Eq. 4.7.1). However, in general, λ^* is not available until the solution has been determined. An augmented Lagrangian method, therefore, must include a procedure for estimating the Lagrange multipliers. Gill,

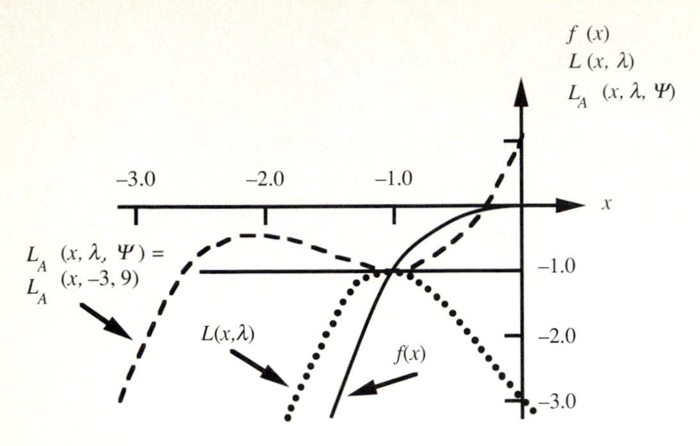

FIGURE 4.7.1
The solid line is the graph of the objective function of Example 4.7.1 $F(x) = x^3$. The dotted line denotes the graph of the Lagrangian function, and the dashed line is the graph of the augmented Lagrangian $L_A(x, \lambda^*, 9)$ (Adapted from Gill, Murray, and Wright, 1981).

Murray, and Wright (1981) present the following algorithm:

Step 0 Select initial estimates of the Lagrange multipliers $\boldsymbol{\lambda}^k = \mathbf{0}$, the penalty parameter ψ, an initial point $\mathbf{x}^k = 0$. Set $k = k + 1$ and set the maximum number of iterations as J.

Step 1 Check to see if $\mathbf{x}^k$ satisfies optimality conditions or if $k > J$. If so, terminate the algorithm.

Step 2 Minimize the augmented Lagrangian function, Minimize $L_A(\mathbf{x}, \boldsymbol{\lambda}, \psi)$, in Eq. (4.7.1). Procedures to consider unboundedness must be considered. The best solution is denoted as $\mathbf{x}^{k+1}$.

Step 3 Update the multiplier estimate by computing $\boldsymbol{\lambda}^{k+1}$.

Step 4 Increase the penalty parameter ψ if the constraint violations at $\mathbf{x}^{k+1}$ have not decreased sufficiently from those at $\mathbf{x}^k$.

Step 5 Set $k = k + 1$ and return to Step 1.

Augmented Lagrangian methods can be applied to inequality constraints. For the set of violated constraints, $\mathbf{g}(\mathbf{x})$ at $\mathbf{x}^k$, the augmented Lagrangian function has discontinuous derivatives at the solution if any of the constraints are active (Gill, Murray, and Wright, 1981). Buys (1972) and Rockafellar (1973*a,b*, 1974) presented the augmented Lagrangian function for inequality-constrained problems

$$L_A(\mathbf{x}, \boldsymbol{\lambda}, \psi) = f(\mathbf{x}) + \sum_{i=1}^{m} \begin{cases} \lambda_i g_i(\mathbf{x}) + \frac{\psi}{2}[g_i(\mathbf{x})]^2, & \text{if } g_i(\mathbf{x}) \le \frac{\lambda_i}{\psi} \\ -\frac{\psi}{2}\lambda_i^2 & \text{if } g_i(\mathbf{x}) > \frac{\lambda_i}{\psi} \end{cases} \tag{4.7.2}$$

Examples of using the augmented Lagrangian procedure for incorporating bound constraints into models for groundwater systems and water distribution systems are described in Chapters 8 and 9.

4.8 CONSTRAINED NONLINEAR OPTIMIZATION: PROJECTED LAGRANGIAN METHOD

The general nonlinear programming problem is now stated by separating out the equality and inequality contraints,

$$\text{Minimize } f(\mathbf{x})$$

subject to

$$\mathbf{g}(\mathbf{x}) = \mathbf{0}$$

$$\mathbf{h}(\mathbf{x}) \geq \mathbf{0}$$

Instead of using the original objective function $f(\mathbf{x})$, the projected Lagrangian method solves the following generalized objective involving Lagrangian multipliers,

$$\text{Minimize } f(\mathbf{x}) - \left[(\boldsymbol{\lambda}_g^k)^T \mathbf{g}(\mathbf{x}) + (\boldsymbol{\lambda}_{\hat{h}}^k)^T \hat{\mathbf{h}}(\mathbf{x})\right] + \left[(\boldsymbol{\lambda}_g^k)^T \mathbf{A}_g^k \mathbf{x} + (\boldsymbol{\lambda}_{\hat{h}}^k)^T \mathbf{A}_{\hat{h}}^k \mathbf{x}\right] \tag{4.8.1}$$

in which $\boldsymbol{\lambda}_g^k$ and $\boldsymbol{\lambda}_{\hat{h}}^k$ are vectors of Lagrange multipliers for equality constraints $\mathbf{g}(\mathbf{x})$ and **active** (at their bounds) inequality constraints $\hat{\mathbf{h}}(\mathbf{x})$ in the kth iteration; $\mathbf{A}_g^k$ and $\mathbf{A}_{\hat{h}}^k$ are the matrices of linearized equality constraints and active inequality constraints in the kth iteration which are determined as

$$\mathbf{A}_g^k = \left[\frac{\partial \mathbf{g}(\mathbf{x})}{\partial \mathbf{x}}\right]_{\mathbf{x}^k}; \qquad \mathbf{A}_{\hat{h}}^k = \left[\frac{\partial \hat{\mathbf{h}}(\mathbf{x})}{\partial \mathbf{x}}\right]_{\mathbf{x}^k} \tag{4.8.2}$$

The constraints of the original model in the projected Lagrangian method is linearized (if the problem is not linearly constrained) as

$$\mathbf{A}_g^k \mathbf{x} = -\mathbf{g}(\mathbf{x}^k) + \mathbf{A}_g^k \mathbf{x}^k \tag{4.8.3}$$

$$\mathbf{A}_{\hat{h}}^k \mathbf{x} = -\hat{\mathbf{h}}(\mathbf{x}^k) + \mathbf{A}_{\hat{h}}^k \mathbf{x}^k \tag{4.8.4}$$

in which $\mathbf{g}(\mathbf{x}^k)$ and $\hat{\mathbf{h}}(\mathbf{x}^k)$ are vectors of constraint values evaluated at the solution points $(\mathbf{x}^k)$ in the kth iteration.

Gill, Murray, and Wright (1981) present a simplified projected Lagrangian algorithm as follows:

Step 0 With $k = 0$, select initial estimates of solution point $\mathbf{x}^k$, Lagrangian multipliers $\boldsymbol{\lambda}_g^k$ for equality constraints $\mathbf{g}(\mathbf{x})$ and $\boldsymbol{\lambda}_{\hat{h}}^k$ for active inequality constraints $\hat{\mathbf{h}}(\mathbf{x})$. Set the maximum number of iterations as J.

Step 1 Check to see if $\mathbf{x}_k$ satisfies optimality conditions or if $k > J$. If so, terminate the algorithm.

Step 2 Use $\mathbf{x}^k$ and solve the following problem

$$\text{Minimize } f(\mathbf{x}) - [(\boldsymbol{\lambda}_g^k)^T \mathbf{g}(\mathbf{x}) + (\boldsymbol{\lambda}_{\hat{h}}^k)^T \hat{\mathbf{h}}(\mathbf{x})] + [(\boldsymbol{\lambda}_g^h)^T \mathbf{A}_g^k \mathbf{x} + (\boldsymbol{\lambda}_{\hat{h}}^k)^T \mathbf{A}_{\hat{h}}^k \mathbf{x}]$$

subject to

$$\mathbf{A}_g^k \mathbf{x} = -\mathbf{g}(\mathbf{x}^k) + \mathbf{A}_g^k \mathbf{x}^k$$

$$\mathbf{A}_{\hat{h}}^k \mathbf{x} = -\hat{\mathbf{h}}(\mathbf{x}^k) + \mathbf{x}^k$$

with appropriate safeguards to cope with unboundedness.

Step 3 Set the optimal solution found in Step 2 to $\mathbf{x}^{k+1}$ and update the Lagrangian multipliers to $\boldsymbol{\lambda}_g^{k+1}$ and $\boldsymbol{\lambda}_{\hat{h}}^{k+1}$. Set $k = k + 1$ and go back to Step 1.

Performance of the above algorithm is sensitive to the initial starting solution. The success of the algorithm to converge to the local minimum requires that the initial starting solution $\mathbf{x}^0$ and $\boldsymbol{\lambda}^0$ are sufficiently close to the optimal values. To improve the model performance, Gill, Murray, and Wright (1981) describe approaches to find a good starting solution, one of which is to replace the objective function Eq. (4.8.1) in Step 1 by the augmented Lagrangian function, described in Section 4.7, with penalty parameter $\psi = 0$.

4.9 NONLINEAR PROGRAMMING CODES

This section briefly introduces nonlinear programming computer codes that have been applied to solve NLP problems. They are: (1) GRG2 (Generalized Reduced Gradient 2) developed by Lasdon and his colleagues (Lasdon et al., 1978; Lasdon and Waren, 1978); (2) GINO (Liebman et al., 1986); (3) MINOS (Modular In-core Nonlinear Optimization System) developed by Murtagh and Saunders (1980, 1983); and (4) GAMS-MINOS.

GRG2 COMPUTER CODE. The GRG2 computer code utilizes the fundamental idea of the generalized reduced gradient algorithm described in Section 4.6. GRG2 requires that the user provide a subroutine GCOMP specifying the objective function and constraints of the nonlinear-programming (NLP) problem. It is optional for the user to provide the subroutine that contains derivatives of the objective function and constraints. If not provided, differentiations are approximated numerically by either forward finite differencing or central finite differencing.

GRG2 provides several alternative ways that can be used to define the search direction. They include the BFGS, the quasi-Newton method and variations of conjugate gradient methods (see descriptions in Section 4.4.3). The default method is the BFGS method.

MINOS COMPUTER CODE. MINOS is a Fortran-based computer code designed to solve large-scale optimization problems. The program solves a linear programming problem by implementing the primal simplex method. When a problem has a nonlinear objective function subject to linear constraints, MINOS uses a reduced-gradient algorithm in conjunction with a quasi-Newton algorithm. In case that the problem involves nonlinear constraints, the projected Lagrangian algorithm (Section 4.8) is implemented. Similar to GRG2, MINOS requires that the user provide subroutine FUNOBJ to specify the objective function and its gradient. Also, subroutine FUNCON is to be supplied by the user to input the constraints and as many of their gradients as possible.

GAMS-MINOS. GAMS-MINOS is a microcomputer version that links GAMS and MINOS. As an illustration the GAMS-MINOS input file for example 4.6.1 is shown in Fig. 4.9.1. Similarities can be found by comparing this input file with the GAMS input file for the LP solution of the manufacturing waste-treatment problem in Appendix 3.A.

GINO. GINO (Liebman et al., 1986) is a microcomputer version of GRG2.

REFERENCES

Buys, J. D.: *Dual Algorithms for Constrained Optimization Problems*, Ph.D. Thesis, University of Leiden, Netherlands, 1972.

Cooper, L. L. and M. W. Cooper: *Introduction to Dynamic Programming*, Pergamon Press, Elmsford, N.Y., 1981.

Chow, V. T., D. R. Maidment, and G. W. Tauxe: "Computer Time and Memory Requirements for DP and DDDP in Water Resource System Analysis," *Water Resources Research*, vol. 11, no. 5, pp. 621–628, Oct. 1971.

Denardo, E. V.: *Dynamic Programming Theory and Applications*, Prentice-Hall, Englewood Cliffs, N.J., 1982.

Dennis, J. E. and R. B. Schnable: *Numerical Methods for Unconstrained Optimization*, Prentice-Hall, Englewood Cliffs, N.J., 1983.

Dreyfus, S. and A. Law: *The Art and Theory of Dynamic Programming*, Academic Press, New York, 1977.

Edgar, T. F. and D. M. Himmelblau: *Optimization of Chemical Processes*, McGraw-Hill, Inc., New York, 1988.

Fletcher, R.: *Practical Methods of Optimization, vol. 1, Unconstrained Optimization*, Wiley, New York, 1980.

Gill, P. E., W. Murray and M. H. Wright: *Practical Optimization*, Academic Press, London and New York, 1981.

Himmelblau, D. M.: *Applied Nonlinear Programming*, McGraw-Hill, Inc., 1972.

Lasdon, L. S., R. L. Fox, and M. W. Ratner: "Nonlinear optimization using the generalized reduced gradient method," *Revue Francaise d'Automatique, Informatique et Recherche Operationnelle*, vol. 3, pp. 73–104, November 1974.

Lasdon, L. S., A. D. Waren, A. Jain, and M. Ratner: "Design and Testing of a Generalized Reduced Gradient Code for Nonlinear Programming," *ACM Transactions on Mathematical Software*, vol. 4, pp. 34–50, 1978.

Lasdon, L. S. and A. D. Waren: "Generalized Reduced Gradient Software for Linearly and Nonlinearly Constrained Problems," in *Design and Implementation of Optimization Software*, H. J. Greenberg (ed.), Sijthoff and Noordhoff, pp. 363–397, 1978.

Liebman, J. S., L. S. Lasdon, L. Schrage and A. Waren: *Modeling and Optimization with GINO*, The Scientific Press, Palo Alto, Calif., 1986.

Luenberger, D. G.: *Introduction to Linear and Nonlinear Programming*, Addison-Wesley, Reading, Mass., 1984.

McCormick, G. P.: *Nonlinear Programming: Theory, Algorithms, and Applications*, Wiley, New York, 1983.

Murtaugh, B. A. and M. A. Saunders: "Large-Scale Linearly Constrained Optimization," *Mathematical Programming*, vol. 14, pp. 41–72, 1978.

```
*    The Problem: find the quantity of goods to be produced and
*    the quantity of waste to be *discharged without treatment
*    which maximize the total net benefit of *the company
*
*        ------- NONLINEAR OBJECTIVE ----------
*        ------- Changed 3 constraints to nonlinear    ------
*-----------------------------------------------SETS-------
     C    constraint /PLANTCAP,DISCLIMIT,MAXWASTE/
     A    amount    /GOODS,WASTE/          ;
*-----------------------------------------
PARAMETER     CC(A)     coefficients of objective function
              /GOODS    -5
              WASTE    -1/

PARAMETER      RHS(C) right-hand-sides of constraints
               /PLANTCAP   10
               DISCLIMIT   4
               MAXWASTE    0/

TABLE   COEFF(C,A)  coefficients in constraints
                    GOODS  WASTE
PLANTCAP              2      -1
DISCIMIT            0.4     0.8
MAXWASTE              2      -1
*-------------------------------------------------------------
VARIABLES    X(A)     decision variables
             MAXBENEFIT
POSITIVE VARIABLE X;

        X.L('GOODS')=2;
        X.L('WASTE')=10;
*-------------------------------------------------------------
EQUATIONS   C1   plant capacity constraint
            C2   waste discharge limit constraint
            C3   no more waste constraint
            PROFIT   profit of company;

C1..COEFF('PLANTCAP','GOODS')*X('GOODS')**1.2 +
    COEFF('PLANTCAP', WASTE')*X('WASTE')=L=RHS('PLANTCAP');

C2..COEFF('DISCLIMIT','GOODS')*X('GOODS')**1.2 +
     COEFF('DISCLIMIT','WASTE')*X('WASTE')=L=RHS ('DISCLIMIT');
C3..COEFF('MAXWASTE','GOODS')*X('GOODS')**1.2 +
    COEFF('MAXWASTE','WASTE')*X('WASTE')=G=RHS ('MAXWASTE');
```

FIGURE 4.9.1
GAMS-MINOS Input for Manufacturing-Waste Treatment Example.

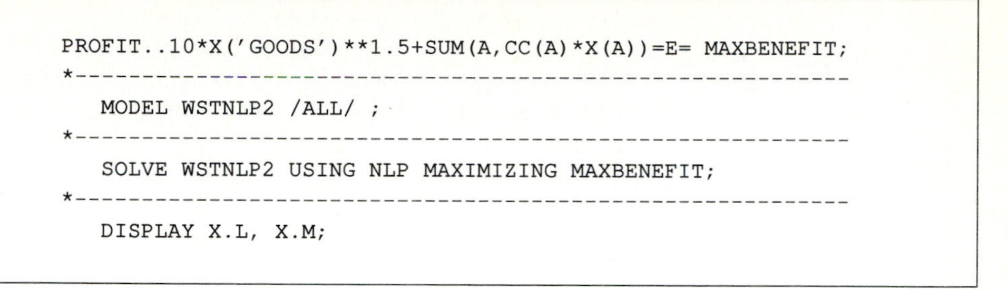

```
PROFIT..10*X('GOODS')**1.5+SUM(A,CC(A)*X(A))=E= MAXBENEFIT;
*---------------------------------------------------------
   MODEL WSTNLP2 /ALL/ ;
*---------------------------------------------------------
   SOLVE WSTNLP2 USING NLP MAXIMIZING MAXBENEFIT;
*---------------------------------------------------------
   DISPLAY X.L, X.M;
```

FIGURE 4.9.1 (*cont.*)

Murtaugh, B. A. and M. A. Saunders: "MINOS/AUGMENTED User's Manual," *Syst. Optimiz. Lab. Tech. Rep. 80–14*, 51 pp., Department of Operations Research, Stanford University, Stanford, Calif., 1980.

Murtaugh, B. A. and M. A. Saunders: "MINOS 5.0 User's Guide," *Syst. Optimiz. Lab. Tech. Rep. 83–20*, 118 pp., Department of Operations Research, Stanford University, Stanford, Calif., 1983.

Rockafellar, R. T.: "A Dual Approach to Solving Nonlinear Programming Problems by Unconstrained Optimization," *Math. Prog.*, vol. 5, 354–373, 1973*a*.

Rockafellar, R. T.: "The Multiplier Method of Hestenes and Powell Applied to Convex Programming," SIAM, *J. Control and Optimization*, vol. 12, 268–285, 1973*b*.

Rockafellar, R. T.: "Augmented Lagrangian Multiplier Functions and Duality in Nonconvex Programming," SIAM, *J. Applied Math*, vol. 12, 555–562, 1974.

PROBLEMS

4.1.1 Solve the project funding allocation Example 4.1.1 using a forward algorithm.

4.1.2 Resolve the project funding allocation Example 4.1.1 with available funding as the state variable using a backward algorithm.

4.1.3 Consider water supply for a region consisting of three cities. The total available water supply is 8 units. Determine the optimal allocation of water to the three cities that maximizes the total economic return from the region. The relationship between the economic return and the quantity of water allocated for each city are given in the following table

Return \ q	City		
	1	2	3
q	$r_1(q)$	$r_2(q)$	$r_3(q)$
0	0	0	0
1	6	5	7
2	12	14	30
3	35	40	42
4	75	55	50
5	85	65	60
6	91	70	70
7	96	75	72
8	100	80	75

4.1.4 Solve the reservoir operation Example 4.1.3 using a DP forward algorithm.

4.1.5 Consider the same reservoir operation problem in Example 4.1.3. Determine the optimal operation policy for maximum annual returns assuming that the reservoir starts with three units of water at the beginning of the year and has three units of water at the end of the year. Use a backward algorithm.

4.1.6 Solve the reservoir operation Problem 4.1.5 using a DP forward algorithm.

4.1.7 Solve the reservoir operations Example 4.1.3 with seasonal inflows of 1, 2, 1, 2 units respectively for the four seasons of the year. Use a DP backward algorithm.

4.2.1 Refer to Problem 4.1.3 and consider that the quantity of water that can be allocated to each of the three cities can be divided into a fraction of a unit. Apply DDDP to refine the optimal solution using the optimal allocation from Problem 4.1.3 as the initial trial trajectory and $\Delta q = 0.5$ units. Use three lattice points (including the trial trajectory) to form the corridor. Terminate the algorithm when $\Delta q = 0.25$ units. Linear interpolation can be used to calculate the economic return.

4.3.1 Rosenbrock's function (Himmelblau, 1972; Fletcher, 1980) is a well known test function for optimization methods given as

$$f(\mathbf{x}) = 100(x_2 - x_1^2)^2 + (1 - x_1)^2$$

Determine the gradient and the Hessian for this function at point $(x_1, x_2) = (1, 1)$.

4.3.2 Determine the convexity of the Rosenbrock's function in Problem 4.3.1 at point $(x_1, x_2) = (1, 1)$ by: (1) method of leading principal minors and; (2) the Eigenvalue method.

4.3.3 Consider the following function

$$f(\mathbf{x}) = 100(x_2 - x_1^3)^2 + (1 - x_1)^2$$

Determine the gradients and Hessian matrices at $\mathbf{x} = (-1, 0)$ and $\mathbf{x} = (0, 1)$. Furthermore, determine the status of the Hessian matrices at the two points.

4.3.4 Consider the following function

$$f(\mathbf{x}) = [1.5 - x_1(1 - x_2)]^2 + [2.25 - x_1(1 - x_2^2)]^2 + [2.625 - x_1(1 - x_2^3)]^2$$

Determine the gradients and Hessian matrices at $\mathbf{x} = (0, 1)$ and $\mathbf{x} = (1, 0)$. Furthermore, determine the status of the Hessian matrices at the two points.

4.4.1 Determine the optimal solution to the following single variable problem by the Newton's method.

$$\text{Minimize } f(x) = x^3 - 6x$$

4.4.2 Perform the tasks below for the problem:

$$f(\mathbf{x}) = x_1^2 + 9x_2^2, \text{ with the initial point } (x_1, x_2) = (5, 1)$$

(*a*) Sketch a few contours of constant value of the function on graph paper.
(*b*) Derive an expression for the optimal step size β^i for the arbitrary initial point $\mathbf{x}^i$.
(*c*) Go through three iterations of the method of steepest descent starting from (5, 1) and plot the results.

4.4.3 Use GINO or GAMS-MINOS to minimize the function

$$f(\mathbf{x}) = x_1^2 + 9x_2^2$$

starting from the initial point (5,1). If GINO is used solve using the BFGS, Fletcher-Reeves, and the 1-step BFGS methods. What is the total number of line searches for each method?

4.4.4 Use GINO to minimize the Rosenbrock function

$$f(\mathbf{x}) = 100(x_2 - x_1^2)^2 + (1 - x_1)^2$$

starting from the initial point $(-1.2, 1)$. Solve using the BFGS, Fletcher-Reeves, and the 1-step BFGS method, and summarize the results. What is the total number of line searches for each method?

4.4.5 Use GINO to minimize the function

$$f(\mathbf{x}) = (x_1 + 10x_2)^2 + 5(x_3 - x_4)^2 + (x_2 - 2x_3)^4 + 10(x_1 - x_4)^4$$

starting from the initial point $(3, -1, 0, 1)$. Solve using the BFGS, Fletcher-Reeves, and the 1-step BFGS methods. Summarize the results. What is the total number of line searches for each method?

4.4.6 Use GINO to minimize the four-dimensional Rosenbrock function

$$f(\mathbf{x}) = 100(x_2 - x_1^2) + (1 - x_1)^2 + 100(x_4 - x_3^2)^2 + (1 - x_3)^2$$

starting from $(-1.2, 1, -1.2, 1)$. Solve using the BFGS, Fletcher-Reeves, and the 1-step BFGS methods. Summarize the results. What is the total number of line searches for each method?

4.5.1 Solve the following minimization problem with one variable using the Lagrange multiplier method.

$$\text{Minimize } f(x) = x^3$$

subject to

$$x + 1 = 0$$

4.5.2 Consider the following minimization problem

$$\text{Minimize } f(\mathbf{x}) = -4x_1 - 6x_2 + 2x_1^2 + 2x_1x_2 + 2x_2^2$$

subject to

$$x_1 + 2x_2 = 2$$

Solve the problem by the Lagrange multiplier method.

4.5.3 Find the optimal solution of the following problem by the Lagrange multiplier method.

$$\text{Minimize } f(\mathbf{x}) = x_1^2 + 4x_2^2 + 5x_3^2$$

subject to

$$x_1 + x_2^2 + x_3 = 5$$

$$x_1 + 5x_2^2 + x_3 = 7$$

4.5.4 Solve the following minimization problem using the Lagrange multiplier method

$$\text{Minimize } f(\mathbf{x}) = x_1^2 + 2x_2^2$$

subject to

$$2x_1 + x_2 = 1$$

4.5.5 Find the radius (r) and the height (h) of a water tank of a closed cylinder shape with minimum surface area to provide a storage volume of 1000m^3. (a) Use the Lagrangian multiplier method, and (b) use the method of substitution.

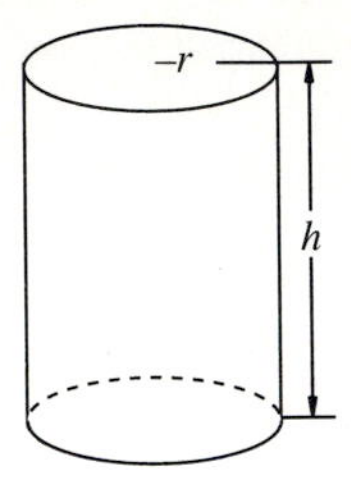

4.6.1 Determine the reduced objective and the reduced gradient for the following programming problem at point $\mathbf{x} = (0.6, 0.4)$:

$$\text{Minimize } f(\mathbf{x}) = (x_1 - 1)^2 + (x_2 - 0.8)^2$$

subject to

$$g_1(\mathbf{x}) = x_1 - x_2 \geq 0$$

$$g_2(\mathbf{x}) = -x_1^2 + x_2 \geq 0$$

$$g_3(\mathbf{x}) = x_1 + x_2 \geq 1$$

$$x_1 \geq 0, 0 \leq x_2 \leq 0.8, x_3 \geq 0$$

4.6.2 Determine the optimal solution to the following problem by the GRG method with the starting point $(x_1, x_2) = (1, 0)$.

$$\text{Minimize } (x_1 - 1)^2 + (x_2 - 2)^2$$

subject to

$$x_1 - 2x_2 \geq 0$$

$$x_1, x_2 \geq 0$$

At each step, indicate the basic and nonbasic variables, reduced objective function, reduced gradient, and the optimal step size. Define the one-dimensional searches using the steepest descent method and show the results graphically.

4.6.3 Solve the following problem using the generalized reduced gradient algorithm.

$$\text{Minimize } f(\mathbf{x}) = x_1^2 + 2x_2^2$$

subject to

$$2x_1 + x_2 \geq 1$$

starting from the point $\mathbf{x}^0 = (x_1^0, x_2^0) = (1, 1)$. At each step, indicate the basic and nonbasic variables, reduced objective function, reduced gradient, and optimal step size. Define the one-dimensional searches using the steepest descent method and show the results graphically.

4.6.4 Solve the following minimization problem using GINO or GAMS-MINOS.

$$\text{Maximize } f(\mathbf{x}) = 3x_1 - 2x_2^2$$

subject to

$$2x_1 - x_2 \geq 0$$

$$x_1 + x_2 \leq 4$$

$$x_1 \geq 0, x_2 \geq 0$$

and the initial point is at $x_1 = 1$ and $x_2 = 2$.

4.6.5 Use GAMS to solve Example 4.6.1.

CHAPTER 5

UNCERTAINTY AND RELIABILITY ANALYSIS OF HYDROSYSTEMS

The first item in discussing risk and reliability for hydrosystem design is to delineate uncertainty and other related terms such as probability and stochasticity. **Uncertainty** could simply be defined as the occurrence of events that are beyond our control. The uncertainty of a hydrosystem is an indeterminable characteristic and is beyond our rigid controls. In the design of hydrosystems, decisions must be made under various kinds of uncertainty.

5.1 REVIEW OF PROBABILITY THEORY

In this section we present a review of some of the basic principles and theories in probability and statistics that are useful to evaluate the reliability of hydrosystems. The numerical evaluation of the reliability for hydrosystems requires the use of probabilistic and statistical models.

5.1.1 Terminology

In probability theory, an **experiment** in general represents the process of observation. The total possible outcomes of the experiment is called the **sample space**. An **event** is any subset of outcomes contained in the sample space. Therefore, an event may be an empty set Ø, or subset of the sample space, or the sample space itself. Since events are sets, the appropriate operators to be used are **union**, **intersection**, and **complement**. The occurrence of event A or event B (implying the union of A and B) is denoted as $A \cup B$ while the joint occurrence of events A and B (implying the intersection of A and B) is denoted as $A \cap B$ or simply (A, B). In this chapter, the complement of event A is denoted as A'. If the two events A and B contain no common elements in the sets, then they are called **mutually exclusive** or **disjoint** which is expressed as $(A, B) = Ø$. If the event A whose occurrence depends on that of event B then this is a **conditional event** denoted by $A|B$.

Probability is a numeric measure of the likelihood of occurrence of an event. In general, probability of the occurrence of an event A can be assessed in two ways: (1) **objective** or **posterior** probabilities based on observations of the occurrence of the event; and (2) **subjective** or **prior** probabilities on the basis of experience and judgement.

5.1.2 Rules of Probability Computations

Three basic axioms of probability that are intuitively understandable: (i) $P(A) \geq 0$ (nonnegativity); (ii) $P(S) = 1$ (totality) with S being the sample space; and (iii) if A and B are mutually exclusive events then $P(A \cup B) = P(A) + P(B)$. From the first two axioms, the value of probability must lie between 0 and 1. The extension of axiom (iii) to any number of mutually exclusive events is

$$P(A_1 \cup A_2 \cup \ldots \cup A_k) = P\left(\bigcup_{i=1}^{k} A_i\right) = \sum_{i=1}^{k} P(A_i) \tag{5.1.1}$$

For two mutually exclusive events A and B, the probability of intersection $P(A \cap B) = P(A, B) = P(Ø) = 0$. After relaxing the requirement of mutual exclusiveness, the probability of the union of two events A and B can be evaluated as

$$P(A \cup B) = P(A) + P(B) - P(A,\ B) \tag{5.1.2}$$

with the following generalization for k events

$$\begin{aligned} P\left(\bigcup_{i=1}^{k} A_i\right) &= \sum_{i=1}^{k} P(A_i) - \sum_{i}^{k} \underset{\neq}{} \sum_{j}^{k} P(A_i,\ A_j) \\ &+ \sum_{i}^{k} \underset{\neq}{} \sum_{j}^{k} \underset{\neq}{} \sum_{l}^{k} P(A_i,\ A_j,\ A_l) - \ldots \\ &+ (-1)^{k+1} P(A_1,\ A_2,\ \ldots,\ A_k) \end{aligned} \tag{5.1.3}$$

If two events are said to be **independent** of each other, the occurrence of one event has no influence on the occurrence of the other. Therefore, events A and B are independent if and only if $P(A, B) = P(A)P(B)$. To generalize this principle, the probability of joint occurrence of k independent events, also referred to as **joint**

probability, is

$$P\left(\bigcap_{i=1}^{k} A_i\right) = \prod_{i=1}^{k} P(A_i) \tag{5.1.4}$$

It should be noted that the mutual exclusiveness of two events does not in general imply independence and vice versa.

Recall the conditional event mentioned above. The probability that a conditional event occurs is called **conditional probability**. The conditional probability $P(A|B)$ can be computed as

$$P(A|B) = P(A, B)/P(B) \tag{5.1.5}$$

where $P(A|B)$ is the probability of occurrence of event A given that event B occurred. In other words, $P(A|B)$ represents our reevaluation of the probability of A in light of the information that event B has occurred. To generalize Eq. (5.1.5), the probability of the joint occurrence of k dependent events can be evaluated as

$$P\left(\bigcap_{i=1}^{k} A_i\right) = P(A_1) P\left(A_2|A_1\right) P\left(A_3|A_2,\ A_1\right) \ldots P\left(A_k|A_{k-1}, \ldots,\ A_1\right) \tag{5.1.6}$$

Sometimes, the probability that event A occurs cannot be determined directly or easily. However, the event A generally occurs along with other **attributes**, C_i, which are other events that cause event A to occur. Referring to Fig. 5.1.1 event A could occur jointly with k mutually exclusive and **collectively exhaustive** attributes C_i, $i = 1, 2, \ldots, k$, where collectively exhaustive refers to the union of all possible events in a sample space. The probability of occurrence of event A, regardless of the cause of the attributes, can be computed as

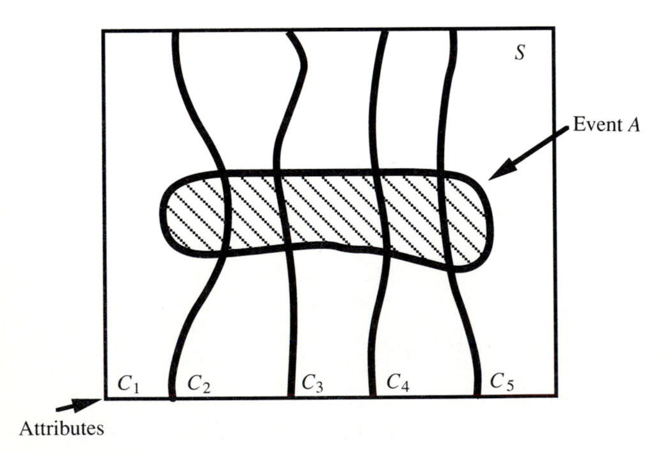

FIGURE 5.1.1
Venn diagram showing event A with attributes.

$$P(A) = \sum_{i=1}^{k} P(A,\ C_i) = \sum_{i=1}^{k} P\left(A|C_i\right) P(C_i) \tag{5.1.7}$$

which defines the **total probability theorem.**

The total probability theorem, states that the occurrence of event A may be affected by a number of attributes C_i, $i = 1, 2, \ldots, k$. In some situations $P(A|C_i)$ is known and one would like to determine the probability that a particular attribute C_i is responsible for the occurrence of event A, that is, $P(C_i|A)$ is required. Based on the definition of conditional probability, Eq. (5.1.5), and the total probability theorem, Eq. (5.1.7), $P(C_i|A)$ can be computed as

$$P\left(C_i|A\right) = \frac{P(C_i, A)}{P(A)} = \frac{P\left(A|C_i\right) P(C_i)}{\sum_{i=1}^{k} P\left(A|C_i\right) P(C_i)} \tag{5.1.8}$$

Equation (5.1.8) is called **Bayes theorem** in which $P(C_i)$ is the **prior probability** representing the initial belief of the probability on the occurence of attribute C_i, $P(A|C_i)$ is the **likelihood function** and $P(C_i|A)$ is the **posterior probability** representing our new evaluation of C_i being responsible in light of the occurrence of event A. Bayes theorem can be used to update and to revise the calculated probability as more information becomes available.

5.1.3 Random Variables and Their Distributions

In analyzing statistical characteristics of hydrosystem performance, many events of interest can be defined by the related random variables. A **random variable** is a real-valued function that is defined on the sample space. A rather standard convention in statistical literature is that a random variable is denoted by the uppercase letter while the lowercase letter represents the **realization**, or actual value, of the corresponding random variable. Following this convention, for instance, Q can be used to represent flow magnitude, a random variable, while q represents the possible value of Q. A random variable can be discrete or continuous. There are many examples of discrete random variables in hydrosystems engineering. This section only considers univariate random variables. Multivariate random variable cases can be found elsewhere (Blank, 1980; Devore, 1987).

The **cumulative distribution function** (CDF), $F(x)$, or simply **distribution function (DF)** of a random variable X is defined as

$$F(x) = P(X \le x) \tag{5.1.9}$$

$F(x)$ is cumulative as its argument or realization, x, increases. Further, as x approaches the lower bound of the random variable X the value of $F(x)$ approaches zero; on the other hand, the value of $F(x)$ approaches one as its argument approaches the upper bound of the random variable X.

For a discrete random variable X, the **probability mass function** (PMF) of X is defined as

$$p(x) = P(X = x) \tag{5.1.10}$$

where $p(x)$ is the **probability mass** which is the probability at a discrete point $X = x$. The PMF of any discrete random variable must satisfy two conditions: (1) $p(x_i) \geq 0$ for all x_i's and (2) $\sum_{\text{all } i} p(x_i) = 1$. The PMF of a discrete random variable and its associated CDF are shown in Figs. 5.1.2*a* and *b*. The CDF of a discrete random variable X appears as a staircase.

For a continuous random variable, the **probability density function** (PDF) is defined as

$$f(x) = \frac{dF(x)}{dx} \tag{5.1.11}$$

in which $F(x)$ is the CDF of X as defined in Eq. (5.1.9). The PDF of a continuous random variable $f(x)$ is the slope of its CDF. Graphical representation of a PDF and CDF for continuous random variables is shown in Figures 5.1.2*c* and *d*. Similar to the discrete case, any PDF of a continuous random variable must satisfy two conditions: (1) $f(x) \geq 0$ and (2) $\int_{-\infty}^{\infty} f(x)\, dx = 1$.

Given the PDF of a continuous random variable X, or the PMF of a discrete random variable its CDF can be obtained using

$$F(x) = \int_{-\infty}^{x} f(x)\, dx \text{ for continuous random variables} \tag{5.2.12a}$$

Discrete Random Variables

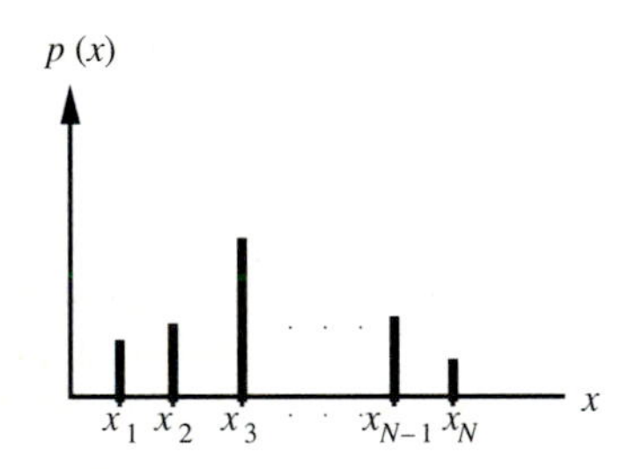

(a) Probability mass function (PMF) of a discrete random variable.

Continuous Random Variables

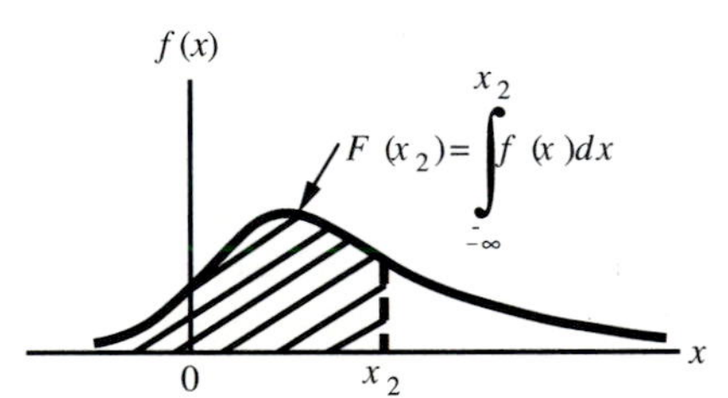

(c) Probability density function (PDF).

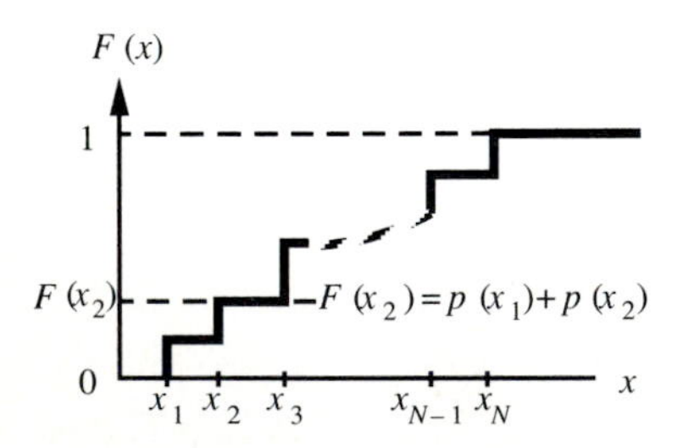

(b) Cumulative distribution function (CDF) of a discrete random variable.

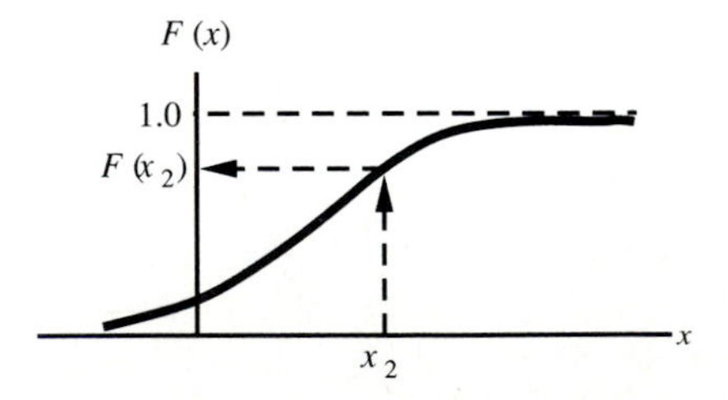

(d) Cumulative distribution function (CDF).

FIGURE 5.1.2
Probability mass function and cumulative distribution of discrete and continuous random variables.

and

$$F(x_n) = \sum_{1 \le i \le n} p(x_i) \text{ for discrete random variables} \tag{5.1.12b}$$

The probability for a continuous random variable to take on a particular value is zero whereas this is not the case for discrete random variables.

5.1.4 Statistical Properties of Random Variables

In statistics the term **population** represents the complete assemblage of all the values representative of a particular random process. A **sample** is any subset of the population.

Descriptors that are commonly used to describe statistical properties of a random variable can be categorized into three types: (1) descriptors showing the central tendency; (2) descriptors showing the dispersion about a central value; and (3) descriptors showing the asymmetry of a distribution. The frequently used descriptors in these three categories are related to the statistical moments of the random variable. The **expected value** of $(X - x_0)^r$ is the rth moment of a random variable X about any reference point $X = x_0$. Mathematically the expected value, $E[(X - x_0)^r]$, for the continuous case is defined as,

$$E\left[(X - x_0)^r\right] = \int_{-\infty}^{\infty} (x - x_0)^r f(x)\, dx \tag{5.1.13a}$$

while for the discrete case,

$$E\left[(X - x_0)^r\right] = \sum_{i=1}^{N} (x_i - x_0)^r p(x_i) \tag{5.1.13b}$$

where $E[\ ]$ is a **statistical expectation operator**. In practice, the first three moments are used to describe the central tendency, variability, and asymmetry of the distribution of a random variable. Without losing generality the following discussion considers that random variables are continuous.

For measuring the central tendency, the **expectation** of a random variable X defined as

$$E[X] = \mu = \int_{-\infty}^{\infty} x f(x)\, dx \tag{5.1.14}$$

frequently is used. This expectation is known as the **mean** of a random variable. Other descriptors or statistical properties for central tendency of a random variable are listed in Table 5.1.1.

Some useful operational properties of expectation:

1. Expectation of the sum of random variables is equal to the sum of the expectation of individual random variables.

$$E\left(\sum_{i=1}^{k} a_i X_i\right) = \sum_{i=1}^{k} a_i E[X_i] \tag{5.1.15a}$$

TABLE 5.1.1
Commonly used statistical properties of a random variable

	Statistical properties
Population	**Sample estimators**
1. *Central Tendency*	
Arithmetic Mean	
$\mu = E(X) = \int_{-\infty}^{\infty} x f(x)\, dx$	$\overline{X} = \frac{1}{n} \sum_{i=1}^{n} X_i$
Median	
x_{md} such that $F(x_{md}) = 0.5$	50th percentile value of data
Geometric Mean	$\left[\prod_{i=1}^{n} X_i \right]^{1/n}$
2. *Variability*	
Variance	
$\sigma^2 = E[(X - \mu)^2]$	$S_x^2 = \frac{1}{n-1} \sum_{i=1}^{n} \left(X_i - \overline{X} \right)^2$
Standard Deviation	
$\sigma = \left[E(X - \mu)^2 \right]^{1/2}$	$S = \left[\frac{1}{n-1} \sum_{i=1}^{n} \left(X_i - \overline{X} \right)^2 \right]^{1/2}$
Coefficient of Variation	
$\Omega = \sigma / \mu$	$C_v = S / \overline{X}$
3. *Symmetry*	
Coefficient of Skewness	
$\gamma = \frac{E(X - \mu)^3}{\sigma^3}$	$G = \frac{n \sum_{i=1}^{n} \left(X_i - \overline{X} \right)^3}{(n-1)(n-2) S^3}$
4. *Correlation*	
Correlation coefficient	
$\rho = \frac{\text{cov}(X, Y)}{\sigma_x \sigma_y}$	$R = \frac{\Sigma \left(X_i - \overline{X} \right) \left(Y_i - \overline{Y} \right)}{\sqrt{\Sigma \left(X_i - \overline{X} \right)^2} \sqrt{\Sigma \left(Y_i - \overline{Y} \right)^2}}$

2. If $X_1, X_2, \ldots, X_k$ are independent random variables, then

$$E \left[\prod_{i=1}^{k} X_i \right] = \prod_{i=1}^{k} E[X_i] \tag{5.1.15b}$$

Two types of moments are commonly used: moments about the origin where $x_o = 0$ and the central moments where $x_o = \mu$. The rth **central moment** is denoted as $\mu_r = E[(X - \mu)^r]$ while the rth moment about the origin is denoted as $\mu'_r = E[X^r]$. The relationships between central moments and moments about the origin of any order

r are

$$\mu_r = \sum_{i=0}^{r} (-1)^i \, {}_rC_i \mu^i \mu'_{r-i} \tag{5.1.16a}$$

$$\mu'_r = \sum_{i=0}^{r} {}_rC_i \mu^i \mu_{r-i} \tag{5.1.16b}$$

where the binomial coefficient ${}_rC_i = r!/[i!(r-i)!]$, μ^i is the mean to the ith power, μ'_{r-i} is the $(r-i)$th order moment about the origin. Equation (5.1.16a) is used to compute the central moments from moments about the origin, while Eq. (5.1.16b) is used to compute the moment about the origin from the central moments.

For measuring the variability, the **variance** of a continuous random variable is defined as

$$\text{Var}[X] = \sigma^2 = E\left[(X-\mu)^2\right] = \int_{-\infty}^{\infty} (x-\mu)^2 f(x)\, dx \tag{5.1.17}$$

which is a second-order central moment. The positive square root of the variance σ^2 is called the **standard deviation**, σ, which is often used as the measure of the degree of uncertainty associated with a random variable. A smaller standard deviation refers to a random variable with less uncertainty. The standard deviation has the same units as the random variable. To compare the degree of uncertainty of two random variables of different units, a nondimensionalized measure $\Omega = \sigma/\mu$, called the **coefficient of variation**, is useful. The following are several important properties of variance:

1.

$$\text{Var}[a] = 0 \tag{5.1.18a}$$

2.

$$\text{Var}[X] = E\left[X^2\right] - E^2[X] \tag{5.1.18b}$$

3.

$$\text{Var}[aX] = a^2 \, \text{Var}[X] \tag{5.1.18c}$$

4. If all random variables, X's, are independent, then

$$\text{Var}\left[\sum_{i=1}^{k} a_i X_i\right] = \sum_{i=1}^{k} a_i^2 \sigma_i^2 \tag{5.1.18d}$$

where a_i is a constant and σ_i is the standard deviation of random variable X_i.

To measure the asymmetry of the PDF of a random variable, the **skew coefficient** γ is used, which is defined as

$$\gamma = E\left[(X-\mu)^3\right] \sigma^3 \tag{5.1.19}$$

The skew coefficient is dimensionless and is related to the third central moment.

The sign of the skew coefficient indicates the extent of symmetry of the probability distribution about its mean. If $\gamma = 0$, the distribution is symmetric about its mean; $\gamma > 0$, the distribution has a long tail to the right; $\gamma < 0$, the distribution has a long tail to the left. Figure 5.1.3 is used to illustrate the shapes of a probability distribution with different skew coefficients and the relative position of the mean μ, the median x_{md}, and the mode x_{mo} are shown in Fig. 5.1.3. The **mode**, x_{mo}, is the value the random variable at the peak of the probability density function.

Statistical moments higher than three are rarely used in practical application because their accuracies decrease rapidly when estimated from a limited sample size. Equations used to compute the sample estimates of the above statistical moments are given in Table 5.1.1.

When two dependent random variables are considered, the degree of *linear* dependence between the two can be measured by the **correlation coefficient** $\rho(X, Y)$ which is defined as

$$\rho(X, Y) = \text{Cov}[X, Y]/\sigma_X \sigma_Y \tag{5.1.20}$$

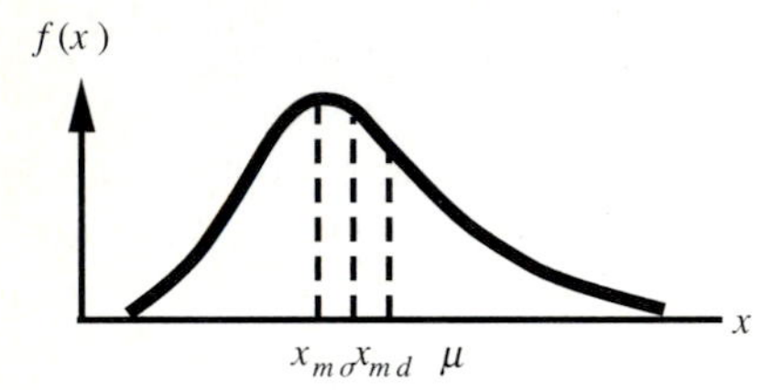

(a) Distribution with positive skew, $\gamma > 0$

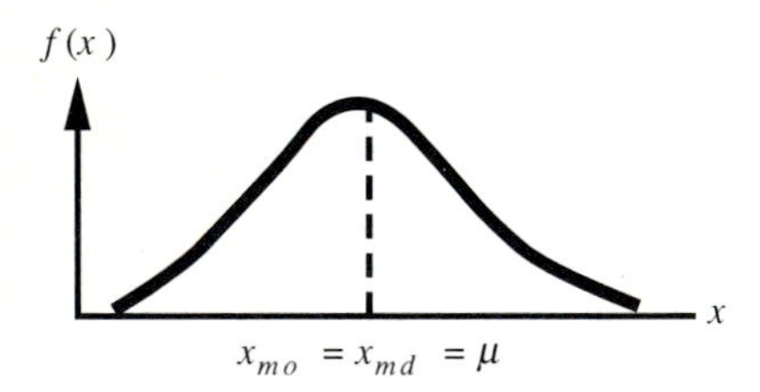

(b) Symmetric distribution, $\gamma = 0$

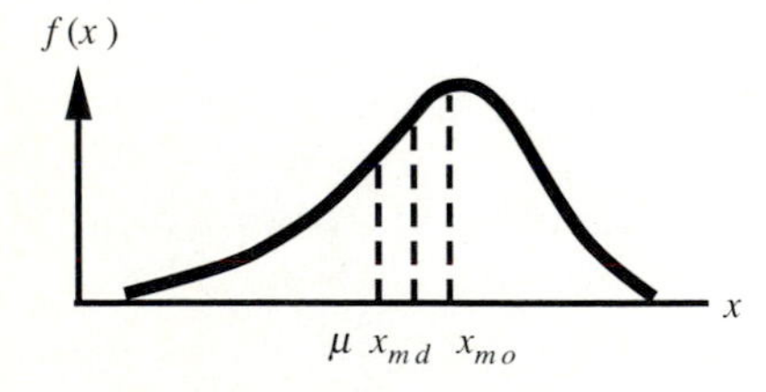

(c) Distribution with negative skew, $\gamma < 0$

FIGURE 5.1.3
Shapes of distribution with different signs of skew coefficient.

where Cov$[X, Y]$ is the **covariance** between random variables X and Y. As an example the correlation coefficient defines the reasonableness of the assumption that values of x and y define a straight line. The covariance is defined as the expected value of the products of $(X - \mu_X)$ and $(Y - \mu_Y)$, which is defined as

$$\text{Cov}[X, Y] = E[(X - \mu_X)(Y - \mu_Y)] = E[XY] - \mu_X \mu_Y \tag{5.1.21a}$$

or

$$\text{Cov}(X, Y) = \frac{1}{N} \sum_{i=1}^{N} (x_i - \bar{x})(y_i - \bar{y}) \tag{5.1.21b}$$

for N pairs of data. The covariance is a measure of tendancy for two variables to vary together. This measure can be zero, negative, or positive referring respectively to uncorrelated variables, negatively correlated variables, or positively correlated variables.

The correlation coefficient must be greater than or equal to -1 and less than or equal to +1, that is, $-1 \leq \rho(X, Y) \leq +1$. The case where $\rho(X, Y) = +1$ means that there is a perfect positive relation between two variables (i.e., all points lie in a straight line) whereas $\rho(X, Y) = -1$ refers to a perfect correlation in opposite directions (i.e., one variable increases as the other decreases). When $\rho(X, Y) = 0$ there is no linear correlation. Figure 5.1.4 illustrates values of correlation. If the two random variables X and Y are independent, then $\rho(X, Y) = \text{Cov}[X, Y] = 0$. However, the reverse is not necessarily true (referring to Fig. 5.1.4*d*). Considering the correlation among the random variables involved, Eq. (5.1.18*d*) can be generalized as

$$\text{Var}\left[\sum_{i=1}^{k} a_i X_i\right] = \sum_{i=1}^{k} a_i^2 \sigma_i^2 + 2 \sum_{i}^{k} \sum_{< \; j}^{k} a_i a_j \text{Cov}\left[X_i, X_j\right] \tag{5.1.22}$$

Example 5.1.1. Consider the mass balance of a surface reservoir over a monthly period where m refers to the mth month. The end-of-month storage S_{m+1} can be computed

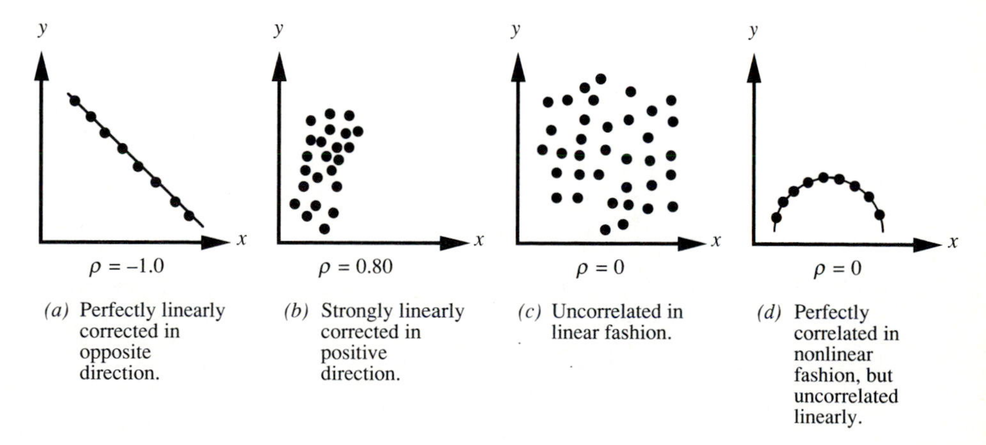

FIGURE 5.1.4
Some examples of the correlation coefficient (adapted from Harr, 1987).

using the conservation of mass

$$ST_{m+1} = ST_m + PP_m + QF_m - EV_m - R_m$$

in which ST_m = initial storage volume in month m, PP_m = precipitation on the reservoir surface during month m, QF_m = surface runoff inflow during month m, EV_m = total monthly evaporation amount during month m, and R_m = controlled monthly release from the reservoir for various purposes. At the beginning of the month, the initial storage volume and the release are known. Further, the monthly total amount of precipitation, surface runoff inflow, and evaporation are uncertain and are assumed to be independent random variables. The means and standard deviations of PP_m, QF_m and EV_m, from historical data of month m are estimated as

$$E(PP_m) = 1 \text{ KAF}, \quad E(QF_m) = 8 \text{ KAF}, \quad E(EV_m) = 3 \text{ KAF},$$

$$\sigma(PP_m) = 0.5 \text{ KAF}, \quad \sigma(QF_m) = 2 \text{ KAF}, \quad \sigma(EV_m) = 1 \text{ KAF}$$

where KAF refers to 1000 acre-feet. Determine the mean and standard deviation of storage volume in the reservoir by the end of the month if the initial storage volume is 20 KAF and designated release for the month is 10 KAF.

Solution. From Eq. (5.1.15*a*), the mean of the end-of-month storage volumes in the reservoir can be determined as

$$E(ST_{m+1}) = ST_m + E(PP_m) + E(QF_m) - R_m$$

$$= 20 + 1 + 8 - 3 - 10 = 16 \text{ KAF}$$

From Eq. (5.1.18*c*), the variance of the end-of-month storage volume in the reservoir can be obtained as

$$\text{Var}(ST_{m+1}) = \text{Var}(PP_m) + \text{Var}(QF_m) + \text{Var}(EV_m)$$

$$= (0.5)^2 + (2)^2 + (1)^2 = 5.25(\text{KAF})^2$$

The standard deviation of ST_{m+1} then is

$$\sigma(ST_{m+1}) = \sqrt{5.25} = 2.29 \text{ KAF}$$

Example 5.1.2. Perhaps the assumption of independence of PP_m, QF_m, and EV_m in Example 5.1.1 may not be quite true in reality. After examining the historical data closely there exist correlations among the three random variables. Analysis of data reveals that $\rho(PP_m, QF_m) = 0.8, \rho(PP_m, EV_m) = -0.4, \rho(QF_m, EV_m) = -0.3$. Recalculate the standard deviation of the end-of-month storage volume.

Solution. By Eq. (5.1.22), the variance of storage volume in the reservoir at the end of the month can be calculated as

$$\begin{aligned}\text{Var}(ST_{m+1}) &= \text{Var}(PP_m) + \text{Var}(QF_m) + \text{Var}(EV_m) + 2\,\text{Cov}(PP_m, QF_m) \\ &\quad - 2\,\text{Cov}(PP_m, EV_m) - 2\,\text{Cov}(QF_m, EV_m) \\ &= \text{Var}(PP_m) + \text{Var}(QF_m) + \text{Var}(EV_m) + 2\rho(PP_m, QF_m)\sigma(PP_m)\sigma(QF_m) \\ &\quad - 2\rho(PP_m, EV_m)\sigma(PP_m)\sigma(EV_m) - 2\,\rho(QF_m, EV_m)\sigma(QF_m)\sigma(EV_m)\end{aligned}$$

$$= (0.5)^2 + (2)^2 + (1)^2 + 2(0.8)(0.5)(2)$$
$$- 2(-0.4)(0.5)(1) - 2(-0.3)(2)(1)$$
$$= 8.45 \text{ (KAF)}^2$$

The corresponding standard deviation of end-of-month storage volume is

$$\sigma(ST_{m+1}) = \sqrt{8.45} = 2.91 \text{ KAF}$$

In Example 5.1.1 the standard deviation was 2.29 KAF. Obviously the assumption of independence resulted in a smaller standard deviation.

5.2 COMMONLY USED PROBABILITY DISTRIBUTIONS

In the reliability analysis of hydrosystems, several probability distributions are frequently used. Based on the nature of the random variable, probability distributions can be classified into discrete distributions and continuous distributions. Two types of discrete distributions are commonly used in reliability analysis: binomial distribution and Poisson distribution. For the continuous random variables, there are several PDFs that are frequently used in reliability analysis. They include the normal, lognormal, Gamma, Weibull, and exponential distributions. Other distributions such as the beta and extremal distributions are sometimes used.

5.2.1 Binomial Distribution

The binomial distribution is applicable to the random processes with only two possible outcomes. The state of components or subsystems in many hydrosystems can be classified as either functioning or failed which is a typical example of binary outcomes. Consider a system involving a total of n independent components each of which has two possible outcomes, say, functioning or failed. For each component, the probability of being functional is p. Then, the probability of having x functioning components in the system can be computed as

$$p(x) = {}_nC_x p^x q^{n-x}, x = 0, 1, 2, \ldots, n \tag{5.2.1}$$

where $q = 1 - p$ and ${}_nC_x$ is a binomial coefficient. A random variable X having a binomial distribution with parameters n and p has the expectation $E(X) = np$ and variance $\text{Var}(X) = npq$. The shape of the PMF of a binomial random variable depends on the values of p and q. The PMF is positively skewed if $p < q$; symmetric if $p = q = 0.5$; and negatively skewed if $p > q$.

Example 5.2.1. The operator of a boat dock has decided to build a new docking facility along a river. In an economic analysis of the situation he decided to have the facility designed to withstand a flood up to 75,000 cubic feet per second (cfs). Furthermore, he has determined that if one flood greater than this occurs in a 5-year period, he can repair his facility and break even on its operation during the 5-year period. If more than one flow in excess of 75,000 cfs occurs, he will lose money. If the annual probability of exceeding 75,000 cfs is 0.15, what is the probability the operator will not lose money?

Solution. Let X be the random variable representing the number of occurrences of floods exceeding 75,000 cfs in the 5-year period. Each year can be treated as a trial in that a flood exceeding 75,000 cfs can either occur or not occur. Therefore, the outcome in each year is binary. The 5-year period can be considered as having 5 trials. The random variable X defined above in the problem has a binomial distribution with parameters $p = 0.15$ and $n = 5$. The operator will not lose money if at most one flood in excess of 75,000 cfs occurs during the 5-year period. The probability of having at most one such flood can be calculated as

$$P \text{ (At most one flood is in excess of 75,000 cfs in 5 years)}$$

$$= P(X \leq 1)$$

$$= P(X = 0) + P(X = 1)$$

$$= {}_5C_0(0.15)^0(1 - 0.15)^5 + {}_5C_1(0.15)^1(1 - 0.15)^4$$

$$= 0.4437 + 0.3915 = 0.8352$$

5.2.2 Poisson Distribution

When $n \to \infty$, and $p \to 0$ while np = constant, the binomial distribution becomes a Poisson distribution with the following PMF

$$p(x) = e^{-\lambda}\lambda^x/x!, \quad x = 0, 1, 2, \ldots \tag{5.2.2}$$

where the parameter $\lambda > 0$ is the mean of the discrete random variable X having a Poisson distribution. The Poisson distribution has been applied widely in modeling the number of occurrences of events within a specified time or space interval. Equation (5.2.2) can be modified to

$$p(x) = e^{-vt}(vt)^x/x!, \quad x = 0, 1, 2, \ldots \tag{5.2.3}$$

in which the parameter v can be interpreted as the average rate of occurrence of an event in time interval $(0, t)$.

Example 5.2.2. Reassess the probability in Example 5.2.1 using the Poisson distribution.

Solution. In Example 5.2.1, it was assumed that a flood in excess of 75,000 cfs cannot occur more than once in each year. Discard this implicit assumption and assume that more than one flood is possible for each year regardless of how small the likelihood. The random variable X follows a Poisson distribution with the parameter $v = np = 5(0.15) = 0.75$. The value 0.75 represents the expected (or average) number of occurrences of a flood in excess of 75,000 cfs over a 5-year period. Therefore, the probability that at most one such flood occurs in a 5-year period can be calculated as

$$P(X \leq 1) = P(X = 0) + P(X = 1)$$

$$= e^{-0.75}(0.75)^0/0! + e^{-0.75}(0.75)^1/1!$$

$$= 0.4724 + 0.3543 = 0.8266$$

Comparing with the probability value, 0.8352, obtained in the previous example, the difference in probability values using the two distributions is less than 1 percent for this example. The difference in probabilities calculated by using a binomial distribution and Poisson distribution would be negligible if the value of p is small. However, the assumption implicitly built into the binomial distribution that only one event occurs in each n would make one prefer to use the Poisson distribution in risk evaluation for most hydrosystem problems.

5.2.3 Normal Distribution

The normal distribution is a well-known probability distribution, also called the Gaussian distribution. Two parameters are involved in a normal distribution, that is, the mean and the variance. A normal random variable having a mean μ and a variance σ^2 is herein denoted as $X \sim N(\mu, \sigma^2)$ with a PDF of

$$f(x) = \frac{1}{\sqrt{2\pi}\sigma} \exp\left[-\frac{1}{2}\left(\frac{x-\mu}{\sigma}\right)^2\right], \qquad \text{for } -\infty < x < \infty \tag{5.2.4}$$

A normal distribution is bell-shaped and symmetric with respect to $x = \mu$. Therefore, the skew coefficient for a normal random variable is zero. A random variable Y that is a linear function of a normal random variable X is also normal. That is, if $X \sim N(\mu, \sigma^2)$ and $Y = aX + b$ then $Y \sim N(a\mu + b, a^2\sigma^2)$. An extension of this theorem is that the sum of normal random variables (independent or dependent) is also a normal random variable with mean and variance that can be computed by Eqs. (5.1.15*a*) and (5.1.22), respectively.

Probability computations for normal random variables are made by first transforming to its standardized variate Z as

$$Z = (X - \mu)/\sigma \tag{5.2.5}$$

in which Z has a zero mean and unit variance. Since Z is a linear function of the random variable X, then, Z is also normally distributed. The PDF of Z, called the **standard normal distribution**, can be expressed as

$$\phi(z) = \frac{1}{\sqrt{2\pi}} \exp\left[-\frac{z^2}{2}\right], \qquad \text{for } -\infty < z < \infty \tag{5.2.6}$$

Tables of the CDF of Z such as Table 5.2.1, can be found in statistics textbooks (Haan, 1977; Blank, 1980; Devore, 1987). Computations of probability for $X \sim N(\mu, \sigma^2)$ can be performed using

$$\begin{aligned} P(X \le x) &= P\left[\frac{X-\mu}{\sigma} \le \frac{x-\mu}{\sigma}\right] \\ &= P[Z \le z] = \Phi(z) \end{aligned} \tag{5.2.7}$$

where $\Phi(z)$ is the CDF of the standard normal random variable Z defined as

$$\Phi(z) = \int_{-\infty}^{z} \phi(s)ds \tag{5.2.8}$$

TABLE 5.2.1
Standard normal curve areas (Devore, 1987) $\Phi(z) = P(Z \leq z)$.

z	0.00	0.01	0.02	0.03	0.04	0.05	0.06	0.07	0.08	0.09
−3.4	0.0003	0.0003	0.0003	0.0003	0.0003	0.0003	0.0003	0.0003	0.0003	0.0002
−3.3	0.0005	0.0005	0.0005	0.0004	0.0004	0.0004	0.0004	0.0004	0.0004	0.0003
−3.2	0.0007	0.0007	0.0006	0.0006	0.0006	0.0006	0.0006	0.0005	0.0005	0.0005
−3.1	0.0010	0.0009	0.0009	0.0009	0.0008	0.0008	0.0008	0.0008	0.0007	0.0007
−3.0	0.0013	0.0013	0.0013	0.0012	0.0012	0.0011	0.0011	0.0011	0.0010	0.0010
−2.9	0.0019	0.0018	0.0017	0.0017	0.0016	0.0016	0.0015	0.0015	0.0014	0.0014
−2.8	0.0026	0.0025	0.0024	0.0023	0.0023	0.0022	0.0021	0.0021	0.0020	0.0019
−2.7	0.0035	0.0034	0.0033	0.0032	0.0031	0.0030	0.0029	0.0028	0.0027	0.0026
−2.6	0.0047	0.0045	0.0044	0.0043	0.0041	0.0040	0.0039	0.0038	0.0037	0.0036
−2.5	0.0062	0.0060	0.0059	0.0057	0.0055	0.0054	0.0052	0.0051	0.0049	0.0048
−2.4	0.0082	0.0080	0.0078	0.0075	0.0073	0.0071	0.0069	0.0068	0.0066	0.0064
−2.3	0.0107	0.0104	0.0102	0.0099	0.0096	0.0094	0.0091	0.0089	0.0087	0.0084
−2.2	0.0139	0.0136	0.0132	0.0129	0.0125	0.0122	0.0119	0.0116	0.0113	0.0110
−2.1	0.0179	0.0174	0.0170	0.0166	0.0162	0.0158	0.0154	0.0150	0.0146	0.0143
−2.0	0.0228	0.0222	0.0217	0.0212	0.0207	0.0202	0.0197	0.0192	0.0188	0.0183
−1.9	0.0287	0.0281	0.0274	0.0268	0.0262	0.0256	0.0250	0.0244	0.0239	0.0233
−1.8	0.0359	0.0352	0.0344	0.0336	0.0329	0.0322	0.0314	0.0307	0.0301	0.0294
−1.7	0.0446	0.0436	0.0427	0.0418	0.0409	0.0401	0.0392	0.0384	0.0375	0.0367
−1.6	0.0548	0.0537	0.0526	0.0516	0.0505	0.0495	0.0485	0.0475	0.0465	0.0455
−1.5	0.0668	0.0655	0.0643	0.0630	0.0618	0.0606	0.0594	0.0582	0.0571	0.0559
−1.4	0.0808	0.0793	0.0778	0.0764	0.0749	0.0735	0.0722	0.0708	0.0694	0.0681
−1.3	0.0968	0.0951	0.0934	0.0918	0.0901	0.0885	0.0869	0.0853	0.0838	0.0823
−1.2	0.1151	0.1131	0.1112	0.1093	0.1075	0.1056	0.1038	0.1020	0.1003	0.0985
−1.1	0.1357	0.1335	0.1314	0.1292	0.1271	0.1251	0.1230	0.1210	0.1190	0.1170
−1.0	0.1587	0.1562	0.1539	0.1515	0.1492	0.1469	0.1446	0.1423	0.1401	0.1379
−0.9	0.1841	0.1814	0.1788	0.1762	0.1736	0.1711	0.1685	0.1660	0.1635	0.1611
−0.8	0.2119	0.2090	0.2061	0.2033	0.2005	0.1977	0.1949	0.1922	0.1894	0.1867
−0.7	0.2420	0.2389	0.2358	0.2327	0.2296	0.2266	0.2236	0.2206	0.2177	0.2148
−0.6	0.2743	0.2709	0.2676	0.2643	0.2611	0.2578	0.2546	0.2514	0.2483	0.2451
−0.5	0.3085	0.3050	0.3015	0.2981	0.2946	0.2912	0.2877	0.2843	0.2810	0.2776
−0.4	0.3446	0.3409	0.3372	0.3336	0.3300	0.3264	0.3228	0.3192	0.3156	0.3121
−0.3	0.3821	0.3783	0.3745	0.3707	0.3669	0.3632	0.3594	0.3557	0.3520	0.3483
−0.2	0.4207	0.4168	0.4129	0.4090	0.4052	0.4013	0.3974	0.3936	0.3897	0.3859
−0.1	0.4602	0.4562	0.4522	0.4483	0.4443	0.4404	0.4364	0.4325	0.4286	0.4247
−0.0	0.5000	0.4960	0.4920	0.4880	0.4840	0.4801	0.4761	0.4721	0.4681	0.4641

5.2.4 Lognormal Distribution

The lognormal distribution is a commonly used continuous distribution when random variables cannot be negative. A random variable X is said to be lognormally distributed if its logarithmic transform $Y = \ln(X)$ has a normal distribution with mean $\mu_{\ln X}$ and variance $\sigma^2_{\ln X}$. The PDF of the lognormal random variable is

$$f(X) = \frac{1}{\sqrt{2\pi} X \sigma_{\ln X}} \exp\left[-\frac{1}{2}\left(\frac{\ln X - \mu_{\ln X}}{\sigma_{\ln X}}\right)^2\right], \qquad \text{for } 0 < X < \infty \qquad (5.2.9)$$

TABLE 5.2.1 continued

z	0.00	0.01	0.02	0.03	0.04	0.05	0.06	0.07	0.08	0.09
0.0	0.5000	0.5040	0.5080	0.5120	0.5160	0.5199	0.5239	0.5279	0.5319	0.5359
0.1	0.5398	0.5438	0.5478	0.5517	0.5557	0.5596	0.5636	0.5675	0.5714	0.5753
0.2	0.5793	0.5832	0.5871	0.5910	0.5948	0.5987	0.6026	0.6064	0.6103	0.6141
0.3	0.6179	0.6217	0.6255	0.6293	0.6331	0.6368	0.6406	0.6443	0.6480	0.6517
0.4	0.6554	0.6591	0.6628	0.6664	0.6700	0.6736	0.6772	0.6808	0.6844	0.6879
0.5	0.6915	0.6950	0.6985	0.7019	0.7054	0.7088	0.7123	0.7157	0.7190	0.7224
0.6	0.7257	0.7291	0.7324	0.7357	0.7389	0.7422	0.7454	0.7486	0.7517	0.7549
0.7	0.7580	0.7611	0.7642	0.7673	0.7704	0.7734	0.7764	0.7794	0.7823	0.7852
0.8	0.7881	0.7910	0.7939	0.7967	0.7995	0.8023	0.8051	0.8078	0.8106	0.8133
0.9	0.8159	0.8186	0.8212	0.8238	0.8264	0.8289	0.8315	0.8340	0.8365	0.8389
1.0	0.8413	0.8438	0.8461	0.8485	0.8508	0.8531	0.8554	0.8577	0.8599	0.8621
1.1	0.8643	0.8665	0.8686	0.8708	0.8729	0.8749	0.8770	0.8790	0.8810	0.8830
1.2	0.8849	0.8869	0.8888	0.8907	0.8925	0.8944	0.8962	0.8980	0.8997	0.9015
1.3	0.9032	0.9049	0.9066	0.9082	0.9099	0.9115	0.9131	0.9147	0.9162	0.9177
1.4	0.9192	0.9207	0.9222	0.9236	0.9251	0.9265	0.9278	0.9292	0.9306	0.9319
1.5	0.9332	0.9345	0.9357	0.9370	0.9382	0.9394	0.9406	0.9418	0.9429	0.9441
1.6	0.9452	0.9463	0.9474	0.9484	0.9495	0.9505	0.9515	0.9525	0.9535	0.9545
1.7	0.9554	0.9564	0.9573	0.9582	0.9591	0.9599	0.9608	0.9616	0.9625	0.9633
1.8	0.9641	0.9649	0.9656	0.9664	0.9671	0.9678	0.9686	0.9693	0.9699	0.9706
1.9	0.9713	0.9719	0.9726	0.9732	0.9738	0.9744	0.9750	0.9756	0.9761	0.9767
2.0	0.9772	0.9778	0.9783	0.9788	0.9793	0.9798	0.9803	0.9808	0.9812	0.9817
2.1	0.9821	0.9826	0.9830	0.9834	0.9838	0.9842	0.9846	0.9850	0.9854	0.9857
2.2	0.9861	0.9864	0.9868	0.9871	0.9875	0.9878	0.9881	0.9884	0.9887	0.9890
2.3	0.9893	0.9896	0.9898	0.9901	0.9904	0.9906	0.9909	0.9911	0.9913	0.9916
2.4	0.9918	0.9920	0.9922	0.9925	0.9927	0.9929	0.9931	0.9932	0.9934	0.9936
2.5	0.9938	0.9940	0.9941	0.9943	0.9945	0.9946	0.9948	0.9949	0.9951	0.9952
2.6	0.9953	0.9955	0.9956	0.9957	0.9959	0.9960	0.9961	0.9962	0.9963	0.9964
2.7	0.9965	0.9966	0.9967	0.9968	0.9969	0.9970	0.9971	0.9972	0.9973	0.9974
2.8	0.9974	0.9975	0.9976	0.9977	0.9977	0.9978	0.9979	0.9979	0.9980	0.9981
2.9	0.9981	0.9982	0.9982	0.9983	0.9984	0.9984	0.9985	0.9985	0.9986	0.9986
3.0	0.9987	0.9987	0.9987	0.9988	0.9988	0.9989	0.9989	0.9989	0.9990	0.9990
3.1	0.9990	0.9991	0.9991	0.9991	0.9992	0.9992	0.9992	0.9992	0.9993	0.9993
3.2	0.9993	0.9993	0.9994	0.9994	0.9994	0.9994	0.9994	0.9995	0.9995	0.9995
3.3	0.9995	0.9995	0.9995	0.9996	0.9996	0.9996	0.9996	0.9996	0.9996	0.9997
3.4	0.9997	0.9997	0.9997	0.9997	0.9997	0.9997	0.9997	0.9997	0.9997	0.9998

which can be derived from the normal PDF, that is, Eq. (5.2.4). Statistical properties of a lognormal random variable of the original scale can be computed from those of the log-transformed variable. To compute the statistical moments of X from those of $\ln X$, the following formulas are useful.

$$\mu_X = \exp(\mu_{\ln X} + \sigma^2_{\ln X}/2) \tag{5.2.10a}$$

$$\sigma^2_X = \mu^2_X \left[\exp(\sigma^2_{\ln X}) - 1\right] \tag{5.2.10b}$$

$$\Omega_X^2 = \exp(\sigma_{\ln X}^2) - 1 \tag{5.2.10c}$$

$$\lambda_X = \Omega_X^3 + 3\Omega_X \tag{5.2.10d}$$

From Eq. (5.2.10*d*) it is obvious that lognormal distributions are always positively skewed because $\Omega_X > 0$. Conversely, the statistical moments of $\ln X$ can be computed from those of X by

$$\mu_{\ln X} = \frac{1}{2} \ln \left[\frac{\mu_X^2}{1 + \Omega_X^2} \right] \tag{5.2.11a}$$

$$\sigma_{\ln X}^2 = \ln(\Omega_X^2 + 1) \tag{5.2.11b}$$

Since the sum of normal random variables is normally distributed, the multiplication of lognormal random variables is also lognormally distributed. Several properties of lognormal random variables are useful:

1. If X is a lognormal random variable and $Y = aX^b$ then, Y has a lognormal distribution with mean $\mu_{\ln Y} = \ln a + b\mu_{\ln X}$ and variance $\sigma^2 \ln Y = b^2 \sigma_{\ln X}^2$.
2. If X and Y are independently lognormally distributed, $W = XY$ has a lognormal distribution with mean $\mu_{\ln W} = \mu_{\ln X} + \mu_{\ln Y}$ and variance $\sigma_{\ln W}^2 = \sigma_{\ln X}^2 + \sigma_{\ln Y}^2$.
3. If X and Y are independent and lognormally distributed then $R = X/Y$ is lognormal with $\mu_{\ln R} = \mu_{\ln X} - \mu_{\ln Y}$ and variance $\sigma_{\ln R}^2 = \sigma_{\ln X}^2 + \sigma_{\ln Y}^2$.

Example 5.2.3. The annual maximum series of flood magnitudes in a river has a lognormal distribution with a mean of 6000 cfs and a standard deviation of 4000 cfs. (a) What is the probability in each year that a flood magnitude would exceed 7000 cfs? (b) Determine the flood magnitude with a return period of 100-years.

Solution. **(a)** Let Q be a random variable representing the annual maximum flood magnitude. Since Q is assumed to follow a lognormal distribution, $\ln(Q)$ is normally distributed with mean and variance that can be computed from Eqs. (5.2.11*a*) and (5.2.11*b*), respectively, with $\Omega_Q = 4000/6000 = 0.667$

$$\mu_{\ln Q} = \frac{1}{2} \ln \left[\frac{6000^2}{1 + 0.667^2} \right] = 8.515$$

$$\sigma_{\ln Q}^2 = \ln(0.667^2 + 1) = 0.368$$

The probability that the flood magnitude exceeds 7000 cfs is

$$\begin{aligned} P(Q \geq 7000) &= P[\ln Q \geq \ln(7000)] \\ &= 1 - P\left[Z \leq \left(\ln(Q) - \mu_{\ln Q}\right)/\sigma_{\ln Q}\right] \\ &= 1 - P\left[Z \leq (\ln 7000 - 8.515)/\sqrt{0.368}\right] \\ &= 1 - P\left[Z \leq (8.85537 - 8.515)/\sqrt{0.368}\right] \\ &= 1 - \Phi(0.558) = 1 - 0.7368 = 0.2632 \end{aligned}$$

(b) A 100-year event in hydrology represents the event that occurs, on the average, once every 100 years. Therefore, the probability in every single year that a 100-year event is equaled or exceeded, is 0.01, i.e., $P(Q \geq q_{100}) = 0.01$ in which q_{100} is the magnitude of the 100-year flood. This part of the problem is to determine q_{100} which is a reverse of part (a).

$$P(Q \leq q_{100}) = 1 - P(Q \geq q_{100}) = 0.99$$

Since

$$P(Q \leq q_{100}) = 1 - P(\ln Q \geq \ln q_{100}) = 0.99$$

$$0.99 = P\left[Z \leq \left(\ln(q_{100}) - \mu_{\ln Q}\right)/\sigma_{\ln Q}\right]$$

$$0.99 = P\left[Z \leq (\ln(q_{100}) - 8.515)/\sqrt{0.368}\right]$$

$$0.99 = \Phi\left\{(\ln(q_{100}) - 8.515)/\sqrt{0.368}\right\}$$

$$0.99 = \Phi(z)$$

From the standard normal probability Table 5.2.1, $z = 2.33$ for $\Phi(2.33) = 0.99$. Solving

$$(\ln(q_{100}) - 8.515)/\sqrt{0.368} = 2.33$$

for q_{100} yields $\ln(q_{100}) = 9.9284$, $q_{100} = 20{,}505$ cfs.

5.3 ANALYSIS OF UNCERTAINTIES

In the design and analysis of hydrosystems there are many quantities of interest that are functionally related to a number of variables of which some are subject to **uncertainty**. For example, hydraulic engineers frequently apply weir flow equations such as $Q = CLH^{1.5}$ to estimate spillway capacity in which the coefficient C and head H are subject to uncertainty. As a result, discharge over the spillway is not certain. A rather straightforward and useful technique for this approximation purpose is the **first-order analysis of uncertainties** or sometimes called the **delta-method**.

The use of first-order analysis of uncertainty is quite popular in many fields of engineering. Such popularity owes to its relative ease in application to a wide array of problems. First-order analysis is used to estimate the uncertainty in a deterministic model formulation involving parameters which are uncertain (not known with certainty). More specifically, first-order analysis enables one to estimate the mean and variance of a random variable which is functionally related to several variables, some of which are random. By using first-order analysis, the combined effect of uncertainty in a model formulation, as well as the use of uncertain parameters, can be assessed. Presentation of first-order analysis to civil engineering problems was made by Benjamin and Cornell (1970), Ang and Tang (1979), and Harr (1987). The method has been applied to various problems in hydraulics, hydrology, and water quality (Burges and Lettenmaier, 1975; Tang et al., 1975; Tung and Mays, 1980, 1981; Brown and Barnwell, 1987; Virjling, 1987; Chow et al., 1988; Tung and Hathhorn, 1988).

Consider a random variable, Y, which is a function of k random variables (multivariate case). Mathematically, Y can be expressed as

$$Y = g(\boldsymbol{X}) \tag{5.3.1}$$

where $\boldsymbol{X} = (X_1, X_2, \ldots, X_k)$ is a vector containing k random variables X_i. Through the use of Taylor's expansion, about the means of k random variables, the first-order approximation of the random variable Y can be expressed as

$$Y = g(\bar{\boldsymbol{x}}) + \sum_{i=1}^{k} \left[\frac{\partial g(\boldsymbol{X})}{\partial X_i}\right]_{\boldsymbol{X}=\bar{\boldsymbol{x}}} (X_i - \bar{x}_i) + \sum_{i=1}^{k} \sum_{j=1}^{k} \left[\frac{\partial^2 g(\boldsymbol{X})}{\partial X_i \partial X_j}\right]_{\boldsymbol{X}=\bar{\boldsymbol{x}}} (X_i - \bar{x}_i)(X_j - \bar{x}_j) + \ldots \tag{5.3.2}$$

in which $\bar{\boldsymbol{x}} = (\bar{x}_1, \bar{x}_2, \ldots, \bar{x}_k)$, a vector containing the means of k random variables. The first-order approximation ignores the second and higher order terms and Eq. (5.3.2) can be simplified as

$$Y \approx g(\bar{\boldsymbol{x}}) + \sum_{i=1}^{k} \left[\frac{\partial g}{\partial X_i}\right]_{\bar{\mathbf{x}}} (X_i - \bar{x}_i) \tag{5.3.3}$$

where $\left[\frac{\partial g}{\partial X_i}\right]_{\bar{\boldsymbol{x}}}$ is called the sensitivity coefficient representing the rate of change of function value $g(\boldsymbol{x})$ at $\boldsymbol{x} = \bar{\boldsymbol{x}}$.

The mean (the expected value) of random variable Y, using Eq. (5.1.15*a*), is approximated as

$$\mu_Y = E[Y] \approx g(\bar{\boldsymbol{x}}) \tag{5.3.4}$$

The variance of Y can be approximated as

$$\text{Var}[Y] = \text{Var}[g(\bar{\boldsymbol{x}})] + \text{Var}\left\{\sum_{i=1}^{k} \left[\frac{\partial g}{\partial X_i}\right] (X_i - \bar{x}_i)\right\}$$

The term $\text{Var}[g(\bar{\boldsymbol{x}})] = 0$ for $g(\bar{\boldsymbol{x}})$ being a constant when the mean values of $\bar{\boldsymbol{x}}$ are used. The above equation reduces to

$$\text{Var}[Y] = 0 + \text{Var}\left\{\sum_{i=1}^{k} a_i (X_i - \bar{x}_i)\right\}$$

where $a_i = \left[\frac{\partial g}{\partial X_i}\right]_{\bar{\boldsymbol{x}}}$. Using Eq. (5.1.22), the variance of Y can be approximated as

$$\sigma_Y^2 = \text{Var}[Y] \approx \sum_{i=1}^{k} a_i^2 \sigma_i^2 + 2 \sum_{i} \sum_{< j} a_i a_j \text{Cov}[X_i, X_j] \tag{5.3.5}$$

where σ_i^2 is the variance corresponding to random variable X_i. If the X_i's are uncorrelated, that is, $\text{Cov}[X_i, X_j] = 0$, then Eq. (5.3.5) reduces to

$$\sigma_Y^2 = \sum_{i=1}^{k} a_i^2 \sigma_i^2 \tag{5.3.6}$$

Equation (5.3.6) can be expressed in terms of the coefficient of variation Ω by dividing both sides by μ_Y^2,

$$\Omega_Y^2 = \sum_{i=1}^{k} a_i^2 \left(\frac{\bar{x}_i}{\mu_Y} \right)^2 \Omega_{X_i}^2 \tag{5.3.7}$$

Equations (5.3.6) or (5.3.7) contain the relative contribution, $a_i^2\sigma_i^2$, of each random component to the total uncertainty of model output Y. Such information can be utilized to design measures to reduce the uncertainties or to minimize the effects of uncertainties.

Example 5.3.1. It is a common practice to use Manning's formula to compute the flow carrying capacity in an open channel. The channel capacity using Manning's formula can be expressed as

$$Q = 1.49 n^{-1} S^{1/2} A^{5/3} P^{-2/3}$$

where P is the wetted perimeter. Because uncertainties exist in estimating roughness coefficients, channel slope, cross-sectional area of flow, and wetted perimeter, the computed channel capacity is also subject to uncertainty. Assume that the degrees of uncertainty in estimating A and P are negligible while uncertainties in the roughness coefficient and the channel slope are significant. Apply first-order analysis of uncertainty to derive the degree of uncertainty in Q in terms of the uncertainties in Manning's roughness n and channel slope S.

Solution. Since A and P are considered to be deterministic without uncertainty, they can be combined into a constant term, $K = 1.49A^{5/3}P^{-2/3}$ in order to express

$$Q = K n^{-1} S^{1/2}$$

The first-order approximation of the mean of Q using Manning's formula can be determined using Eq. (5.3.3)

$$\begin{aligned} Q &\approx \overline{Q} + \left[\frac{\partial Q}{\partial n}\right]_{(\bar{n},\bar{S})} (n - \bar{n}) + \left[\frac{\partial Q}{\partial S}\right]_{(\bar{n},\bar{S})} (S - \bar{S}) \\ &= \overline{Q} + \left[-K\bar{n}^{-2}\bar{S}^{1/2}\right](n - \bar{n}) + \left[0.5\, K\bar{n}^{-1}\bar{S}^{-1/2}\right](S - \bar{S}) \end{aligned}$$

where $\overline{Q} = K\bar{n}^{-1}\bar{S}^{1/2}$. The variance of channel capacity can be obtained by applying the variance operator to the equation, assuming n and S are independent random variables,

$$\sigma_Q^2 = \left[\frac{\partial Q}{\partial n}\right]_{(\bar{n},\bar{S})}^2 \sigma_n^2 + \left[\frac{\partial Q}{\partial S}\right]_{(\bar{n},\bar{S})}^2 \sigma_S^2$$

where $\left[\frac{\partial Q}{\partial n}\right]$ and $\left[\frac{\partial Q}{\partial S}\right]$ are sensitivity coefficients. Alternatively, the uncertainty of Q in terms of the coefficient of variation can be derived by using Eq. (5.3.7) with $X_1 = n$ and $X_2 = S$ as

$$\begin{aligned}
\Omega_Q^2 &= \sum_{i=1}^{2} \left[\frac{\partial Q}{\partial X_i}\right]^2 \left[\frac{\bar{x}_i}{\bar{Q}}\right]^2 \Omega_{x_i}^2 \\
&= \left[\frac{\partial Q}{\partial n}\right]^2 \left[\frac{\bar{n}}{\bar{Q}}\right]^2 \Omega_n^2 + \left[\frac{\partial Q}{\partial S}\right]^2 \left[\frac{\bar{S}}{\bar{Q}}\right]^2 \Omega_S^2 \\
&= \left[\frac{-K\bar{S}^{1/2}}{\bar{n}^2}\right]^2 \left[\frac{\bar{n}}{\bar{Q}}\right]^2 \Omega_n^2 + \left[\frac{0.5K}{\bar{n}S^{1/2}}\right]^2 \left[\frac{\bar{S}}{\bar{Q}}\right]^2 \Omega_S^2 \\
&= \left[\frac{-K\bar{S}^{1/2}}{\bar{Q}}\right]^2 \left[\frac{1}{\bar{n}^2}\right] \Omega_n^2 + [0.5]^2 \left[\frac{K}{\bar{n}\bar{S}^{1/2}}\right]^2 \left[\frac{\bar{S}}{\bar{Q}}\right]^2 \Omega_S^2 \\
&= \left[\bar{n}^2\right] \left[\frac{1}{\bar{n}^2}\right] \Omega_n^2 + 0.25 \left[\frac{1}{\bar{S}}\right] \left[\bar{S}\right] \Omega_S^2 \\
&= \Omega_n^2 + 0.25\Omega_S^2
\end{aligned}$$

5.4 RELIABILITY COMPUTATIONS USING LOAD-RESISTANCE ANALYSIS

Practically all hydrosystems are designed to be placed in natural environments that are subject to various external stresses. The **resistance** or strength of a hydrosystem is its ability to accomplish the intended mission satisfactorily without failure when subjected to external stresses or loadings. Loadings or stresses tend to cause failure of the system. When the strength of the system is exceeded by the stress, failure occurs. The resistance of a hydrosystem can be the flow carrying capacity while the loading could be the magnitude of flow passing through the system. From the previous discussions of the existence of uncertainties in the design, analysis and modeling of hydrosystems, the resistance or strength of the system and the imposed loadings or stresses, more often than not, are random and subject to uncertainty. The **reliability** of the system can be evaluated by studying the interaction of load and resistance.

The **reliability** of a hydrosystem is defined as the probability that the capacity of the system (i.e., the resistance) exceeds or equals the loading. On the other hand, the **risk** is the probability of the loading exceeding the resistance. Consider the random variables L and R which represent the loading and resistance, respectively. The reliability of a hydrosystem can be expressed mathematically as

$$\alpha = P(L \leq R) \tag{5.4.1}$$

The relationship between reliability (α) and risk (α') is

$$\alpha' = P(L > R) = 1 - \alpha \tag{5.4.2}$$

Computations of reliability or risk using Eqs. (5.4.1) and (5.4.2) does not consider the time dependence of the loading. It is generally applied when the performance of the system subject to a single worst loading event is evaluated. From the reliability computation viewpoint, this is referred to as a **static reliability model**.

5.4.1 Direct Integration Method

From Eqs. (5.4.1) and (5.4.2), the computation of risk and reliability require knowledge of the probability distributions of loading and resistance. In terms of the joint PDF of the loading and resistance, $f_{L,R}(l, r)$, Eq. (5.4.1) can be expressed as

$$\alpha = \int_0^\infty \int_0^r f_{L,R}(l, r)\, dl\, dr \tag{5.4.3}$$

If the loading L and resistance R are independent, Eq. (5.4.3) can be written as

$$\alpha = \int_0^\infty f_R(r) \left[\int_0^r f_L(l)\, dl \right] dr = \int_0^\infty f_R(r) F_L(r)\, dr \tag{5.4.4}$$

in which $f_L(l)$ and $f_R(r)$ are the PDF's of loading L and resistance R, respectively and $F_L(r)$ is the CDF of the loading evaluated at $L = r$. This computation of reliability is called **load-resistance interference** which can be schematically shown in Fig. 5.4.1. The computation steps involved for reliability determination using Eq. (5.4.4) is shown in Fig. 5.4.1. The method of direct integration in general is analytically tractable for only a very few special combinations of probability distributions. Numerical integrations are performed typically for reliability determination.

5.4.2 Methods of Using Safety Margin and Safety Factor

The **safety margin** (SM) is defined as the difference between the resistance and the anticipated loading, that is, $SM = R - L$. From Eq. (5.4.1), the reliability of a system can be expressed in terms of the safety margin as

$$\alpha = P(R - L \geq 0) = P(SM \geq 0) \tag{5.4.5}$$

Using the safety margin for computing reliability requires knowing the probability distribution of SM. If that is the case, the reliability can be obtained by $\alpha = 1 - F_{\text{SM}}(0)$ in which $F_{\text{SM}}()$ is the CDF of the safety margin SM. Under some special cases, the distribution of SM can be easily assessed without mathematical manipulations. For example, from the discussion of the normal distribution (Section 5.2.3), the distribution of SM is normally distributed with the mean μ_{SM} and variance σ^2_{SM} if the loading and resistance are both normal random variables. The mean and the variance of SM can be obtained, based on Eqs. (5.1.15a) and (5.1.22), as

$$\mu_{\text{SM}} = \mu_R - \mu_L \tag{5.4.6}$$

$$\sigma^2_{\text{SM}} = \sigma^2_R + \sigma^2_L - 2\text{Cov}(L, R) \tag{5.4.7}$$

If loading and resistance are independent then the covariance term in Eq. (5.4.7) is zero so that $\sigma_{\text{SM}} = \left(\sigma^2_R + \sigma^2_L\right)^{1/2}$. Under the normality conditions, the reliability of

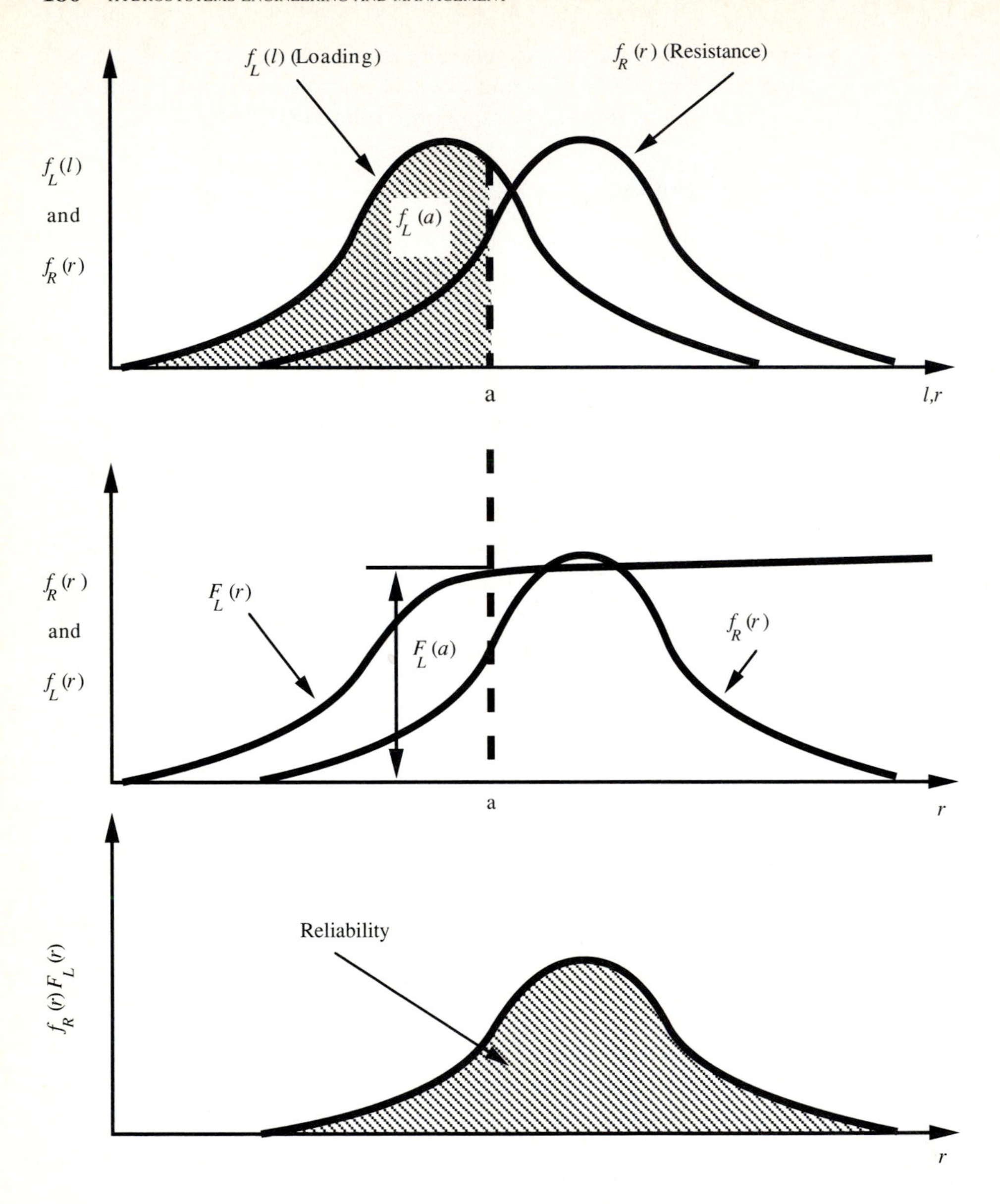

FIGURE 5.4.1
Graphical illustration of the steps involved in reliability computation by Eq. (5.4.4).

the system, can be determined by subtracting μ_{SM} from both sides of the inequality in Eq. (5.4.5) and dividing both sides by σ_{SM} to obtain

$$\alpha = P\left(\frac{\text{SM} - \mu_{SM}}{\sigma_{SM}} \geq \frac{-\mu_{SM}}{\sigma_{SM}}\right)$$
$$= P\left(Z \geq -\mu_{SM}/\sigma_{SM}\right) = \Phi\left(\mu_{SM}/\sigma_{SM}\right) \tag{5.4.8}$$

where $Z = (SM - \mu_{\text{SM}})/\sigma_{\text{SM}}$ is a standard normal random variate, and $\Phi(\ )$ is the standard normal CDF.

Example 5.4.1. A city's estimated annual water demand is three units, with a standard deviation of one unit. It is also known that the city's water supply system has an estimated mean capacity of five units with a standard deviation of 0.75 units. Calculate the reliability or probability of supply exceeding the demand using safety margin as the performance criterion. Assume that both the demand and supply are independent normal random variables.

Solution. From the problem statement, the demand is L and supply is R with $\mu_L = 3$, $\sigma_L = 1$; $\mu_R = 5$, $\sigma_R = 0.75$. The mean and variance of safety margin SM can be calculated using Eqs. (5.4.6) and (5.4.7) as $\mu_{\text{SM}} = 5 - 3 = 2$ and $\sigma^2_{\text{SM}} = (0.75)^2 + (1)^2 = 1.5625$. Since both demand and supply are normal random variables, therefore SM is also a normal random variable (refer to Section 5.2.3). The reliability of the water supply system is the probability of being able to meet the demand which can be calculated as

$$\alpha = P(\text{SM} \geq 0) = \Phi\left(\mu_{\text{SM}}/\sigma_{\text{SM}}\right) = \Phi\left(2/\sqrt{1.5625}\right) = \Phi(1.60) = 0.945$$

The risk, the probability of not being able to meet the demand, of the water supply system is $\alpha' = 1 - \alpha = 1 - 0.945 = 0.055$.

Although exact solutions for reliability computation is desirable, it might not be practical because the exact PDF of the safety margin cannot be easily derived. In such circumstances the use of the normal approximation based on the mean and variance of safety margin might be a viable alternative. Ang (1973) indicated that, provided $\alpha < 0.99$, reliability is not greatly influenced by the choice of distribution for L and R and the assumption of a normal distribution for SM is quite satisfactory. However, for a reliability higher than this (e.g. $\alpha = 0.999$), the shape of the tails of the distribution become very critical in which case accurate assessment of the distribution of SM or a direct integration procedure should be used to evaluate the reliability or risk.

The **safety factor** (SF) is defined as the ratio of resistance to loading, R/L. Because safety factor SF is the ratio of two random variables, consequently, it is also a random variable. The reliability then can be written as $P(SF \geq 1)$. Several safety factor measures and their usefulness in hydraulic engineering design and analysis are discussed by Yen (1979). Similar to the safety margin, reliability computations using safety factor requires knowing the PDF of SF.

The simplest case is when both loading L and resistance R have lognormal distributions. The logarithmic transform of SF leads to the difference of $\ln(R)$ and $\ln(L)$ which are both normally distributed. The reliability computation can then be proceeded as the safety margin case

$$\alpha = P\left(\frac{R}{L} \geq 1\right) = P(\text{SF} \geq 1) = P[\ln(\text{SF}) \geq 0]$$

$$= P[\ln(R/L) \geq 0] = P[(\ln R - \ln L) \geq 0]$$

$$= P\left[\frac{(\ln R - \ln L) - (\mu_{\ln R} - \mu_{\ln L})}{\sqrt{\sigma^2_{\ln R} + \sigma^2_{\ln L}}} \geq \frac{0 - (\mu_{\ln R} - \mu_{\ln L})}{\sqrt{\sigma^2_{\ln R} + \sigma^2_{\ln L}}}\right]$$

$$= P\left[Z \leq \mu_{\ln \mathrm{SF}}/\sigma_{\ln \mathrm{SF}}\right]$$

$$= \Phi\left[\mu_{\ln \mathrm{SF}}/\sigma_{\ln \mathrm{SF}}\right] \tag{5.4.9}$$

in which $\mu_{\ln \mathrm{SF}} = (\mu_{\ln R} - \mu_{\ln L})$ and $\sigma_{\ln \mathrm{SF}} = (\sigma^2_{\ln R} + \sigma^2_{\ln L})^{1/2}$, with $\mu_{\ln R}$, $\mu_{\ln L}$, $\sigma_{\ln R}$, and $\sigma_{\ln L}$ derived by using Eqs. (5.2.11*a*) and (5.2.11*b*). After some algebraic manipulations, Eq. (5.4.9) can also be expressed in terms of the statistical properties of L and R directly as (Chow, Maidment, and Mays, 1988)

$$\alpha = 1 - \Phi\left\{\frac{-\ln\left[\frac{\mu_r}{\mu_L}\sqrt{\frac{1+\Omega^2_L}{1+\Omega^2_R}}\right]}{\sqrt{\ln\left[\left(1+\Omega^2_L\right)\left(1+\Omega^2_R\right)\right]}}\right\} \tag{5.4.10}$$

Example 5.4.2. Solve Example 5.4.1 assuming capacity (R) and demand (L) are both lognormally distributed.

Solution. From Example 5.4.1 the coefficients of variation of demand and supply are $\Omega_L = 1/3 = 0.333$; $\Omega_R = 0.75/5 = 0.15$. Applying Eq. (5.4.10), the reliability of the water supply system to meet the demand can be calculated as

$$\alpha = 1 - \Phi\left\{\frac{-\ln\left[\frac{5}{3}\sqrt{\frac{1+(0.333)^2}{1+(0.15)^2}}\right]}{\sqrt{\ln\left[\left(1+0.15^2\right)\left(1+0.333^2\right)\right]}}\right\}$$

$$= 1 - \Phi(-1.5463) = 1 - 0.061 = 0.939$$

5.4.3 First-Order Second-Moment Methods

Reliability can be expressed in terms of a **performance function**, such as the safety factor or safety margin, which describes the system performance. A system performance can be described in terms of the loading $L = g(\boldsymbol{X})$ and the resistance $R = h(\boldsymbol{Y})$ as $W(\boldsymbol{X}, \boldsymbol{Y})$ which could be one of the following forms:

$$W_1(\boldsymbol{X}, \boldsymbol{Y}) = R - L = h(\boldsymbol{Y}) - g(\boldsymbol{X}) = \mathrm{SM} \tag{5.4.11}$$

$$W_2(\boldsymbol{X}, \boldsymbol{Y}) = (R/L) - 1 = [h(\boldsymbol{Y})/g(\boldsymbol{X})] - 1 = \mathrm{SF} - 1 \tag{5.4.12}$$

$$W_3(\boldsymbol{X}, \boldsymbol{Y}) = \ln(R/L) = \ln[h(\boldsymbol{Y})] - \ln[g(\boldsymbol{X})] = \ln(\mathrm{SF}) \tag{5.4.13}$$

where **X** and **Y** are vectors of the uncertain parameters in defining the loading and resistance. Equation (5.4.11) is identical to the safety margin while Eqs. (5.4.12) and

(5.4.13) are based on safety factor representation. Therefore, the reliability is the probability that the performance function is greater than or equal to zero.

MEAN-VALUE FIRST-ORDER SECOND-MOMENT (MFOSM) METHOD. Identical to the first-order analysis of uncertainty described in Section 5.3.2, the MFOSM method estimates the mean (μ_W) and standard deviation (σ_W) of the performance variable W by Eqs. (5.3.4) and (5.3.5), respectively. Once the mean and variance of W are estimated, a **reliability index** β is computed as

$$\beta = \mu_W / \sigma_W \tag{5.4.14}$$

and the reliability can be computed as

$$\alpha = P(W \geq 0) = 1 - P(W < 0) = 1 - F_W(0) = 1 - F_{W'}(-\beta) \tag{5.4.15}$$

in which $F_W(\ )$ is the CDF of the performance variable W and W' is the standardized performance variable defined as $W' = (W - \mu_W)/\sigma_W$. The normal distribution is commonly used for W in which case the reliability can be simply computed as

$$\alpha = 1 - \Phi(-\beta) = \Phi(\beta) \tag{5.4.16}$$

in which $\Phi(\ )$ is the standard normal CDF (see Table 5.2.1).

Example 5.4.3. Consider a section of man-made open channel having concrete sides with a gravel bottom. Assume that the uncertainties in channel cross-section area (A) and wetted perimeter (P) are negligible. The values of channel cross-section area (A) and wetted perimeter (P) are 90 ft^2 and 35 ft, respectively. However, the Manning roughness coefficient (n) and channel slope (S) are subject to uncertainty. The means of the roughness coefficient and the slope are 0.017 and 0.0016 ft/ft, respectively. The coefficients of variation of n and S are 20- and 30 percent, respectively. Determine the reliability that the channel has a conveyance capacity of 350 cfs.

Solution. Based on Manning's equation, the flow rate of the open channel is

$$\begin{aligned} Q &= 1.49A^{5/3}P^{-2/3}n^{-1}S^{1/2} \\ &= (1.49)(90)^{5/3}(35)^{-2/3}n^{-1}S^{1/2} \\ &= 251.7n^{-1}S^{1/2} \end{aligned}$$

By Eq. (5.3.4), the mean of the conveyance capacity of the open channel for $\bar{n} = 0.017$ and $\bar{S} = 0.016$ is

$$\mu_Q \approx 251.7(0.017)^{-1}(0.0016)^{1/2} = 592.3 \text{ cfs}$$

The uncertainty of conveyance capacity, in terms of the coefficient of variation, can be computed by Eq. (5.3.7), or referring to Example 5.3.2, as

$$\Omega_Q^2 = \Omega_n^2 + 0.25\Omega_S^2 = (0.2)^2 + 0.25(0.3)^2 = 0.0625$$

Hence, the standard deviation of channel conveyance capacity is

$$\sigma_Q = \mu_Q \Omega_Q = 592.3(0.0625)^{0.5} = 148.1 \text{ cfs}$$

Assuming a normal distribution for the channel conveyance capacity, the reliability that the channel is able to deliver the required flow rate of 350 cfs can be estimated as

$$\begin{aligned}\alpha = P(Q \geq 350) &= P[Z \geq (350 - 592.3)/148.1] \\ &= P[Z \geq -1.636] \\ &= 1 - \Phi(-1.636) \\ &= \Phi(1.636) = 0.949\end{aligned}$$

Although the MFOSM method is simple and straightforward in use, it, however, possesses some weaknesses which include: (1) inability to handle distributions with a large skew coefficient; (2) generally poor estimation of the mean and variance of nonlinear functions; and (3) sensitivity of the computed risk to the formulation of performance variables (Ang and Tang, 1984; Yen et al., 1986). To reduce the effect of nonlinearity, one way is to include the second-order terms in a Taylor expansion. This would increase the burden of analysis by having to compute the second-order partial derivatives and higher-order statistical moments, which may not be so easily and reliably obtained.

5.4.4 Dynamic (Time-Dependent) Reliability Model

In some situations, reliability evaluation of hydrosystems is made with respect to a specified time framework. For example, one might be interested in the risk of overflow of a urban storm water detention basin in the summer season when convective thunderstorms prevail. Loadings to most hydrosystems are caused by the occurrence of hydrological events such as floods, storms, or droughts which are random by nature. **Dynamic** or **time-dependent** reliability analysis considers repeated applications of loading and also can consider the change of the distribution of resistance with time. The objective of the reliability computations for the dynamic models is to determine the system reliability over a specified time interval in which the number of occurrences of loadings is also a random variable.

Repeated loadings on a hydrosystem are characterized by the time each load is applied and the behavior of time intervals between the application of loads. From a reliability theory viewpoint, the uncertainty about the loading and resistance variables may be classified into three categories: **deterministic**, **random-fixed**, and **random-independent** (Kapur and Lamberson, 1977). For the deterministic category, the loadings assume values that are exactly known a priori. For the random-fixed case, the randomness of loadings varies in time in a known manner. For the random-independent case, the loading is not only random but the successive values assumed by the loading are statistically independent.

Reliability computations for dynamic models can be made for deterministic and random cycle times. The development of a model for deterministic cycles will be given below which naturally leads to the model for random cycle times. Consider a hydrosystem with a capacity $R = r$ subject to n repeated loadings $L_1, L_2, \ldots, L_n$.

When the system capacity r is fixed, the system reliability after n loadings, $\alpha(n,r)$ is

$$\begin{aligned}\alpha(n,r) &= P[(L_1 \le r) \cap (L_2 \le r) \cap \ldots \cap (L_n \le r)] \\ &= P[L_{\max} \le r]\end{aligned} \qquad (5.4.17)$$

in which $L_{\max} = \max\{L_1,\ L_2, \ldots,\ L_n\}$, the maximum loading, is also a random variable. If all random loadings are independent and each with its own distribution, Eq. (5.4.17) is simplified to

$$\alpha(n,r) = \prod_{i=1}^{n} F_{L_i}(r) \qquad (5.4.18)$$

where $F_{L_i}(\ )$ is the CDF of the ith loading. In case that all loadings are generated by the same statistical population, that is, all L's are identically distributed, Eq. (5.4.18) can be further reduced to

$$\alpha(n,r) = [F_L(r)]^n \qquad (5.4.19)$$

Reliability under the fixed number of loadings n can be expressed as

$$\alpha(n) = \int_0^{\infty} \alpha(n,r) f_R(r)\, dr \qquad (5.4.20)$$

Since the loadings to many hydrosystems are related to hydrological events, the occurrence of the number of loadings, in general, is uncertain. The reliability of the system under random loading in the specified time interval $[0,t]$ can be expressed as

$$\alpha(t) = \sum_{n=0}^{\infty} \pi(t|n)\alpha(n) \qquad (5.4.21)$$

in which $\pi(t|n)$ is the probability of n loadings occurring in the time interval $[0,t]$. A Poisson distribution can be used to describe the probability of the number of events occurring in a given time interval. In fact, the Poisson distribution has been found to be an appropriate model for the number of occurrences of hydrological events (Todorovic and Yevjevich, 1969; Zelenhasic, 1970; Rousselle, 1972; Fogel and Duckstein, 1982). Referring to Eq. (5.2.3), $\pi(t|n)$ can be expressed as

$$\pi(t|n) = \frac{e^{-\nu t}(\nu t)^n}{n!} \qquad (5.4.22)$$

where ν is the mean rate of occurrence of the loading in $[0,t]$ which can be estimated from historical data.

Substituting Eq. (5.4.22) into Eq. (5.4.21), the time-dependent reliability for the random independent loading and random-fixed resistance can be expressed as

$$\begin{aligned}\alpha(t) &= \sum_{n=0}^{\infty} \frac{e^{-\nu t}(\nu t)^n}{n!} \int_0^{\infty} f_R(r)[F_L(r)]^n\, dr \\ &= \int_0^{\infty} f_R(r) e^{-\nu t[1-F_L(r)]}\, dr\end{aligned} \qquad (5.4.23)$$

Figure 5.4.2 illustrates dynamic risk-safety factor curves generated by Lee and Mays (1984) for levee design on the Guadalupe River near Victoria, Texas.

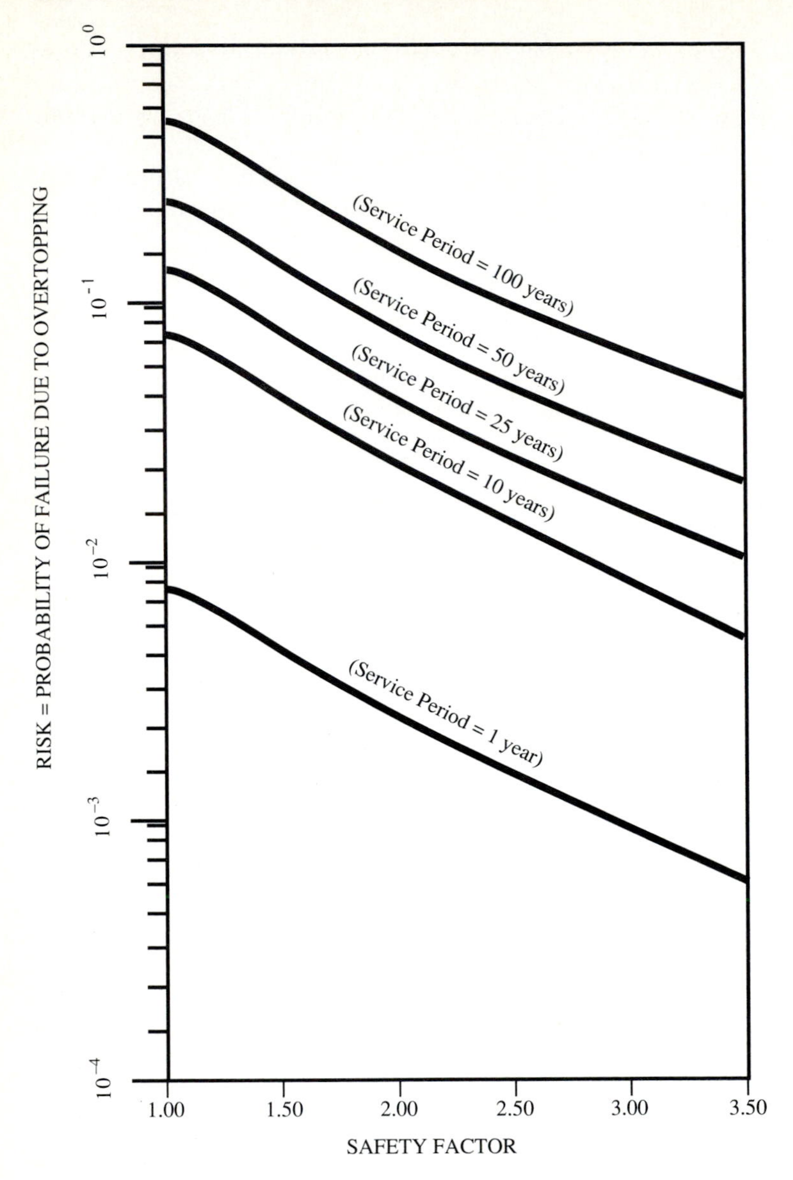

FIGURE 5.4.2
Dynamic risk-safety factor curves for levee design using log-Pearson III loading distribution, T_r = 100 years on the Guadalupe River near Victoria, Texas.

5.5 RELIABILITY USING TIME-TO-FAILURE ANALYSIS

Instead of considering detailed interactions of resistance and loading over time, a system or its components can be treated as a black box and their performance observed over time. This reduces the reliability analysis to a one dimensional problem involving only time as the random variable. In such cases, **time-to-failure** (T) of a system or component of the system is the random variable with the PDF, $f_T(t)$, called the **failure density function**. Time-to-failure analysis is particularly suitable for assessing the reliability of systems involving components which are repairable.

5.5.1 Failure Density Function

The failure density function serves as the common thread in the reliability computations using time-to-failure analysis. Using the failure density function, the reliability of a system or a component within a time interval $[0, t]$ can be expressed as

$$\alpha(t) = \int_t^{\infty} f_T(t)\, dt \tag{5.5.1}$$

which represents the probability that the system experiences no failure within $[0, t]$. The risk or **unreliability**, based on Eq. (5.5.1), then can be expressed as

$$\alpha'(t) = 1 - \alpha(t) = \int_0^t f_T(t)\, dt \tag{5.5.2}$$

Schematically, the reliability and unreliability are shown in Fig. 5.5.1.

5.5.2 Failure Rate and Hazard Function

The probability of failure of a hydrosystem in some time interval $(t, t + \Delta t)$ can be expressed in terms of the reliability as

$$\begin{aligned}\int_t^{t+dt} f_T(t)\, dt &= \int_t^{\infty} f_T(t)\, dt - \int_{t+\Delta t}^{\infty} f_T(t)\, dt \\ &= \alpha(t) - \alpha(t + \Delta t)\end{aligned}$$

which define the difference in reliability at time t, $\alpha(t)$ and at time $t + \Delta t$, $\alpha(t + \Delta t)$. The rate at which a failure occurs in the time interval $(t, t + \Delta t)$ is called the **failure rate**. Failure rate is the probability that a failure per unit time occurs in the time

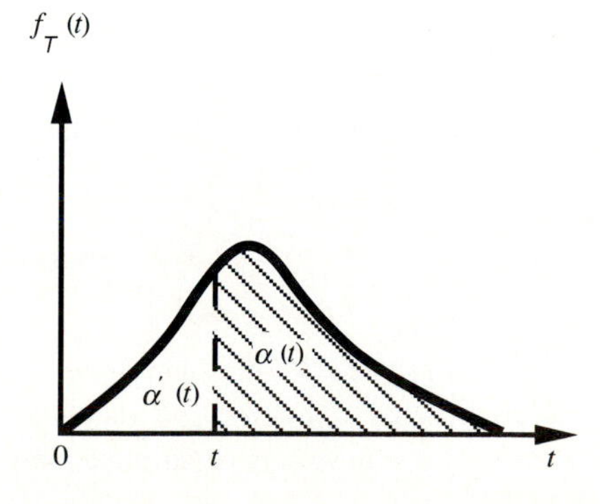

FIGURE 5.5.1
Schematic diagram of reliability and unreliability for time-to-failure analysis.

interval, given that a failure has not occurred prior to time t. The failure rate in the time interval from t to $t + \Delta t$ is then

$$\frac{\alpha(t) - \alpha(t + \Delta t)}{\Delta t \alpha(t)}$$

A **hazard function** is the limit of the failure rate as Δt approaches 0, that is,

$$m(t) = \lim_{\Delta t \to 0} \left[\frac{\alpha(t) - \alpha(t + \Delta t)}{\Delta t \alpha(t)} \right] = \frac{1}{\alpha(t)} \left[-\frac{d[\alpha(t)]}{dt} \right] \tag{5.5.3a}$$

$$= \frac{-1}{\alpha(t)} \frac{d\alpha(t)}{dt} = \frac{f_T(t)}{\alpha(t)} \tag{5.5.3b}$$

The hazard function, therefore, is the instantaneous failure rate.

The quantity $m(t)dt$ is the probability that a component fails during the time interval $(t, t + dt)$. The hazard function indicates change in the failure rate over the life of a hydrosystem or components. The practical usefulness of the hazard function $m(t)$ is that the failure density function $f_T(t)$ and reliability function $\alpha(t)$ can be derived directly from it. From Eq. (5.5.3a), the reliability function $\alpha(t)$ can be expressed in terms of the hazard function $m(t)$ by integrating both sides of $d\alpha(t)/\alpha(t) = -m(t)dt$. The result then is

$$\alpha(t) = \exp \left[- \int_0^t m(t)\, dt \right] \tag{5.5.4}$$

Substituting Eq. (5.5.4) into Eq. (5.5.3b), the failure density function, $f_T(t)$, can be expressed, in terms of hazard function $m(t)$, as (Kapur and Lamberson, 1977)

$$f_T(t) = m(t) \exp \left[- \int_0^t m(\tau)\, d\tau \right] \tag{5.5.5}$$

For an exponential distribution $f_T(t) = \lambda e^{-\lambda t}$ the corresponding reliability function is $\alpha(t) = e^{-\lambda t}$ then the hazard function is

$$m(t) = \frac{f_T(t)}{\alpha(t)} = \lambda$$

which is a constant. The relationships of $f_T(t)$, $\alpha(t)$, and $m(t)$ for other types of failure density functions, can be found elsewhere (Henley and Kumamoto, 1981).

5.5.3 Mean-Time-To-Failure

The **mean-time-to-failure** (MTTF) is the expected value of time to failure stated mathematically as

$$\text{MTTF} = \int_0^\infty t f_T(t)\, dt \tag{5.5.6}$$

which has a unit of time (e.g., minutes, hours, etc.).

Example 5.5.1. Because of its ability to define failure time and its relative simplicity for performing reliability computations, the exponential distribution is perhaps the most widely used failure density function. Suppose the time to failure of a pump in a water distribution system is assumed to follow an exponential distribution with the parameter $\lambda = 0.0008$/hr (i.e., 7 failures/yr). Determine the reliability of the pump in the period of (0,100 hrs.), the failure rate, and the MTTF.

Solution. The failure density function of the pump can be expressed as

$$f_T(t) = \lambda e^{-\lambda t} = 0.0008\ e^{-0.0008t}, t \geq 0$$

The reliability of the pump at any time $t > 0$ is calculated, according to Eq. (5.5.1), as

$$\alpha(t) = \int_t^{\infty} f_T(t)\ dt = \int_t^{\infty} \lambda e^{-\lambda t} dt = e^{-\lambda t} = e^{-0.0008t}, t \geq 0$$

The reliability of the pump for $t = 100$ is $\alpha(t = 100) = \exp(-0.08) = 0.9231$ and the associated risk is $\alpha'(t = 100) = 1 - a(t = 100) = 0.0769$. Based on Eq. (5.5.3) the failure rate or hazard function $m(t)$ for the exponential failure density function is

$$m(t) = \frac{f_T(t)}{\alpha(t)} = \lambda = 0.0008$$

The MTTF of the pump, by Eq. (5.5.6), is

$$\text{MTTF} = \int_0^{\infty} t(0.0008\ e^{-0.0008t})\ dt = 1/0.0008 = 1250 \text{ hrs.}$$

5.5.4 Repair Density Function, Repair Rate, and Mean-Time-To-Repair

Similar to the failure density function, the **repair density function**, $g(t)$, describes the random characteristics of the time required to repair a failed component when failure occurs at time zero. The **probability of repair**, $G(t)$, is the probability that the component repair is completed before time t, given that the component failed at $t = 0$. In other words, $G(t)$ is the CDF of the repair time having the repair density function $g(t)$.

Repair rate, $r(t)$, similar to the failure rate through the hazard function, is the probability that the component is repaired per unit time given that the component failed at time zero and is still not repaired at time t. The quantity $r(t)dt$ is the probability that a component is repaired during the time interval $(t, t + dt)$ given that the component fails at time t. Similar to Eq. (5.5.3b) the relation between repair density function, $g(t)$, repair rate, $r(t)$, and repair probability, $G(t)$, is

$$r(t) = g(t)/[1 - G(t)] \tag{5.5.7}$$

which can be derived in the same manner or analogous to the failure rate and hazard function.

Given a repair rate $r(t)$, the repair density function $g(t)$ and the repair probability $G(t)$ are determined, respectively, as

$$g(t) = r(t) \exp\left[-\int_0^t r(\tau)\ d\tau\right] \tag{5.5.8}$$

$$G(t) = 1 - \exp\left[-\int_0^t r(\tau)\, d\tau\right] \tag{5.5.9}$$

The **mean-time-to-repair** (MTTR) is the expected value of the time to repair a failed component, defined as

$$\text{MTTR} = \int_0^\infty t g(t)\, dt \tag{5.5.10}$$

which has a unit of time.

5.5.5 Mean-Time-Between-Failure and Mean-Time-Between-Repair

The **mean-time-between-failure** (MTBF) is the expected value of time between two consecutive failures. For a repairable component, the MTBF is the sum of MTTF and MTTR, i.e.,

$$\text{MTBF} = \text{MTTF} + \text{MTTR} \tag{5.5.11}$$

The **mean-time-between-repair** (MTBR) is the expected value of the time between two consecutive repairs and equals the MTBF.

Example 5.5.2. Consider the pump in Example 5.5.1, which has an exponential repair density function with parameter $\eta = 0.02$/hr; determine the MTTR for the pump system.

Solution. The repair density function can be written as

$$g(t) = \eta e^{-\eta t} = 0.02\, e^{-0.02t}, t \geq 0$$

so that the MTTR is

$$\text{MTTR} = \int_0^\infty t\eta e^{-\eta t} dt = \frac{1}{\eta} = \int_0^\infty t(0.02\, e^{-0.02t})\, dt = 1/0.02 = 50 \text{ hrs.}$$

5.5.6 Availability and Unavailability

The reliability of a component is a measure of the probability that the component would be continuously functional without interruption through the entire period $[0, t]$. This measure is appropriate if a component is **nonrepairable** and has to be discarded when the component fails. In case that components in the system are repairable, a measure that has a broader meaning than that of the reliability is needed.

The **availability** $A(t)$ of a component is the probability that the component is in operating condition at time t, given that the component was as good as new at time zero. The reliability generally differs from the availability because reliability requires the continuation of the operational state over the specified interval $[0, t]$. Subcomponents contribute to the availability but not to the reliability if the subcomponent that failed before time t is repaired and is then operational at time t. As a result, the availability is always larger than or equal to the reliability, that is, $A(t) \geq \alpha(t)$. For a repairable component, $A(t) > \alpha(t)$ while for a nonrepairable component, $A(t) = \alpha(t)$.

As shown in Fig. 5.5.2, the availability of a nonrepairable component decreases to zero as t gets larger, whereas the availability of a repairable component converges to a nonzero positive value.

The **unavailability** $U(t)$ at time t is the probability that a component is in the failed state at time t, given that it started in the operational state at time zero. Availability and unavailability of the component are complimentary events, therefore,

$$A(t) + U(t) = 1 \tag{5.5.12}$$

Knowing $A(t) \geq \alpha(t)$, it can be concluded that $U(t) \leq \alpha'(t)$.

Using an exponential failure and repair density function, the resulting failure rate and repair rate, according to the definitions given in Eqs. (5.5.3) and (5.5.7), are constants equal to their respective parameters. For a constant failure rate and a constant repair rate the analysis of the entire process can be simplified to analytical solutions. Henley and Kumamoto (1981) use **Laplace transforms** to derive the unavailability as

$$U(t) = \frac{\lambda}{\lambda + \eta}\left[1 - e^{(\lambda+\eta)t}\right] \tag{5.5.13}$$

and availability as

$$A(t) = 1 - U(t) = \frac{\eta}{\lambda + \eta} + \frac{\lambda}{\lambda + \eta} e^{(\lambda+\eta)t} \tag{5.5.14}$$

The **steady state** or **stationary unavailability**, $U(\infty)$, and the **stationary availability**, $A(\infty)$, when $t \to \infty$, are respectively

$$U(\infty) = \frac{\text{MTTR}}{\text{MTTF} + \text{MTTR}} = \frac{\lambda}{\lambda + \eta} \tag{5.5.15}$$

and

$$A(\infty) = \frac{\text{MTTF}}{\text{MTTF} + \text{MTTR}} = \frac{\eta}{\lambda + \eta} \tag{5.5.16}$$

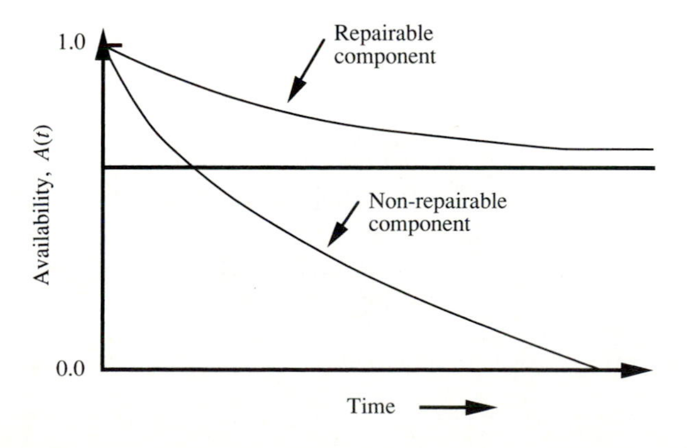

FIGURE 5.5.2
Availability for repairable and nonrepairable components.

Example 5.5.3. Referring to Examples 5.5.1 and 5.5.2, calculate the availability and unavailability of the pump for t = 100 hrs. Then compute the stationary availability and stationary unavailability.

Solution. Substituting λ = 0.0008 and η = 0.02 into Eqs. (5.5.13) and (5.5.14), the corresponding unavailability and availability of the pump at time t = 100 hrs. are 0.0336 and 0.9667, respectively.

The stationary unavailability and availability for the pump can be calculated, using MTTF = 1250 hrs. and MTTR = 50 hrs., then by Eqs. (5.5.15) and (5.5.16), $U(\infty)$ = 0.03846 and $A(\infty)$ = 0.96154, respectively.

Example 5.5.4. Regression equations can be developed for the break rates of water mains using data from specific water distribution systems. As an example, Walski and Pelliccia (1982) developed break-rate regression equations for the system in Binghamton, New York. These equations are:

Pit Cast Iron:

$$N(t) = 0.02577\ e^{0.0207t}$$

Sandspun Cast Iron:

$$N(t) = 0.0627\ e^{0.0137t}$$

where $N(t)$ = break rate (in breaks/mile/year) and t = age of the pipe (in years). The break rates versus the age of pipes for the above two cast iron pipes are shown in Fig. 5.5.3. Walski and Pelliccia (1982) also developed a regression equation for the time required to repair pipe breaks: $t_r = 6.5d^{0.285}$ where t_r = time to repair in hours and d = pipe diameter in inches. Derive the expressions for the failure rate, reliability, and failure density function for a 5-mile water main of sandspun cast iron pipe.

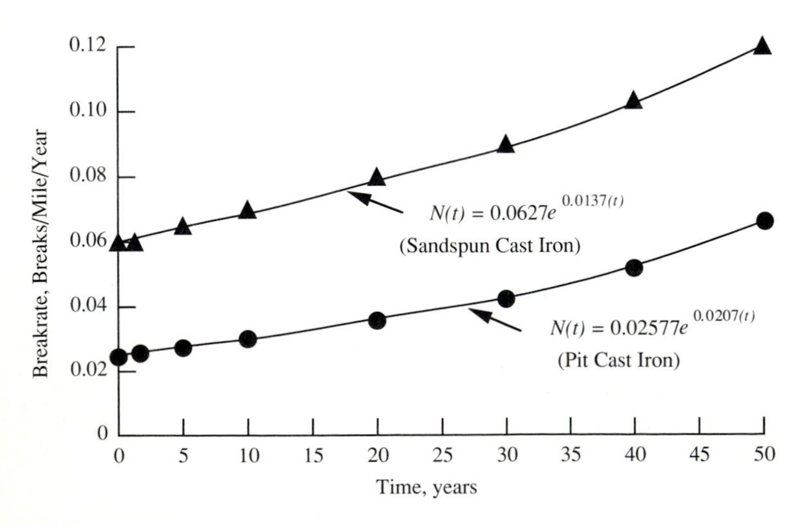

FIGURE 5.5.3
Break-rate curves for pit cast-iron and sandspun cast-iron pipes.

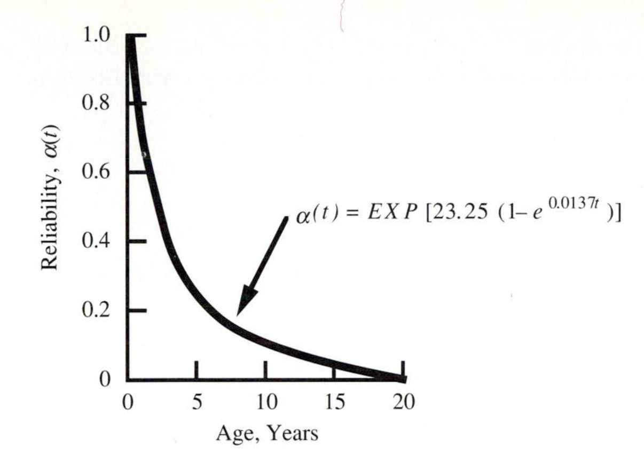

FIGURE 5.5.4
Reliability curves for pipe evaluation example.

Solution. The break rate per year (i.e., failure rate or hazard function for the 5-mile water main) can be calculated as

$$m(t) = 5 \text{ miles} \times N(t) = 0.3185\ e^{0.0137t}$$

The reliability of this 5-mile water main then can be computed using Eq. (5.5.5) as

$$\alpha(t) = \exp\left[-\int_0^t 0.3185\ e^{0.0137s}\ ds\right] = \exp\left[23.25\left(1 - e^{0.0137t}\right)\right]$$

The failure density $f_T(t)$ can be calculated, using Eq. (5.5.4), as

$$f_T(t) = 0.3185\ e^{0.0137t} \exp\left[23.25\left(1 - e^{0.0137t}\right)\right]$$

The reliability of the 5-mile main for various mission times is plotted as Fig. 5.5.4.

5.6 RELIABILITY ANALYSIS OF SIMPLE SYSTEMS

Most systems are composed of several subsystems. The reliability of a system depends on how the components are interconnected. Methods for computing reliability of simple series and parallel systems are presented in this section.

SERIES SYSTEMS. The simplest type system is a **series system** in which every component must function if the system is to function. Considering the time-to-failure T_i for the ith component as a random variable, then the system reliability over the period $(0, t)$ for a system of n components is

$$\alpha_s(t) = \prod_{i=1}^{n} P(T_i \geq t) = \prod_{i=1}^{n} \alpha_i(t) \tag{5.6.1}$$

where $\alpha_s(t)$ is the system reliability and $\alpha_i(t)$ is the reliability for the ith component. The system reliability for a series system is less than the individual component reliability. For a component that has failure times exponentially distributed (with constant failure rates) so that the ith component reliability is $\exp(-\lambda_i t)$, then the system reliability is

$$\alpha_s(t) = \exp\left(-\sum_{i=1}^{n} \lambda_i t\right) \tag{5.6.2}$$

The MTTF is

$$\text{MTTF} = \int_0^\infty \exp\left(-\sum_{i=1}^{n} \lambda_i t\right) dt = \frac{1}{\sum_{i=1}^{n} \lambda_i} \tag{5.6.3}$$

Example 5.6.1. Consider two different pumps in series, both of which must operate to pump the required quantity. The constant failure rates for the pumps are $\lambda_1 = 0.0003$ failures/hr and $\lambda_2 = 0.0002$ failures/hr. Determine the system reliability for a 2000-hr. mission time and the MTTF of the system.

Solution. For a 2,000-hr mission time, the system reliability is

$$\alpha_s(t) = e^{-(0.0003+0.0002)(2000)} = 0.90484$$

and

$$\text{MTTF} = \frac{1}{0.0003 + 0.0002} = 2{,}000 \text{ hrs.}$$

PARALLEL SYSTEMS. A **parallel system** is defined as one which will fail if and only if all units in the system fail or malfunction. The pure parallel system is one in which all components are initially activated, and any component can maintain the system operation. The system reliability is then expressed as

$$\alpha_s(t) = 1 - \prod_{i=1}^{n} [1 - \alpha_i(t)] \tag{5.6.4}$$

For a system with an exponentially distributed time-to-failure and a constant failure rate for each component of the system, the system reliability is

$$\alpha_s(t) = 1 - \prod_{i=1}^{n} \left(1 - e^{-\lambda_i t}\right) \tag{5.6.5}$$

and the MTTF for a system with identical components is

$$\text{MTTF} = \frac{1}{\lambda} \sum_{i=1}^{n} \frac{1}{i} \tag{5.6.6}$$

Hence, the system reliability for a parallel system is higher than the individual component reliability and, accordingly, the MTTF of a parallel system is longer than those of individual components.

Example 5.6.2. Consider two identical pumps operating in a redundant configuration so that either pump could fail and the peak discharge could still be delivered. Both pumps have a failure rate of $\lambda = 0.0005$ failures/hr. and both pumps start operating at $t = 0$. Determine the system reliability over the time period (0,1000 hrs.) and the MTTF of the system.

Solution. The system reliability for a mission time of $t = 1{,}000$ hr. is

$$\alpha_s(t) = 2e^{-\lambda t} - e^{-2\lambda t} = 2e^{-(0.0005)(1000)} - e^{-2(0.0005)(1000)}$$

$$= 1.2131 - 0.3679 = 0.8452$$

The MTTF is

$$\text{MTTF} = \frac{1}{\lambda}\left(\frac{1}{1} + \frac{1}{2}\right) = \frac{3}{2}\frac{1}{\lambda} = 3{,}000 \text{ hrs.}$$

5.7 OPTIMIZATION OF RELIABILITY

5.7.1 Reliability Design with Redundancy

Consider the design of a hydrosystem consisting of n main subsystems that are arranged in series so that the failure of one subsystem will cause failure of the entire system. Reliability (probability of no failure) of the hydrosystem can be improved by installing standby units in each subsystem. A design may require the use of standby units, which means that each subsystem may include up to K units in parallel. The total capital available for the hydrosystem is C and the cost for the ith subsystem is $C_i(k_i)$ where k_i is the number of parallel units in subsystem i.

The objective of this type of problem is to determine the number of parallel units k_i in subsystem i that maximizes the reliability of the system, α_s, without exceeding the available capital. System reliability would be defined as the product of the individual subsystem reliabilities $\alpha_i(k_i)$ for a system of n subsystems in series and k_i parallel units in each subsystem. This optimization problem is stated as

$$\max \alpha_s = \prod_{i=1}^{n} \alpha_i(k_i) \tag{5.7.1}$$

$$\text{S.T.} \quad \sum_{i=1}^{n} C_i(k_i) \le C \tag{5.7.2}$$

where k_i is the decision variable satisfying $0 \le k_i \le K$.

For this problem, the DP approach can be used to solve the above optimization model. The DP backward recursive equation would then be multiplicative as

$$f_i(b_i) = \begin{cases} \max\limits_{k_i} [\alpha_i(k_i)] & i = n \quad (5.7.2a) \\ \max\limits_{k_i} \{\alpha_i(k_i) \cdot f_{i+1}[b_i - C_i(k_i)]\} & i = 1, \ldots, n-1 \quad (5.7.2b) \end{cases}$$

where the state b_i is the total capital available for components i, $i+1$, ..., n. The stages are the subsystems $i = 1$, ..., n. The above problem is very similar to the capital budgeting problem (see Chapter 4).

5.7.2 Reliability Apportionment

Consider a hydrosystem consisting of $i = 1, \ldots, n$ subsystems, each of which is to be developed independently and are in series. The objective is to quantify a reliability goal, α_i, for each subsystem i so that a system reliability goal $\bar{\alpha}_s$, is obtained with minimum expenditure of capital. This optimization problem can be stated as

$$\text{Min} \sum_{i=1}^{n} C_i(\alpha_i, \bar{\alpha}_i) \tag{5.7.3}$$

subject to

$$\prod_{i=1}^{n} \bar{\alpha}_i \geq \bar{\alpha}_s \tag{5.7.4}$$

and

$$0 \leq \alpha_i \leq \bar{\alpha}_i \leq 1 \qquad i = 1, \ldots, n \tag{5.7.5}$$

where α_i is the reliability of each subsystem at the present state, $\bar{\alpha}_i$ is the reliability goal set for each subsystem ($\alpha_i \leq \bar{\alpha}_i \leq 1$) and $\bar{\alpha}_s$ is the system reliability goal ($0 < \bar{\alpha}_s < 1$).

This reliability optimization problem, which is nonlinear, can be converted to a DP problem where each subsystem is a stage; the state variable is the reliability allocated to each stage $\bar{\alpha}_i$ to meet the reliability goal; and the decisions are also the reliability allocated to each stage. The DP recursive equation for this problem is

$$f_i(\bar{\alpha}_i) = \underset{\bar{\alpha}_i}{\text{Min}} \{C_i(\alpha_i, \bar{\alpha}_i) + f(\bar{\alpha}_{i-1}\} \tag{5.7.6}$$

5.8 CHANCE-CONSTRAINED MODELS

In system modeling, one frequently is required to use quantities that cannot be assessed with certainty. Some of the parameters in an optimization model that are uncertain can be treated as random variables. In an LP model, some of the constraint coefficients a_{ij} and/or RHS coefficients b_i could be subject to uncertainty. Because of the uncertainty, the compliance of constraints for a given solution cannot be ensured. In other words, it is possible that, for any solution $\mathbf{x}$, there is a certain probability that constraints will be violated. Under such circumstances, it is logical to replace the original constraint by a probabilistic statement in the form of a **chance-constraint** as

$$P\left\{\sum_{j=1}^{n} a_{ij}x_j \leq b_i\right\} \geq \alpha_i, i = 1, 2, \ldots, m \tag{5.8.1}$$

in which α_i is the specified reliability of compliance for the ith constraint.

A probability statement is not mathematically operational for algebraic solution. For this reason, the so-called **deterministic equivalent** of Eq. (5.8.1) must be derived. There are three cases in which the random elements in equation (5.8.1) could occur:

(1) only the RHS coefficients, b_i, are random; (2) only elements a_{ij} on the LHS of the constraints are random; and (3) both a_{ij} and b_i are random.

5.8.1 Right-Hand Side Coefficients Random

The simplest of the three cases is case (1) in which only the RHS coefficient b_i is random. The derivation of the deterministic equivalent of a chance-constraint for this case is described as follows. For clarity, replace the random RHS coefficient b_i by B_i to denote that it is a random variable. Then, Eq. (5.8.1) is

$$P\left[\sum_{j=1}^{n} a_{ij}x_j \leq B_i\right] \geq \alpha_i \tag{5.8.2}$$

in which the random RHS coefficient B_i has a CDF, $F_{B_i}(b)$, with mean μ_{B_i} and standard deviation σ_{B_i} (see Fig. 5.8.1a). Equation (5.8.2) is equivalent to

$$P\left[B_i \leq \sum_{j=1}^{n} a_{ij}x_j\right] \leq 1 - \alpha_i \tag{5.8.3}$$

which can also be expressed in terms of the CDF of the random RHS coefficient, B_i, as

$$F_{B_i}\left(\sum_{j=1}^{n} a_{ij}x_j\right) \leq 1 - \alpha_i \tag{5.8.4}$$

Using the standardized variate of the random RHS coefficient, that is, $Z_{B_i} = (B_i - \mu_{B_i})/\sigma_{B_i}$, Eqs. (5.8.3) and (5.8.4) can be expressed, respectively, as

$$P\left[Z_{B_i} \leq \frac{\sum_{j=1}^{n} a_{ij}x_j - \mu_{B_i}}{\sigma_{B_i}}\right] \leq 1 - \alpha_i \tag{5.8.5}$$

and

$$F_{Z_{B_i}}\left(\frac{\sum_{j=1}^{n} a_{ij}x_j - \mu_{B_i}}{\sigma_{B_i}}\right) \leq 1 - \alpha_i \tag{5.8.6}$$

The deterministic equivalent of the original chance-constrained Eq. (5.8.2) is the inverse of Eq. (5.8.6)

$$\frac{\sum_{j=1}^{n} a_{ij}x_j - \mu_{B_i}}{\sigma_{B_i}} \leq F_{Z_{B_i}}^{-1}(1 - \alpha_i) \tag{5.8.7}$$

which can be rewritten as

$$\sum_{j=1}^{n} a_{ij}x_j \leq \mu_{B_i} + z_{B_i,\,1-\alpha_i}\sigma_{B_i} \tag{5.8.8}$$

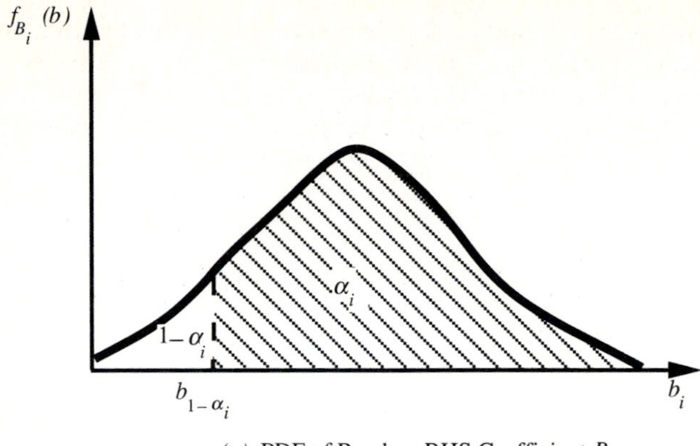

(a) PDF of Random RHS Coefficient B_i

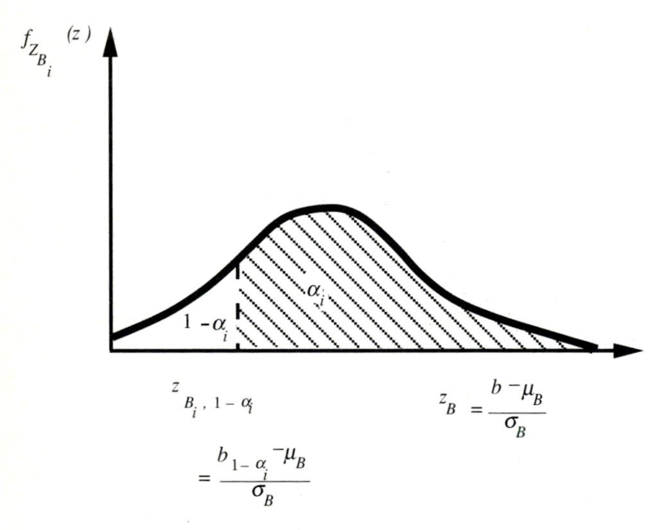

(b) PDF of Standardized Random RHS Coefficient Z_{B_i}

FIGURE 5.8.1
Probability density function (PDF) for chance constraint models.

where the specific value of $z_{B_i,\ 1-\alpha_i}$ is $F^{-1}_{Z_{B_i}}(1-\alpha_i)$, which is the $(1-\alpha_i)$th quantile of the standardized RHS coefficient B_i (see Fig. 5.8.1*b*). Knowing the PDF of the random RHS coefficient B_i and the required constraint compliance reliability, α_i, the specific value $z_{B_i,1-\alpha_i}$ can be determined. For example, if B_i has a normal distribution, then $Z_{B_i,1-\alpha_i}$ (referring to Section 5.2.3) is the standardized normal variate satisfying $\Phi(z_{B_i,1-\alpha_i}) = 1 - \alpha_i$ where $\Phi()$ is the standard normal CDF. Equation (5.8.8) illustrates that, when only the RHS coefficient B_i in an LP model is subject to uncertainty, the deterministic equivalent of the chance-constraint Eq. (5.8.1) is still linear. Furthermore, as the reliability requirement α_i decreases, the value of $z_{B_i,1-\alpha_i}$ increases and, consequently, the RHS of Eq. (5.8.8) increases.

Example 5.8.1. For the manufacturing-waste treatment problem of Example 3.1.1 the treatment capacity of 10 units is uncertain. The uncertain treatment capacity is assessed to follow a normal distribution with a mean value of 10 units and standard deviation of 1 unit. Formulate a chance-constraint such that the waste to be treated does not exceed the treatment capacity with a 95 percent reliability. Then derive its deterministic equivalent of the chance-constraint.

Solution. Let C be the random treatment capacity having a normal distribution with mean ($\mu_C = 10$ units) and standard deviation ($\sigma_C = 1$ unit). The constraint that the treated waste ($2x_1 - x_2$) does not exceed the treatment capacity with a 95 percent reliability can be stated as

$$P[2x_1 - x_2 \leq C] \geq 0.95$$

The deterministic equivalent of the above chance-constraint, based on Eq. (5.8.8) is

$$2x_1 - x_2 \leq \mu_C + z_{(1-0.95)}\sigma_C$$

where $z_{(1-0.95)} = z_{0.05} = -z_{0.95} = -1.645$ (from Table 5.2.1). The RHS is $\mu_C + z_{(1-0.95)}\sigma_C = 10 + (-1.645)(1) = 8.355$. The resulting deterministic equivalent of the chance-constraint is

$$2x_1 - x_2 \leq 8.355$$

5.8.2 Technological Coefficients Random

In the case that only the technological coefficients a_{ij} are random (replaced by A_{ij} to denote randomness) in an LP model with known CDF, $F_{A_{ij}}(a)$, the chance-constraint (Eq. 5.8.1) can be expressed as

$$P\left[\sum_{j=1}^{n} A_{ij}x_j \leq b_i\right] \geq \alpha_i \tag{5.8.9}$$

Let $A_i = \sum_{j=1}^{n} A_{ij}x_j$, then Eq. (5.8.9) can be written as

$$P[A_i \leq b_i] \geq \alpha_i \tag{5.8.10a}$$

or equivalently as

$$P\left[Z_{A_i} = \frac{A_i - \mu_{A_i}}{\sigma_{A_i}} \leq \frac{b_i - \mu_{A_i}}{\sigma_{A_i}}\right] \geq \alpha_i \tag{5.8.10b}$$

in which μ_{A_i} and σ_{A_i} are the mean and standard deviation of the random variable A_i. Since A_i is a linear combination of random variables A_{ij}, the mean of A_i can be derived by Eq. (5.1.15a) as

$$\mu_{A_i} = \sum_{j=1}^{n} \mu_{A_{ij}}x_j \tag{5.8.11}$$

and variance, by Eq. (5.1.22), as

$$\sigma_{A_i}^2 = \sum_{j=1}^{n} \sigma_{A_{ij}}^2 x_j^2 + 2 \sum_{j=1}^{n-1} \sum_{j'=j+1}^{n} x_j x_{j'} \text{Cov}\left(A_{ij}, A_{ij'}\right) \tag{5.8.12}$$

in which $\mu_{A_{ij}}$ and $\sigma_{A_{ij}}$ are the mean and standard deviation of the random technological coefficients in an LP model and Cov$(A_{ij}, A_{ij'})$ is the covariance between different technological coefficients in the ith constraint. If all random technological coefficients are uncorrelated, the variance of the random LHS term (A_i) reduces to (refer to Eq. (5.1.18*d*))

$$\sigma_{A_i}^2 = \sum_{j=1}^{n} \sigma_{A_{ij}}^2 x_j^2 \tag{5.8.13}$$

In case that all random technological coefficients follow a normal distribution, then, based on the properties of the normal random variables described in Section 5.2.3, the random LHS (A_i) is a normal random variable with mean and variance given by Eqs. (5.8.11) and (5.8.12), respectively.

Equation (5.8.10*b*) can be expressed as

$$F_{A_i}\left(\frac{b_i - \mu_{A_i}}{\sigma_{A_i}}\right) \geq \alpha_i \tag{5.8.14}$$

which is equal to

$$\frac{b_i - \mu_{A_i}}{\sigma_{A_i}} \geq z_{A_i,\alpha_i} \tag{5.8.15}$$

in which $z_{A_i,\alpha_i} = F_{A_i}^{-1}(\alpha_i)$. Substituting expressions for μ_{A_i} and σ_{A_i} in Eq. (5.8.15), the deterministic equivalent of the chance-constraint Eq. (5.8.9) is

$$\sum_{j=1}^{n} \mu_{A_{ij}} x_j + z_{A_i,\alpha_i} \sqrt{\sum_{j=1}^{n} \sigma_{A_{ij}}^2 x_j^2 + 2 \sum_{j < j'} \sum x_j x_{j'} \text{Cov}(A_{ij}, A_{ij'})} \leq b_i \tag{5.8.16}$$

for correlated random technological coefficients and

$$\sum_{j=1}^{n} \mu_{A_{ij}} x_j + z_{A_i,\alpha_i} \sqrt{\sum_{j=1}^{n} \sigma_{A_{ij}}^2 x_j^2} \leq b_i \tag{5.8.17}$$

for uncorrelated random technological coefficients. Under the normality condition, z_{A_i,α_i} is the α_ith quantile of the standardized normal random variable. Note that in Eqs. (5.8.16) and (5.8.17), the second terms of the constraint involve the square root of the quadratic function of the decision variables. Therefore, when the technological coefficients on the LHS of an LP model are random, the deterministic equivalent of the chance-constraint equation is no longer linear. Procedures for solving an optimization model involving a nonlinear objective function and/or nonlinear constraints are under the realm of nonlinear programming which is described in Chapter 4.

Example 5.8.2. For the manufacturing-waste treatment Example 3.1.1 the treatment capacity of 10 units is certain. However, the quantity of waste to be generated during the manufacturing process is uncertain and assumed to follow a normal distribution with a mean of 2 units and a standard deviation of 0.4 units for every unit of product manufactured. Formulate the chance-constraint such that the waste to be treated does not exceed the treatment capacity of 10 units with a reliability of 95 percent. Also, derive the deterministic equivalent of the chance-constraint.

Solution. The quantity of waste produced per unit of product (A_{11}) is considered a random variable. The chance-constraint for the water treatment capacity can be written as

$$P\left[A_{11}x_1 - x_2 \leq 10\right] \geq 0.95$$

in which A_{11} follows a normal distribution with mean $\mu_{A_{11}} = 2$ and standard deviation of $\sigma_{A_{11}} = 0.4$. Let $A_1 = A_{11}x_1 - x_2$ which involves only one random variable A_{11} because x_2 is the quantity of waste discharged directly with no treatment (a decision variable without uncertainty). Because A_{11} is a normal random variable, then A_1 is also a normal random variable with mean $\mu_{A_1} = 2x_1 - x_2$ and variance $\sigma^2_{A_1} = (0.4x_1)^2$. Based on Eq. (5.8.17), the deterministic equivalent of this chance-constraint is

$$2x_1 - x_2 + z_{A_1,0.95}\sqrt{(0.4x_1)^2} \leq 10$$

From the normal probability table (Table 5.2.1), the value of $z_{A_1,0.95} = 1.645$. The above inequality can be rewritten as

$$2.658x_1 - x_2 \leq 10$$

The deterministic equivalent is linear because there is only one random technologic coefficient in the waste treatment capacity constraint.

5.8.3 Right-Hand Side and Technological Coefficents Random

Finally, when both the RHS coefficient and some of the technological coefficients in a given constraint of an LP model are random, the chance-constrained equation can be expressed as

$$P\left[\sum_{j=1}^{n} A_{ij}x_j - B_i \leq 0\right] \geq \alpha_i \qquad (5.8.18)$$

Following the same procedure as described previously, the deterministic equivalent of Eq. (5.8.18) can be derived as

$$\sum_{j=1}^{n} \mu_{A_{ij}}x_j + Z_{T_i,\alpha_i} \sqrt{\sum_{j=1}^{n}\sigma^2_{A_{ij}}x_j^2 + \sigma^2_{B_i} + 2\sum_{j}\sum_{<j'}x_j x_{j'}\text{Cov}\left(A_{ij}, A_{ij'}\right) + 2\sum_{j=1}^{n}x_j\text{Cov}\left(A_{ij}, B_i\right)} \leq \mu_{B_i} \qquad (5.8.19)$$

in which $T_i = \sum_{j=1}^{n} A_{ij}x_j - B_i$ and Cov(A_{ij}, B_i) is the covariance between random technological coefficient A_{ij} and the RHS coefficient B_i for the ith constraint.

REFERENCES

Ang, A. H-S.: "Structural Risk Analysis and Reliability-Based Design," *Journal of Structural Engineering Division*, ASCE, vol. 99, no. ST9, 1973.

Ang, A. H-S and W. H. Tang: *Probability Concepts in Engineering Planning and Design, Volume I: Basic Principles*, Wiley, New York, 1979.

Ang, A. H-S., and Tang, W. H.: *Probability Concepts in Engineering and Design, Volume II: Decision, Risk, and Reliability*, Wiley, New York, 1984.

Benjamin, J. R. and C. A. Cornell: *Probability, Statistics, and Decisions for Civil Engineers*, McGraw-Hill, Inc., New York, 1970.

Blank, L.: *Statistical Procedures for Engineering, Management, and Science*, McGraw-Hill, Inc., 1980.

Brown, L. C. and T. O. Barnwell: "The Enhanced Stream Water Quality Models QUAL2E and QUAL2E-UNCAS: Documentation and User Manual," *Report*, EPA/600/3-87/007, U.S. Environmental Protection Agency, 1987.

Burges, S. J. and D. P. Lettenmaier: "Probabilistic Methods in Stream Quality Management," *Water Resources Bulletin*, vol. 11, 1975.

Cheng, S. T.: "Overtopping Risk Evaluation of an Existing Dam," Ph.D. Dissertation, Department of Civil Engineering, University of Illinois, Urbana-Champaign, Ill., 1982.

Chow, V. T., D. R. Maidment, and L. W. Mays: *Applied Hydrology*, McGraw-Hill, Inc., New York, 1988.

Devore, J. L.: *Probability and Statistics for Engineering and Sciences*, 2d ed., Brooks/Cole, Monterey, Calif., 1987.

Fogel, M., and L. Duckstein: "Stochastic Precipitation Modeling for Evaluating Non-point Source Pollution," in *Statistical Analysis of Rainfall and Runoff*, V. P. Singh, ed., Water Resources Publication, Littleton, Colo., 1982.

Haan, C. T.: *Statistical Methods in Hydrology*, Iowa State University Press, 1977.

Harr, M.: *Reliability-Based Design in Civil Engineering*, McGraw-Hill, Inc., New York, 1987.

Henley, E. J. and H. Kumamoto: *Reliability Engineering and Risk Assessment*, Prentice-Hall, Englewood Cliffs, N.J., 1981.

Kapur, K. C. and L. R. Lamberson: *Reliability in Engineering Designs*, Wiley, New York, 1977.

Lee, H. L. and L. W. Mays: "Improved Risk and Reliability Model for Hydraulic Structures," *Water Resources Research*, AGU, vol. 19, no. 5, pp. 1415–1422, 1984.

Mays, L. W. ed.: *Reliability Analysis of Water Distribution Systems*, ASCE, New York, 1989.

Rousselle, J.: "On Some Problems of Flood Analysis," Ph.D. Dissertation, Colorado State University, 1970.

Tang, W. H., L. W. Mays and B. C. Yen: "Optimal Risk-Based Design of Storm Sewer Networks," *Journal of Environmental Engineering Division*, ASCE, vol. 103, no. EE3, June, 1975.

Todorovic, P. and V. Yevjevich: "Stochastic Process of Precipitation," Hydrology Paper, No. 35, Colorado State University, 1969.

Tung, Y. K. and L. W. Mays: "Risk Analysis for Hydraulic Structures," *Journal of Hydraulics Division*, ASCE, vol. 106, no. HY5, May, 1980.

Tung, Y. K. and L. W. Mays: "Risk Models for Levee Design," *Water Resources Research*, AGU, vol. 17, no. 4, August, 1981.

Tung, Y. K. and W. E. Hathhorn: "Assessment of Probability Distribution of Dissolved Oxygen Deficit," *Journal of Environmental Engineering*, ASCE, vol. 114, no. 6, December, 1988.

Virjling, J. K.: "Probabilistic Design of Water Retaining Structures," in *Engineering Reliability and Risk in Water Resources*, L. Duckstein and E. J. Plate, eds., Martinus Nijhoff, Dordrecht, The Netherlands, 1987.

Walski, T. M. and A. Pelliccia: "Economic Analysis of Water Main Breaks," *Journal of the American Water Works Association*, March, 1982.

Yen, B. C.: "Safety Factor in Hydrologic and Hydraulic Engineering Design," *Reliability in Water Resources Management*, E. A. McBean, K. W. Hipel, and T. E. Unny, eds., Water Resources Publications, Littleton, Colo., 1979.

Yen, B. C., Cheng, S. T., and C. S. Melching: "First-Order Reliability Analysis," in *Stochastic and Risk Analysis in Hydraulic Engineering*, B. C. Yen, ed., Water Resources Publications, Littleton, Colo., 1986.

Zelenhasic, E.: "Theoretical Probability Distribution for Flood Peaks," Hydrology Paper, no. 42, Colorado State University, 1970.

PROBLEMS

5.1.1 Figure 5.P.1 shows that two tributaries 1 and 2 merge to a main channel 3. Assume that tributaries 1 and 2 have equal bankful capacities, however, hydrological characteristics of the two corresponding drainage basins is somewhat different. During storm events, $P(E_1) = P$ (Bankful capacity of tributary 1 is exceeded) $= 0.10$, $P(E_2) = P$ (Bankful capacity of tributary 2 is exceeded) $= 0.20$. Also, we know that $P(E_1|E_2) = 0.50$ and $P(E_2|E_1) = 1.00$.

(*a*) If the bankful capacity of the main channel 3 is the same as tributaries 1 and 2, what is the probability that bankful capacity of the main channel is exceeded? (Assume that when tributaries 1 and 2 are carrying less than their bankful capacities, flow in main channel 3 exceeding its bankful capacity is 0.2).

(*b*) If the bankful capacity of main channel 3 is twice that of tributaries 1 and 2, what is the probability that bankful capacity of the main channel is exceeded? (Assume that, if only tributaries 1 or 2 exceeds its bankful capacity, flow in main channel 3 exceeding its bankful capacity is 0.15).

5.1.2 A detention basin is designed to accommodate excessive surface runoff temporarily during storm events. The detention basin should not have overflowed, if possible, to prevent potential pollution of the stream or other receiving bodies. For simplicity, the amount of daily rainfall is categorized as heavy, moderate, and light (including none). With the present storage capacity, the detention basin is capable of accommodating runoff generated by two consecutive days of heavy rainfall or three consecutive days of at least moderate rainfall. The daily rainfall amounts around the detention basin site are not entirely independent. In other words, the amount of daily rainfall at a given day would affect the daily rainfall amount in the next day. Let random variable X_t represent the amount of rainfall in any day t. The transition probability matrix indicating the conditional probability of rainfall amount in a given day t conditioned on the rainfall amount of the previous day is shown in the following table.

		X_{t+1}		
		H	M	L
$X_t =$	H	0.3	0.5	0.2
	M	0.3	0.4	0.3
	L	0.1	0.3	0.6

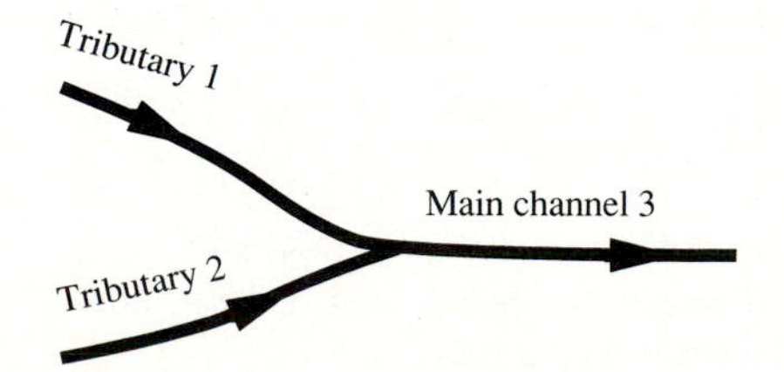

FIGURE 5.P.1

For a given day the amount of rainfall is light, what is the probability that the detention basin will overflow in the next three days?

5.1.3 Referring to Problem 5.1.2, compute the probability that the detention basin will overflow in the next three days. Assume that, at any given day of the month, the probabilities for the various rainfall amounts are: $P(H) = 0.1$, $P(M) = 0.3$, $P(L) = 0.6$.

5.1.4 Factories 1, 2, and 3 occasionally release lethal waste into a river (denoted as events F_i, $i = 1, 2, 3$) with the probabilities of 0.25, 0.50 and 0.25, respectively. Probabilities of fish-killing ingredients being in the waste (as event A) of the three factories are: $P(A|F_1) = 0.75$, $P(A|F_2) = 0.25$, and $P(A|F_3) = 0.05$ (Haan, 1977).

(*a*) When a waste release has occurred, what is the probability of causing fish killing?

(*b*) When fish killing is observed, what is the probability of each factory being responsible?

5.1.5 Before a section of concrete pipe of special order can be accepted for installation in a culvert project, the thickness of the pipe is inspected by State Highway Department personnel for specification compliance by ultrasonic reading. For this particular project, the required thickness of concrete pipe wall must be at least 3 inches. The inspection is done by arbitrarily selecting a point on the pipe surface and measure the thickness at that point. The pipe is accepted if the thickness from the ultrasonic reading exceeds 3 inches; otherwise, the entire section of the pipe is rejected. Suppose, from past experience, that 90 percent of all pipe sections manufactured by the factory were found to be in compliance with specifications. However, the ultrasonic thickness determination is only 80 percent reliable.

(*a*) What is the probability that a particular pipe section is well manufactured and will be accepted by the Highway Department?

(*b*) What is the probability that a pipe section is poorly manufactured and will be accepted on the basis of ultrasonic test?

5.1.6 A quality-control inspector is testing sample output from a manufacturing process for concrete pipes for a storm sewer project wherein 95 percent of the items are satisfactory. Three pipes are chosen randomly for inspection. The successive quality events may be considered as independent. What is the probability that: (a) none of the three pipes inspected are satisfactory; and (b) exactly two are satisfactory.

5.1.7 Consider the possible failure of a water supply system to meet the demand during any given summer day.

(*a*) Determine the probability that the supply will be insufficient if the probabilities shown in the table are known.

	Demand Level	***P* (level)**	***P* (Inadequate supply/level)**
D1	100,000 gpd	0.6	0.0
D2	150,000 gpd	0.3	0.1
D3	200,000 gpd	0.1	0.5

(*b*) Find the probability that a demand level of 150,000 gpd was the *cause* of the system's failure to meet demand if an inadequate supply is observed.

(*c*) The likelihood of a pump failing and causing the system to fail is 0.02 regardless of the demand level, what is the probability of system failure in this case?

5.1.8 Assume that the Manning's roughness has a triangular distribution with the lower bound, a, the mode, b, and the upper bound, c, as shown in Fig. 5.P.2.

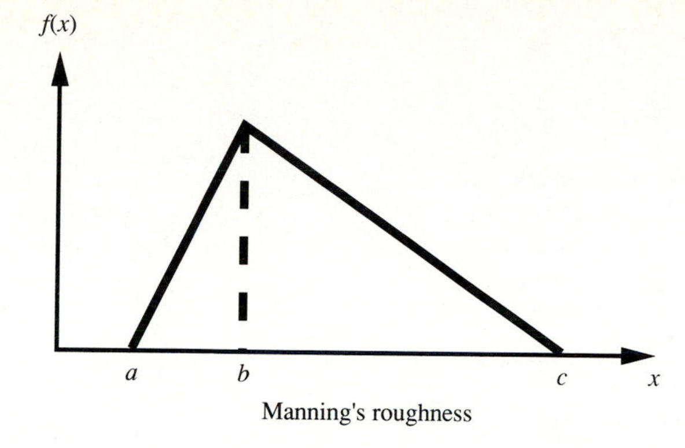

FIGURE 5.P.2
Triangular distribution for Manning's roughness.

(*a*) Derive the PDF of Manning's roughness.
(*b*) Derive the expression for the mean and variance of Manning's roughness.

5.2.1 The well-known Thiem equation can be used to compute drawdown in a confined and homogeneous aquifer as

$$s_{ij} = \frac{\ln\left(r_{oj}/r_{ij}\right)}{2\pi T} Q_j = \xi_{ij} Q_j$$

in which s_{ij} = drawdown at observation location i resulting from a pumpage of Q_j at the jth production well; r_{oj} = radius of influence of production well j; r_{ij} = distance between observation point i and production well j; and T = transmissivity of aquifer. The overall effect of aquifer drawdown at any observation point i, when more than one production well is in operation, can be obtained, by the principle of linear superposition, as the sum of responses caused by all production wells in the field, that is,

$$s_i = \sum_{j=1}^{N} s_{ij} = \sum_{j=1}^{N} \xi_{ij} Q_j$$

where N = total number of production wells in operation.

Consider a system consisting of two production wells and one observation well. The locations of the three wells, the pumping rates of the two production wells and their zone of influence are shown in the Fig. 5.P.3. It is assumed that the transmissivity of the aquifer has a lognormal distribution with the mean μ_T = 4000 gpd/ft. and standard deviation σ_T = 2000 gpd/ft.

(*a*) Prove that the total drawdown in the aquifer field is also lognormally distributed.
(*b*) Compute the exact values of the mean and variance of total drawdown at the observation point.
(*c*) Compute the probability that the resulting drawdown at the observation point does not exceed 2 feet.
(*d*) If the maximum allowable probability of the total drawdown exceeding 2 feet is 0.10, find out the maximum allowable total pumpage from the two production wells.

5.2.2 A company plans to build a production factory by a river. You are hired by the company as a consultant to analyze the flood risk of the factory site. It is known that the magnitude of an annual flood has a lognormal distribution with a mean of 30,000 cfs and standard deviation of 25,000 cfs. It is also known that, from the field investigation, the stage-discharge relationship of the channel reach is $Q = 1500H^{1.4}$ where Q = flow rate in

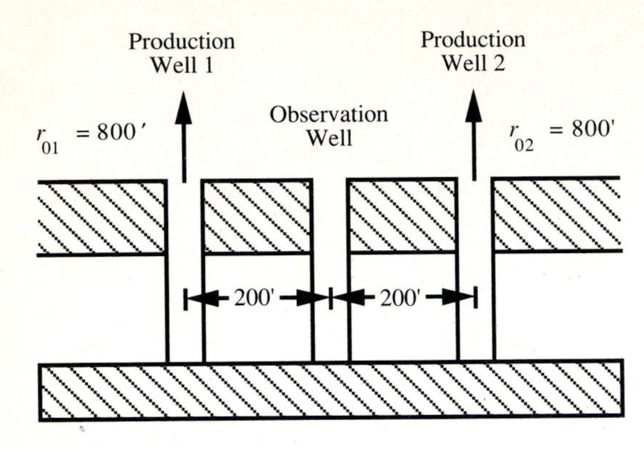

FIGURE 5.P.3

cfs and H = water surface elevation in feet above a given datum. The elevation of the tentative location for the factory is 15 feet above the datum.

(*a*) What is the annual risk that the factory site will be flooded?

(*b*) If the factory is to be operated over a 25-year period, what is the flood risk in this period?

(*c*) At this particular plant site, it is also known that the flood damage function can be approximated as

$$\text{Damage (in \$1,000)} = \begin{cases} 0, & \text{if } H \leq 15' \\ 40(\ln H + 8)(\ln H - 2.7), & \text{if } H > 15' \end{cases}$$

What is the annual expected flood damage? (Use an appropriate numerical approximation technique for calculation.)

5.3.1 Referring to Problem 5.2.1, apply first-order analysis to each individual drawdown by the two production wells

(*a*) to estimate the mean and variance of the total drawdown at the observation point and to compare the result from part (b) of Problem 5.2.1;

(*b*) Would the inclusion of second order terms in Taylor expansion improve the estimation of the mean of total drawdown?

(*c*) Repeat part (c) of Problem 5.2.1 using the mean and variance obtained in part (a) by the first-order analysis. Assume that the distribution of the total drawdown is log normal.

5.3.2 In the design of storm sewer systems, the rational formula

$$Q_L = CiA$$

is frequently used in which Q_L is the surface inflow resulting from a rainfall of intensity i falling on the contributing drainage area of A, and C is the runoff coefficient. On the other hand, Manning's formula, for full pipeflow,

$$Q_C = 0.463n^{-1}S^{1/2}D^{8/3}$$

is commonly used to compute the flow carrying capacity of storm sewers in which D is the diameter of sewer pipe, and n and S are Manning's roughness and pipe slope, respectively.

Consider that all the parameters in the rational formula and Manning's equation are independent random variables with their respective mean μ_X and standard deviation

σ_X where X represents a random variable.

(*a*) Apply first-order analysis to derive the expressions for the mean and variance of safety margin.

(*b*) Given the following data on the the parameters, compute the reliability of a 36-inch pipe using the normal distribution for safety margin.

Parameter	**Mean**	**Standard deviation**
C	0.825	0.058575
i (in./hr.)	4.000	0.6
A (ac)	10.000	0.5
n	0.015	0.00083
D (ft.)	3.0	0.03
S_0	0.005	0.00082

5.3.3 Repeat Problem 5.3.1 by considering both normal and lognormal distributions for the safety factor as the performance criterion.

5.3.4 Consider a water distribution system (see Fig. 5.P.4) consisting of a storage tank serving as the source, and a 1-ft diameter cast-iron pipe of 1-mile long, leading to a user. The head elevation at the source is maintained at a constant level of 100 feet above the user. It is also known that, at the user end, the required pressure head is fixed at 20 psi with variable demand on flow rate. Assume that the demand in flow rate is random, having a lognormal distribution with the mean 3 cfs and standard deviation 0.3 cfs. Because of the uncertainty in pipe roughness and pipe diameter, the supply to the user is not certain. We know that the pipe has been installed for about 3 years. Therefore, our estimation of the pipe roughness in the Hazen-Williams equation is about 130 with some error of ± 20. Furthermore, knowing the manufacturing tolerance, the 1-ft pipe has an error of ± 0.05 ft. Again, we assume that both the pipe diameter and the Hazen-William's C coefficient has lognormal distributions with the means of 1 ft. and 130 and standard deviations of 0.05 ft and 20, respectively. Compute the reliability that the demand of the user can be satisfied.

5.3.5 Refer to Problem 5.3.4 and use safety margin as the performance criterion to estimate the mean and variance of the safety margin by first-order analysis. Further, compute the reliability using the normal distribution for the safety margin. Compare the exact reliability computed in Problem 5.3.4.

5.3.6 Repeat Problem 5.3.5 using safety factor as the performance criterion. Compare the reliability computed in this problem with those computed in Problems 5.3.4 and 5.3.5.

5.4.1 Suppose that, at a given dam site, the flood flows follow a triangular distribution with lower bound (l_a = 0), mode (l_b = 2000 cfs), and upper bound (lc = 5000 cfs). The spillway capacity of the dam, due to various hydraulic uncertainties described in Chapter 1, is also a random variable with a triangular distribution having a lower bound (r_a = 2500 cfs), mode (r_b = 3500 cfs), and upper bound (r_c = 4500 cfs). Use a direct integration method to calculate the reliability that the spillway capacity is able to convey the flood flow.

5.4.2 Refer to Problem 5.4.1. The flood flow distribution remains the same whereas the probability distribution of spillway capacity is simplified to a piecewise uniform distribution with the following PDF

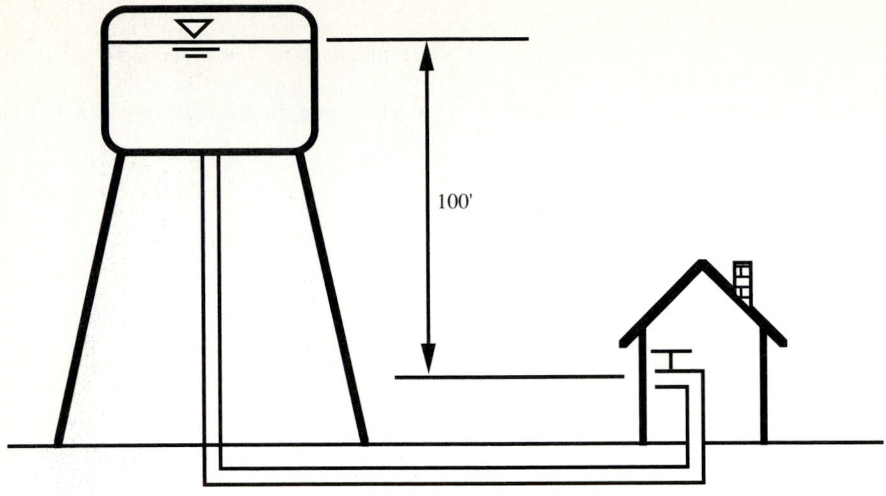

FIGURE 5.P.4

$$f_R(r) = \begin{cases} 0.0004, 2500 \text{ cfs} \leq r < 3000 \text{ cfs} \\ 0.0006, 3000 \text{ cfs} \leq r < 4000 \text{ cfs} \\ 0.0004, 4000 \text{ cfs} \leq r < 4500 \text{ cfs}. \end{cases}$$

Determine the spillway reliability by direct integration method.

5.4.3 Refer to Problem 5.4.1 and assume that flood flow and spillway capacity are independent random variables. Compute the mean and variance of the safety margin and use the normal distribution for the safety margin to compute the approximate reliability.

5.4.4 Resolve Problem 5.4.3 using the piecewise uniform distribution for spillway capacity given in Problem 5.4.2.

5.4.5 Resolve Problem 5.4.3 using the MFOSM method using performance function Eq. (5.4.12).

5.4.6 Resolve Problem 5.4.3 using the MFOSM method using performance function Eq. (5.4.13).

5.5.1 The failure rate having a form of power function as

$$m(t) = \frac{\beta t^{\beta-1}}{\theta^\beta}, \beta > 0$$

is very versatile. The shape of $m(t)$ can be a constant, increasing, or decreasing by choosing different values of β.

(*a*) Let $\theta = 1$, construct the failure function curves for $\beta = 1/2$, 1, 2, and 4.

(*b*) Prove that the failure rate with the above power function form has the failure density function $f_T(t)$

$$f_T(t) = \frac{\beta t^{\beta-1}}{\theta^\beta} \exp\left[-\left(\frac{t}{\theta}\right)^\beta\right], \quad t > 0$$

and reliability function

$$\alpha(t) = \exp\left[-\left(\frac{t}{\theta}\right)^\beta\right], \quad t > 0$$

5.5.2 Generally, after a system or a component such as a pump is assembled and placed in operation, the initial failure rate is higher than that encountered later. The initial failures may be due to various manufacturing and assembling defects that escape detection by the quality control system. As the defective parts are replaced with new ones, the reliability improves. After this break-in period, the failure rate would stay rather constant for some time before it gradually increases again due to wear out. This bathtube type of failure can be approximated as a piecewise linear function as shown graphically in Fig. 5.P.5.
(*a*) Show that the corresponding failure density function, $f_T(t)$, is

$$f_T(t) = \begin{cases} (c_0 + m_0 - c_1 t)\exp\left\{-\left[(c_0+m_0)t - c_1\left(\frac{t^2}{2}\right)\right]\right\}, & 0 \le t \le \frac{c_0}{c_1} \\ m_0 \exp\left\{-m_0 t + \left(\frac{c_0^2}{2c_1}\right)\right\}, & \frac{c_0}{c_1} < t \le t_0 \\ [c(t-t_0)+m_0]\exp\left\{-\left[\left(\frac{c}{2}\right)(t-t_0)^2 + \left(\frac{c_0^2}{2c_1}\right) + m_0 t\right]\right\}, & t > t_0 \end{cases}$$

(*b*) Show that the reliability function is

$$\alpha(t) = \begin{cases} \exp\left\{-\left[(c_0+m_0)t - c_1\left(\frac{t^2}{2}\right)\right]\right\}, & 0 \le t \le \frac{c_0}{c_1} \\ \exp\left\{-\left(m_0 t + \frac{c_0^2}{2c_1}\right)\right\} & \frac{c_0}{c_1} < t \le t_0 \\ \exp\left\{-\left[\left(\frac{c}{2}\right)(t-t_0)^2 + m_0 t + \left(\frac{c_0^2}{2c_1}\right)\right]\right\}, & t > t_0 \end{cases}$$

(*c*) Consider a pump having a piecewise linear bathtube failure rate function with parameters $c_0 = 0.0024$ failures/hr., $m_0 = 0.0008$ failures/hr., $c_1 = 0.000008$ failures/hr., and $c = 0.00001$ failures/hr. Construct the diagrams for failure rate, failure density, and reliability functions.

5.5.3 Compute (a) the MTTF and (b) the most likely time to failure for the failure density function in Problem 5.5.1.

5.5.4 Studying the historical pipe-break records, it is found that the number of breaks can be estimated as

$$N(t) = N(t_0) - t_0)^b$$

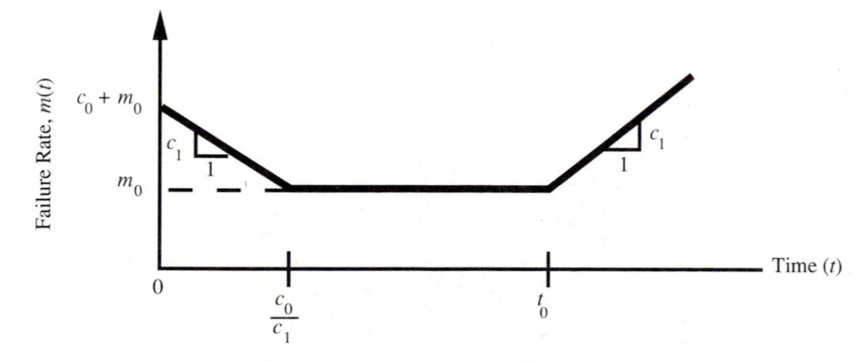

FIGURE 5.P.5
Piecewise linear bathtube failure rate function.

in which $N(t)$ = number of breaks per 1000-ft length of pipe in year t; t = time in years; t_0 = base year when pipe was installed; and b = constant, $b \geq 1$.

(*a*) Derive the reliability function and failure density function for the system having a 1000-ft-long pipe.

(*b*) Derive the expression for the MTTF for the system with 1000-ft-long pipe.

5.6.1 Consider a pump station consisting of n identical pumps in series each of which has a failure rate of power function form as defined in Problem 5.5.1. Assume that each pump performs independently.

(*a*) Derive the expression for the reliability of the pump station.

(*b*) Derive the failure density function for the pump station.

5.6.2 Resolve Problem 5.6.1 assuming that the n pumps in the pump station are arranged in parallel.

5.6.3 Refer to Figs. 5.P.6 and 5.P.7 for two systems. Assume that all components are identical and perform independently. The reliability of each component is α.

(*a*) Derive the expressions for system reliability of the two systems.

(*b*) Which system configuration gives a higher system reliability?

5.7.1 In a pump station, three components are arranged in series. Different components will install different types of pumps. The reliability of the station can be improved by

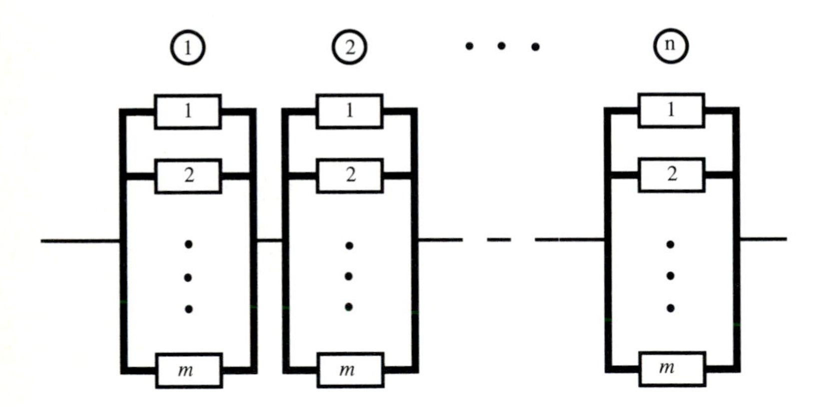

FIGURE 5.P.6

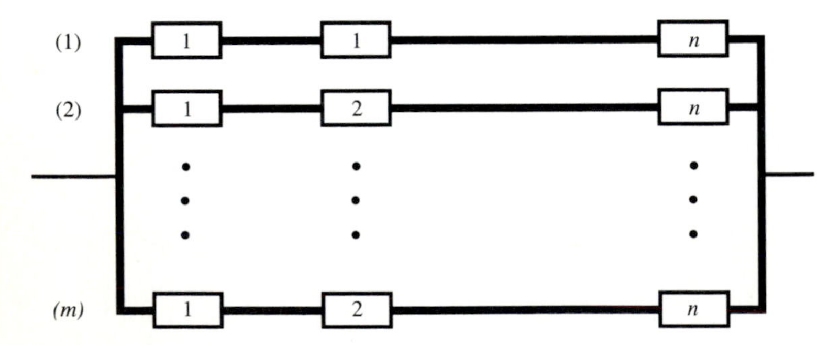

FIGURE 5.P.7

installing standby pump units on each component. Because of space limitations, only three parallel pump units in each component can be installed. The total capital available for installing standby units for the whole pump station is $100,000. The standby pump unit to be installed for components 1, 2, and 3 has a reliability of 0.90, 0.95, 0.99 with cost $15,000, $20,000, and $30,000, respectively. Formulate a model and determine the number of standby pump units for each component that maximizes system reliability by dynamic programming approach.

5.8.1 Refering to Example 5.8.2, derive the deterministic equivalents for the constraints on waste discharge limit and Constraint 3 in the manufacturing-waste treatment example. Use 95 percent for constraint compliance requirement and assume quantity of waste generated per unit good produced is a normal random variable.

5.8.2 Combine the derived chance constraints in Problem 5.8.1 and Example 5.8.2 to reformulate the optimal manufacturing-waste treatment example by maximizing the expected total net benefit.

5.8.3 Resolve Problem 5.8.2 by using $\alpha = 0.99$ and discuss the results as compared with those for Problem 5.8.2.

5.8.4 Reformulate the optimal manufacturing-waste treatment problem by considering both treatment capacity and quantity of waste generated per unit good produced are independent normal random variables simultaneously. Again, the objective function is to maximize the expected total net benefit. Use information contained in Examples 5.8.1 and 5.8.2 along with α = 95 percent for constraint compliance requirement. Furthermore, solve the model by appropriate technique.

5.8.5 Resolve Problem 5.8.4 by changing the objective function to maximize the probability that the total net benefit exceeds $25 K.

PART II

WATER SUPPLY ENGINEERING AND MANAGEMENT

CHAPTER 6

WATER DEMAND FORECASTING

6.1 WATER USE AND FORECASTING

Water use can be divided into two categories, **consumptive use**, in which water is an end to itself, and **nonconsumptive use**, in which water is a means to an end. Consumptive use includes municipal, agriculture, industry and mining. Nonconsumptive use includes instream uses such as hydropower, transportation and recreation. From an economic viewpoint, we have the greatest ability to model consumptive uses. Consumptive uses are modeled using consumptive functions and nonconsumptive uses are modeled using production functions. **Water use** refers to the amount of water applied to achieve various ends so that it is a descriptive concept. **Water demand** is the scheduling of quantities that consumers use per unit of time for particular prices of water, which is an analytical concept.

Municipal water use can be divided into categories of residential (houses and apartments), commercial (businesses and stores), institutional (schools and hospitals), industrial, and other water use (park watering, swimming pools, fire-fighting). To the water delivered for these uses (or **consumption**), must be added to that lost due to leakage from the distribution system to determine the amount of treated water (or **production**); then, adding the amount consumed by the treatment processes, yields the water withdrawn from all sources, or water **supply**, for the city. Unlike water use in agriculture, where water is an input into a production system, municipal water use is mostly for meeting human needs without direct economic consequences.

Water use for light industry is included in municipal water use, but some industries are such intensive water users that they must be identified and studied separately. Chief among these is cooling water for steam-electric power generation. Other major water-using industries are petroleum refining, chemicals and steel manufacturing, textiles, food processing, and pulp and paper mills. Agricultural water use includes water use for irrigating fields and for the drinking and care of animals. Irrigation, which is the largest user, may be further classified into flood irrigation, spray irrigation, and trickle irrigation, according to the method of application.

The ability to manage and operate existing water supply facilities and then to plan and design new water supply facilities is directly tied to the ability to describe both present and future water use. The future could refer to hours, days, weeks, months, or years, depending upon the particular problem. **Demand forecasting** is used to forecast water use in the future based upon previous water use and the socioeconomic and climate parameters of past and present water use. Water demand or water use may exhibit hourly, daily, monthly, seasonal and annual variations. Fig. 6.1.1 illustrates the temporal distribution of water use for three cities in the U.S. Various cities have differing factors affecting the water use. Austin, Texas (Fig. 6.1.1*a*) shows a definite similarity of the seasonal variation in water use. Boca Raton, Florida (Fig. 6.1.1*b*) shows less seasonal variation compared with Austin, Texas; however, the growth through time is evident in the Florida data. The trend for Allentown, Pennsylvania (Fig. 6.1.1*c*) is different; a rising trend up to 1976–1977 is apparent, followed by a decrease, possibly reflecting economic and population trends in the state.

All water supply is practically derived from two sources: surface water (e.g., streams, lakes, etc.) and groundwater. It is estimated that over four trillion gallons of precipitation fall on the 48 contiguous states daily. Of that total, Americans use about 450 to 700 billion gallons daily which is a little over 10 percent of the total amount (Dzurick, 1988). The majority of total precipitation, about 65 percent of it, returns to the atmosphere through evaporation and transpiration. Considering the total volumes of surface water and groundwater available, the United States can be considered as a water-abundant nation.

A **forecast** is an estimate of the future state of a parameter that has four dimensions: quantity, quality, time, and space. In the context of water-demand forecasting, the parameter of interest could be the daily average use, daily maximum use, and others. In water project design and planning, the major factors determining the project cost are the quantity of water that must be supplied, treated, distributed, and of waste water to be collected, treated, and disposed of each year. The character, size, and timing of engineering works for water facilities in the future largely depend on the future water use which must be forecasted. Therefore, the ability to manage and operate existing water supply facilities and then to plan and design new water supply facilities is directly tied to the ability to describe both present and future water use.

Future in forecasting could refer to hours, days, weeks, months or years, depending upon the particular problem. Because of the size and capital intensiveness of most water projects, the time scale in water demand forecasting generally is years with 15–25 years for medium-range forecasting and 50 years for long-range forecast-

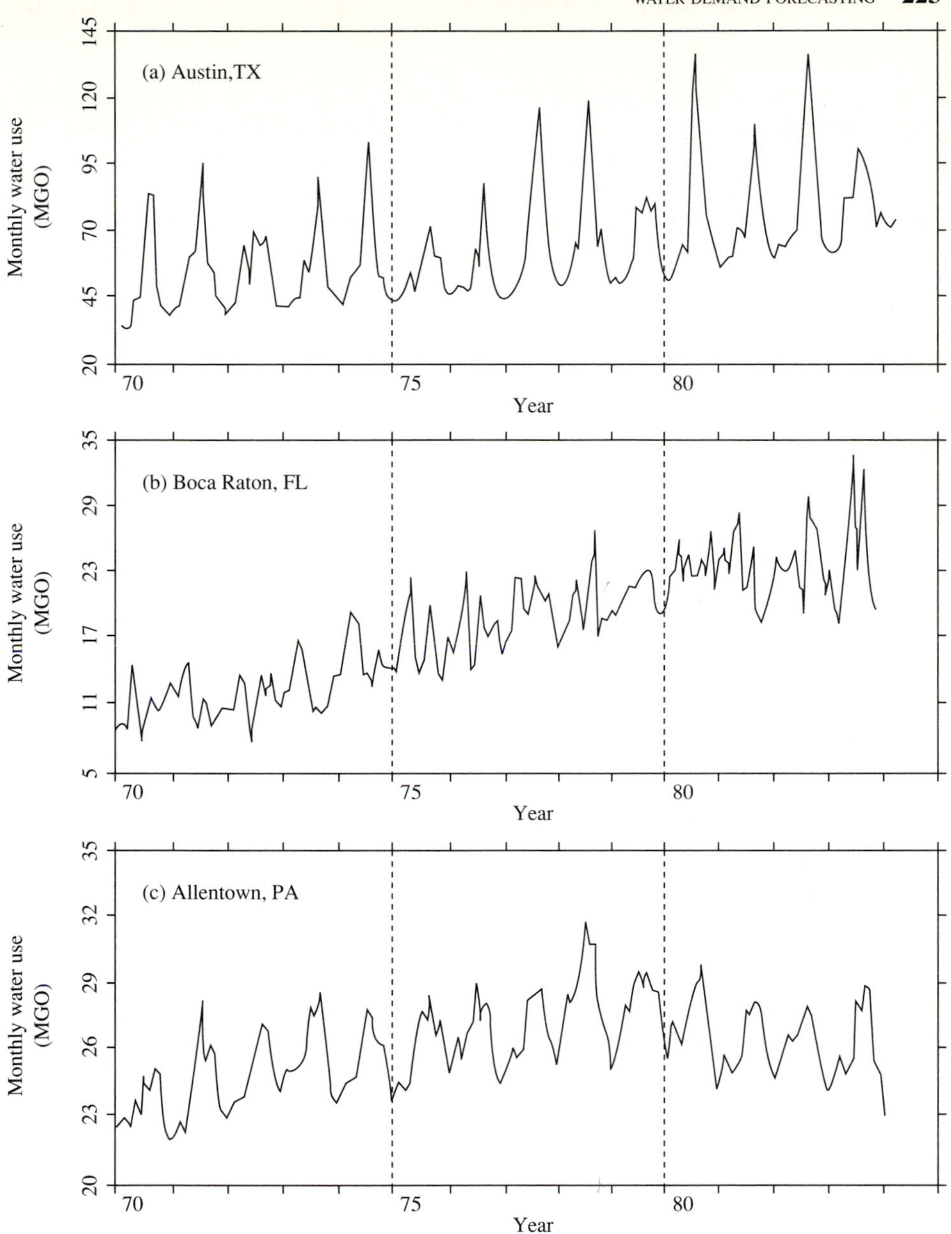

FIGURE 6.1.1
Monthly water use trends in three U.S. cities (Maidment et al., 1985).

ing. Forecasting cannot strictly be a scientific procedure, since the future, properly speaking, does not exist (Encel et al., 1976).

Water demand is defined in economic terms that are related to its price. It differs from the concept of water requirement used in engineering analysis. Forecasts of water demand should also reflect technological changes in production processes, product outputs, raw materials, water handling and waste treatment methods, social taste, and public policies with respect to water use and development. Explicit inclusion of these factors is important in medium and long-range forecasts. Otherwise, forecast results would be of limited value to decision-makers. Therefore, simplistic methods such as linear extrapolation of past water demand (called **projection**) are generally not appropriate for long-term forecasting. However, the methods remain appropriate to assist in managing water during a *crisis period*, during which the forecast-time horizon is short.

Due to the ever changing nature of social, economic, and political environments in a region, there exist numerous uncertainties in any forecast. Errors in water use forecasts may arise from inappropriate or unintended assumptions made in determining the parameters of forecast. These include future population, industry mix, and relationships between the values of model parameters and level of water use. Whatever the cause, errors in forecasting produce excess economic and environmental costs. Such costs may be avoided through the use of improved forecasting approaches. In addition, improved methodologies for forecasting water demands are needed to account for: (1) growing number of conflicts among water uses and water users; (2) increasing realization of interrelationships among the different outputs from water resource systems; and (3) increasing scope and scale of water resources development.

Aggregate water use is the sum of uses by many individual users for many individual purposes. Aggregate methods tend to conceal all but the least common denominator among trends. The advantage of using disaggregate methods is that the likely effect on the total demand due to change in any sector can be assessed.

6.2 FORECASTING MUNICIPAL AND INDUSTRIAL WATER USE

6.2.1 Classification of Methods

Forecasting municipal water demand is an important task for water utility agencies, involving three interrelated activities. The first activity is **supply management** which refers to forecasting water demand so that investments in new supply facilities can be scaled, sequenced, and timed. A second interrelated activity is **demand management** to determine the impact of water meter installation, leak detection, leak control, price changes, conservation measures, and rationing. The third activity is **demand-supply management** which uses water use forecasts to integrate and coordinate supply and demand management policies.

Boland et al. (1981) classify forecasting approaches into three types: (a) **single coefficient methods** which have only one explanatory variable; (b) **multiple coefficient methods** with more than one explanatory variable; and (c) **probabilistic methods** or **contingency tree methods**. Table 6.2.1 provides a comparison of the various forecasting approaches. **Explanatory variables** are variables that are used to explain the demand for water, such as population, price, income, and annual precipitation.

TABLE 6.2.1

Comparison of forecasting approaches (Boland et al., 1983)

	Single coefficient methods			Multiple coefficient methods		Probabilistic methods
	Per capita	**Per connection**	**Unit use coefficient**	**Requirements model**	**Demand model**	**Contingency tree**
Facilitates forecasts consistent with Principles & Guidelines	No	No	When used in disaggregate forecasts	When used in disaggregate forecasts	When used in disaggregate forecasts	Yes
Facilitates evaluation of water conservation measures	No	No	When used in disaggregate forecasts	When used in disaggregate forecasts	When used in disaggregate forecasts	Yes
Suitable for preliminary or reconnaissance studies	Yes	Yes	Yes	No, too complex	No, too complex	No, too complex
Suitable for project planning applications	No	No	When used in disaggregate forecasts	When used in disaggregate forecasts	When used in disaggregate forecasts	Yes
Quantity of data needed	Very little	Very little	Moderate	Moderate to large	Moderate to large	Depends upon application
Difficulty of obtaining needed data	Low	Low	Low to moderate	Moderate to high	Moderate to high	Depends upon application

Source: Adapted from Table III-2, p. III-35, in the *Forecasting Assessment*.

SINGLE COEFFICIENT METHODS. These methods include per capita, per connection, and unit use coefficient methods. Frequently only water production data for a city as a whole are available, along with an estimate of the city's population so that the per capita method is used. In these circumstances, an approximate relation is used to estimate municipal water use:

$$Q(t) = u(t)\text{POP}(t) \tag{6.2.1}$$

where $Q(t)$ is the mean daily water use, POP(t) is population in period t, and $u(t)$ is the mean daily water use per capita (gal./cap./day or liters/cap./day). Using Eq. (6.2.1), the population POP(t) and the mean per capita use $u(t)$ must be estimated or predicted. To obtain the peak daily water use and the peak hourly water use, the average use rates obtained from Eq. (6.2.1) are multiplied by peak-to-average ratios ranging from 1.5 to 3.0.

The per capita approach is widely used; however, it has serious shortcomings for most forecasting applications (Boland et al., 1981) because it ignores many factors that affect water use. These include housing type, household size, climate, commercial activity, income, price, etc. Goals of the U.S. Water Resources Council Principles and Guidelines (1983) state that these additional factors should be included and that a **sectorally disaggregated basis** should be used in preparation of forecasts, which refers to separation by major use sector. Neither of the above goals can be accomplished with the per capita method. To refine the per capita method, the mean daily per capita water use coefficient $u(t)$ can be developed for the various water use categories. The coefficients can also be disaggregated by geographic region and by season.

As a variation of the per capita method, the water supply for a city or community may be estimated using the **per connection method** by

$$Q(t) = \frac{1}{\eta}\sum_{i=1}^{n} C_i(t)W_i(t) \tag{6.2.2}$$

where $W_i(t)$ is the water use per person or per water connection in water use category i, $C_i(t)$ is the number of connections to the water distribution system in category i, $i = 1, 2, \ldots, n$, and η is an efficiency ($0 \leq \eta \leq 1$) representing leakage and water lost in treatment. For a well-managed system, $\eta = 0.9$.

The per connection (number of connections or customers) method usually has the advantage that historical data on the number of connections is readily available. Basically, this method does have the advantages and disadvantages of the per capita method. Other methods which use only one explanatory variable are referred to as **unit use coefficient methods**. Examples of unit use coefficients are water use per employee in industrial and commercial water demand forecasting and water use per unit land of a specific crop in agricultural water forecasting. These methods may be refined and consistent with the Principles and Guidelines (US WRC, 1983) when used in a disaggregate forecast provided that significant explanatory variables are not omitted. In all cases, single coefficient methods rely on the prediction of a key variable and the unit use coefficient in the future. The methods provide reasonably reliable short-term forecasts and become increasingly questionable for long-term forecasts.

MULTIPLE COEFFICIENT METHODS. These methods can be categorized into requirements models and demand models. **Requirements models** include physical/psychological variables that are correlated with water use but do not necessarily include water price and household or per capita income. **Demand models** are based on economic reasoning and include only variables which are correlated significantly with water use and which are expected to be causally related to water use (Boland et al., 1981). As an example, demand models include price and income in addition to other explanatory variables.

A demand model for annual water use could be of the form

$$\text{Annual Water Use} = a_0 + a_1\,(\text{Population}) + a_2\,(\text{Price}) + a_3\,(\text{Income}) + a_4\,(\text{Annual Precipitation}) \qquad (6.2.3)$$

or of a nonlinear form

$$\text{Annual Water Use} = b_0\,(\text{Population})^{b_1}\,(\text{Price})^{b_2}\,(\text{Income})^{b_3}\,(\text{Annual Precipitation})^{b_4} \qquad (6.2.4)$$

The number and types of explanatory variables used in demand models or requirements models vary greatly from one application to another. Data availability, required accuracy, and local conditions all have an effect on the number and types of explanatory variables. When applied in disaggregate forecasts, multiple coefficient methods meet the requirements of the Principles and Guidelines of the U.S. WRC (1983). Again, to predict future water use by a multiple coefficient method, future values of casual factors must be predicted by other means.

CONTINGENCY TREE METHODS. These methods consider uncertain factors in water use forecast. Typically, a base forecast is prepared by one of the methods discussed above and then the forecast is modified to reflect a combination of the uncertain factors. A joint probability of each of the combinations is determined for association with the forecasted water use.

In selecting a forecasting method, consideration must be given to the specific application of the forecast. Most methods could be used for a forecast of average annual aggregate water use. However, if the forecast is to be used in the design of a water treatment and distribution system, much more reliable estimates of maximum day water use are required. The design of a reservoir for water supply may require forecasts of seasonal or monthly water use in addition to annual water use. Other forecast applications may be to show the effect of various levels of economic development on water use. For this type of application, the model must include the needed economic variables. In summary, forecasting methods must make the best use of available data in order to provide the required water use information for the planning and design process.

6.2.2 General Form of Models

A general water demand model can be expressed as

$$Q = f(x_1, x_2, \ldots, x_k) + \epsilon \qquad (6.2.5)$$

where f is the function of explanatory variables $x_1, x_2, \ldots, x_k$, and ϵ is a stochastic error (random variable) describing the joint effect on Q of all the factors not explicitly considered by the explanatory variables. The stochastic error ϵ is assumed to have an expected value of zero; a constant variance; and uncorrelated errors, that is, the expected value of the product $\epsilon_i \cdot \epsilon_j$ is zero.

For practical applications, water demand models do not have an a priori analytical form but are commonly assumed to be additive, multiplicative, or a combination of these. Demand models typically are of the linear, logarithmic, or semi-logarithmic form given respectively as

$$Q = a_0 + a_1 x_1 + \cdots + a_k x_k + \epsilon \tag{6.2.6}$$

or

$$\ln Q = b_0 + b_1 \ln x_1 + \cdots + b_k \ln x_k + \epsilon \tag{6.2.7}$$

or

$$Q = c_0 + c_1 \ln x_1 + \cdots + c_k \ln x_k + \epsilon \tag{6.2.8}$$

These forms of demand models allow for easy estimation of the model parameters, $a_0, \ldots, a_k$, $b_0, \ldots, b_k$, and $c_0, \ldots, c_k$, by use of regression analysis, assuming that the explanatory variables are determined independently of the dependent variable. If the explanatory variables include other dimensions of water demand, then the above condition of independence is not satisfied. For example, if the quantity of water demanded and the price of water were determined in the same process, this is a simultaneous determination problem. In Section 6.3, the theories of regression analysis using the least squares method for estimating model parameters based on a given data set are presented.

6.2.3 Data Availability

In order to develop the outlined water demand models, various types of data for the explanatory variables are required. In effect, the type of data available dictates the variables to be used in a forecasting model. Table 6.2.2 lists some of the typical data types and possible sources.

6.3 REGRESSION MODELS FOR WATER USE FORECASTING

6.3.1 Regression Concepts for Water Use Forecasting

Many empirical studies in hydrosystem problems involve relating system responses of interest to a number of contributing factors. For example, hydrologists often relate surface runoff characteristics such as peak flow, or runoff volume to meteorological and physiographical characteristics of a basin such as rainfall volume, precipitation intensity, watershed size, extent of urbanization, etc. In water demand forecasting, water resource engineers wish to develop a model that relates water use to various

TABLE 6.2.2

Data types and possible sources (Boland et al., 1983)

Data category	Specific data items	Possible data sources
1. Population	a. Population; household size b. Population projections	U.S. Census of Population, housing; local and state planning agencies; city and county data books; state demographer; local and state economic development agency; econometric firms; state and national statistical abstracts; OBERS regional projections
2. Housing	a. Number of housing units by type and by market value; housing density; average lot sizes; assessed valuations b. Housing unit projections by type	U.S. Census of Population, housing; U.S. Census Metropolitan Housing Characteristics; local and state planning agencies; real estate assessment agencies; state demographer; state financial agencies; local and state economic development agency; local zoning commission; econometric firms; OBERS regional projections
3. Employment	a. Total employment by major industry sectors; employment disaggregated by 3- and 4-digit SIC categories; local historical employment growth rates b. Aggregate and disaggregate (by SIC)* employment projections	U.S. Census of Population: Detailed Socioeconomic Characteristics; U.S. Census: County Business Patterns; U.S. Census of Manufacturers, Services, Wholesale Trade, etc.; local and state planning agencies; U.S. Bureau of Labor Statistics; U.S. Department of Commerce; Monthly Labor Review; employment security divisions; local and state economic development agencies; state financial agencies; econometric firms; local manufacturing directory; local service directory; Chambers of Commerce employment listings; OBERS regional projections
4. Other economic variables	a. Consumer Price Index; construction cost index; personal and household income b. Income projections	U.S. Census of Population: Detailed Socioeconomic Characteristics; local and state planning agencies; state financial agencies; Department of Commerce; Monthly Labor Review; state and national statistical abstracts; U.S. Bureau of Labor Statistics; econometric firms; OBERS regional projections
5. Climate	Local weather patterns: rainfall, temperature, evapotranspiration rates, moisture deficit (normal and temporal conditions)	National Oceanic and Atmospheric Administration (NOAA); National Weather Service; university experiment stations; soil and water conservation districts; local airports
6. Land use	Land use patterns; zoning ordinances	U.S. Census of Agriculture; local and state planning agencies; city zoning commissions; city directories; U.S. Census: Block Statistics Reports
7. Water statistics	Water/wastewater prices and rate structures; historical monthly water use by customer class; historical monthly number of accounts by customer class; historical data on unaccounted losses; scope of self-supplied users	Local water supply agency; engineering reports; state water surveys; customer surveys; state regulatory agency; local government ordinances
8. Conservation	Implemented conservation measures; future conservation alternatives; measurements of reduction, coverage of effectiveness of measures; social acceptability; institutional framework; water-using appliances	Local water supply agencies; state regulatory agency; local ordinances and regulatory statistics; local and state planning agencies; interviews of government officials and general public; consumer (satisfaction evaluation) reports; manufacturer (water-using appliance) specifications; literature studies

*SIC refers to standard industrial classification.

apparent social, demographic, economic and hydrologic factors. The adjective "apparent" is used because, in many empirical studies, the exact cause-effect relationship is not entirely understood. However, development of such empirical relations help analysts to gain insight and understanding concerning the system behavior.

Regression analysis is the most frequently used statistical technique to investigate and model such empirical relationships. It is applied widely practically in every field of engineering and science. To illustrate the concept of the regression analysis, consider that the engineers for the city of Austin, Texas are interested in developing a model for predicting water use in the future. The first task is to identify the factors that might affect the water use which could be, but are not limited to, population size, price of water, average income, and annual precipitation. Once the factors are identified, data are collected for the various factors over time as shown in Table 6.3.1. One can imagine that the total annual water use for Austin is a result of the combined effect of many factors. For purposes of illustration, the per capita approach is used to examine the relation between the annual total water use and its population.

Naturally, one's intuition is that the annual water use would increase as its population grows. However, a more useful question is how does water use vary with respect to the population size. If such a relationship can be established, engineers and city planners can predict the amount of water use for the anticipated population

TABLE 6.3.1
Water demand data for Austin, Texas

Year	Population	Price ($/1000 gal)	Income ($/person)	Annual precipitation (in.)	Water use (ac-ft)
1965	216,733	0.98	5,919	40.57	39,606
1966	223,334	0.95	5,970	25.19	40,131
1967	230,135	1.20	6,521	33.54	45,667
1968	237,144	1.15	7,348	40.43	40,780
1969	244,366	1.10	7,965	33.59	45,330
1970	251,808	1.05	8,453	30.64	50,683
1971	259,900	1.00	8,713	24.95	56,600
1972	268,252	1.20	9,286	26.07	57,157
1973	276,873	1.13	9,694	40.46	57,466
1974	285,771	1.06	9,542	36.21	63,263
1975	294,955	0.98	9,684	36.81	57,357
1976	304,434	0.93	10,152	39.17	51,163
1977	314,217	0.87	10,441	22.14	68,413
1978	324,315	0.81	10,496	30.97	69,994
1979	334,738	1.10	10,679	37.50	65,204
1980	345,496	1.05	10,833	27.38	78,564
1981	354,401	0.96	11,060	45.73	76,339
1982	368,135	0.91	11,338	26.63	87,309
1983	383,326	0.87	11,752	33.98	82,120
1984	399,147	0.84	12,763	26.30	97,678
1985	424,120	1.41	12,748	32.49	97,708

growth so that adequate water supply can be provided. The first step to establish such a relationship is to plot the population size (on the horizontal axis) versus the corresponding water use (on the vertical axis) in the form of a scatter diagram as shown Fig. 6.3.1. The points in Fig. 6.3.1 show some degree of scatter, however, a rather clear upward trend is visible.

A next step could be to develop a mathematical function to describe this upward trend of water use with respect to the population size. It may be hypothesized that the water use, Q, for the city of Austin is linearly related to its population, POP, which can be described by the following equation

$$Q = \beta_0 + \beta_1 \text{ POP} \tag{6.3.1}$$

in which β_0 is the intercept and β_1 is the slope of the line. Equation (6.3.1) is a deterministic model in which water use (Q) is uniquely determined by population (POP). In fact, data points shown in Fig. 6.3.1 do not exactly fall on a straight line, so Eq. (6.3.1) is modified as

$$Q = \beta_0 + \beta_1 \text{ POP} + \epsilon \tag{6.3.2a}$$

or

$$y = \beta_0 + \beta_1 x + \epsilon \tag{6.3.2b}$$

in which y is Q, x is POP, and ϵ is the error term denoting the discrepancy between the observed water use (Q or y) and that estimated by the straight line equation

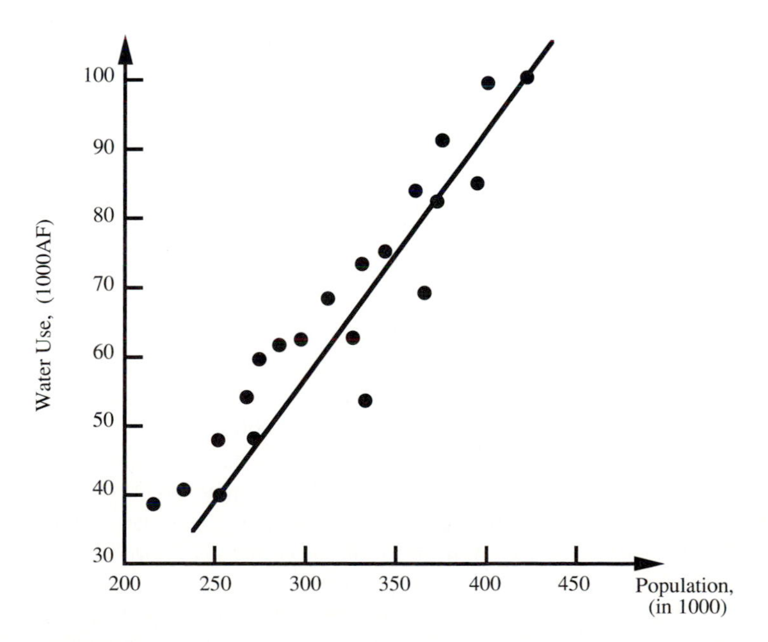

FIGURE 6.3.1
Plot of annual water use versus population for city of Austin, Texas.

($\beta_0 + \beta_1$ POP). In other words, the error (ϵ) accounts for the failure of the proposed model to exactly fit the observed data. Sources that contribute to such error could arise from measurement errors, inadequate model, and the effect of other factors on water use.

Equation (6.3.2) is a **linear regression model**. The variable y is referred to as the **dependent variable** (or **response**) while variable x is the **independent variable** (or **regressor**). More precisely, Eq. (6.3.2) is referred to as a **simple linear regression** model because there is only one independent variable. A general extension of Eq. (6.3.2) involving more than one independent variable is called the **multiple linear regression model** which can be expressed as

$$y = \beta_0 + \beta_1 x_1 + \cdots + \beta_k x_k + \epsilon \tag{6.3.3}$$

in which β represents the parameters of the model called **regression coefficients**. It should be made clear that the adjective "linear" used for Eqs. (6.3.2) and (6.3.3) indicates that the dependent variable y is linear with respect to regression coefficients, β, not because y is linearly related to the independent variable x.

Once the form of the model is hypothesized, the next phase of the regression analysis is to estimate the regression coefficients, which are unknown model parameters, using data observations for dependent and independent variables. This task of fitting a model to data requires adoption of a criterion for measuring the degree of goodness-of-fit. The most commonly used criterion is based on the least squares principle, which is described later. Following the phase of parameter estimation, the analyst should investigate the appropriateness of the model and quality of fit. This step is necessary to ensure the practical usefulness and theoretical validity of the resulting regression model. Checking the model adequacy is an important step for identifying any potential flaws in the model and can lead to modification and refinement of the model.

Developing a model relating a dependent variable and independent variables (or regressors) by regression analysis is an iterative process. The success in developing a reasonable model depends largely on the analyst's ability to interpret the resulting model and to correlate model behavior to the process under investigation.

Before discussing the technical aspects, there are several important concepts in regression analysis that one should recognize. Regression models generally do not imply a cause and effect relationship between variables involved. A causal relationship between variables should be supported by evidence other than what is indicated from a data set alone. In many empirical investigations, engineers frequently have a sufficient understanding of the process and are able to select independent variables that are causal factors contributing to the response. In such circumstances, regression analysis can be used to provide confirmation to such causal relationships. Equations developed by the regression analysis are only an approximation to more complex relationships that exist in the real-life processes. Furthermore, all regression models are only valid within the region of data used in the analysis. Serious error can be made if one applies a regression equation outside the bounds of the regressors in the data set. Hence, to enhance the applicability of a regression model, analysts should collect data which are representative of the process under study.

6.3.2 Linear Regression

Referring to Eq. (6.3.3), the expression for individual data observations can be expressed as

$$y_i = \beta_0 + \beta_1 x_{il} + \cdots + \beta_k x_{ik} + \epsilon_i, \qquad i = 1, 2, \ldots, n \tag{6.3.4}$$

where y_i is the ith observation of the dependent variable, x_{ik} is the ith observation of the kth independent variable, ϵ_i is the corresponding error, and n is the total number of observations in the data set. In matrix form, the linear regression model represented by Eq. (6.3.4) can be written as

$$\mathbf{y} = \mathbf{X}\boldsymbol{\beta} + \boldsymbol{\epsilon} \tag{6.3.5}$$

in which

$$\mathbf{y} = \begin{bmatrix} y_1 \\ y_2 \\ \vdots \\ y_n \end{bmatrix} \qquad \mathbf{X} = \begin{bmatrix} 1 & x_{11} & x_{12} & \ldots & x_{1k} \\ 1 & x_{21} & x_{22} & \ldots & x_{2k} \\ \vdots & \vdots & \vdots & \vdots & \vdots \\ 1 & x_{n1} & x_{n2} & \ldots & x_{nk} \end{bmatrix}$$

$$\boldsymbol{\beta} = \begin{bmatrix} \beta_0 \\ \beta_1 \\ \vdots \\ \beta_k \end{bmatrix} \qquad \boldsymbol{\epsilon} = \begin{bmatrix} \epsilon_1 \\ \epsilon_2 \\ \vdots \\ \epsilon_n \end{bmatrix}$$

In regression analysis, the typical assumptions are

1. The random error associated with each observation is normally distributed with mean zero and unknown variance σ^2, that is, $E(\epsilon_i) = 0$ and Var $(\epsilon_i) = \sigma^2$ for $i = 1, 2, \ldots, n$.
2. All random errors are statistically independent. That is, $\text{Cov}(\epsilon_i, \epsilon_j) = 0$ for $i \neq j$.
3. The values of the independent variables $x_1, x_2, \ldots, x_k$ are measured without error.
4. The dependent variable y is a random variable with mean and variance, respectively, described as

$$E(y|x_1, x_2, \ldots, x_k) = \beta_0 + \beta_1 x_1 + \cdots + \beta_k x_k \tag{6.3.6a}$$

$$\text{Var}(y|x_1, x_2, \ldots, x_k) = \sigma^2 \tag{6.3.6b}$$

From Eqs. (6.3.6*a*, *b*), the mean of y is a linear function of independent variables while the variance of y is a constant which does not depend on the values of independent variables.

In general, the independent variables, **x**, in Eq. (6.3.3) each can be different variables. A special case is when the model is a polynomial function of the same independent variable given as

$$y = \beta_0 + \beta_1 x + \beta_2 x^2 + \cdots + \beta_k x^k + \epsilon \tag{6.3.7}$$

in which k is the order of polynomial function. Other commonly used multiple linear regression models, Eqs. (6.2.7) and (6.2.8) which, referring to Fig. 6.3.2, are respectively the logarithmic transforms of

$$y = e^{\beta_0} x_1^{\beta_1} x_2^{\beta_2} \ldots x_k^{\beta_k} e^{\epsilon} \tag{6.3.8}$$

and

$$e^y = e^{\beta_0} x_1^{\beta_1} x_2^{\beta_2} \ldots x_k^{\beta_k} e^{\epsilon} \tag{6.3.9}$$

respectively.

The important task in regression analysis is to estimate the unknown regression coefficients β based on an observed data set. Referring to Fig. 6.3.1, it is graphically possible to draw many different lines through the data points. The question is "which is the best straight line fitting the data?" In regression analysis, unknown regression coefficients are determined using the **least square principle** by which the sum of squares of differences between the observed and the computed dependent variable y is minimized. Mathematically, the least squares criterion in a linear regression can be stated as

$$\begin{aligned} \underset{\hat{\beta}_0, \hat{\beta}_1 \ldots, \hat{\beta}_k}{\text{Minimize}} \; D &= \sum_{i=1}^{n} \epsilon_i^2 \\ &= \sum_{i=1}^{n} (y_i - \hat{y}_i)^2 \\ &= \sum_{i=1}^{n} [y_i - (\hat{\beta}_0 + \hat{\beta}_1 x_{i1} + \cdots + \hat{\beta}_k x_{ik})]^2 \end{aligned} \tag{6.3.10}$$

in which D is the measure of goodness-of-fit, y_i is the observed value of the ith response, and $\hat{y}_i$ is the computed value of the ith response associated with the particular values of $\hat{\beta}_0, \hat{\beta}_1, \ldots, \hat{\beta}_k$, given as $\hat{y}_i = \hat{\beta}_0 + \hat{\beta}_1 x_{i1} + \cdots + \hat{\beta}_{ik} x_{ik}$.

The determination of regression coefficients, in essence, is an unconstrained minimization problem with Eq. (6.3.10) as the objective function. The unknown regression coefficients β can be determined by the optimization techniques discussed in Section 4.4. Using the optimality condition of Eq. (4.4.2), the values of β that minimize D must satisfy the following necessary conditions:

$$\left[\frac{\partial D}{\partial \beta_0}\right]_{\hat{\beta}_0, \hat{\beta}_1 \ldots, \hat{\beta}_k} = -2 \sum_{i=1}^{n} (y_i - \hat{\beta}_0 - \hat{\beta}_1 x_{i1} - \cdots - \hat{\beta}_k x_{ik}) = 0 \tag{6.3.11a}$$

$$\left[\frac{\partial D}{\partial \beta_j}\right]_{\hat{\beta}_0, \hat{\beta}_1, \ldots, \hat{\beta}_k} = -2 \sum_{i=1}^{n} (y_i - \hat{\beta}_0 - \hat{\beta}_1 x_{i1} - \cdots - \beta_k x_{ik}) x_{ij} = 0, \quad j = 1, 2, \ldots, k \tag{6.3.11b}$$

Equations (6.3.11*a*, *b*) can be simplified to the following **normal equations** with $k+1$ unknowns and $k + 1$ equations.

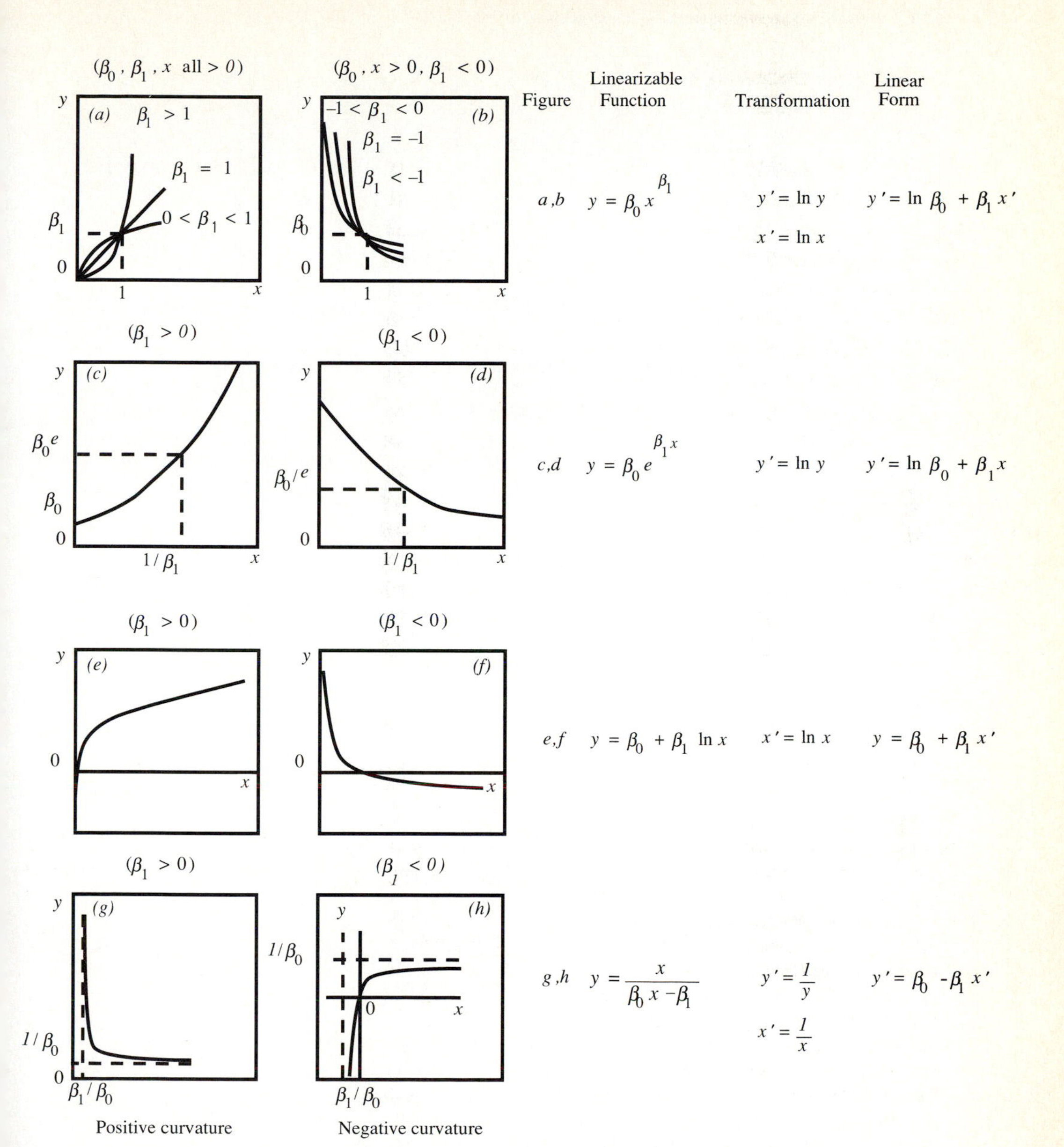

FIGURE 6.3.2
Linearizable functions. From Daniel and Wood, 1980.

$$
\begin{aligned}
n\hat{\beta}_0 &+ \sum_{i=1}^{n} x_{i1}\hat{\beta}_1 + \sum_{i=1}^{n} x_{i2}\hat{\beta}_2 + \cdots + \sum_{i=1}^{n} x_{ik}\hat{\beta}_k = \sum_{i=1}^{n} y_i \\
\sum_{i=1}^{n} x_{i1}\hat{\beta}_0 &+ \sum_{i=1}^{n} x_{i1}^2\hat{\beta}_1 + \sum_{i=1}^{n} x_{i1}x_{i2}\hat{\beta}_2 + \cdots + \sum_{i=1}^{n} x_{i1}x_{ik}\hat{\beta}_k = \sum_{i=1}^{n} y_i x_{i1} \\
&\vdots \\
\sum_{i=1}^{n} x_{ik}\hat{\beta}_0 &+ \sum_{i=1}^{n} x_{ik}x_{i1}\hat{\beta}_1 + \sum_{i=1}^{n} x_{ik}x_{i2}\hat{\beta}_2 + \cdots + \sum_{i=1}^{n} x_{ik}^2\hat{\beta}_k = \sum_{i=1}^{n} y_i x_{ik}
\end{aligned}
\tag{6.3.12}
$$

in which x_{ij} represents the ith observation of the jth independent variable, $i = 1, 2, \ldots, n$ and $j = 1, 2, \ldots, k$.

In matrix form, the normal equations (6.3.12) can be expressed as

$$(\mathbf{X}^T\mathbf{X})\hat{\boldsymbol{\beta}} = \mathbf{X}^T\mathbf{y} \tag{6.3.13}$$

in which T indicates the transpose of a matrix or a vector. Equation (6.3.13) can be solved for $\hat{\boldsymbol{\beta}}$ to obtain the following mathematical expression of the ordinary least square regression coefficients,

$$\hat{\boldsymbol{\beta}} = (\mathbf{X}^T\mathbf{X})^{-1}\mathbf{X}^T\mathbf{y} \tag{6.3.14}$$

in which $\hat{\boldsymbol{\beta}} = (\hat{\beta}_0, \hat{\beta}_1, \ldots, \hat{\beta}_k)^T$. The $(\mathbf{X}^T\mathbf{X})^{-1}$ is a square symmetric $(k+1) \times (k+1)$ matrix whose inverse always exists if the independent variables are linearly independent; that is, no column of the $\mathbf{X}$ matrix is a linear combination of the other columns.

Once the regression coefficients have been estimated the dependent variable, $\hat{y}_i$, can be computed using

$$\hat{y}_i = \hat{\beta}_0 + \hat{\beta}_1 x_{i1} + \cdots + \hat{\beta}_k x_{ik} \tag{6.3.15}$$

Referring to Eq. (6.3.2*a*), since there is only one independent variable, that is, population, then $k = 1$. By letting $x = x_1$, the solutions to the normal Eqs. (6.3.12) are

$$\hat{\beta}_0 = \overline{y} - \hat{\beta}_1\overline{x} \tag{6.3.16}$$

and

$$\hat{\beta}_1 = S_{xy}/S_{xx} \tag{6.3.17}$$

where the averages of y_i and x_i, are

$$\overline{y} = \frac{1}{n}\sum_{i=1}^{n} y_i \qquad \text{and} \qquad \overline{x} = \frac{1}{n}\sum_{i=1}^{n} x_i$$

and

$$S_{xx} = \sum_{i=1}^{n}(x_i - \overline{x})^2 = \sum_{i=1}^{n} x_i^2 - n\overline{x}^2 \tag{6.3.18}$$

$$S_{xy} = \sum_{i=1}^{n} y_i(x_i - \overline{x}) = \sum_{i=1}^{n} y_i x_i - n\overline{xy} \tag{6.3.19}$$

Example 6.3.1. Based on the data given in Table 6.3.1, develop the simple linear regression equation relating annual water use in ac-ft (AF) to population for the city of Austin, Texas that fits through the data points shown in Fig. 6.3.1.

Solution. From Table 6.3.1, compute n = 21; Σx_i = 6,341,600; Σy_i = 1,328,538; Σx_i^2 = 1,989,270,175,744; $\Sigma x_i y_i$ = 21,170,978,816; $\overline{x}$ = 301,980.9; and $\overline{y}$ = 63,263.7. From Eqs. (6.3.18) and (6.3.19), compute S_{xx} = 74,228,039,680, S_{xy} = 21,170,978,816. Using Eqs. (6.3.16) and (6.3.17), the unknown regression coefficients are obtained as $\hat{\beta}_1 = S_{xy}/S_{xx} = .2852$ and $\hat{\beta}_0 = -22{,}865.9$. The resulting regression model is $\hat{y} = -22{,}865.9 + 0.2852x$ in which $\hat{y}$ is the annual total water use in ac-ft (AF) and x is the population size (in persons). The line represented by the resulting regression equation is shown as the solid line in Fig. 6.3.1.

6.3.3 Accuracy of Regression Models

In regression analysis, there are other quantities in addition to the regression coefficients that are useful and need to be estimated. These are the **residuals** and the variance term associated with the regression model. The residual is the difference between the observed and predicted value of the dependent variable which can be estimated using

$$e_i = y_i - \hat{y}_i = y_i - (\hat{\beta}_0 + \hat{\beta}_1 x_{i1} + \cdots + \hat{\beta}_k x_{ik}), \qquad i = 1, 2, \ldots, n \tag{6.3.20}$$

The residuals play an important role in checking the adequacy of the resulting regression model. The variance associated with the regression model s^2 is

$$s^2 = \frac{\sum_{i=1}^{n} e_i^2}{n - (k+1)} = \frac{\mathbf{e}^T\mathbf{e}}{n - k - 1} \tag{6.3.21}$$

in which $\mathbf{e} = (e_1, e_2, \ldots, e_n)^T$ is the vector of errors. The denominator of Eq. (6.3.21) is called the **degrees of freedom**, which is obtained by subtracting the number of unknown regression coefficients in the model from the number of data observations. The square root of the model variance (s) is called the **standard error of estimate**.

Furthermore, the value of the **coefficient of determination** between the dependent variable y and the regressors can be computed as

$$r^2 = 1 - \frac{\sum_{i=1}^{n} e_i^2}{\sum_{i=1}^{n} (y_i - \overline{y})^2} = 1 - \frac{SS_e}{S_{yy}} \tag{6.3.22}$$

where $SS_e = \sum_{i=1}^{n} e_i^2$ is called the **residual sum of squares** and $S_{yy} = \sum_{i=1}^{n}(y_i - \overline{y})^2$ is called the **total sum of squares** about the mean of the dependent variable. The coefficient of determination has a value bounded between 0 and 1 representing the percentage of variation in the dependent variable y explained by the resulting regression equation. Therefore, the term (SS_e/S_{yy}) in Eq. (6.3.22) indicates the portion of variation in y not explained by the regression model.

TABLE 6.3.2
Computations for Example 6.3.2

i (1)	y_i (2)	$\hat{y}_i$ (3)	e_i (4)
1	39,606.0	38,949.7	656.3
2	40,131.0	40,832.4	−701.4
3	45,667.0	42,772.1	2,894.9
4	40,780.0	44,771.2	−3,991.2
5	45,330.0	46,831.0	−1,501.0
6	50,680.0	48,953.6	1,726.4
7	56,600.0	51,261.6	5,338.4
8	57,157.0	53,643.7	3,513.3
9	57,466.0	56,102.5	1,363.5
10	63,263.0	58,640.4	4,622.6
11	57,357.0	61,259.8	−3,902.8
12	51,163.0	63,963.4	−12,800.4
13	68,413.0	66,753.6	1,659.4
14	69,994.0	69,633.7	360.3
15	65,204.0	72,606.5	−7,402.5
16	78,564.0	75,674.9	2,889.1
17	76,339.0	78,214.7	−1,875.7
18	87,309.0	82,131.9	5,177.1
19	82,120.0	86,464.6	−4,344.6
20	97,687.0	90,977.0	6,710.0
21	97,708.0	98,099.7	−391.7

Example 6.3.2. Using the results from Example 6.3.1, compute the standard error of estimate associated with the regression model for Austin and the coefficient of determination between water use and population size.

Solution. Table 6.3.2 illustrates the computations for this example. Column (2) contains the observed annual water uses. Based on the regression model developed in Example 6.3.1, the predicted water use in different years are given in Column (3) of Table 6.3.2. The differences between the observed and the predicted values are the errors listed in Column (4).

Using the error values in Column (4), Eq. (6.3.21) is used to compute the variance

$$s^2 = \Sigma e_i^2/(n-2) = 43{,}374{,}300.2/(21-2) = 22{,}842{,}664.4$$

The standard error is then

$$s = \sqrt{22{,}842{,}664.4} = 4779.4 \text{ AF}$$

The coefficient of determination is computed using Eq. (6.3.22)

$$r^2 = 1 - \frac{\Sigma e_i^2}{S_{yy}} = 1 - \frac{434{,}019{,}040}{21{,}170{,}978{,}816} = 0.9329$$

which indicates that 93.3 percent of the variability in the water use-population data for Austin from 1965–1985 are explained by the regression equation obtained in Example 6.3.1.

An important application of the regression model is to predict a new observation corresponding to a specified level of the regressor x. Note that the β in Eq. (6.3.15) are estimators of the unknown regression coefficients, so they are in fact random variables. As a result, for a given regressor $\mathbf{x}_0 = (x_{01}, x_{02}, \ldots, x_{0k})$, the prediction of the future response by $\hat{y}_0 = \hat{\beta}_0 + \hat{\beta}_1 x_{01} + \cdots + \hat{\beta}_k x_{0k}$ is also subject to error. In addition to estimating the future value of the dependent variable, $\hat{y}_0$, it is also informative to determine the **prediction interval** which has a high probability of capturing the actual future value.

6.3.4 General Comments about Regression Analysis

Several assumptions involved in least squares regression analysis were previously discussed. After a regression model is developed, it is important to check its adequacy with respect to those assumptions. In a case where the assumptions are violated, the validity of the model will be doubtful. Generally, the degree of adequacy of a regression model cannot be revealed by the summary statistics such as r^2. Although examining the model adequacy by formal statistical tests is good, it is often more informative to perform such tasks visually by having residual plots.

The first step is to test if the observations of the dependent variable y follow a normal distribution. Because the validity of normality ensures the correctness of statistical inferences including the derivations of confidence and prediction intervals. The second test is to check model adequacy by examining the validity of the constant variance assumption. This is done by plotting the model residuals against the dependent variable y and individual independent variables x.

Some error patterns in regression analysis that one commonly encountered are shown in Fig. 6.3.3. Strictly speaking, the least square procedure described above is called the **ordinary least square** (OLS) method which abides the condition of constant variance (see Fig. 6.3.3*a*) in model parameter estimation. One could also encounter error variances decreasing with increasing level of dependent or independent variable. To explicitly account for the case of unequal variance such as shown in Fig. 6.3.3*c*, the **weighted** or **generalized least square** method can be applied (Montgomery and Peck, 1982; Neter, Wasserman, and Kutner, 1983).

Because of the cost associated with the effort to collect data on regressors and dependent variables, it is desirable to use the least number of regressors as possible to yield the most representative regression model. An important issue in regression analysis is to determine the best subset of regressors. In general, the contributions of such independent variables are functions of the inclusion of other variables in the model and, therefore, cannot be assessed a priori. There are many statistical procedures such as step-wise regression developed for selecting the best subset of regressors. For more comprehensive discussions of various topics in regression analysis, refer to Draper and Smith (1981), Montgomery and Peck (1982), and Neter, Wasserman, and Kutner (1983).

There are many computer-software packages that can be used to perform a rather comprehensive regression analysis. Some of the well known ones are SAS (Statistical Analysis System, 1988) and SPSS (Statistical Packages for Social Sciences, 1986).

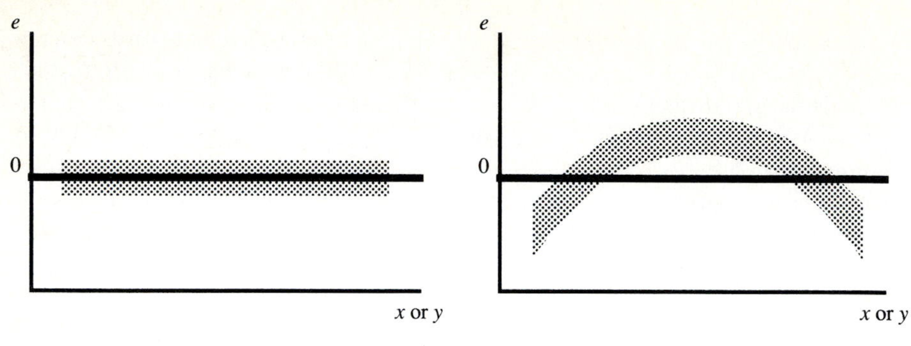

(a) Constant error variance

(b) Indication of nonlinear terms of x or y should be included in the model

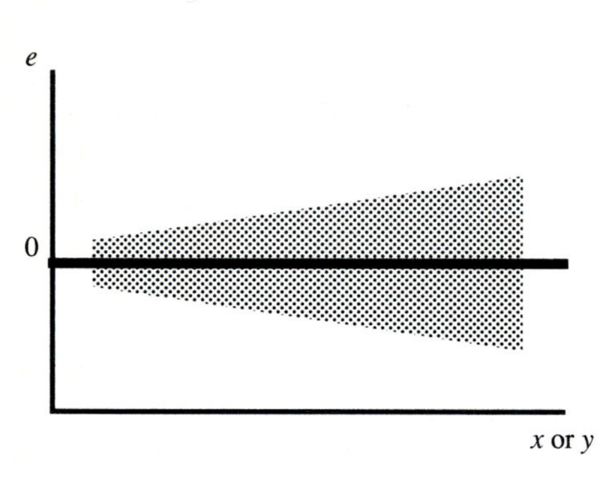

(c) Error variance increase as x or y increases

FIGURE 6.3.3
Prototype residual plots (after Neter, Wasserman, and Kutner, 1983). If the constant variance assumption is valid, the residual plot should appear as Figure 6.3.3*a*. In cases that the residual plot appears as other configurations, Figure 6.3.3*b* or *c*, the regression model should be modified. Figure 6.3.3*c* shows that a larger variability of residual is associated with a larger value of the dependent or independent variable.

These statistical packages are capable of handling large data sets, estimating regression parameters, computing confidence/prediction intervals, perform various tests, and selecting the best independent variables, etc.

Finally, let us briefly discuss alternative curve fitting criteria other than the least square criterion. When using the least square criterion, analysts place much higher weights to larger deviations than to small ones. To reduce such heavy emphasis of large deviations on curve fitting, the criterion of minimizing the sum of absolute deviation can be used. In general, the result of parameter estimation based on the minimum sum of absolute deviation is less sensitive to the presence of extraordinarily large or small observations (outliers) in the data set than that of using a least squares criterion. As the third curve fitting criterion, if one is only concerned with the worst estimation, the objective function that minimizes the largest deviation can be adopted.

Note that the parameter estimation in regression using the least squares criterion is a nonlinear unconstrained minimization such that the unknown regression coefficients can be any real number. In some situations, physical processes dictate the sign and the range of values of unknown regression coefficients. This could make the solutions from solving the normal Eq. (6.3.14) infeasible. When bounds on regression coefficients exist, the problem can be solved by a quadratic programming technique. Adopting the criteria of minimizing the sum of absolute deviations or of minimizing the largest deviation, the regression coefficients in Eq. (6.3.3) can be determined by linear programming. The reason that the least squares criterion is overwhelmingly used in practice is because statistical inferences have been extensively developed. Explorations of statistical properties of regression coefficient estimators based on the other two criteria are still lacking.

6.4 CASCADE MODELS FOR WATER USE FORECASTING

Time series of monthly municipal water use may have a historical pattern of variation that can be separated into long memory components and short run components (Maidment and Parzen, 1984*a*, *b*). **Long memory components** are (1) a trend which reflects the year to year effect of slow changes in population, water price, and family income, and (2) seasonality which reflects the cyclic pattern of variation in water use within a year. **Short term components** could be (1) autocorrelation which reflects the perpetuation of departures of water use from the long term pattern of variation, and (2) climate correlation which reflects the effect on water use of abnormal climatic events such as no rainfall or a lot of rainfall. The cascade model gets its name from the fact that, in the model development, four sequential steps are involved in transforming the monthly water demand time series. A detailed description of the components in the development of a cascade model is presented in Fig. 6.4.1.

An original monthly municipal water use time series can be expressed as, $Q(t)$ for $t = 1, \ldots, T$ or $Q_a(m, y)$ for $m = 1, \ldots, 12$ and $y = 1, \ldots, Y$, where t is the index of months from the beginning of the series to the last month T; m is a monthly index within each year; and y is the index of years where Y is the total number of years. Such a time series can be expressed further as the sum of a long-term memory component, $Q_L(t)$ or $Q_L(m, y)$, and a short-term memory component, $Q_S(t)$ or $Q_S(m)$, as

$$Q(t) = Q_L(t) + Q_S(t) \tag{6.4.1a}$$

or

$$Q_a(m, y) = Q_L(m, y) + Q_S(m) \tag{6.4.1b}$$

with

$$Q(t) = Q_a(m, y), \qquad t = 12(y - 1) + m \tag{6.4.1c}$$

Figure 6.4.2 illustrates the partitioning of the monthly water use time series for Canyon City, Texas into a long-term-memory component and a short-term memory component.

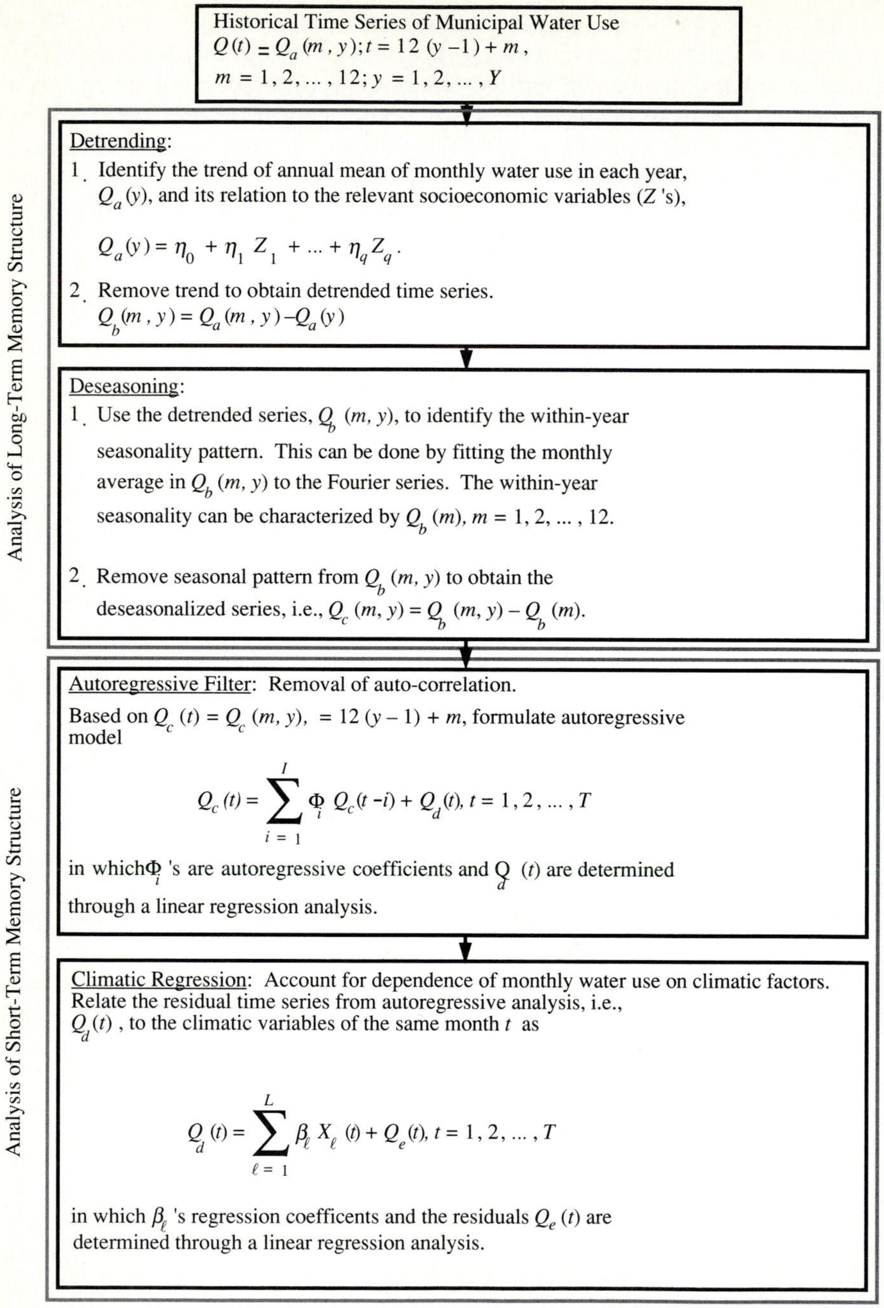

FIGURE 6.4.1
Cascade transformation for monthly water use time series.

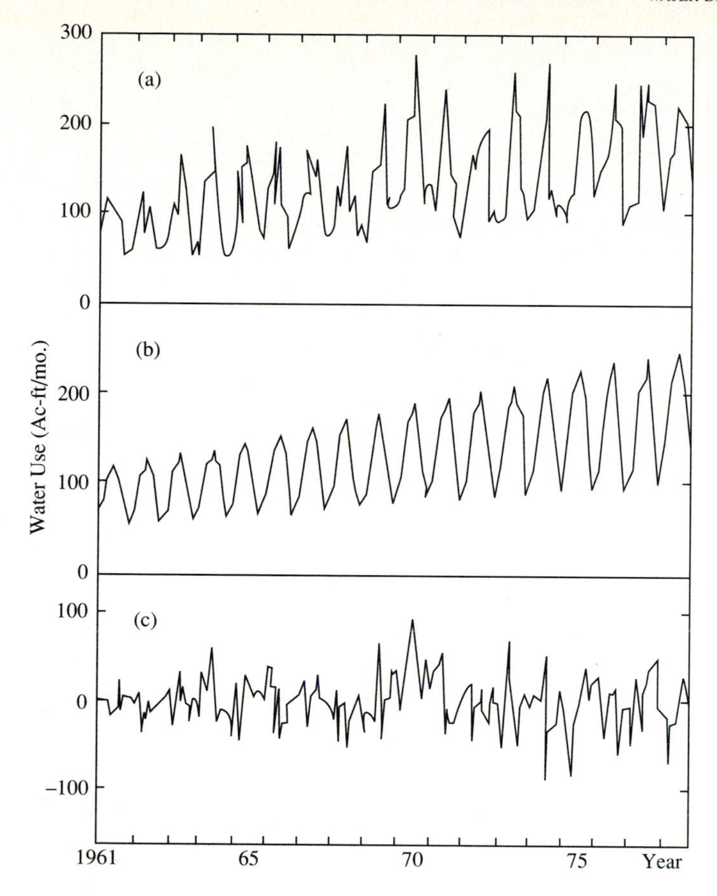

FIGURE 6.4.2
Partition of water use time series into (*a*) long memory component and (*b*) short memory component (*c*) $a = b + c$. Data from Canyon, Texas, 1961–1978 (Maidment and Parzen, 1984*a*).

6.4.1 Detrending

The first step in developing a cascade model is to identify the over-year water use trend in the historical time series and then remove it. The over-year water use trend is contained in $Q_a(y)$ representing the annual mean of monthly water use in year y. Therefore, the main task in this step is to identify the functional form of $Q_a(y)$. From historical data the annual mean of monthly water use in each year y can be estimated as

$$Q_a(y) = \frac{1}{12} \sum_{m=1}^{12} Q_a(m, y) \qquad y = 1, \ldots, Y \tag{6.4.2}$$

Alternatively, the effects of socioeconomic factors such as water price, population, number of water connections, and household income could be accounted for in the annual mean of monthly water use by relating it to the relevant socioeconomic factors as

$$Q_a(y) = \eta_0 + \eta_1 Z_1(y) + \eta_2 Z_2(y) + \cdots + \eta_q Z_q(y), \qquad y = 1, 2, \ldots, Y \tag{6.4.3}$$

where η represents coefficients, $Z_j(y)$ is the jth socioeconomic factor considered to be relevant in year y and the $j = 1, \ldots, q$ socioeconomic factors. The coefficients η in Eq. (6.4.3) can be estimated by regression analysis using historical data of $Q_a(y)$ and $Z_j(y)$. The use of Eq. (6.4.3) in defining the annual water use trend is especially plausible for forecasting water use and is related to future socioeconomic conditions.

Once the annual water use trend $Q_a(y)$ is identified, it is removed (or subtracted) from the historical monthly water use time series, $Q_a(m, y)$, to obtain the detrended time series, $Q_b(m, y)$, as

$$Q_b(m, y) = Q_a(m, y) - Q_a(y) \tag{6.4.4}$$

in which $Q_a(y)$ can be estimated by Eq. (6.4.2) or preferably from Eq. (6.4.3).

Example 6.4.1. Table 6.4.1 contains monthly water use for the city of Austin, Texas from 1977 to 1981. Remove the long-term annual trend in the data.

Solution. The monthly water use time series for Austin, Texas, plotted in Fig. 6.4.3 shows a mild increasing trend in water use. The long-term annual water use trend could be identified through its relationship to socioeconomic factors such as population, water price, and average annual income, etc, such as those contained in Table 6.3.1 for the

TABLE 6.4.1
Monthly water use time series (in 1000 AF), $Q_a(m, y)$, for the city of Austin, Texas

y	m	$Q_a(m, y)$	y	m	$Q_a(m, y)$	y	m	$Q_a(m, y)$
1977	1	7.3154	1978	9	5.2363	1980	5	5.3235
	2	5.8856		10	6.2953		6	9.2917
	3	4.7828		11	4.5265		7	12.0619
	4	3.9762		12	4.3172		8	9.4198
	5	5.0147	1979	1	4.5076		9	6.8067
	6	6.6388		2	3.8646		10	6.0580
	7	7.9304		3	4.3843		11	5.4162
	8	7.8839		4	4.6256		12	4.8631
	9	6.4274		5	5.0220	1981	1	5.1559
	10	3.7504		6	6.6136		2	4.8765
	11	4.4824		7	6.5667		3	5.4689
	12	4.3244		8	7.1248		4	6.2575
1978	1	4.2920		9	6.4226		5	6.3392
	2	3.8670		10	6.9475		6	5.8842
	3	4.9031		11	4.7260		7	8.2813
	4	5.5918		12	4.3980		8	7.5630
	5	5.4707	1980	1	4.3540		9	5.8916
	6	7.1933		2	4.2739		10	6.7701
	7	10.3885		3	5.1617		11	5.9962
	8	7.9107		4	5.5340		12	6.2181

city of Austin, Texas. In the process of identifying the long-term water use trend, the monthly average water use for each year, that is, annual total water use divided by 12, is used as the dependent variable which is regressed against other socioeconomic factors. For purposes of illustration, consider that long-term water use in Austin is related to its population, unit water price, and average income as

$$Q_a(y) = \eta_0 + \eta_1 Z_1(y) + \eta_2 Z_2(y) + \eta_3 Z_3(y), \qquad y = 1965, 1966, \ldots, 1985.$$

in which $Q_a(y)$ is the monthly average water use (in 1000 AF) in a year y, $Z_1(y)$ is the population (in 10,000 people) in year y, $Z_2(y)$ is the unit water price (in dollars) in year y, $Z_3(y)$ is the average income (in \$1000) in year y, and η are model parameters to be estimated by regression analysis. Since three independent variables are involved in the above water use trend model, multiple regression analysis should be used. Based on the 21 years of observations, the unknown model parameters can be estimated by using Eq. (6.3.14) as

$$\hat{\boldsymbol{\eta}} = (\mathbf{X}^T\mathbf{X})^{-1}\mathbf{X}^T\mathbf{Q}_a$$

in which $\hat{\boldsymbol{\eta}}$ is a vector of estimated model parameters, $\hat{\boldsymbol{\eta}} = (\hat{\eta}_0, \hat{\eta}_1, \hat{\eta}_2, \hat{\eta}_3)^T$, $\mathbf{Q}_a$ is the column vector of monthly average water use, and $\mathbf{X}$ is the matrix containing observations of independent variables, namely,

$$\begin{array}{cccc} & \mathbf{Z}_1 & \mathbf{Z}_2 & \mathbf{Z}_3 \end{array}$$

$$\mathbf{X} = \begin{bmatrix} 1.0 & 21.6733 & 0.98 & 5.919 \\ 1.0 & 22.3334 & 0.95 & 5.970 \\ 1.0 & 23.0135 & 1.20 & 6.521 \\ 1.0 & 23.7144 & 1.15 & 7.348 \\ 1.0 & 24.4366 & 1.10 & 7.965 \\ 1.0 & 25.1808 & 1.05 & 8.453 \\ 1.0 & 25.9900 & 1.00 & 8.713 \\ 1.0 & 26.8252 & 1.20 & 9.286 \\ 1.0 & 27.6873 & 1.13 & 9.694 \\ 1.0 & 28.5711 & 1.06 & 9.542 \\ 1.0 & 29.4955 & 0.98 & 9.684 \\ 1.0 & 30.4434 & 0.93 & 10.152 \\ 1.0 & 31.4217 & 0.87 & 10.441 \\ 1.0 & 32.4315 & 0.81 & 10.496 \\ 1.0 & 33.4738 & 1.10 & 10.679 \\ 1.0 & 34.5496 & 1.05 & 10.833 \\ 1.0 & 35.4401 & 0.96 & 11.060 \\ 1.0 & 36.8135 & 0.91 & 11.338 \\ 1.0 & 38.3326 & 0.87 & 11.752 \\ 1.0 & 39.9147 & 0.84 & 12.763 \\ 1.0 & 42.4120 & 1.41 & 12.748 \end{bmatrix}; \quad \mathbf{Q}_a = \begin{bmatrix} 39.606 \\ 40.131 \\ 45.667 \\ 40.780 \\ 45.330 \\ 50.683 \\ 56.600 \\ 57.157 \\ 57.466 \\ 63.263 \\ 57.357 \\ 51.163 \\ 68.413 \\ 69.994 \\ 65.203 \\ 78.563 \\ 76.339 \\ 87.309 \\ 82.120 \\ 97.678 \\ 97.708 \end{bmatrix}$$

The results of matrix algebraic manipulation or from applying a statistical package for regression are

$$\hat{\boldsymbol{\eta}} = \begin{bmatrix} -1.9430 \\ 0.2313 \\ 0.0331 \\ 0.0204 \end{bmatrix}$$

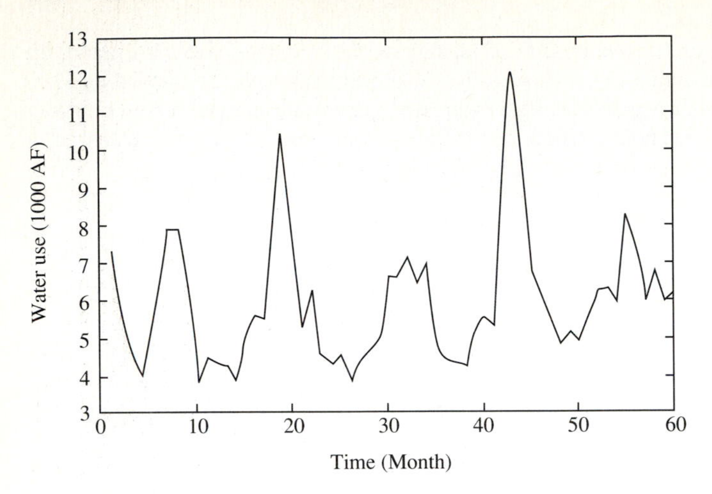

FIGURE 6.4.3
Monthly water use time series, $Q_a(m, y)$, for the city of Austin, Texas, from 1977 to 1981.

Once the model parameters are estimated, they can be used to estimate the monthly average water use (in 1000 AF) for 1977–1981 as

$$\mathbf{Q}_a = \mathbf{X}_{77-81}\hat{\eta} = \begin{bmatrix} Q_a(77) \\ Q_a(78) \\ Q_a(79) \\ Q_a(80) \\ Q_a(81) \end{bmatrix} = \begin{bmatrix} 1.0 & 31.4217 & 0.87 & 10.4410 \\ 1.0 & 32.4315 & 0.81 & 10.4960 \\ 1.0 & 33.4738 & 1.10 & 10.6790 \\ 1.0 & 34.5496 & 1.05 & 10.8330 \\ 1.0 & 35.4401 & 0.96 & 11.0600 \end{bmatrix} \begin{bmatrix} -1.9430 \\ 0.2313 \\ 0.0331 \\ 0.0204 \end{bmatrix} = \begin{bmatrix} 5.5673 \\ 5.8000 \\ 6.0545 \\ 6.3048 \\ 6.5124 \end{bmatrix}$$

Then, the detrended monthly water use time series, $Q_b(m, y)$, can be obtained by subtracting the above monthly average water use from the original monthly water use data, $Q_a(m, y)$, using Eq. (6.4.4), i.e.,

$$Q_b(m, y) = Q_a(m, y) - Q_a(y) \text{ for } y = 1977, 1978, \ldots, 1981$$

The detrended monthly water use time series is given in Table 6.4.2 and plotted in Fig. 6.4.4.

6.4.2 Deseasonalization

The detrended monthly water use time series, $Q_b(m, y)$, generally contains within-year seasonal water use patterns which should be removed. To identify the within-year or monthly seasonal water use pattern, the detrended monthly time series $Q_b(m, y)$ can be used to compute the arithmetic average $Q_b(m)$ for each month within a year that contains a seasonality component, using

$$Q_b(m) = \frac{1}{Y} \sum_{y=1}^{Y} Q_b(m, y), \qquad m = 1, 2, \ldots, 12 \tag{6.4.5}$$

Because the presence of seasonality could result in the occurrence of periodicity in the monthly water use time series, it is possible to model such periodicity by using a **Fourier series** consisting of the summation of sine and cosine terms.

TABLE 6.4.2
Detrended monthly water use time series, $Q_b(m,y)$, in 1000 AF

y	m	$Q_b(m,y)$	y	m	$Q_b(m,y)$	y	m	$Q_b(m,y)$
1977	1	1.7481	1978	9	−0.5637	1978	9	−0.9813
	2	0.3183		10	0.4953		10	2.9869
	3	−0.7845		11	−1.2735		11	5.7571
	4	−1.5911		12	−1.4828		12	3.1150
	5	−0.5526	1979	1	−1.5469	1979	1	0.5019
	6	1.0715		2	−2.1899		2	−0.2468
	7	2.3631		3	−1.6702		3	−0.8886
	8	2.3166		4	−1.4289		4	−1.4417
	9	0.8601		5	−1.0325		5	−1.3565
	10	−1.8169		6	0.5591		6	−1.6359
	11	−1.0849		7	0.5122		7	−1.0435
	12	−1.2429		8	1.0703		8	−0.2549
1978	1	−1.5080		9	0.3681		9	−0.1732
	2	−1.9330		10	0.8930		10	−0.6282
	3	−0.8969		11	−1.3285		11	1.7689
	4	−0.2082		12	−1.6565		12	1.0506
	5	−0.3293	1980	1	−1.9508	1980	1	−0.6208
	6	1.3933		2	−2.0309		2	0.2577
	7	4.5885		3	−1.1431		3	−0.5162
	8	2.1107		4	−0.7708		4	−0.2943

A Fourier series can be used to determine the within-year or monthly seasonal water use pattern using (Salas et al., 1980)

$$Q_b(m) = \sum_{k=0}^{6} a_k \cos\left(\frac{2\pi km}{12}\right) + b_k \sin\left(\frac{2\pi km}{12}\right), \qquad m = 1, \ldots, 12 \qquad (6.4.6)$$

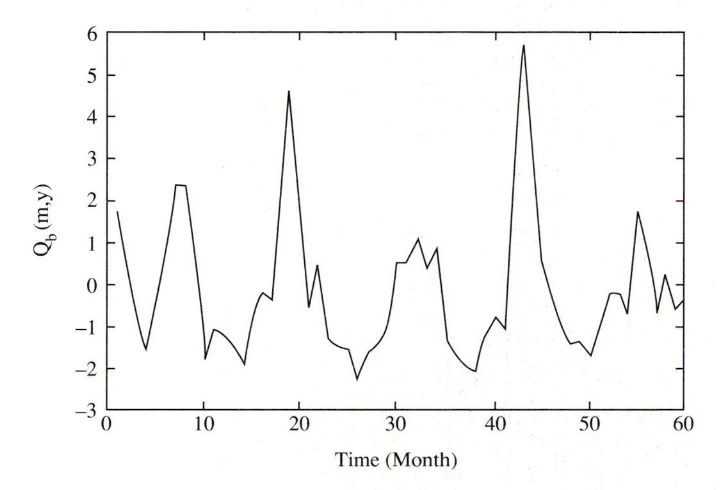

FIGURE 6.4.4
Detrended monthly water use time series, $Q_b(m,y)$, for the city of Austin, Texas, 1977–1981.

where the Fourier coefficients a_0, a_k, and b_k are determined using

$$a_0 = \frac{1}{12}\sum_{m=1}^{12}\left[\frac{1}{Y}\sum_{y=1}^{Y} Q_b(m,y)\right] \tag{6.4.7a}$$

$$a_k = \frac{1}{6}\sum_{m=1}^{12}\left[\frac{1}{Y}\sum_{y=1}^{Y} Q_b(m,y)\right]\cos\left(\frac{2\pi km}{12}\right) \qquad \text{for } k = 1, 2, \ldots, 6 \tag{6.4.7b}$$

$$b_k = \frac{1}{6}\sum_{m=1}^{12}\left[\frac{1}{Y}\sum_{y=1}^{Y} Q_b(m,y)\right]\sin\left(\frac{2\pi km}{12}\right) \qquad \text{for } k = 1, 2, \ldots, 6 \tag{6.4.7c}$$

Computed Fourier coefficients are not necessarily all statistically significant (Maidment and Parzen, 1984*b*), so that only those Fourier coefficients that pass the appropriate statistical significance test are used. The seasonality of water use represented by $Q_b(m)$ is subtracted from the detrended water use series, $Q_b(m, y)$, resulting in the **deseasonalized water use series**, $Q_c(m, y)$, as

$$Q_c(m,y) = Q_b(m,y) - Q_b(m), \qquad \text{for } m = 1, 2, \ldots, 12;\ y = 1, 2, \ldots, Y \tag{6.4.8}$$

in which $Q_b(m)$ can be estimated by either Eq. (6.4.5) or (6.4.6).

Example 6.4.2. Examining Fig. 6.4.4 the detrended monthly water use time series $Q_b(m, y)$ exhibits periodicity or cyclic pattern. Remove the periodicity or seasonality present in the detrended monthly water use.

Solution. The periodicity or cyclic pattern in Fig. 6.4.4 can be identified by fitting the detrended monthly water use time series to a Fourier series defined by Eqs. (6.4.6) and (6.4.7) using the average water use for each month of the detrended data. The average water use (in 1000 AF) for each month $Q_b(m)$ can be calculated by Eq. (6.4.5) using the detrended monthly time series $Q_b(m, y)$, in Table 6.4.2.

$$\begin{bmatrix} Q_b(m=1) \\ Q_b(m=2) \\ Q_b(m=3) \\ Q_b(m=4) \\ Q_b(m=5) \\ Q_b(m=6) \\ Q_b(m=7) \\ Q_b(m=8) \\ Q_b(m=9) \\ Q_b(m=10) \\ Q_b(m=11) \\ Q_b(m=12) \end{bmatrix} = \begin{bmatrix} -0.92282 \\ -1.49428 \\ -1.10764 \\ -0.85078 \\ -0.61378 \\ 1.07652 \\ 2.99796 \\ 1.93264 \\ 0.10912 \\ -0.08354 \\ -1.01834 \\ -1.22364 \end{bmatrix}$$

The variation of $Q_b(m)$ within a year is plotted as the solid curve shown in Fig. 6.4.5.

The $Q_b(m)$ obtained above are used in Eq. (6.4.6) and the Fourier coefficients a_k and b_k are determined by Eqs. (6.4.7*a–c*). The Fourier coefficient a_1 can be computed using Eq. (6.4.7*b*) with $k = 1$ as

$$a_1 = Q_b(1)\cos\left(\frac{\pi}{6}\right) + Q_b(2)\cos\left(\frac{2\pi}{6}\right) + \cdots + Q_b(12)\cos\left(\frac{12\pi}{6}\right)$$

The results of these computations for the Fourier coefficients using Eqs. (6.4.7*a–c*) are

$$\begin{bmatrix} a_0 \\ a_1 \\ a_2 \\ a_3 \\ a_4 \\ a_5 \\ a_6 \\ b_0 \\ b_1 \\ b_2 \\ b_3 \\ b_4 \\ b_5 \\ b_6 \end{bmatrix} = \begin{bmatrix} -0.099882 \\ -1.229319^* \\ 0.220158^* \\ 0.059920 \\ -0.186528^* \\ 0.019309 \\ -0.014597 \\ 0 \\ -1.101187^* \\ 0.733226^* \\ -0.383243^* \\ 0.336968^* \\ 0.109563 \\ 0 \end{bmatrix}$$

The regenerated $Q_b(m)$ can be obtained by substituting the above Fourier coefficients Eq. (6.4.6). Among the fourteen Fourier coefficients obtained above only those with * are significantly different from zero. To reduce computational burden only those Fourier

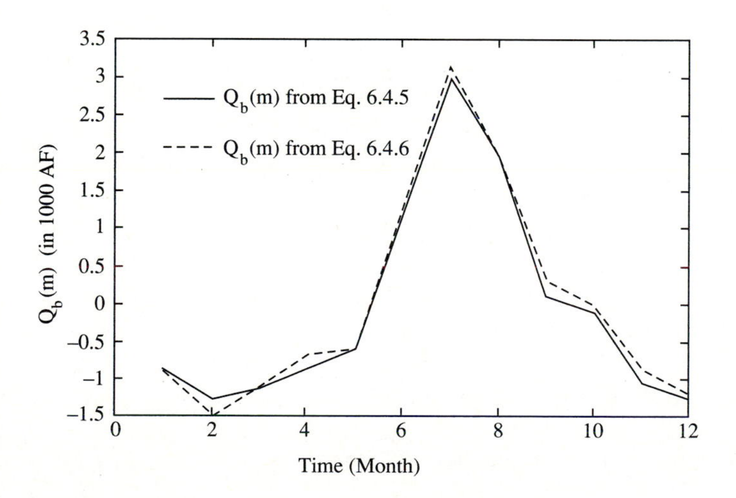

FIGURE 6.4.5
Average monthly detrended water use, $Q_b(m)$. The $Q_b(m)$ values plotted as a solid line are based upon an arithmetic average of $Q_b(m, y)$ values using Eq. (6.4.5). The $Q_b(m)$ values plotted as a dashed line are based upon a Fourier series developed using the $Q_b(m)$ from the arithmetic average.

coefficients that are significantly different from zero are retained in the model leading to

$$\begin{aligned}Q_b(m) = & -1.229319\cos(m\pi/6) + 0.220158\cos(2m\pi/6) \\ & -0.186528\cos(4m\pi/6) - 1.101187\sin(m\pi/6) \\ & +0.733226\sin(2m\pi/6) - 0.383243\sin(3m\pi/6) \\ & +0.336968\sin(4m\pi/6)\end{aligned}$$

Based on the above reduced model the regenerated values of $Q_b(m)$ are computed as

$$\begin{bmatrix} Q_b(m=1) \\ Q_b(m=2) \\ Q_b(m=3) \\ Q_b(m=4) \\ Q_b(m=5) \\ Q_b(m=6) \\ Q_b(m=7) \\ Q_b(m=8) \\ Q_b(m=9) \\ Q_b(m=10) \\ Q_b(m=11) \\ Q_b(m=12) \end{bmatrix} = \begin{bmatrix} -0.868296 \\ -1.241950 \\ -1.124620 \\ -0.698981 \\ -0.592700 \\ 1.262929 \\ 3.128603 \\ 1.894670 \\ 0.311267 \\ -0.020979 \\ -0.854253 \\ -1.195689 \end{bmatrix}$$

The regenerated $Q_b(m)$ are plotted as the dashed curve in Fig. 6.4.5. With only half of the number of the original model, the regenerated $Q_b(m)$ fit the observed one quite well. Then, the deaseasonalized monthly water use time series can be obtained, using Eq. (6.4.8),

$$Q_c(y, m) = Q_b(y, m) - Q_b(m), \qquad m = 1, 2, \ldots, 12;\ y = 1977, \ldots, 1981.$$

Because the long-term trend and seasonality are removed, the identity of month and year in $Q_c(m, y)$ can be replaced by $Q_c(t)$ with $t = 1, 2, \ldots, 60$ for the deseasonalized water use time series. The deseasonalized monthly water use time series $Q_c(t)$ is given in Table 6.4.3 and is plotted in Fig. 6.4.6. The periodicity in the presence of $Q_b(y, m)$ in Fig. 6.4.4 vanishes.

6.4.3 Autoregressive Filtering

The detrended and deseasonalized monthly water use time series $Q_c(m, y)$ now does not possess over-year trend and within-year seasonality, and can be represented as $Q_c(t) = Q_c(m, y)$ with $t = 12(y-1) + m$. Still the detrended and deseasonalized monthly water use time series $Q_c(t)$ may possess, like many hydrologic time series, autocorrelation representing a short memory process. The third step of the cascade model development is to identify such short memory structure in $Q_c(t)$ as

$$Q_c(t) = \sum_{i=1}^{I} \phi_i Q_c(t-i) + Q_d(t), \qquad t = 2, \ldots, T \tag{6.4.9}$$

in which $Q_c(t-i)$ is the detrended and deaseasonalized monthly water demand series lagged by i months; I is the maximum number of lags; ϕ's are the unknown autore-

TABLE 6.4.3
Deseasonalized monthly water use time series, $Q_c(t)$

t	$Q_c(t)$	t	$Q_c(t)$	t	$Q_c(t)$
1	2.61639	21	−0.87496	41	−0.38859
2	1.56025	22	0.51627	42	1.72397
3	0.34011	23	−0.41924	43	2.62849
4	−0.89211	24	−0.28711	44	1.22032
5	0.04010	25	−0.67860	45	0.19063
6	−0.19142	26	−0.94794	46	−0.22582
7	−0.76550	27	−0.54558	47	−0.03434
8	0.42192	28	−0.72991	48	−0.24601
9	0.54883	29	−0.43979	49	−0.48820
10	−1.79592	30	−0.70382	50	−0.39394
11	−0.23064	31	−2.61640	51	0.08111
12	−0.04721	32	−0.82437	52	0.44408
13	−0.63970	33	0.05683	53	0.41950
14	−0.69104	34	0.91397	54	−1.89112
15	0.22771	35	−0.47424	55	−1.35970
16	0.49078	36	−0.46081	56	−0.84407
17	0.26340	37	−1.08250	57	−0.93206
18	0.13037	38	−0.78894	58	0.27867
19	1.45989	39	−0.01848	59	0.33805
20	0.21602	40	−0.07181	60	0.90138

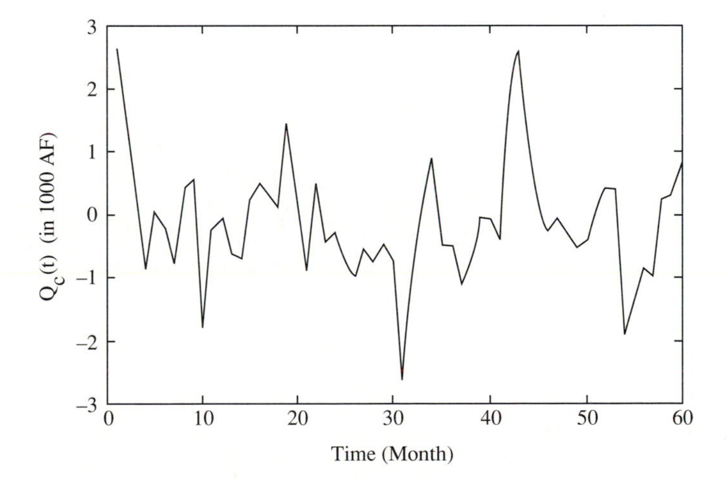

FIGURE 6.4.6
Deseasonalized monthly water use time series, $Q_c(t)$.

gressive model parameters; and $Q_d(t)$ is the residual representing the water use time series after autocorrelation within $Q_c(t)$ has been removed.

The autoregressive parameters ϕ_i and $Q_d(t)$ are estimated using a linear regression analysis of the following equations

$$
\begin{array}{lll}
t = 2 & Q_c(2) = \phi_1 Q_c(1) & + Q_d(2) \\
t = 3 & Q_c(3) = \phi_1 Q_c(2) + \phi_2 Q_c(1) & + Q_d(3) \\
t = 4 & Q_c(4) = \phi_1 Q_c(3) + \phi_2 Q_c(2) + \phi_3 Q_c(1) & + Q_d(4) \\
\vdots & \vdots & \vdots
\end{array}
$$

$$
t = T \quad Q_c(T) = \phi_1 Q_c(T-1) + \phi_2 Q_c(T-2) + \cdots + \phi_1 Q_c(T-I) + Q_d(T) \quad (6.4.10)
$$

which are from Eq. (6.4.9) with $Q_c(t) = Q_c(m, y)$ where $t = 12(y-1) + m$. The residuals $Q_d(t)$ represent the water use time-series removal of autocorrelation are determined using Eq. (6.4.9).

Example 6.4.3. Based on the deseasonalized monthly water use data from Table 6.4.3, perform autoregressive filtering to remove autocorrelation within the monthly water use data.

Solution. Autocorrelation in a time series is a measure of persistence of the observation, the existence of which can be identified by computing the autocorrelation coefficients of the different time lags. For example, the lag-1 autocorrelation in the deseasonalized water use time series, in a simple regression framework, is the correlation coefficient of the following regression equation,

$$
Q_c(t) = \phi_0 + \phi_1 Q_c(t-1) \qquad \text{for } t = 2, 3, \ldots, 60.
$$

determined by using Eq. (6.3.22) with $\{Q_c(2), Qc(3), \ldots, Q_c(60)\}$ as the observations for the dependent variable and $\{Q_c(1), Q_c(2), \ldots, Q_c(59)\}$ for the independent variable. Alternatively, the lag-1 autocorrelation coefficient can be computed using Eq. (5.1.26) by letting $\mathbf{y} = \{Q_c(t)\}$ and $\mathbf{x} = \{Q_c(t-1)\}$. In a similar fashion, the autocorrelation for lag-k is measured by the correlation coefficient between two subseries $\{Q_c(k+1), Q_c(k+2), \ldots, Q_c(60)\}$ and $\{Q_c(1), Q_c(2), \ldots, Q_c(60-k)\}$. Based on the deseasonalized data from Example 6.4.2, the autocorrelation coefficients up to lag-8 are

lag	correl. coeff.	lag	correl. coeff.
1	0.4622	5	−0.1281
2	0.0516	6	−0.0840
3	−0.0036	7	0.0312
4	−0.0363	8	0.0052

From this table it is observed that, for this data set, the value of the autocorrelation coefficient decreases rapidly as the lag increases. Although there are some fluctuations in the autocorrelation coefficients, it can be judged that only lag-1 is significantly different from zero for this deseasonalized time series. Therefore, the autoregressive relationship for the deseasonalized water use time series can be modeled by the lag-1 regression equation as indicated above. If more than one lag is believed or is tested to be nonzero, those lags should be incorporated in the general autoregressive model given by Eq. (6.4.9). Based on the deseasonalized water use data, shown in Fig. 6.4.6, the lag-1 autoregressive model determined by a regression analysis is

$$Q_c(t) = -0.0954167 + 0.4322Q_c(t-1)$$

The residuals $Q_d(t) = Q_c(t) - (-0.0959167 + 0.4322Q_c(t-1))$ given in Table 6.4.4 are then the time series free from the long-term trend, seasonality, and self-persistence. The plot of residual time series, $Q_d(t)$, is shown in Fig. 6.4.7.

6.4.4 Climatic Regression

The final step of the cascade model development is to account for the dependence of the monthly water use time series $Q_d(t)$ on the concurrent climatic factors: rainfall, evaporation, and temperature. Such dependency relationship can be modeled as

$$Q_d(t) = \sum_{\ell=1}^{L} \beta_\ell X_\ell(t) + Q_e(t), t = 1, 2, \ldots, T \tag{6.4.11}$$

in which $X_\ell(t)$ is the ℓth climatic factor, L is the total number of climatic factors considered, and $Q_e(t)$ are the residuals that follow a pure random error series with

TABLE 6.4.4
Monthly water use time series, $Q_d(t)$, after autoregression has been removed

t	$Q_d(t)$	t	$Q_d(t)$	t	$Q_d(t)$
1	0	21	−0.87291	41	−0.26214
2	0.52486	22	0.98985	42	1.98734
3	−0.23880	23	−0.54696	43	1.97881
4	−0.94370	24	−0.01049	44	0.17971
5	0.52109	25	−0.45909	45	−0.24137
6	−0.11334	26	−0.55924	46	−0.21279
7	−0.58735	27	−0.04045	47	0.15866
8	0.84819	28	−0.39870	48	−0.13574
9	0.46189	29	−0.02891	49	−0.28646
10	−1.93771	30	−0.41833	50	−0.08753
11	0.64096	31	−2.21679	51	0.34680
12	0.14789	32	0.40185	52	0.50443
13	−0.52388	33	0.50854	53	0.32298
14	−0.31915	34	0.98483	54	−1.97702
14	0.61280	35	−0.77385	55	−0.44694
16	0.48777	36	−0.16042	56	−0.16098
17	0.14670	37	−0.78792	57	−0.47184
18	0.11194	38	−0.22567	58	0.77693
19	1.49896	39	0.41792	59	0.31302
20	−0.31952	40	0.03158	60	0.85069

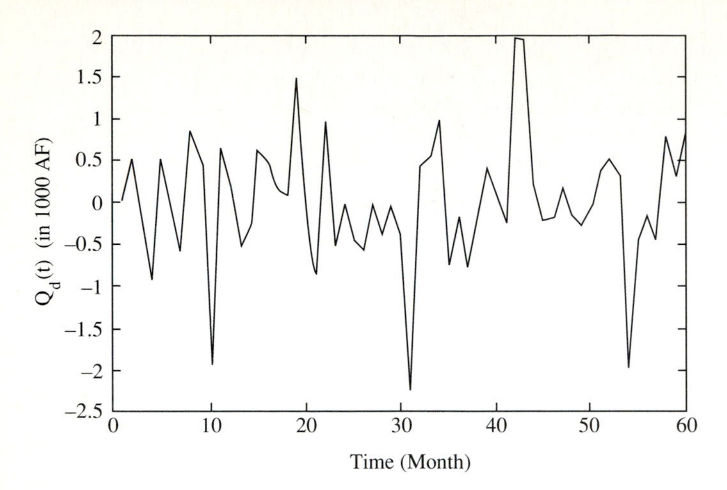

FIGURE 6.4.7
Residual monthly water use time series from autoregressive filtering, $Q_d(t)$.

zero mean and constant variance. The residual $Q_e(t)$ is completely uncorrelated. Unknown model parameters β in Eq. (6.4.11) can be estimated by linear regression analysis using the historical data for the climatic variables $X_\ell(t)$ and the known values of $Q_d(t)$.

The regression equations are

$$
\begin{array}{ll}
t = 1 & Q_d(1) = \beta_1 X_1(1) + \beta_2 X_2(1) + \cdots + \beta_L X_L(1) + Q_e(1) \\
t = 2 & Q_d(2) = \beta_1 X_1(2) + \beta_2 X_2(2) + \cdots + \beta_L X_L(2) + Q_e(2) \\
\vdots & \vdots \qquad \vdots \qquad \vdots \qquad \vdots \\
t = T & Q_d(T) = \beta_1 X_1(T) + \beta_2 X_2(T) + \cdots + \beta_L X_L(T) + Q_e(T)
\end{array}
\tag{6.4.12}
$$

in which the only unknowns are the $\beta_1, \ldots, \beta_L$ and the residuals $Q_e(1), \ldots, Q_e(T)$.

Example 6.4.4. Using the residual time series from the autoregressive filtering obtained in Example 6.4.3, perform climatic regression.

Solution. In this example, the climatic data considered are monthly precipitation (X_1) and maximum temperature (X_2) listed in Table 6.4.5. The model for the climatic regression is

$$Q_d(t) = \hat{\beta}_0 + \hat{\beta}_1 X_1 + \hat{\beta}_2 X_2 + Q_e(t), \qquad t = 1, 2, \ldots, 60$$

The results of a regression analysis yield the following model parameters

$$\hat{\boldsymbol{\beta}} = \begin{bmatrix} \hat{\beta}_0 \\ \hat{\beta}_1 \\ \hat{\beta}_2 \end{bmatrix} = \begin{bmatrix} -0.654815 \\ -0.138738 \\ 0.013205 \end{bmatrix}$$

The residual time series, $Q_e(t)$, from the climatic regression is the pure random error shown in Fig. 6.4.8.

TABLE 6.4.5

Monthly precipitation (in inches) and maximum temperature (in °F) from 1977–1981 for the city of Austin, Texas

t	Precipitation	Temperature	t	Precipitation	Temperature	t	Precipitation	Temperature
1	2.25	50.77	21	4.44	86.73	41	5.43	83.74
2	2.58	66.00	22	1.38	81.29	42	0.31	95.00
3	2.18	72.39	23	5.48	69.60	43	0.28	100.10
4	6.08	77.43	24	2.84	60.13	44	1.18	96.48
5	1.24	83.32	25	2.11	48.10	45	5.66	90.70
6	1.22	92.03	26	3.54	58.04	46	1.29	80.03
7	0.21	95.71	27	3.76	71.03	47	3.41	68.33
8	0.06	97.68	28	2.98	76.03	48	1.24	63.71
9	3.10	94.17	29	7.29	81.06	49	1.61	60.35
10	1.19	82.52	30	0.83	88.90	50	1.18	63.61
11	1.69	72.50	31	10.54	92.13	51	3.05	68.87
12	0.34	65.10	32	0.61	92.35	52	0.81	81.80
13	0.88	48.77	33	1.40	88.27	53	9.02	84.35
14	1.95	53.89	34	0.45	86.03	54	14.96	89.60
15	0.84	70.42	35	0.59	67.77	55	3.39	94.00
16	1.72	79.60	36	3.40	64.16	56	0.91	95.94
17	5.78	87.06	37	0.85	60.32	57	2.65	88.87
18	2.98	92.40	38	2.33	65.00	58	7.04	82.94
19	1.19	97.29	39	3.20	72.87	59	0.72	76.23
20	1.49	94.68	40	2.20	79.43	60	0.39	67.84

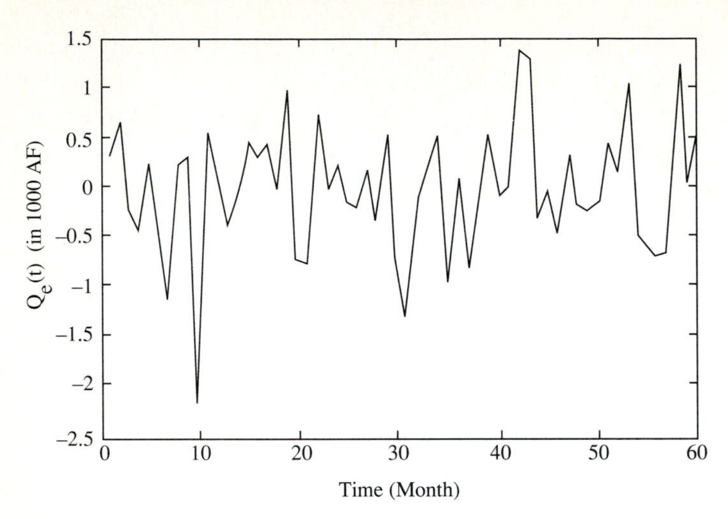

FIGURE 6.4.8
Random residual monthly water use, $Q_e(t)$, from climatic regression.

6.4.5 Application for Monthly Water Use Forecast

Following through the process of transforming historical monthly water use data as described above and in Fig. 6.4.1, let us now assume that the cascade model has been developed with all its unknown coefficients estimated. That is, the water use in future time ($t > T$) can be predicted by the developed cascade model as

$$\hat{Q}_a(t) = \hat{Q}_a(m, y) = \hat{Q}_a(y) + \hat{Q}_b(m) + \hat{Q}_c(t) + \hat{Q}_d(t), \qquad \text{for } t > T \tag{6.4.13}$$

in which '^' represents the predicted values with each term on the RHS calculated by

$$\hat{Q}_a(y) = \tilde{\eta}_0 + \tilde{\eta}_1 \hat{Z}_1(y) + \cdots + \tilde{\eta}_q \hat{Z}_q(y) \tag{6.4.14}$$

$$\hat{Q}_b(m) = \sum_{k=0}^{6} \tilde{a}_k \cos\left(\frac{2\pi km}{12}\right) + \tilde{b}_k \sin\left(\frac{2\pi km}{12}\right) \tag{6.4.15}$$

$$\hat{Q}_c(t) = \sum_{i=1}^{I} \tilde{\phi}_i \hat{Q}_c(t - i), \qquad t = T + 1, T + 2, \ldots \tag{6.4.16}$$

$$\hat{Q}_d(t) = \sum_{\ell=1}^{L} \tilde{\beta}_\ell \hat{X}_\ell(t), \qquad t = T + 1, T + 2, \ldots \tag{6.4.17}$$

in which "~" represents the estimated values ($\tilde{\eta}, \tilde{a}_k, \tilde{b}_k, \tilde{\phi}$, and $\tilde{\beta}$) from the regression analysis.

To forecast future values for $\hat{Q}_a(y)$ forecast on the future socioeconomic factors, $Z(y)$, must be made beforehand. The seasonality component, $\hat{Q}_b(m)$, however, depends only on the month being considered. To compute future values of deseasonalized water use time series, $\{\hat{Q}_c(t), t > T\}$, reliance on the last I historical time series $\{Q_c(t), T - I \leq t \leq T\}$ is needed to perform the recursive calculations. Finally, to compute the contribution to water use from climatic variables, forecasts of the future values of climatic variables are necessary. The future total water use is then the sum of all four components. A flow chart of using the cascade model for future water use forecast is shown in Fig. 6.4.9.

6.5 ECONOMETRIC MODELS FOR WATER DEMAND FORECASTING

The conventional regression analysis (Section 6.3) is for one-way, **unidirectional causality**. The variables on the RHS of equations are assumed to have impacts on the response variable on the LHS. There is no feedback in the other direction. In reality it is a fact that everything depends on everything else. The outcomes of an economic system depend on the equilibrium of many interrelated variables and factors (Judge et al., 1985). Therefore, strictly speaking, the regression equation using price only on the RHS does not completely capture the true interaction among economic variables. To account for the feedback nature of some variables in economic systems, simultaneous-equation systems are more appropriate than those single-equation systems as in conventional regression analysis.

In water demand forecasting, the quantity of water used in a given year would depend on the price of water, population, income, precipitation, and others. On the other hand, the price of water may depend on the quantity of water to be supplied. At equilibrium, the demand and supply of water will be equal. This interaction and feedback of variables in water demand forecasting can be expressed as

$$\mathbf{Q}_d = \mathbf{P}\gamma + \mathbf{POP}\delta_1 + \mathbf{I}\delta_2 + \mathbf{PREC}\delta_3 + \mathbf{e}_1 \tag{6.5.1}$$

$$\mathbf{P} = \mathbf{Q}_s\gamma + \mathbf{e}_2 \tag{6.5.2}$$

$$\mathbf{Q}_d = \mathbf{Q}_s \tag{6.5.3}$$

in which $\mathbf{Q}_d$ and $\mathbf{Q}_s$ are T by 1 vectors of quantity of water demand and supply, respectively; $\mathbf{P}$ is a T by 1 vector of price over T periods, **POP**, **I**, **PREC** are T by 1 vectors of population, income, and precipitation, respectively, and **e** are vectors of disturbances; and γ, δ_1, δ_2, and δ_3 are coefficients that must be determined.

Broadly speaking, economic variables can be classified into two types: **endogenous variables** and **exogenous variables**. Endogenous variables are those whose outcome values are determined through joint interaction with other variables within the system. In other words, the impact of endogenous variables is two-way. For example, price, water demand, and supply are endogenous variables. On the other hand, exogenous variables effect the outcome of endogenous variables, but their values are determined from outside the system. There is no feedback between the exogenous variables. Exogenous variables in the above water demand model are population, income, and precipitation.

A system of simultaneous equations may include (1) **behavior equations**, (2) **technical equations**, (3) **institutional equations**, (4) **accounting** or **definitional**

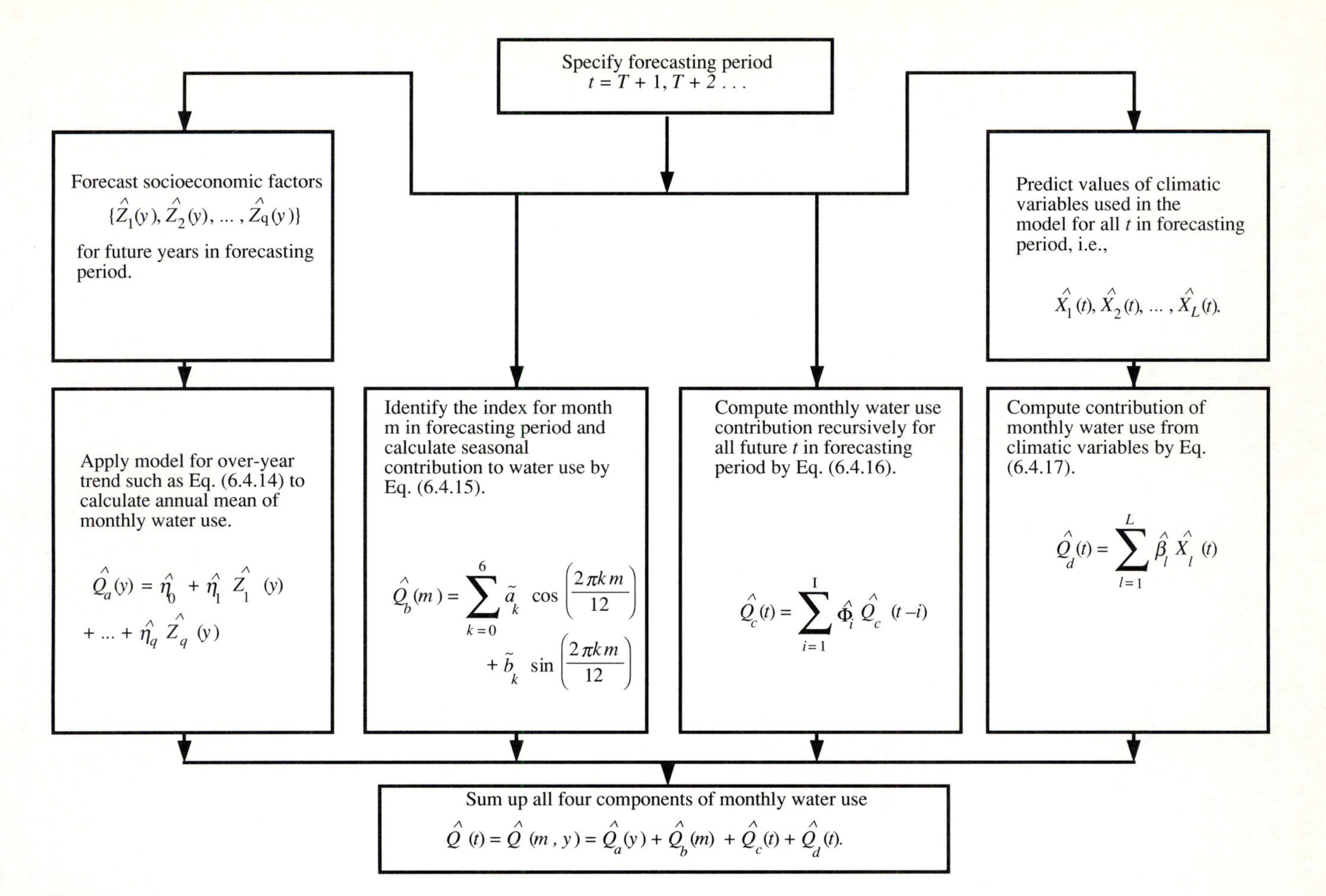

FIGURE 6.4.9
Application of cascade model for monthly water use forecasting.

equations, and (4) **equilibrium conditions** (Judge et al., 1982). Behavior equations describe the behavior of consumers and producers in an economic system. Referring to the water demand forecasting model presented, behavior equations are demand and supply functions such as Eqs. (6.5.1) and (6.5.2). Technical equations depict input-output relations such as production functions (see Chapter 2). Example of institutional equations are tax rules and regulations set by the government. Definitional equations reflect economic relations among variables involved, such as income equals to consumption plus investment and government expenditure. Equilibrium conditions specify the conditions under which the outcomes of variables are to be determined in the market. For example, under the perfect competitive market, an equilibrium price is obtained if the quantity demanded equals to the quantity supplied, such as Eq. (6.5.3). In econometrics, these equations define the structure of an economic system under study and therefore are called **structural equations.**

It should be noted that institutional and definitional equations, and equilibrium conditions are deterministic without stochastic terms and unknown parameters. They provide important feedback relations for the joint determined variables. On the other hand, behavior and technical equations specify possible relationships among endogenous and exogenous variables and thus contain stochastic terms and unknown parameters.

6.5.1 Simultaneous Equations

Consider a system of simultaneous equations containing G endogenous variables (y) and K exogenous variables (x) over T periods of time:

$$\sum_{j=1}^{G} y_{tj}\gamma_{ji} + \sum_{k=1}^{K} x_{tk}\delta_{ki} + e_{ti} = 0, \qquad i = 1, 2, \ldots G;\ t = 1, 2, \ldots, T \tag{6.5.4}$$

in which γ_{ji} and δ_{ki} are called **structural parameters** which are to be estimated. For a particular t, the G simultaneous equations can be expressed as

$$\begin{array}{c}
(y_{t1}\gamma_{11} + y_{t2}\gamma_{21} + \cdots + y_{tG}\gamma_{G1}) + (x_{t1}\delta_{11} + x_{t2}\delta_{21} + \cdots + x_{tK}\delta_{K1}) + e_{t1} = 0 \\
(y_{t1}\gamma_{12} + y_{t2}\gamma_{22} + \cdots + y_{tG}\gamma_{G2}) + (x_{t1}\delta_{12} + x_{t2}\delta_{22} + \cdots + x_{tK}\delta_{K2}) + e_{t2} = 0 \\
\vdots \qquad \vdots \qquad \vdots \qquad \vdots \qquad \vdots \qquad \vdots \qquad \vdots \qquad \vdots \ \ \vdots \\
(y_{t1}\gamma_{1G} + y_{t2}\gamma_{2G} + \cdots + y_{tG}\gamma_{GG}) + (x_{t1}\delta_{1G} + x_{t2}\delta_{2G} + \cdots + x_{tK}\delta_{KG}) + e_{tG} = 0
\end{array} \tag{6.5.5}$$

With all T observations, Eq. (6.5.5) can be expressed in matrix form as

$$\mathbf{Y}_{T\times G}\mathbf{\Gamma}_{G\times G} + \mathbf{X}_{T\times K}\mathbf{\Delta}_{K\times G} + \mathbf{E}_{T\times G} = \mathbf{0}_{T\times G} \tag{6.5.6}$$

A system is complete if the number of endogenous variables equals to the number of equations. The simultaneous equations system, Eq. (6.5.6), can be visualized as the linear regression in a multivariate setting in which more than one response variable must be considered simultaneously.

6.5.2 Reduced Form Equations

In principle, the reduced form equations are obtained by expressing G endogenous variables in terms of K exogenous variables. Equation (6.5.6) can be written in the equivalent form,

$$\mathbf{Y} = \mathbf{X\Pi} + \mathbf{V} \tag{6.5.7}$$

where $\mathbf{\Pi} = -\mathbf{\Delta\Gamma}^{-1}$, a K by G matrix, and $\mathbf{V} = -\mathbf{E\Gamma}^{-1}$, a T by G matrix. More specifically, the ith endogenous variable, $\mathbf{y}_{.i}$ (corresponding to the ith column of matrix $\mathbf{Y}$), can be expressed as

$$\mathbf{y}_{.i} = \mathbf{X}\boldsymbol{\pi}_{.i} + \mathbf{V}_{.i} \tag{6.5.8}$$

or algebraically,

$$y_{ti} = \sum_{j=1}^{K} x_{tj}\pi_{ji} + v_{ti} \tag{6.5.9}$$

From Eq. (6.3.14), the OLS estimators of the unknown parameters in the reduced form equations are

$$\hat{\mathbf{\Pi}} = (\mathbf{X}^T\mathbf{X})^{-1}\mathbf{X}^T\mathbf{Y} \tag{6.5.10}$$

which minimize $\hat{\mathbf{V}}^T\hat{\mathbf{V}}$ where $\hat{\mathbf{V}}^T = \mathbf{Y} - \mathbf{X}\hat{\mathbf{\Pi}}$. The OLS estimator $\hat{\mathbf{\Pi}}$ can be shown to be consistent and unbiased for the reduced form equation. Once the parameters of the reduced form are estimated, the parameters matrix of the original model $\mathbf{\Gamma}$ can be estimated as

$$\hat{\mathbf{\Gamma}} = -\hat{\mathbf{\Pi}}^{-1}\mathbf{\Delta} \tag{6.5.11}$$

Knowing the relationship between the reduced parameters and the structural parameters, that is, $\mathbf{\Pi\Gamma} = -\mathbf{\Delta}$, one would think that the values of structural parameters can be derived from those of the reduced parameters. However, to obtain the solutions to structural parameters in terms of the estimated reduced form parameters depends on the characteristics of $\mathbf{\Pi\Gamma} = -\mathbf{\Delta}$. This leads to the topic of the identification problem of the simultaneous equations which is discussed in great detail by Intriligator (1978), Judge et al. (1985), and Fomby et al. (1988).

There are three possible outcomes from the identification of each individual equation in the structural equations. That is, an equation could be **exactly identified**, **over-identified**, or **under-identified**. The structural parameters in an exactly identified equation can be uniquely determined from the estimated reduced-form parameters. Nonuniqueness of structural parameters will occur when the equation is over-identified while the equation is undistinguishable with other equations if it is under-identified.

Example 6.5.1. Use the first 5 years of data for Austin, Texas, in Table 6.3.1 to estimate the unknown parameters $\pi_1, \ldots, \pi_4$ in the reduced-form Eqs. (6.5.7) for an econometric model to forecast water demand.

Solution. The system of three simultaneous equations (6.5.1–6.5.3), can be reduced to two simultaneous equations by substituting Eq. (6.5.3). To demonstrate the estimation procedure the model is simplified by assuming that the price and population impact the water demand while, on the other hand, demand and income impact the water price. The hypothesized simultaneous equations then reduce to the following.

$$-Q + P\gamma_{21} + POP\delta_{11} + e_1 = 0 \qquad (a)$$

$$Q\gamma_{12} - P + I\delta_{22} + e_2 = 0 \qquad (b)$$

where $\gamma_{11} = -1$, $\gamma_{22} = -1$, $\delta_{12} = 0$, and $\delta_{12} = 0$. The two endogenous variables are water demand (in 10000 AF) and price, $\mathbf{Y} = (\mathbf{Q}, \mathbf{P})$, while the two exogenous variables are population (in 10000) and income (in \$1000), that is, $\mathbf{X} = (\mathbf{POP}, \mathbf{I})$. The simultaneous equations in terms of $\mathbf{X}$ and $\mathbf{Y}$ can be expressed by Eq. (6.5.6) as

$$\mathbf{Y}_{5\times 2}\mathbf{\Gamma}_{2\times 2} + \mathbf{X}_{5\times 2}\mathbf{\Delta}_{2\times 2} + \mathbf{E}_{5\times 2} = \mathbf{0}_{5\times 2}$$

where

$$\mathbf{Y} = \begin{bmatrix} 3.9606 & 0.98 \\ 4.0131 & 0.95 \\ 4.5667 & 1.20 \\ 4.0780 & 1.15 \\ 4.5330 & 1.10 \end{bmatrix} \quad \text{and} \quad \mathbf{X} = \begin{bmatrix} 2.16733 & 5.919 \\ 2.23334 & 5.970 \\ 2.30135 & 6.521 \\ 2.37144 & 7.348 \\ 2.44366 & 7.965 \end{bmatrix}$$

$$\mathbf{\Gamma} = \begin{bmatrix} \gamma_{11} & \gamma_{12} \\ \gamma_{21} & \gamma_{22} \end{bmatrix} = \begin{bmatrix} -1 & \gamma_{12} \\ \gamma_{21} & -1 \end{bmatrix} \quad \text{and} \quad \mathbf{\Delta} = \begin{bmatrix} \delta_{11} & \delta_{12} \\ \delta_{21} & \delta_{22} \end{bmatrix} = \begin{bmatrix} \delta_{11} & 0 \\ 0 & \delta_{22} \end{bmatrix}$$

Using Eq. (6.5.10), first calculate $\mathbf{X}^T\mathbf{X} = \begin{bmatrix} 26.5765 & 78.0577 \\ 78.0577 & 230.6332 \end{bmatrix}$ and the corresponding inverse $(\mathbf{X}^T\mathbf{X})^{-1} = \begin{bmatrix} 6.3300 & -2.1424 \\ -2.1424 & 0.7294 \end{bmatrix}$. By Eq. (6.5.10), the unknown reduced form parameters are

$$\hat{\mathbf{\Pi}} = (\mathbf{X}^T\mathbf{X})^{-1}\mathbf{X}^T\mathbf{Y} = \begin{bmatrix} \pi_{11} & \pi_{12} \\ \pi_{21} & \pi_{22} \end{bmatrix} = \begin{bmatrix} 2.0305 & 0.4178 \\ -0.0661 & 0.0169 \end{bmatrix}$$

6.5.3 Estimation of Structural Parameters

Note that the objective of estimation is to obtain the unknown structural parameters $\mathbf{\Gamma}$ and $\mathbf{\Delta}$ in the original structural equations. Because of the biasness and inconsistency of the OLS estimators as applied directly to the original structural equations, the reduced form equations are solved instead to obtain unbiased and consistent estimators. Once the unknown parameters $\mathbf{\Pi}$ in the reduced form equations are estimated, the unknown structural parameters $\mathbf{\Gamma}$ and $\mathbf{\Delta}$, can be estimated through the relation $\mathbf{\Pi\Gamma} = -\mathbf{\Delta}$. As mentioned above, if the individual equation in a simultaneous equations system is identified exactly, the structural parameters in the original simultaneous equations can be determined uniquely from the estimated reduced form parameters. Such a procedure is called the **indirect least squares (ILS) method**. In general conditions in which the simultaneous equations system is not identified, more elaborate methods such as **2-stage least squares (2SLS)** or **3-stage least squares (3SLS)** must be used (Judge et al., 1985).

Example 6.5.2. Determine the structural parameters to complete the econometric model for water demand forecasting in Austin, Texas (Example 6.5.1) using the estimated reduced form parameters.

Solution. Use $\mathbf{\Pi\Gamma} = -\mathbf{\Delta}$ and, substitute $\hat{\mathbf{\Pi}}$ estimated in Example 6.5.1, in the relation, then

$$\hat{\mathbf{\Pi}}\mathbf{\Gamma} = \begin{bmatrix} 2.0305 & 0.4178 \\ -0.0661 & 0.0169 \end{bmatrix} \begin{bmatrix} -1 & \gamma_{12} \\ \gamma_{21} & -1 \end{bmatrix} = \begin{bmatrix} \delta_{11} & \delta_{12} \\ \delta_{21} & \delta_{22} \end{bmatrix}$$

$$\begin{bmatrix} -2.0305 + 0.4178\gamma_{21} & 2.0305\gamma_{12} - 0.4178 \\ 0.0661 + 0.0169\gamma_{21} & -0.0661\gamma_{12} - 0.0169 \end{bmatrix} = \begin{bmatrix} -\delta_{11} & 0 \\ 0 & -\delta_{22} \end{bmatrix}$$

To estimate the structural parameters (i.e., γ_{21} and δ_{11}) in the first equation (a) of Example 6.5.1, solve the following system of equations

$$-2.0305 + 0.4178\gamma_{21} = -\delta_{11}$$

$$.0661 + 0.0169\gamma_{21} = 0$$

from which $\hat{\gamma}_{21} = -3.9112$ and $\hat{\delta}_{11} = 3.6646$. Similarly, the structural parameters in the second equation (b) of Example 6.5.1 can be estimated by solving

$$2.0305\gamma_{12} - 0.4178 = 0$$

$$-0.0661\gamma_{12} - 0.0169 = -\delta_{22}$$

from which $\hat{\gamma}_{12} = 0.2058$ and $\hat{\delta}_{22} = 0.0305$. The resulting simultaneous equations derived by the ILS method are

$$Q + 3.9112P = 3.6646 \text{ POP}$$

$$-0.2058\ Q + P = 0.0305I$$

6.6 IWR—MAIN WATER USE FORECASTING SYSTEM

The IWR—MAIN (Institute for Water Resources—Municipal and Industrial Needs) water use forecasting system (Davis et al., 1988) is a software package for personal computers containing a number of forecasting models, parameter generating procedures, and data management techniques. Major elements of this system are shown in Fig. 6.6.1. Urban water uses are divided into four major sectors: residential, commercial/institutional, industrial and public/unaccounted, with sectors further disaggregated into a number of categories (Table 6.6.1). Most forecasts will utilize approximately 130 specific categories of water use, with the maximum number of categories being 284 (Baumann, 1987).

Model use for a given urban area is dependent upon verification of the empirical equations for estimating water use. Water use estimates for one or more historical years are used for verification for which the first year of the time series of water use estimates is the base year and the following years are the forecast years. IWR—MAIN generates estimates of water use for the base year and all historical years in the forecast, and are subsequently compared with recorded water use to determine required

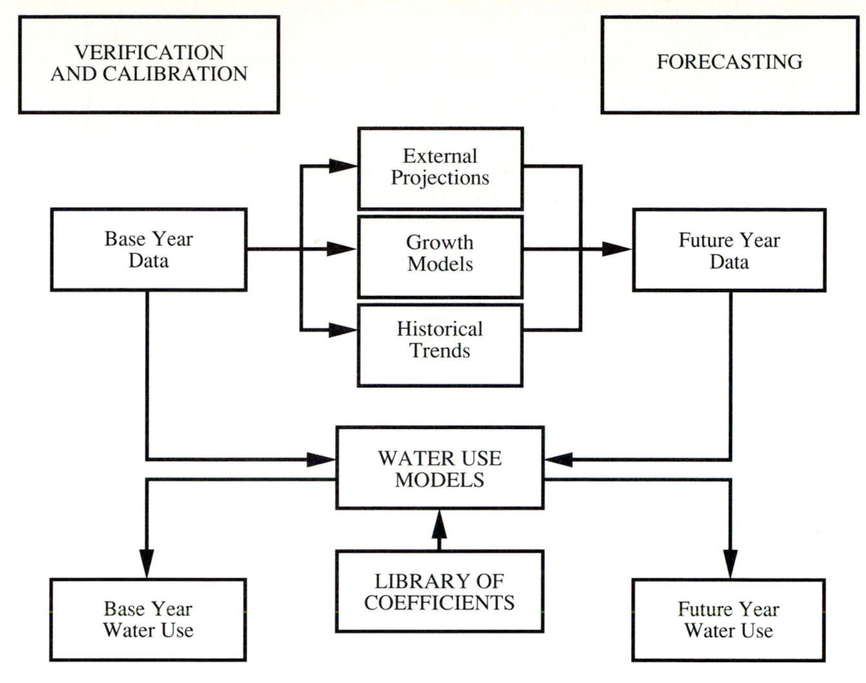

FIGURE 6.6.1
Major components of IWR—MAIN system (Baumann, 1987).

adjustments in the water use model. Three methods (Fig. 6.6.1) are used to project any base year data to any forecast year: (1) using external projections introduced by the user; (2) using the internal growth models; and (3) extrapolating local historical data provided by the user. Typically, all three methods are used; however, extremes would be to externally provide all water use parameters for input to generating all the water use parameters internally in the model.

Input data requirements depend upon the forecast method selected with the minimum input for a water use forecast being the total resident population, total employment, and median household income. Fig. 6.6.2 shows the data requirements and procedure for deriving disaggregate water use estimates.

Water use is estimated for each of the four major sectors: residential, commercial/institutional, industrial, and public/unaccounted. Residential water use is first disaggregated into the four housing types listed in Table 6.6.1, then divided into as many as 25 market value ranges with each type of housing.

ESTIMATING RESIDENTIAL WATER USE. IWR—MAIN estimates residential water use by means of econometric models that link water use to price of water and disposable income. Residential water use equations are applied to estimate water use for winter, summer, average annual, and maximum day for each housing type. Forecasting of residential water use is performed by projecting the number of housing units for each value range of each housing type. The number of units in each range

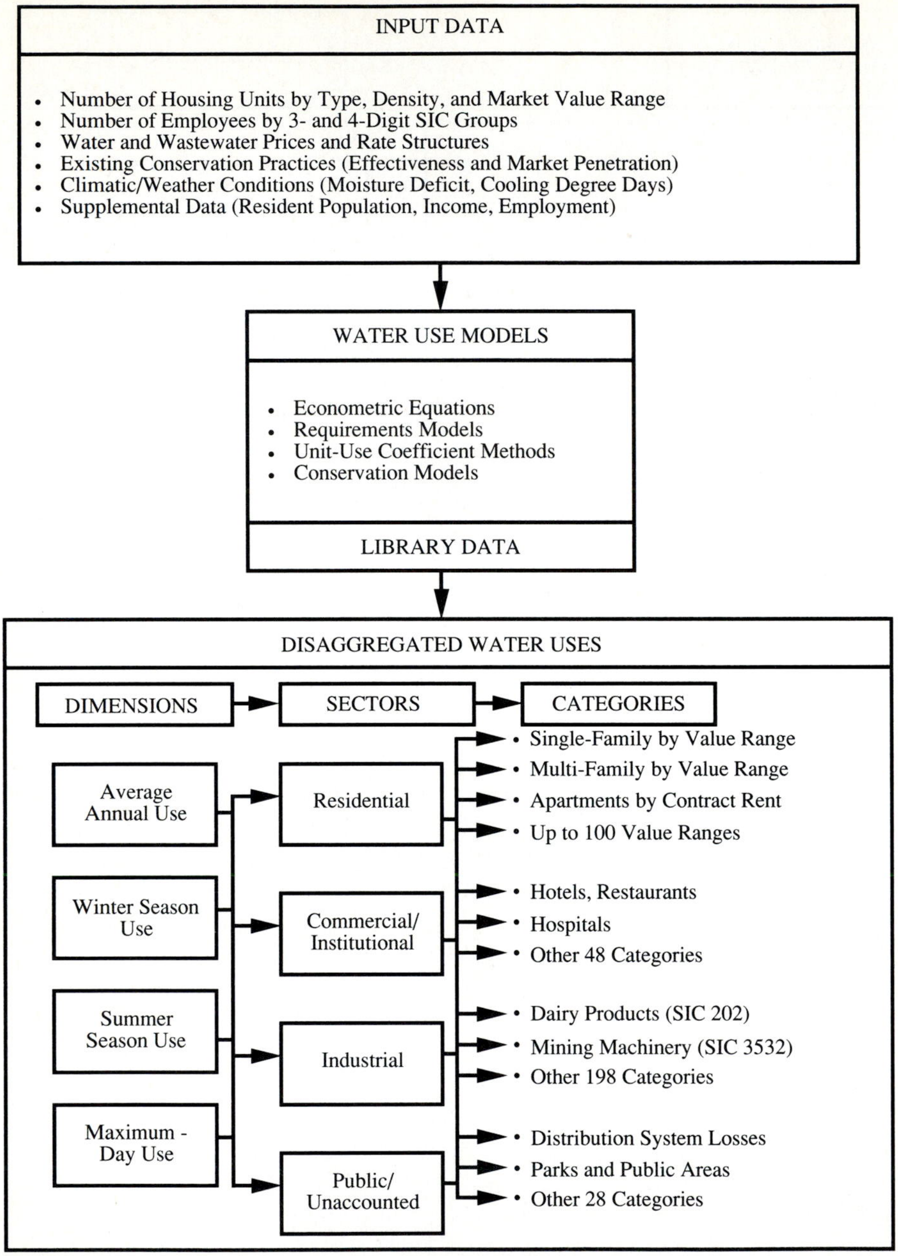

INPUT DATA

- Number of Housing Units by Type, Density, and Market Value Range
- Number of Employees by 3- and 4-Digit SIC Groups
- Water and Wastewater Prices and Rate Structures
- Existing Conservation Practices (Effectiveness and Market Penetration)
- Climatic/Weather Conditions (Moisture Deficit, Cooling Degree Days)
- Supplemental Data (Resident Population, Income, Employment)

WATER USE MODELS

- Econometric Equations
- Requirements Models
- Unit-Use Coefficient Methods
- Conservation Models

LIBRARY DATA

DISAGGREGATED WATER USES

DIMENSIONS → SECTORS → CATEGORIES

Average Annual Use

Winter Season Use

Summer Season Use

Maximum - Day Use

Residential
- Single-Family by Value Range
- Multi-Family by Value Range
- Apartments by Contract Rent
- Up to 100 Value Ranges

Commercial/ Institutional
- Hotels, Restaurants
- Hospitals
- Other 48 Categories

Industrial
- Dairy Products (SIC 202)
- Mining Machinery (SIC 3532)
- Other 198 Categories

Public/ Unaccounted
- Distribution System Losses
- Parks and Public Areas
- Other 28 Categories

FIGURE 6.6.2
IWR—MAIN procedure for estimating water use (Dziegielewski, 1987).

TABLE 6.6.1
Water use forecasting approaches in the IWR—MAIN system (Baumann, 1987)

Sector	Water use category	Forecast method
Residential	Metered and sewered residences	Econometric demand models
	Flat rate and sewered residences	Multiple coefficient requirements models
	Flat rate and unsewered residences	Multiple coefficient requirements models
	Master-metered apartments	Multiple coefficient requirement models
Commercial/ Institutional	Up to 50 user categories, defined as groups of four-digit SIC codes	Unit use coefficients (per employee)
Industrial	Up to 200 user categories, presently including 198 manufacturing categories, defined by three-digit and four-digit SIC codes	Unit use coefficients (per employee)
Public/Unaccounted	Up to 30 user categories, such as distribution system losses and free service	Unit use coefficients or per capita requirements

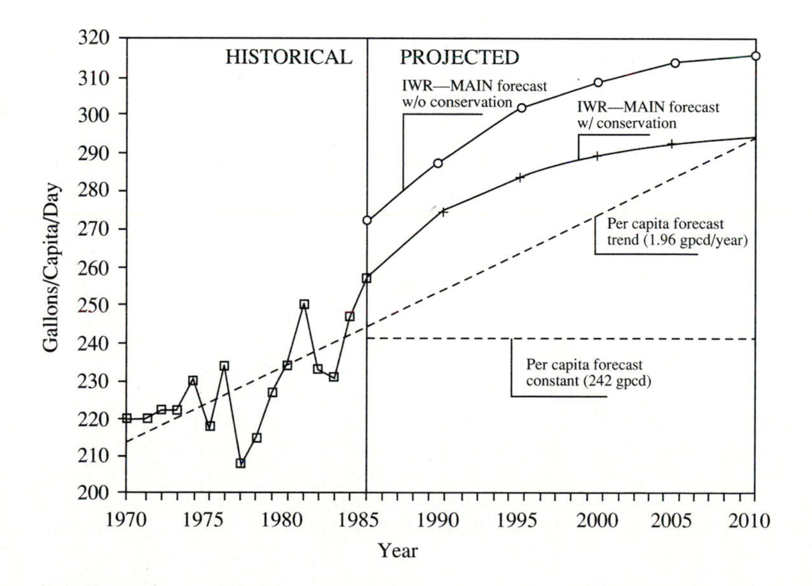

FIGURE 6.6.3
Comparison of per capita method and IWR—MAIN method applied to Anaheim, California (Dziegielewski and Boland, 1989).

and each type is multiplied by the water use per house unit. Econometric equations are used in absence of external housing projections by type and value group.

NONRESIDENTIAL WATER USE. IWR—MAIN can consider up to 280 categories of nonresidential water use within the three major sectors, that is, industrial, commercial/institutional, and public/unaccounted. Water use for each category is predicted by multiplying the number of employees by the water use per employee per day in that category. The SIC in Fig. 6.6.2 refers to standard industrial classification. The categorization of commercial and institutional users in public/unaccounted categories are service and distribution losses. Service is estimated using resident population as the water use parameter with a default value of 5.2 gallons per person per day. Distribution system losses can be estimated using resident population assuming losses of 14.9 gallons per person per day or as a percent of total municipal use.

Nonresidential water use forecasts require that the employment in each category be projected to each forecast year. Employment can be: (a) projected externally and input into IWR—MAIN; (b) generated internally based upon historical employment trends for each category; or (c) estimated from projections of total employment provided for the base year and all forecast years. Fig. 6.6.3 illustrates the comparison of the per capita method with the IWR—MAIN method for application to Anaheim, California. Two forecasts are shown, with and without conservation.

REFERENCES

Baumann, D. D.: "Demand Management and Urban Water Supply Planning," *Proc. Am. Soc. Civ. Eng. Conf., The Role of Social and Behavioral Sciences in Water Resources Planning and Management*, D. D. Baumann and Y. Y. Haimes, eds., May 1987.

Boland, J. J., Baumann, D. D., and B. Dziegielewski: *An Assessment of Municipal and Industrial Water Use Forecasting Approaches*, Contract Report 81-C05, U.S. Army Engineer Institute for Water Resources, Fort Belvoir, Va., 1981.

Boland, J. J., W.-S. Moy, R. C. Steiner, and J. L. Pacey: *Forecasting Municipal and Industrial Water Use: A Handbook of Methods*, Contract Report 83C-01, U.S. Army Engineer Institute for Water Resources, Fort Belvoir, Va., July 1983.

Daniel, C. and F. S. Wood: *Fitting Equations to Data*, 2d ed., Wiley, New York, 1980.

Davis, W. U., D. Rodrigo, E. Opitz, B. Dziegielewski, D. D. Baumann, and J. Boland: *IWR—Main Water Use Forecasting System, Version 5.1: User's Manual and System Description.* IWR-Report 88-R-6, U.S. Army Corps of Engineers Institute for Water Resources, Fort Belvoir, Va., 1988.

Draper, N. R., and H. Smith: *Applied Regression Analysis*, Wiley, New York, 1981.

Dziegielewski, B.: "The IWR—MAIN Disaggregate Water Use Model," *Proceedings of the 1987 UCOWR Annual Meeting*, UCOWR Executive Director's Office, Carbondale, Ill., August 1987.

Dziegielewski, B., and J. J. Boland: "Forecasting Urban Water Use: The IWR—MAIN model," *Water Resources Bulletin*, American Water Resources Association, vol. 25, no. 1, pp. 101–109, February 1989.

Dzurick, A. A.: "Water Use and Public Policy," in *Civil Engineering Practice 5/Water Resources/Environmental*, P. N. Cheremisinoff, N. P. Cheremisinoff, and S. L. Cheng, eds., Technomics Publishing Co., Inc., Lancaster, Pa, 1988.

Encel, S., P. K. Pauline, and W. Page, ed.: *The Art of Anticipation: Values and Methods in Forecasting*, Pica Press, New York, 1976.

Fomby, T. B., R. C. Hill, and S. R. Johnson: *Advanced Econometric Methods*, Springer-Verlag, New York. 1988.

Intriligator, M. D.: *Econometric Models, Techniques, and Applications*, Prentice-Hall, Englewood Cliffs, N.J. 1978.

Judge, G. G., R. C. Hill, W. E. Griffths, H. Lutkepohl, and T. C. Lee: *Introduction to the Theory and Practices Econometrics*, Wiley, New York, 1982.

Judge, G. G., W. E. Griffths, R. C. Hill, H. Lutkepohl, and T. C. Lee: *The Theory and Practices Econometrics*, Wiley, New York, 1985.

Maidment, D. R. and E. Parzen: *A Cascade Model of Monthly Municipal Water Use*, Texas Engineering Experiment Station, Texas A&M University, College Station, 1981.

Maidment, D. R. and E. Parzen: "Time Patterns of Water Use in Six Texas Cities," *Journal of Water Resources Planning and Management*, ASCE, vol. 110, no. 1, pp. 90–106, January 1984*a*.

Maidment, D. R. and E. Parzen: "Cascade Model of Monthly Municipal Water Use," *Water Resources Research*, vol. 20, no. 1, pp. 15–23, 1984*b*.

Maidment, D. R., S.-P. Miaou, D. N. Nvule, and S. G. Buchberger: *Analysis of Daily Water Use in Nine Cities, Technical Report 201*, Center for Research in Water Resources, The University of Texas, Austin, Tex., February 1985.

Montgomery, D. C., and E. A. Peck: *Introduction to Linear Regression Analysis*, Wiley, New York, 1982.

Neter, J., W. Wasserman, and M. H. Kutner: *Applied Linear Regression Models*, Irwin, Homewood, Ill., 1983.

Salas, J. D., J. W. Delleur, V. Yevjevich, and W. L. Lane: *Applied Modeling of Hydrologic Time Series*, Water Resources Publication, Littleton, Colo. 1980.

SAS Institute Inc.: *SAS/STAT User's Guide*, Release 6.03 edition, Cary, N.C. SPSS, Inc., *SPSS* User's Guide*, 2d ed., Chicago, 1986.

U.S. Water Resources Council: *Economic and Environmental Principles and Guidelines for Water and Related Land Resources Implementation Studies*, Washington, D.C., U.S. Government Printing Office, March 1983.

PROBLEMS

6.2.1 Based on the first sixteen (16) years of water use data (1965–1980) for Austin, Texas as contained in Table 6.3.1, forecast the water use in 1981–1985 using the per capita method and compare with observed water use.

6.2.2 Repeat Problem 6.2.1 using the first 15 years of water use data, 1966–1980, for Bastrop, Texas (in Table 6.P.1).

6.3.1 Based on the first 16 years of annual water use (1965–1980) for Austin given in Table 6.3.1 develop a model using only the element of time, that is, years, to project the water use in 1981–1985.

6.3.2 Among the four factors, that is, population (X_1), water price (X_2), average income (X_3) and annual precipitation (X_4), that potentially might affect water use in Austin, how do you determine which would be the best two factors that you want to include in developing the linear regression model having the form of Eq. (6.2.6)? Explain your rationale and show the necessary computations.

6.3.3 Consider that population and average income are to be included in the following water use forecasting model for Austin,

$$Y = a_0 + a_1 X_1 + a_2 X_2$$

in which Y = annual water use, X_1 = population size, and X_2 = average income. Determine the three unknown coefficients (a_0, a_1, a_2) by regression analysis using the first 16-years of data (1965–1980) from Table 6.3.1.

TABLE 6.P.1
Water demand data for Bastrop, Texas

Year	Population	Price ($/1000 gal)	Income ($/person)	Annual precipitation (in.)	Water use (ac-ft)
1966	3067	0.68	4592	30.0	472
1967	3078	0.66	4831	33.9	545
1968	3089	0.64	5156	56.7	453
1969	3100	0.61	5405	35.7	510
1970	3112	0.58	5342	34.7	578
1971	3173	0.78	5808	27.1	878
1972	3236	0.75	6121	33.5	546
1973	3301	0.71	6487	45.8	737
1974	3366	0.64	6669	42.5	739
1975	3433	0.59	6505	41.5	963
1976	3502	0.56	6973	53.3	809
1977	3571	0.52	6766	31.0	877
1978	3642	0.49	8161	34.0	1189
1979	3715	0.45	8470	43.7	919
1980	3789	0.69	8492	27.5	868
1981	4063	1.06	9137	52.7	518
1982	4326	1.00	8952	29.1	755
1983	4616	0.96	9049	37.7	784
1984	4927	0.93	9473	30.3	790
1985	5280	1.07	9449	37.2	816

6.3.4 Based on the water use forecast model developed in Problem 6.3.3 for Austin, forecast the annual water use in 1981–1985. Compare the accuracy of water use by the forecast model and the projection model developed in Problem 6.3.1.

6.3.5 Repeat Problem 6.3.3 by adopting the model form of Eq. (6.2.7), that is,

$$\ln Y = b_0 + b_1 \ln X_1 + b_2 \ln X_2$$

in which Y = annual water use, X_1 = population size, and X_2 = average income. Also, compare the performance of the two forecast models the one in Problem 6.3.3 and the current model considered.

6.3.6 Consider a water forecasting model having the following form

$$Y = c_0 + c_1 \ln X_1 + c_2 \ln X_2$$

in which Y = annual water use, X_1 = population size, and X_2 = average income. Determine the coefficients by regression analysis using the first 16 years (1965–1980) data in Table 6.3.1.

6.3.7 Use the annual water demand data in the Table 6.P.1 for Bastrop, Texas to determine the coefficients of a demand model for annual water use using the form of Eq. (6.2.6). Consider population (X_1) and average income (X_2) as the two independent variables in the model.

6.3.8 Repeat Problem 6.3.7 to determine the coefficients of a demand model for annual water use using the form of Eq. (6.2.7).

6.3.9 Repeat Problem 6.3.7 to determine the coefficients of a demand model for annual water use using the form of Eq. (6.2.8).

6.3.10 Using population (X_1), averaged income (X_2), and the linear water demand forecast model

$$Y = a_0 + a_1 X_1 + a_2 X_2$$

as in Problem 6.3.3, develop a linear programming model which determines a_0, a_1, and a_2 by minimizing the sum of absolute errors between the predicted and observed water use. Furthermore, solve the LP model using the first 16 years (1965–1980) of data in Table 6.3.1.

6.3.11 Repeat Problem 6.3.10 by changing the objective to minimize the largest error between the observed and predicted water use.

6.4.1 Solve Example 6.4.1 using the last four years (1978–1981) of monthly data.

6.4.2 Solve Example 6.4.2 using the last four years (1978–1981) of monthly data.

6.4.3 Solve Example 6.4.3 using the last four years (1978–1981) of monthly data.

6.4.4 Solve Example 6.4.4 using the last four years (1978–1981) of monthly data.

6.5.1 Adopting the econometrics model used in Example 6.5.1, estimate the unknown parameters in the reduced form equations using the first 16 years (1965–1980) of data in Table 6.3.1.

6.5.2 Based on the reduced form model developed in Problem 6.5.1, determine the structural parameters in the original econometrics model in Example 6.5.1.

6.5.3 Apply the developed econometric model in Problem 6.5.2 to forecast the annual water use and price for 1981–1985 based on the information contained in Table 6.3.1.

6.5.4 Develop an econometric model having the same form as in Example 6.5.1 based on the first 15 years (1966–1980) of water use data for Bastrop, Texas given in Table 6.P.1.

CHAPTER 7

SURFACE WATER SYSTEMS

7.1 SURFACE WATER RESERVOIR SYSTEMS

The primary function of reservoirs is to smooth out the variability of surface water flow through control and regulation and make water available when and where it is needed. The use of reservoirs for temporary storage would result in an undesirable increase in water loss through seepage and evaporation. However, the benefits that can be derived through regulating the flow for water supplies, hydropower generation, irrigation, recreational uses, and other activities would offset such losses. The net benefit associated with any reservoir development project is dependent on the size and operation of the reservoir, as well as the various purposes of the project.

Reservoir systems may be grouped into two general operation purposes: **conservation** and **flood control**. Conservation purposes include water supply, low-flow augmentation for water quality, recreation, navigation, irrigation and hydroelectric power, and any other purpose for which water is saved for later release. Flood control is simply the retention or detention of water during flood events for the purpose of reducing downstream flooding. This chapter focuses only on surface water reservoir systems for conservation. The flood control aspect of reservoir system operation is discussed in Chapter 13.

Generally, the total reservoir storage space in a multipurpose reservoir consists of three major parts (see Fig. 7.1.1): (1) the **dead storage zone**, mainly required for sediment collection, recreation, or hydropower generation; (2) the **active storage**, used

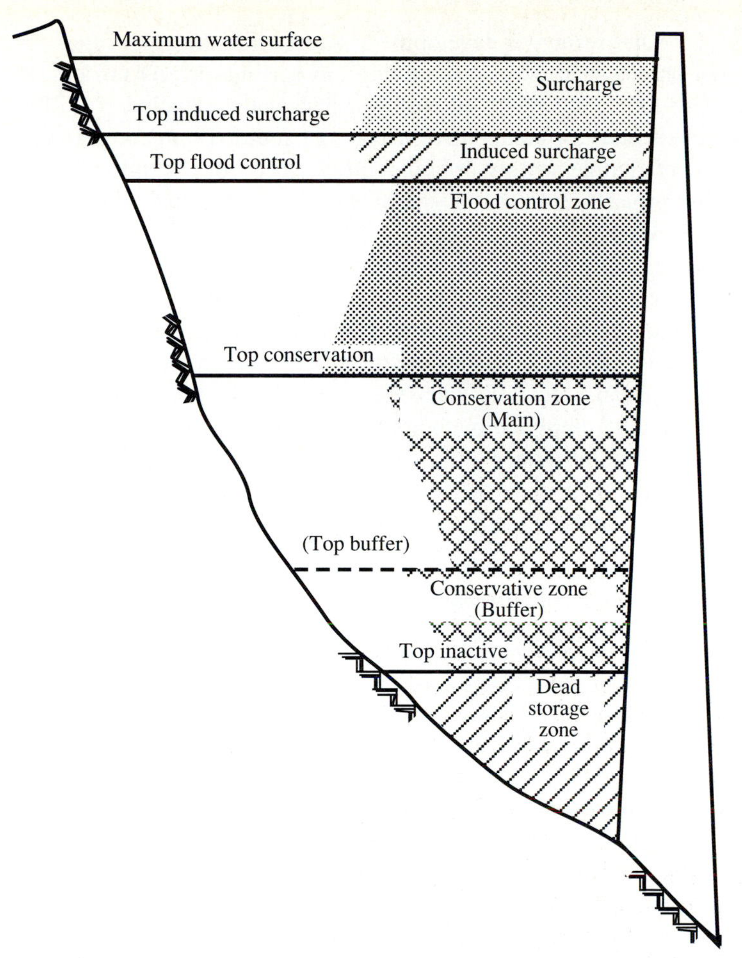

FIGURE 7.1.1
Reservoir storage allocation zones (U.S. Army Corps of Engineers, 1977).

for conservation purposes, including water supplies, irrigation, navigation, etc.; (3) the **flood control storage** reserved for storing excessive flood volume to reduce potential downstream flood damage. In general, these storage spaces could be determined separately and combined later to arrive at a total storage volume for the reservoir.

7.2 STORAGE-FIRM YIELD ANALYSIS FOR WATER SUPPLY

The determination of storage-yield relationships for a reservoir project is one of the basic hydrologic analyses associated with the design of reservoirs. Two basic problems in storage-yield studies (U.S. Army Corps of Engineers—HEC, 1977) are: (1) determination of storage required to supply a specified yield; and (2) determination of yield for a given amount of storage. The former is usually encountered in the planning and early design phases of a water resources development study while the latter often occurs in the final design phases or in the reevaluation of an existing project for a more comprehensive analysis. Other objectives of storage-yield analysis include: (1) the determination of complementary or competitive aspects of multiple-project development; (2) determination of complementary or competitive aspects of multiple-purpose development in a single project; and (3) analysis of alternative operation rules for a project or group of projects.

The procedures used to develop a storage-yield relationship include (U.S. Army Corps of Engineers, 1977): (1) simplified analysis; and (2) detailed sequential analysis. The simplified techniques are satisfactory when the study objectives are limited to preliminary or feasibility studies. Detailed methods that include simulation and optimization analysis are usually required when the study objectives advance to the design phase. The objective of simplified methods is to obtain a reasonably good estimate of the results which can be further improved by a detailed sequential analysis. Factors affecting the selection of method for analysis are: (1) study requirements; (2) degrees of accuracy required; and (3) the basic data required and available.

7.2.1 Firm-Yield Analysis Procedures

Firm yield is defined as the largest quantity of flow or flow rate that is dependable at the given site along the stream at all times. More specifically, Chow et al. (1988) define the firm yield of a reservoir as the mean annual withdrawal rate that would lower the reservoir to its minimum allowable level just once during the critical drought of record. The most commonly used method to determine the firm yield of an unregulated river is to construct a **flow-duration curve**, which is a graph of the discharge as a function of the percent of time that the flow is equalled or exceeded as shown in Fig. 7.2.1 for the monthly flows in Table 7.2.1.

The flow-duration curve can be developed for a given location on the stream by arranging the observed flow rates in the order of descending magnitude. From this, the percentage of time for each flow magnitude to be equalled or exceeded can be computed. Then, this percentage of time of exceedence is plotted against the flow magnitude to define the flow-duration relationship. The firm yield is the flow magnitude that is equalled and exceeded 100 percent of the time for a historical sequence of flows. Flow-duration curves are used in the determination of the water-supply potential in the planning and design of water resources projects, in particular the hydropower plants.

MASS-CURVE ANALYSIS. To increase the firm yield of an unregulated river, surface impoundment facilities are constructed to regulate the river. Two methods, **mass-curve analysis** and **sequent-peak analysis**, can be used to develop storage-yield relationships for specific locations along a river. A **mass curve** is a plot of the cumulative flow volumes as a function of time. Mass-curve analysis was first developed by Ripple in 1883. The method uses historical or synthetic stream flow sequences over a time interval, $[0, T]$. Implicitly the analysis assumes that the time interval includes the **critical period** which is the time period over which the flows have reached

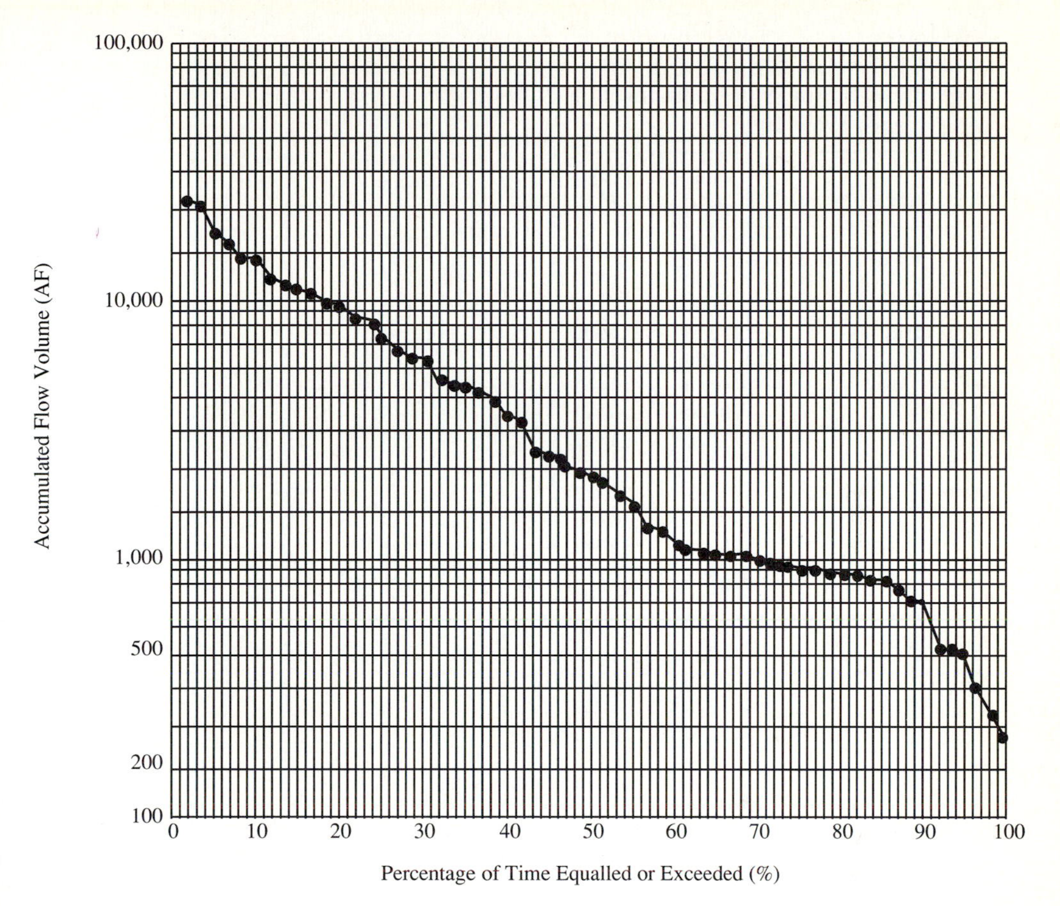

FIGURE 7.2.1
Flow-duration curve for the Little Weiser River near Indian Valley, Idaho (1966–1970).

a minimum causing the greatest drawdown of a reservoir. Mass-curve analysis can be implemented using graphical procedures. Two graphical procedures are described here. The first is the original **Ripple method** (shown in Fig. 7.2.2) in which the value of cumulative reservoir inflows over time from Table 7.2.1 are plotted.

The Ripple method is applicable when the release is constant. In cases where the releases vary, however, it is easier to compute the difference between the cumulative reservoir inflows and cumulative reservoir releases. The required active storage volume is the maximum difference. Of course, this alternative approach can be applied to a constant release case and can be implemented graphically. The procedures can be applied repeatedly by varying releases to derive the storage-yield curve at a given reservoir site.

The assumption implicitly built into the mass-curve method is that the total release over the time interval of analysis does not exceed the total reservoir inflows. In mass-curve analysis, the critical sequence of flows might occur at the end of the

TABLE 7.2.1
Monthly flows in the Little Weiser River near Indian Valley, Idaho, for water years 1966–1970

t	Year	Month	Flow (AF)	ΣQF_t AF
1	1965	10	742	742
2		11	1,060	1,802
3		12	1,000	2,802
4	1966	1	1,500	3,302
5		2	1,080	4,382
6		3	6,460	10,842
7		4	10,000	20,842
8		5	13,080	33,922
9		6	4,910	38,832
10		7	981	39,813
11		8	283	40,096
12		9	322	40,398
13		10	404	40,822
14		11	787	41,609
15		12	2,100	43,709
16	1967	1	4,410	48,119
17		2	2,750	50,869
18		3	3,370	54,239
19		4	5,170	59,409
20		5	19,680	79,089
21		6	19,630	98,719
22		7	3,590	102,309
23		8	710	103,019
24		9	518	103,537
25		10	924	104,461
26		11	1,020	105,481
27		12	874	106,355
28	1968	1	1,020	107,375
29		2	8,640	116,015

streamflow record. When this occurs, the period of analysis is doubled from $[0, T]$ to $[0, 2T]$ with the inflow sequence repeating itself in the second period, and the analysis proceeds. If the required total release exceeds the total historical inflow over the recorded period, the mass-curve analysis does not yield a finite reservoir capacity.

SEQUENT-PEAK ANALYSIS. The sequent-peak method computes the cumulative sum of inflows QF_t minus the reservoir releases R_t, that is, $\Sigma_t u_t = \Sigma_t(QF_t - R_t)$, for all time periods t over the time interval of analysis $[0, T]$. To solve this problem graphically, the cumulative sum of u_t is plotted against t. The required storage for the interval is the vertical difference between the first peak and the low point before the sequent peak. The method has the same two assumptions as the mass-curve analysis.

TABLE 7.2.1
continued

t	Year	Month	Flow (AF)	ΣQF_t AF
30		3	6,370	122,385
31		4	6,720	129,105
32		5	13,290	142,395
33		6	9,290	151,685
34		7	1,540	153,225
35		8	915	154,140
36		9	506	154,646
37		10	886	155,532
38		11	3,040	158,572
39		12	2,990	161,562
40	1969	1	8,170	169,732
41		2	2,800	172,532
42		3	4,590	177,122
43		4	21,960	199,082
44		5	30,790	229,872
45		6	14,320	244,192
46		7	2,370	246,562
47		8	709	247,271
48		9	528	247,799
49		10	859	248,658
50		11	779	249,437
51		12	1,250	250,687
52	1970	1	11,750	262,437
53		2	5,410	267,849
54		3	5,560	273,407
55		4	5,610	279,017
56		5	24,330	303,347
57		6	32,870	336,217
58		7	7,280	343,497
59		8	1,150	344,647
60		9	916	345,563

Algebraically, the sequent-peak method can be implemented using the following equation recursively.

$$K_t = \begin{cases} R_t - QF_t + K_{t-1}, & \text{if positive} \\ 0, & \text{otherwise} \end{cases} \tag{7.2.1}$$

where K_t is the required storage capacity at the beginning of period t. The initial value of K_t at $t = 0$ is set to zero. In general, the method using Eq. (7.2.1) is applied repeatedly, up to twice the length of the recorded time span to account for the possibility that the critical flow sequence occurs at the end of the streamflow record. The maximum value of the calculated K_t is the required active reservoir storage capacity for the flow sequence and the considered releases.

In reality, the hydrological components, precipitation, evaporation and seepage in addition to the streamflow inflows determine the storage volume in a reservoir.

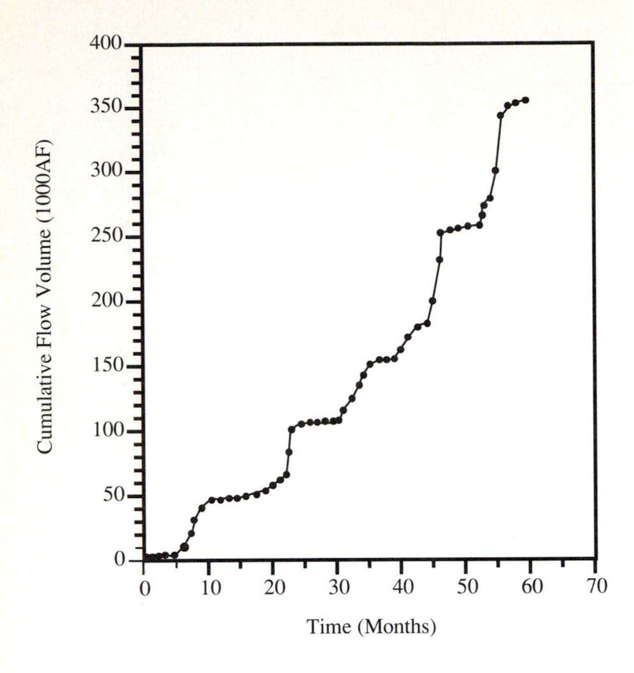

FIGURE 7.2.2
Mass curve for the Little Weiser River near Indian Valley, Idaho, based on monthly total from 1966–1970.

Precipitation that directly falls on the reservoir surface contributes to the storage volume. Evaporation and seepage result in losses to the available water in active reservoir storage. Depending on the location and the geological conditions of the reservoir site, the total loss from evaporation and seepage are an important influence on the mass balance of the reservoir system. Neglecting such factors would result in serious overestimation of the water availability and, consequently, underestimation of the required reservoir storage capacity to support the desired releases. In arid and semiarid areas, such as the southwestern United States, the quantity of water loss through evaporation may be large enough to significantly lessen the positive effects of impounding the water.

The amount of water loss through evaporation and seepage is a function of storage, impounding surface area, and geological and meteorological factors. The net inflows to a reservoir must be adjusted and used in the mass-curve and the sequent-peak methods. The adjusted reservoir inflow QF_{ta} in period t can be estimated as

$$QF_{ta} = QF_t + PP_t - EV_t - SP_t \tag{7.2.2}$$

in which PP_t is the precipitation amount on the reservoir surface, EV_t is the evaporation, and SP_t is the seepage loss during period t. The elements on the RHS of Eq. (7.2.2) depend on the storage and reservoir surface area during time period t which is, in turn, a function of those hydrological components.

Example 7.2.1. Assume that the average monthly evaporation loss and precipitation for several years are as follows:

Month	10	11	12	1	2	3	4	5	6	7	8	9
EV(AF)	270	275	280	350	470	450	400	350	370	330	300	290
PP(AF)	3	5	5	10	30	50	100	150	70	10	2	3

The loss through seepage is negligible at the site. Determine the required active storage for producing 2000 AF/month firm yield.

Solution. Computation by the sequent-peak method considering other hydrologic components are shown in Table 7.2.2. Columns (2)–(5) contain data for monthly required release, surface inflow, precipitation, and evaporation, respectively. Columns (3)–(5) are used to compute the adjusted inflow according to Eq. (7.2.2). The adjusted inflow for each month is used in Eq. (7.2.1) to compute K_t. The active storage required is 8840 AF as indicated in Table 7.2.2. The presence of evaporation loss results in an increase in required active storage. It should be pointed out that, in this example, the monthly precipitation and evaporation amounts are constants and are assumed independent of storage. In actuality, monthly values for PP_t and EV_t are functions of storage which is an unknown quantity in the exercise. To accurately account for the values of PP_t and EV_t as storage changes a trial-and-error procedure is needed to determine the required K_a for a given firm yield.

7.2.2 Optimization Procedures for Firm-Yield Analysis

Mass-curve analysis and the sequent-peak method are used in the planning stages to determine the capacity of a single-surface reservoir for a specified release pattern. It enables engineers to develop storage-yield curves for the reservoir site under consideration. However, the ability of the two methods to analyze a reservoir system involving several reservoirs is severely restricted. Furthermore, active storage capacity of a reservoir depends on various hydrologic elements whose contributions to the mass balance, in turn, are a function of unknown reservoir storage. Such an implicit relationship cannot be accounted for directly by the mass-curve and sequent-peak analysis.

Optimization models, on the other hand, can explicitly consider such implicit relationships which can be solved directly by appropriate methods. In addition, systems consisting of several multiple-purpose reservoirs can be modeled and their interrelationships accounted for in an optimization model.

For illustration consider a reservoir designed solely for the purpose of water supply. The essential feature of an optimization model for reservoir capacity determination is the mass balance equation,

$$ST_{t+1} = ST_t + PP_t + QF_t - R_t - EV_t \tag{7.2.3}$$

where ST_t is the reservoir storage at the beginning of time period t.

A model to determine the minimum active storage capacity (K_a) for a specified firm yield release (R^*) can be formulated as

$$\text{Minimize } K_a \tag{7.2.4a}$$

TABLE 7.2.2

Computations of sequent-peak method considering other hydrologic components

t (month) (1)	R_t (AF/mon) (2)	QF_t (AF/mon) (3)	PP_t (AF/mon) (4)	EV_t (AF/mon) (5)	K_{t-1} (AF/mon) (6)	K_t (AF/mon) (7)
1	2000.	742.	3.	270.	0.	1525.
2	2000.	1060.	5.	275.	1525.	2735.
3	2000.	1000.	5.	280.	2735.	4010.
4	2000.	1500.	10.	350.	4010.	4850.
5	2000.	1080.	30.	470.	4850.	6210.
6	2000.	6460.	50.	450.	6210.	2150.
7	2000.	10000.	100.	400.	2150.	0.
8	2000.	13080.	150.	350.	0.	0.
9	2000.	4910.	70.	370.	0.	0.
10	2000.	981.	10.	330.	0.	1339.
11	2000.	283.	2.	300.	1339.	3354.
12	2000.	322.	3.	290.	3354.	5319.
13	2000.	404.	3.	270.	5319.	7182.
14	2000.	787.	5.	275.	7182.	8665.
15	2000.	2100.	5.	280.	8665.	8840.
16	2000.	4410.	10.	350.	8840.	6770.
17	2000.	2750.	30.	470.	6770.	6460.
18	2000.	3370.	50.	450.	6460.	5490.
19	2000.	5170.	100.	400.	5490.	2620.
20	2000.	19680.	150.	350.	2620.	0.
21	2000.	19630.	70.	370.	0.	0.
22	2000.	3590.	10.	330.	0.	0.
23	2000.	710.	2.	300.	0.	1588.
24	2000.	518.	3.	290.	1588.	3357.
25	2000.	924.	3.	270.	3357.	4700.
26	2000.	1020.	5.	275.	4700.	5950.
27	2000.	874.	5.	280.	5950.	7351.
28	2000.	1020.	10.	350.	7351.	8671.
29	2000.	8640.	30.	470.	8671.	2471.

subject to

a. conservation of mass in each time period t,

$$ST_t - ST_{t+1} - \hat{R}_t = R^* - QF_t - PP_t + EV_t, \qquad t = 1, \ldots, T \qquad (7.2.4b)$$

where $\hat{R}_t$ is the amount of release in excess of the specified firm release R^*.

b. reservoir capacity cannot be exceeded during any time period,

$$ST_t - K_a \leq 0, \qquad t = 1, \ldots, T \qquad (7.2.4c)$$

Decision variables are ST_t, $\hat{R}_t$, and K_a. The model is linear if the hydrological variables QF_t, PP_t, and EV_t are known quantities. The required minimum active storage K_a^* to yield the specified firm release R^* can be determined by an LP

TABLE 7.2.2
continued

t (month) (1)	R_t (AF/mon) (2)	QF_t (AF/mon) (3)	PP_t (AF/mon) (4)	EV_t (AF/mon) (5)	K_{t-1} (AF/mon) (6)	K_t (AF/mon) (7)
30	2000.	6370.	50.	450.	2471.	0.
31	2000.	6720.	100.	400.	0.	0.
32	2000.	13290.	150.	350.	0.	0.
33	2000.	9290.	70.	370.	0.	0.
34	2000.	1540.	10.	330.	0.	780.
35	2000.	915.	2.	300.	780.	2163.
36	2000.	506.	3.	290.	2163.	3944.
37	2000.	886.	3.	270.	3944.	5325.
38	2000.	3040.	5.	275.	5325.	4555.
39	2000.	2990.	5.	280.	4555.	3840.
40	2000.	8170.	10.	350.	3840.	0.
41	2000.	2800.	30.	470.	0.	0.
42	2000.	4590.	50.	450.	0.	0.
43	2000.	21960.	100.	400.	0.	0.
44	2000.	30790.	150.	350.	0.	0.
45	2000.	14320.	70.	370.	0.	0.
46	2000.	2370.	10.	330.	0.	0.
47	2000.	709.	2.	300.	0.	1589.
48	2000.	528.	3.	290.	1589.	3348.
49	2000.	859.	3.	270.	3348.	4756.
50	2000.	779.	5.	275.	4756.	6247.
51	2000.	1250.	5.	280.	6247.	7272.
52	2000.	11750.	10.	350.	7272.	0.
53	2000.	5410.	30.	470.	0.	0.
54	2000.	5560.	50.	450.	0.	0.
55	2000.	5610.	100.	400.	0.	0.
56	2000.	24330.	150.	350.	0.	0.
57	2000.	32870.	70.	370.	0.	0.
58	2000.	7280.	10.	330.	0.	0.
59	2000.	1150.	2.	300.	0.	1148.
60	2000.	916.	3.	290.	1148.	2519.

solution algorithm. To derive the firm yield-storage relationship, the model can be solved repeatedly by varying the firm release R^*.

Example 7.2.2. Using the monthly streamflow data shown in Table 7.2.1 for the Little Weiser River near Indian Valley, Idaho, formulate an LP model to determine the minimum required active storage to produce 2000 AF/month of firm yield. Assume that the reservoir is initially full with $ST_1 = K_a$.

Solution. The objective function for the problem is given in Eq. (7.2.4*a*), that is, Minimize K_a. The first set of constraints is the flow mass-balance equations for each month. With an initial storage volume of $ST_1 = K_a$, for the first month ($t = 1$), the mass balance is $ST_1 - ST_2 - \hat{R}_1 = R^* - QF_1 - PP_1 + EV_1$ in which $ST_1 = K_a$, $R^* = 2000$ AF/mon, $QF_1 = 742$ AF/mon, and, from Example 7.2.1, $PP_1 = 3$ AF/mon, and EV_1 = 270 AF/mon. The resulting mass balance constraint for month $t = 1$ is $K_a - ST_2 - \hat{R}_1 = 1525$. Similarly, the mass balance constraint for the month $t = 2$ is $ST_2 - ST_3 - \hat{R}_2 = R^* - QF_2 - PP_2 + EV_2 = 1210$. The process is repeated for each month over the entire period of analysis, that is, $T = 60$ months.

The second set of constraints is to ensure that the monthly storage ST_t does not exceed the reservoir active capacity, that is, Eq. (7.2.4*c*). Specifically, this set of constraints can be expressed as $ST_2 - K_a \leq 0$; $ST_3 - K_a \leq 0$; ...; $ST_{60} - K_a \leq 0$; and $ST_{61} - K_a \leq 0$. The LP model for this example consists of a total of $2T + 1 = 121$ unknown decision variables and $2T = 120$ constraints with T mass balance constraints and T storage constraints.

Alternatively, the following optimization model can be used to determine the firm yield-storage relationship. This model maximizes the firm release (R) for a specified active storage (K_a^*).

$$\text{Maximize } Z = R \qquad (7.2.5a)$$

subject to

$$ST_t - ST_{t+1} - \hat{R}_t - R = -QF_t - PP_t + EV_t, \qquad t = 1, \ldots, T \qquad (7.2.5b)$$

$$ST_t \leq K_a^*, \qquad t = 1, \ldots, T \qquad (7.2.5c)$$

The firm-yield storage relationship can be constructed by repeatedly solving this model for different values of specified active storage K_a^*.

In case that evaporation and precipitation volumes are a function of surface area of the reservoir which, in turn, depends on the reservoir storage, one could incorporate the storage-area relationship in the optimization model. The storage-area relationship can be derived from conducting a topographical survey that determines the storage volume and surface area for a given elevation (see Fig. 7.2.3). To incorporate the storage-area relationship in the optimization model, the model formulation represented by Eqs. (7.2.4*a–c*) can be modified as

$$\text{Minimize } Z = K_a \qquad (7.2.6a)$$

subject to

$$ST_t - ST_{t+1} + p_t \cdot A_t(ST_t) - e_t \cdot A_t(ST_t) - \hat{R}_t = R^* - QF_t, \qquad t = 1, \ldots, T \qquad (7.2.6b)$$

$$ST_t - K_a \leq 0, \qquad t = 1, \ldots, T \qquad (7.2.6c)$$

where p_t and e_t are the depths of precipitation and evaporation per unit area during period t, respectively. For almost all reservoir sites, the relationship between storage volume and surface area is nonlinear. Therefore, the model represented by Eqs. (7.2.6*a–c*) is a nonlinear optimization model. The nonlinear storage-area relation in Eq. (7.2.6*b*) can be approximated by a linear function to facilitate implementation of linear programming.

Note that in the above single reservoir sizing and operation model, Eqs. (7.2.5*a–c*), the decision variable is the firm yield while unknown storage volumes ST_t are state variables, and are functions of the decision variable R. Therefore, the problem

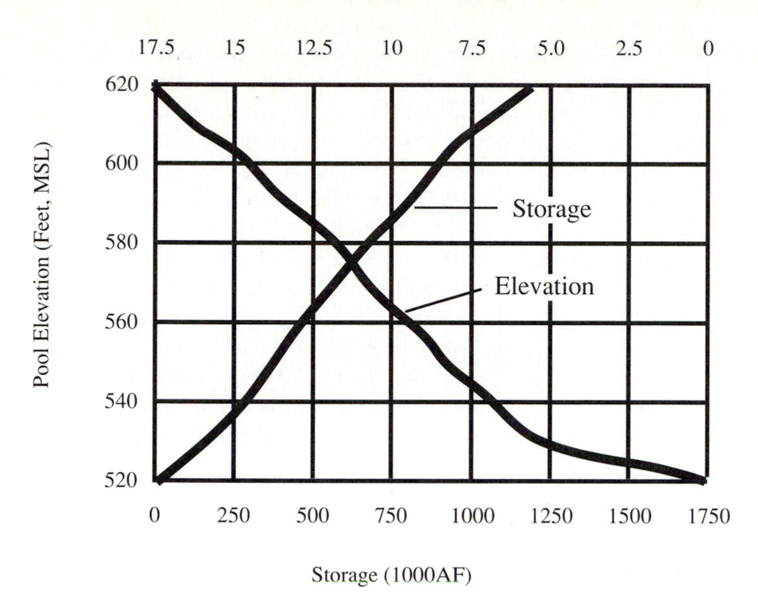

FIGURE 7.2.3
Storage-elevation and area-elevation curves (U.S. Army Corps of Engineers, 1977).

in effect is one dimensional. A one-dimensional search scheme such as the golden section method described in Section 4.2.2 can be applied. Consider the problem of determining the firm yield, R^*, for a specified K_a described by Eqs. (7.2.5*a–c*). If the firm yield assumed is greater than the minimum required, the resulting storage computed by mass-balance Eq. (7.2.5*b*) in certain periods would be negative. Therefore, the firm yield that results in the minimum of the absolute values of the storages is the solution to the reservoir model stated by Eqs. (7.2.5*a–c*). This constrained reservoir model can be transformed to an unconstrained one with the following objective function

$$\text{Minimize } Z = |ST_1, ST_2, \ldots, ST_T| \tag{7.2.7}$$

The problem can be solved by the golden-section method in which the mass-balance constraint Eq. (7.2.5*b*) is incorporated into the optimization technique in the form of reservoir routing. The original characteristics of a storage-surface area relation can be used as it does not require any manipulation to fit the solution procedure.

Example 7.2.3. Describe a model, based upon the golden section method, to determine the firm yield for the Little Weiser River near Indian Valley, Idaho, for a reservoir with an active storage of 8840 AF. The monthly data are listed in Table 7.2.1.

Solution. As described in Section 4.4.2 the golden-section method requires specification of an initial interval of firm yield. Assume that this interval is (1000 AF/mon, 10,000 AF/mon). Furthermore, if the required accuracy for the final interval is to be less than 50

AF/mon, the needed function evaluations, each involving reservoir routing of 60 monthly flows, can be computed using $(0.618)^N(10000 - 1000) < 50$. Solving for the needed function evaluations $N > 10.79$. Therefore, eleven function evaluations of objective function Eq. (7.2.7) are needed such that the firm yield by the golden-section method does not deviate from the actual required firm yield by more than 50 AF/mon.

Following the golden-section algorithm described in Section 4.4.2, first determine the two interior trial firm yields, $R_{a,1}$ and $R_{a,2}$, within the interval (1000 AF/mon, 10,000 AF/mon); They are $R_{a,1} = 1000 + 0.382(10000 - 1000) = 4438$ AF/mon and $R_{a,2} = 10000 - 0.382(10000 - 1000) = 6562$ AF/mon. For $R_{a,1} = 4438$ AF/mon, reservoir flow routing based on the continuity Eq. (7.2.3) results in a minimum end-of-month storage of −21934 AF. This corresponds to the objective function value of 21934 AF. Repeatedly, using $R_{a,2} = 6562$ AF/yr, the second function evaluation results in a minimum end-of-month storage of −101425 AF yielding the objective function value of 101425 AF. Since the objective function value of 101425 AF is larger than 21934 AF, so the interval of (1000 AF/mon, 4438 AF/mon) is eliminated. Using the shortened interval (1000 AF/mon, 4438 AF/mon), two interior points are determined in the same fashion and the search procedure is repeated.

7.3 STORAGE-FIRM ENERGY ANALYSIS

7.3.1 Concepts of Water Supply for Power Generation

Power available from a river is directly proportional to the flow rate that passes through the turbines and the potential head available to operate the turbines. The turbine units in a power plant convert the potential and kinetic energy of water into mechanical energy and ultimately into electrical energy. Hydroelectrical power HP (in terms of horsepower, hp) that can be generated by a turbine is

$$\mathrm{HP} = \frac{e_t \gamma Q h_e}{550} \tag{7.3.1}$$

where Q is the flow rate (or discharge) in cfs through the turbine, γ is the specific weight of water in lb/ft^3, and h_e is the effective (or net) potential head available in feet, and e_t is the turbine efficiency of the power generating units. A commonly used metric unit for power is kilowatt (kW). One horsepower is equal to 0.7457 kW. Since **power** is the rate of energy, **energy** produced by a power generating unit is equal to the power multiplied by the time period of production. The commonly used units for energy are kilowatt-hour (kWh) or megawatt-hour (mWh).

It should be pointed out that the net (or effective) head in Eq. (7.3.1) is the actual head available for power generation. It can be obtained by subtracting various losses through friction, entrance conditions, and other hydraulic losses from the gross head. The **gross head** is the difference in elevation between the upstream water surface and the point where the water passes through the turbine. The **hydraulic efficiency** (e_h) of a hydroelectric plant is defined as the ratio of the net head to the gross head. The term e_t in Eq. (7.3.1) is the efficiency of the power generating units resulting from energy losses through machine operation. Therefore, the overall efficiency of a hydropower plant e_p can be obtained by multiplying hydraulic efficiency and turbine efficiency.

In general, the overall **hydropower plant efficiency** ranges from 60 percent to 70 percent. Equation (7.3.1) can be used to determine the horsepower by replacing the turbine efficiency e_t with e_p and replacing the effective head with the gross head available (h_g) for the hydropower plant operation.

The maximum power that can be generated at a hydroelectric power plant under conditions of normal head and full flow is called **plant capacity**. **Firm power** is the amount of power that can be generated and produced with very little or no interruption and **firm energy** is the corresponding energy. Firm power is typically thought of as being available 100 percent of the time. In general, a hydropower plant during the year will produce a substantial amount of power in excess of the firm power. The power generated in excess of firm power is called **secondary** (or **surplus, interruptable**) **power**. The supply of this secondary power cannot be relied upon and, therefore, the rate of secondary power is generally well below that of the firm power. The secondary power is interruptable but is available more that 50 percent of the time. The third type of power is called **dump power** which is much less reliable and is available less than 50 percent of the time.

7.3.2 Determination of Firm Energy

Stream flow varies with respect to time and space. The hydropower potential at a given site on a stream depends on the flow rate and head available. In particular, to determine the firm power for a run-of-the-river hydropower plant, which has very little or no storage, statistical analysis of the streamflow sequence is essential. The commonly used method is the flow-duration curve. Because a run-of-the-river plant has no storage and the available head is fixed, the variability of energy produced by such a plant is directly proportional to the flow variability. In other words, the amount of hydropower energy that can be delivered 90 percent of the time would be generated by the flow rate that is equalled or exceeded 90 percent of the time. Customarily, the firm energy for a run-of-the-river plant is assumed on the basis of flow that is available 90 to 97 percent of the time.

Example 7.3.1. Referring to the monthly flow data for the Little Weiser River in Table 7.2.1, a run-of-the-river hydropower plant is proposed at the site. The head available at the site is 30 feet and the plant efficiency is about 0.70. Determine the firm energy and dump energy that are expected if the plant is to be constructed.

Solution. For 1 cfs of flow passing through the proposed run-of-the-river plant, the power output is determined, using Eq. (7.3.1) as,

$$\frac{0.7 \times 62.4 \times 1 \times 30}{550} \times 0.7457 = 1.777 \text{ kW}/1 \text{ cfs}$$

for $e_p = 0.7$, $\gamma = 62.4$ 1b/ft^3, $Q = 1$ cfs, and $h_g = 30$ ft. For simplicity, assume that each month has 30 days. The average flow rate corresponding to 1 AF/month is

$$\frac{43560 \text{ ft}^3/\text{ac-ft}}{\frac{30 \text{ days}}{\text{month}} \times \frac{24 \text{ hrs}}{\text{days}} \times \frac{60 \text{ sec}}{\text{min}} \times \frac{60 \text{ min}}{\text{hr}}} = 0.0168 \text{ cfs}/\text{AF}/\text{month}$$

Therefore, 1 AF/month of flow volume would produce energy of 0.0168 cfs/AF/month $\times$ 1.777 kW/cfs $\times$ 720 hrs/month = 21.502 kWh/AF/month. The firm energy that could be produced at the site, based on the firm yield of 283 AF/month found in Example 7.2.1, is 283 $\times$ 21.502 = 6085 kWh. The dump power (or energy) is the power available at least 50 percent of the time. From Fig. 7.2.1, the flow volume that is equalled or exceeded 50 percent of the time is 2800 AF/month. Therefore, the corresponding dump energy is 60,206 kWh.

The greater the firm yield, the greater the hydropower that can be generated when the head is fixed. However, the notion of energy includes the element of time, in addition to flow rate and head. Therefore, the amount of energy that can be generated by a hydropower plant is also limited by the volume of water available. To determine the storage requirement for a specified firm energy, it involves two interrelated variables: flow rate and head.

Consider a storage reservoir to be designed solely for hydropower generation. The problem is to determine the storage required to produce a specified firm energy. Similar to the problem of storage-firm yield analysis, the storage required to produce the specified firm energy can be determined by a mass-curve analysis and the sequent-peak method provided that the mass curve of energy demand is available in terms of flow rate over the recorded period. However, the complexity of determining the storage volume in firm energy determination arises from the fact that firm energy, flow rate, head, and storage are all interrelated. There are practically an infinite number of possible combinations of flow rate and head that can produce the specified firm energy. Therefore, sequential analysis by flow routing through the reservoir must be performed to determine the required storage. During the course of routing, the head available for power generation is determined from a storage-elevation relationship. The procedure can be repeated to determine the required reservoir storage volume associated with different specified firm energy levels to construct the storage-firm energy relationship.

In investigating the hydropower potential at a given site for a specified storage volume, one requires computation of the availability of secondary power. In such a case, one has to route historical or synthesized flow sequences through the reservoir to obtain a time history of reservoir elevation and outflow from which the time series of hydropower energy can be computed. Then an energy-duration curve can be produced to determine the percentage of time the secondary power will be available.

Example 7.3.2. Using the storage required for producing 2000 AF/month firm yield from Example 7.2.1, determine the firm energy and dump energy at the site. For simplicity, the equation representing the storage-water elevation relation at the reservoir site is given as follows

$$h_w = \begin{cases} \frac{6\ ST}{650}, & 0 \le ST \le 6500 \text{ AF} \\ 55.6 + \frac{2}{2950} ST, & ST \ge 6500 \text{ AF} \end{cases}$$

in which h_w is the depth of water (in feet) behind the dam corresponding to a storage volume ST (in AF). Assume that the initial reservoir is full with 8840 AF and the corresponding water surface elevation of 78.58 ft. Uncontrolled spills are considered available for power generation. The monthly average precipitation and evaporation are used (as in Example 7.2.1) to simplify the computations.

Solution. The required active storage volume for 2000 AF/month firm yield is 8840 AF from Example 7.2.1. Computations for flow routing through the reservoir using inflows from Column 3, precipitation (Column 4) and evaporation (Column 5) are shown in Table 7.3.1. The results of flow routing are the amount of overflow and end-of-period storage given in Columns (7) and (8), respectively. The overflow due to a spill in month t, $(\hat{R}_t)$, is computed by

$$\hat{R}_t = \begin{cases} ST_{t+1} - ST_{\text{full}}, & \text{if } ST_{t+1} > ST_{\text{full}} \\ 0, & \text{otherwise} \end{cases}$$

in which ST_{full} is the available storage volume when the reservoir is full; $ST_{\text{full}} =$ 8840 AF. The energy (in kWh) produced in each month t (Column 10) is calculated using

$$ER_t = 0.7457 \frac{e_p (43{,}560) Q_t \bar{h}_t}{550 \times 3600} = 0.7166\ Q_t \bar{h}_t$$

where $Q_t = R + \hat{R}_t$ is the total reservoir release (in AF) during month t, that is, Q_t is the total reservoir release (in AF) during month t, which is the sum of controlled firm yield R and uncontrolled overflow spill, $\hat{R}_t$; e_p is the plant efficiency, 0.70; $\bar{h}_t$ is the head available, computed by the sum of the head drop from the top of the dam to the turbine (h_o = 30 ft) and the average depth of water behind the dam. That is,

$$\bar{h}_t = h_o + h_w \left(\frac{ST_t + ST_{t+1}}{2} \right)$$

The average available head ($\bar{h}_t$) in each month and the electrical energy produced are shown in the Columns (9) and (10), respectively. Based on the last column of Table 7.3.1, the firm energy is 44,153 kWh which is the minimum ER_t in Column (10). The dump energy is 130,497 kWh or less (see Fig. 7.3.1 for energy-duration curve) which is the 30th largest value of the 60 values. A comparison of an energy-duration curve with and without storage volume is also shown in Fig. 7.3.1.

7.4 RESERVOIR SIMULATION

7.4.1 Operation Rules

The purpose of **operating rules** (policies) for water resource systems is to specify how water is managed throughout the system. These rules are specified to achieve system stream flow requirements and system demands in a manner that maximizes the study objectives which may be expressed in the form of benefits. System demands may be expressed as minimum desired and minimum required flows to be met at selected locations in the system. Operation rules may be designed to vary seasonally in response to the seasonal demands for water and the stochastic nature of supplies. Operating rules, often established on a monthly basis, prescribe how water is to be regulated during the subsequent month (or months) based on the current state of the system.

TABLE 7.3.1

Flow routing for determining the firm energy based on monthly flow at Little Weiser River near Indian Valley, Idaho (1966–1970)

t (month) (1)	ST_t (AF) (2)	QF_t (AF/mon) (3)	PP_t (AF/mon) (4)	EV_t (AF/mon) (5)	R_t (AF/mon) (6)	$\hat{R}_t$ (AF/mon) (7)	ST_{t+1} (AF) (8)	$\bar{h}_t$ (ft) (9)	ER_t (kWh) (10)
1	8840.	742.	3.	270.	2000.	0.	7315.	91.1	130528.
2	7315.	1060.	5.	275.	2000.	0.	6105.	90.1	129200.
3	6105.	1000.	5.	280.	2000.	0.	4830.	80.5	115327.
4	4830.	1500.	10.	350.	2000.	0.	3990.	70.7	101337.
5	3990.	1080.	30.	470.	2000.	0.	2630.	60.6	86784.
6	2630.	6460.	50.	450.	2000.	0.	6690.	73.0	104644.
7	6690.	10000.	100.	300.	2000.	5550.	8840.	92.7	501777.
8	8840.	13080.	150.	350.	2000.	10880.	8840.	95.3	879414.
9	8840.	4910.	70.	370.	2000.	2610.	8840.	92.5	305498.
10	8840.	981.	10.	330.	2000.	0.	7501.	91.1	130619.
11	7501.	283.	2.	300.	2000.	0.	5486.	89.9	128900.
12	5486.	322.	3.	290.	2000.	0.	3521.	71.6	102573.
13	3521.	404.	3.	270.	2000.	0.	1658.	53.9	77253.
14	1658.	787.	5.	275.	2000.	0.	175.	38.5	55120.
15	175.	2100.	5.	280.	2000.	0.	0.	30.8	44153.**
16	0.	4410.	10.	350.	2000.	0.	2070.	39.6	56755.
17	2070.	2750.	30.	470.	2000.	0.	2380.	50.5	72430.
18	2380.	3370.	50.	450.	2000.	0.	3350.	56.4	80897.
19	3350.	5170.	100.	300.	2000.	0.	6220.	74.2	106298.
20	6220.	19680.	150.	350.	2000.	14860.	8840.	95.7	1156728.
21	8840.	19630.	70.	370.	2000.	17330.	8840.	97.5	1350089.
22	8840.	3590.	10.	330.	2000.	1270.	8840.	92.0	215634.
23	8840.	710.	2.	330.	2000.	0.	7252.	91.1	130498.
24	7252.	518.	3.	290.	2000.	0.	5483.	88.8	127233.
25	5483.	924.	3.	270.	2000.	0.	4140.	74.4	106648.
26	4140.	1020.	5.	275.	2000.	0.	2890.	62.4	89496.
27	2890.	874.	5.	280.	2000.	0.	1489.	50.2	71961.
28	1489.	1020.	10.	350.	2000.	0.	169.	37.7	53962.
29	169.	8640.	30.	470.	2000.	0.	6369.	60.2	86242.

In reservoir operation, the benefit function used should indicate that shortages cause severe adverse consequences while surplus may enhance benefits only moderately. It is common practice to define operating rules in terms of a minimum yield or target value. If water supply to all demand points was rigidly constrained when droughts occurred, it would be impossible to satisfy all demands.

Reservoir storage is commonly divided into different zones, as shown in Fig. 7.1.1. **Rule curves** indicate the boundary of storage of various zones (see Fig. 7.4.1) throughout the year. In developing rule curves for a multipurpose reservoir consideration must be given to whether or not conflicts in serving various purposes occur. When a number of reservoirs serve the same purpose, system rule curves should be developed.

TABLE 7.3.1
continued

t (month) (1)	ST_t (AF) (2)	QF_t (AF/mon) (3)	PP_t (AF/mon) (4)	EV_t (AF/mon) (5)	R_t (AF/mon) (6)	$\hat{R}_t$ (AF/mon) (7)	ST_{t+1} (AF) (8)	$\bar{h}_t$ (ft) (9)	ER_t (kWh) (10)
30	6369.	6370.	50.	450.	2000.	1499.	8840.	91.3	228829.
31	8840.	6720.	100.	300.	2000.	4420.	8840.	93.1	428267.
32	8840.	13290.	150.	350.	2000.	11090.	8840.	95.4	894420.
33	8840.	9290.	70.	370.	2000.	6990.	8840.	94.0	605319.
34	8840.	1540.	10.	330.	2000.	0.	8060.	91.3	130890.
35	8060.	915.	2.	300.	2000.	0.	6677.	90.6	129839.
36	6677.	506.	3.	290.	2000.	0.	4896.	83.4	119547.
37	4896.	886.	3.	270.	2000.	0.	3515.	68.8	98631.
38	3515.	3040.	5.	275.	2000.	0.	4285.	66.0	94590.
39	4285.	2990.	5.	280.	2000.	0.	5000.	72.9	104412.
40	5000.	8170.	10.	350.	2000.	1990.	8840.	91.0	260089.
41	8840.	2800.	30.	470.	2000.	360.	8840.	91.7	155104.
42	8840.	4590.	50.	450.	2000.	2190.	8840.	92.3	277238.
43	8840.	21960.	100.	300.	2000.	19660.	8840.	98.3	1525085.
44	8840.	30790.	150.	350.	2000.	28590.	8840.	101.3	2220204.
45	8840.	14320.	70.	370.	2000.	12020.	8840.	95.7	961132.
46	8840.	2370.	10.	330.	2000.	50.	8840.	91.6	134576.
47	8840.	709.	2.	300.	2000.	0.	7251.	91.1	130497.
48	7251.	528.	3.	290.	2000.	0.	5492.	88.8	127286.
49	5492.	859.	3.	270.	2000.	0.	4084.	74.2	106337.
50	4084.	779.	5.	275.	2000.	0.	2593.	60.8	87161.
51	2593.	1250.	5.	280.	2000.	0.	1568.	49.2	70519.
52	1568.	11750.	10.	350.	2000.	2138.	8840.	87.9	260658.
53	8840.	5410.	30.	470.	2000.	2970.	8840.	92.6	329789.
54	8840.	5560.	50.	450.	2000.	3160.	8840.	92.7	342635.
55	8840.	5610.	100.	300.	2000.	3310.	8840.	92.7	352789.
56	8840.	24330.	150.	350.	2000.	22130.	8840.	99.2	1713476.
57	8840.	32870.	70.	370.	2000.	30570.	8840.	102.0	2379577.
58	8840.	7280.	10.	330.	2000.	4960.	8840.	93.3	465202.
59	8840.	1150.	2.	300.	2000.	0.	7692.	91.2	130711.
60	7692.	916.	3.	290.	2000.	0.	6321.	90.4	129488.

**Firm energy

It is essential that operating rules are formulated with information that will be available at the time when operation decisions are made. If forecasts are used in operation, the degree of reliability should be taken into account in deriving operating rules. Likewise, all physical, legal, and other constraints should be considered in formulating and evaluating operation rules. Further, uncertainties associated with the rule curves, and changes in physical and legal conditions should be incorporated in developing the rule curves, if possible.

Rule curves are developed to provide guidance on what operational policy is to be employed at a reservoir or dam site. The operational decision is based on the current state of the system and the time of year which accounts for the seasonal variation of

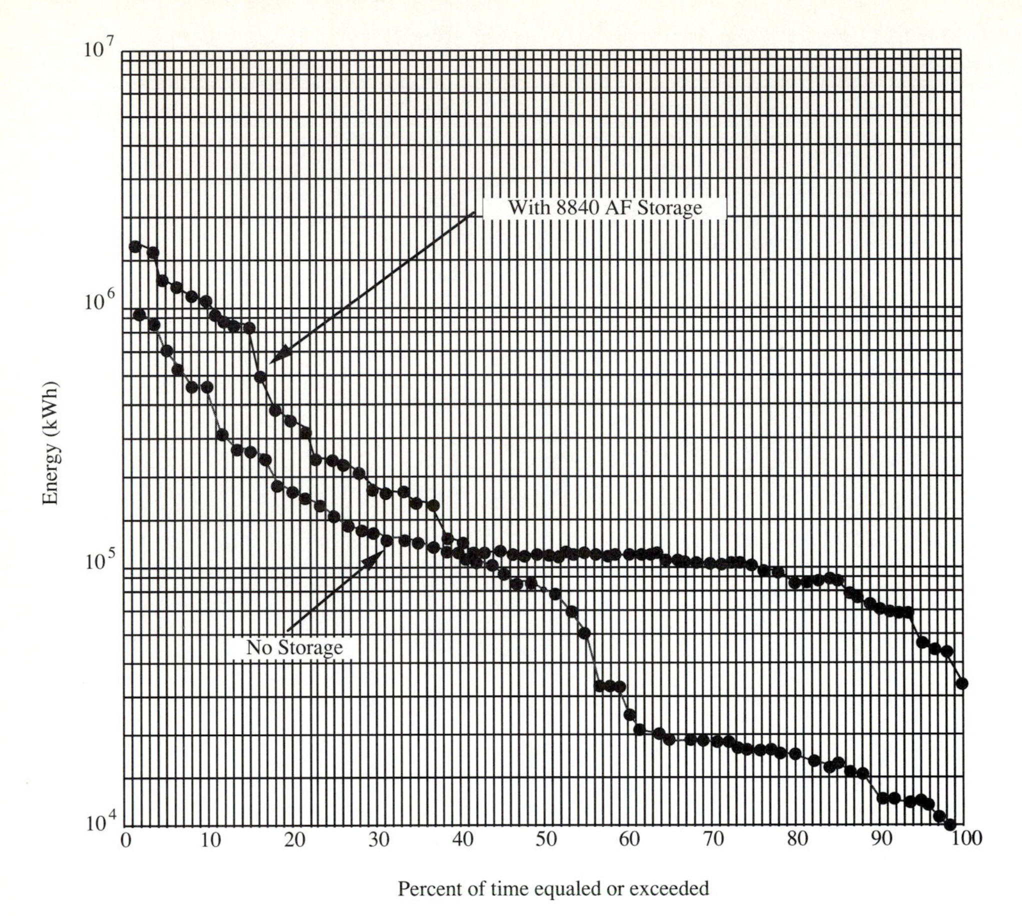

FIGURE 7.3.1
Energy-duration curves for Little Weiser River.

reservoir inflows. A simple rule curve may specify the next period's release based solely on the storage level in the current month. A more complicated rule curve might consider storage at other reservoirs, specifically at downstream control points, and perhaps a forecast of future expected inflows to the reservoir.

Three basic methods have been used in planning, design and operation of reservoir systems: (a) simplified methods such as nonsequential analysis; (b) simulation analysis; and (c) optimization analysis. Simple methods are generally used for analyzing systems involving one reservoir with one purpose using data for only a critical flow period. Simulation models can handle much more complex system configurations and can preserve much more fully the stochastic, dynamic characteristics of reservoir systems. It is not generally the intent to find the optimal alternative in design and planning when simulation models are used. The search for an optimal alternative is dependent on the engineer's ability to manipulate design variables and operating policies in an efficient manner. There may be no guarantee that a globally-optimal

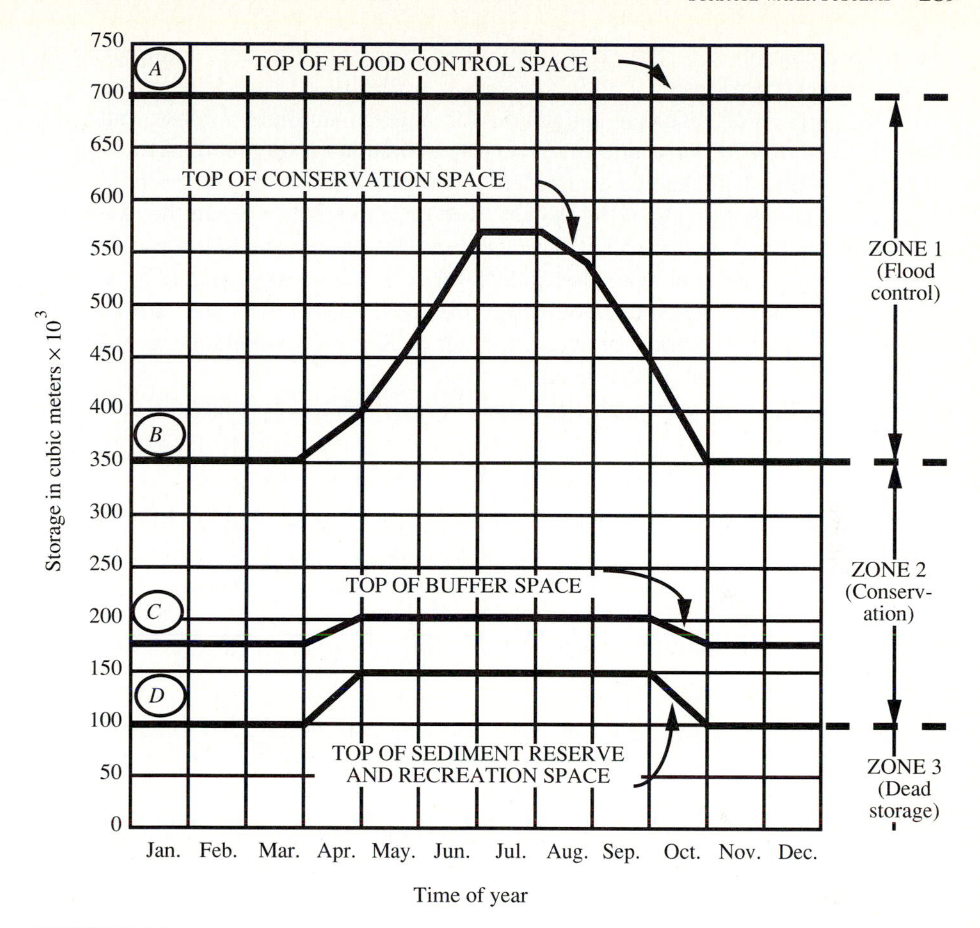

FIGURE 7.4.1
Example of seasonally varying storage boundaries for multipurpose reservoir (U.S. Army Corps of Engineers, 1977).

alternative is found. Optimization models may have a greater number of assumptions and approximations than the simulation models that are generally needed to make the model mathematically tractable.

7.4.2 Conservation Simulation

Reservoir simulation refers to the mathematical simulation of a river network with reservoir(s). The planning and operation of a reservoir system requires the simulation of these systems to determine if demands can be met for water supply (municipal, industrial, and/or agricultural users, hydropower, instream flow maintenance for water quality, and flood control.) For purposes of discussion here water uses are considered in two categories: flood control and conservation; and where conservation use refers to all non-flood control uses. This section is limited to conservation uses and Chapter 13 discusses flood control use.

The purpose of reservoir simulation for a given multiple-purpose, multiple-reservoir system is to determine the reservoir operation (reservoir releases) over a given time period with known streamflows at input points to these reservoirs and other control points throughout the system. The objective is to operate the reservoirs so as to best meet flow demands for water uses. Reservoir simulation can be used to determine if a reservoir operation policy for a particular system can be used to meet demands. Reservoir simulation could also be used in a trial-and-error fashion to develop reservoir operation strategies (policies). Reservoir simulation is also used for determining reservoir storage requirements.

The routing of flow through a reservoir is accomplished by using the continuity equation

$$ST_{t+1} = ST_t + QF_t - R_t - E_t$$

where t is a time period, ST_t is the reservoir storage at the beginning of time period t, and QF_t is the inflow during time period t, R_t is the release during time period t and E_t is the net evaporation $(EV_t - PP_t)$ from the reservoir surface.

The general procedure for conducting a reservoir system analysis for conservation purposes using simulation involves (U.S. Army Corps of Engineers, 1977):

1. identifying the system;
2. determining the study objectives and specifying the criteria used to measure the objectives;
3. examining the availability of the system data;
4. formulating a model which is mathematically and quantitatively representative of the system's components, hydrology, and operating criteria;
5. validating the model;
6. organizing and solving the model; and
7. analyzing and evaluating the results according to how well they achieve the objectives of the study.

7.4.3 HEC-5 Simulation Model

One of the most widely used reservoir simulation models is the U.S. Army Corps of Engineers, Hydrologic Engineering Center, HEC-5 computer program (U.S. Army Corps of Engineers, 1982), which was developed to simulate the operation of multipurpose, multireservoir systems. This model can be used to simulate the operation of a system of reservoirs in a river network for flood control, water supply, hydropower, and instream flow maintenance for water quality. HEC-5 can be used to determine both reservoir storage requirements and operational strategies for flood control and/or conservation purposes. Conservation or flood control reservoir storage and operation may be determined by iteratively analyzing the performance of reservoir(s) using different reservoir sizes (storage volumes) and control strategies. For flood control purposes the performance can be measured in terms of flow violations or expected annual flood damages and net benefits. Benefit calculations can also be performed for hydropower. Individual reservoir storages for conservation (non flood control) demands can be determined to meet a specified demand or the maximum reservoir yield that can be obtained from a specified storage.

The major capabilities of HEC-5 are summarized as follows:

- Flood control operation (including the computation of expected annual damages).
- Determination of firm yield for a single reservoir.
- Hydropower system simulation.
- Multiple-purpose, multiple-reservoir system operation and analysis.
- Simulate operation of an on-line or an off-line pumped storage project.

Table 7.4.1 outlines the reservoir operation criteria used in HEC-5 and Table 7.4.2 lists the operational priorities of HEC-5.

Index levels for each reservoir are assigned by the program user for use in determining the priority of releases among reservoirs. A reservoir system is operated to meet specified operation constraints first and then to keep the reservoir in balance. A reservoir system is in balance when all reservoirs are at the same index level. The priority for releases in the balancing of reservoir levels is governed by the index levels. The reservoirs at the highest levels at the end of a current time period, assuming no releases, are given first priority for the current time period.

The concept of **equivalent reservoirs** is used in determining the priority of reservoir levels among parallel reservoirs or other subsystems of a reservoir system that have **tandem reservoirs**. Tandem reservoirs are reservoirs that are operated in conjunction with each other. The level of each reservoir in a subsystem is weighted by the storage in the reservoir to develop a storage weighted level for the subsystem of reservoirs.

7.5 OPTIMAL SIZING AND OPERATION OF A SINGLE MULTIPLE-PURPOSE RESERVOIR

Optimization models for reservoir systems can be classified into two categories: (1) optimization models for planning purposes; and (2) optimization models for real-time operations (Yeh, 1982). The typical constraints in reservoir optimization models are shown in Table 7.5.1. They primarily include a mass-balance equation, maximum and minimum storage levels, maximum and minimum releases, flow-carrying capacities of hydraulic structures such as penstock, contractual, legal, and institutional requirements for the various purposes of the system. In a reservoir design and planning study, determination of optimal operating policy is frequently sought in conjunction with the search for the optimal reservoir capacity.

Consider that a reservoir is designed for water supply, irrigation, power generation, maintaining instream flow, and storage for recreation. The problem is to determine both the capacity and operation of a multiple-purpose conservation reservoir

TABLE 7.4.1

Reservoir operation criteria used in HEC-5 (U.S. Army Corps of Engineers, 1982)

A. Reservoirs are operated to satisfy constraints at individual reservoirs, to maintain specified flows at downstream control points, and to keep the system in balance. Constraints at individual reservoirs with gated outlets are as follows:

(1) When the level of a reservoir is between the top of conservation pool and the top of flood pool, releases are made to attempt to draw the reservoir to the top of the conservation pool without exceeding the designated channel capacity at the reservoir or at downstream control points for which the reservoir is being operated.

(2) Releases are made equal to or greater than the minimum desired flows when the reservoir storage is greater than the top of buffer storage, and equal to the minimum required flow if between level one and the top of buffer pool. No releases are made when the reservoir is below level one (top of inactive pool). Releases calculated for hydropower requirements will override minimum flows if they are greater than the controlling desired or required flows.

(3) Releases are made equal to or less than the designated channel capacity at the reservoir until the top of flood pool is exceeded, and then all excess flood water is dumped if sufficient outlet capacity is available. If insufficient capacity exists, a surcharge routing is made. Input options permit channel capacity releases (or greater) to be made prior to the time that the reservoir level reaches the top of the flood pool if forecasted inflows are excessive.

(4) Rate of change criteria specifies that the reservoir release cannot deviate from the previous period release by more than a specified percentage of channel capacity at the dam site, unless the reservoir is in surcharge operation.

B. Operational criteria for gated reservoirs for specified downstream control points are as follows:

(1) Releases are not made (as long as flood storage remains) which would contribute to flooding at one or more specified downstream locations during a predetermined number of future periods. The number of future periods considered is the lesser of the number of reservoir release routing coefficients or the number of local flow forecast periods.

(2) Releases are made, where possible, to exactly maintain downstream flows at channel capacity (for flood operation) or for minimum desired or required flows (for conservation operation). In making a release determination, local (intervening area) flows can be multiplied by a contingency allowance (greater than 1 for flood control and less than 1 for conservation) to account for uncertainty in forecasting these flows.

C. Operation criteria for keeping a gated flood control reservoir system in balance are as follows:

(1) Where two or more reservoirs are in parallel operation for a common control point, the reservoir that is at the highest index level assuming no releases for the current time period, will be operated first to try to increase the flows in the downstream channel to the target flow. Then the remaining reservoirs will be operated in a priority established by index levels to attempt to fill any remaining space in the downstream channel without causing flooding during any of a specified number of future periods.

(2) If one of two parallel reservoirs has one or more reservoirs upstream whose storage should be considered in determining the priority of releases from the two parallel reservoirs, then an equivalent index level is determined for the tandem reservoirs based on the combined storage in the tandem reservoirs.

(3) If two reservoirs are in tandem, the upstream reservoir can be operated for control points between the two reservoirs. In addition, when the upstream reservoir is being operated for the downstream reservoir, an attempt is made to bring the upper reservoir to the same index level as the lower reservoir based on index levels at the end of the previous time period.

D. Parallel conservation operation procedures are utilized when one or more gated reservoirs are operated together to serve some common downstream flow requirements. The following steps are utilized by HEC-5 to determine the reservoir releases necessary for the downstream location MY:

TABLE 7.4.1
continued

(1) Determine all reservoirs operating for downstream location (MY)

(2) Determine priorities of reservoirs operating for MY based on index levels (for flood control operation only).

(3) Calculate table of releases to bring all other parallel reservoirs to level of each reservoir in turn.

(4) Calculate release to bring all parallel reservoirs to each target storage level. Also determine sum of releases to bring system to top of conservation and top of buffer pools.

(5) If no upstream parallel reservoir has been operated for flood control or water supply at MY and no requirement for low flow exists and no flooding will occur at MY within forecast period, skip operation for MY.

(6) Check for future flooding at MY within forecast period. If flooding occurs, operate for flood control.

(7) If no flooding, determine conservation releases for each parallel reservoir to bring system reservoirs to some appropriate level as follows:

(a) If release to satisfy minimum desired flow is less than discharge to bring system to top of buffer level—release at each reservoir is based on MIN DESIRED Q at MY.

(b) If not, and release required to satisfy minimum required flow is greater than discharge to bring system to top of buffer level—release at each reservoir is based on MIN REQUIRED Q at MY.

(c) ELSE—release flow required to bring system to top of buffer level (more than required flow but less than desired).

(d) If release for minimum required flow exceeds discharge to bring system to level 1, only release to level 1.

E. Tandem conservation system operational procedures are utilized when one or more upstream reservoirs are operated for a downstream reservoir in order to balance the conservation storage in the system based on storage target levels. The procedures are designed to balance the system storage levels based on the previous period's storage levels without causing any release of water which exceeds the downstream requirements except during periods of high flows when the conservation pools are full. When the upstream reservoir, for the previous time period, is at an index level below that of the downstream reservoir and both are below the index level for the top conservation pool, releases from the upstream project are made to satisfy the upstream project's minimum flow requirement and to, at least, bring the upstream reservoir down to the index level of the downstream reservoir. When the upstream reservoir's index level, for the previous time period, is greater than the index level for the downstream reservoir, the upstream reservoir is operated to bring the upstream reservoir down to the level of the downstream reservoir for the previous time period. Two additional criteria must also be satisfied. First, the release from the upstream reservoir must not be allowed to cause the lower reservoir to spill or waste water just due to balancing levels. Second, the downstream reservoir must not be required to empty all of its conservation storage in meeting its requirements if there is still water in the upstream projects. This condition could occur without special routines due to the use of the previous time period for the balancing level. It is necessary to use the previous period's index level because the reservoir release for the downstream project, for the current time period, is not known when the upstream reservoir's release is being calculated.

F. Reservoir operational priority for different purposes is shown in Table 7.4.2.

to maximize the annual net benefit. The schematic diagram of a single multiple-purpose reservoir is shown in Fig. 7.5.1 in which releases for water supply and irrigation are made through different conveyance structures. The primary decision variables are the reservoir storage and releases during different time periods to various users. The objective function of the problem can be expressed as

TABLE 7.4.2
Reservoir operation priority in HEC-5 (U.S. Army Corps of Engineers, 1982)

Condition	Normal Priority	Optional Priority
During flooding at downstream location:	No release for power requirements	Release for primary power
If primary power releases can be made without increasing flooding downstream:	Release down to top of buffer pool	Release down to top of inactive pool (level 1)
During flooding at downstream location:	No releases for minimum flow	Release minimum desired flow
If minimum *desired* flows can be made without increasing flooding downstream:	Release min flow between top of conservation and top of buffer pool	Same as normal
If minimum *required* flows can be made without increasing flooding downstream:	Release min flow between top of conservation and top of inactive pool	Same as normal
Diversions from reservoirs (except when diversion is a function of storage):	Divert down to top of buffer pool	Divert down to top of inactive pool (level 1)

$$\text{Maximize } NB = \sum_i \left\{ B_i(T_i) - \sum_t \left[L_{i,t}(D_{i,t}) - G_{i,t}(E_{i,t}) \right] \right\} - C(K) \quad (7.5.1a)$$

where NB is the total annual net benefit, $B_i(T_i)$ is the benefit from an annual target allocation T_i to the ith user, that is, water supply, irrigation, power generation, and recreation; $D_{i,t}$ and $E_{i,t}$ are the deficit and excess, respectively, with respect to the target allocation T_i for user i in period t; $L_{i,t}(\ )$ and $G_{i,t}(\ )$ are the loss and gain functions, respectively, corresponding to $D_{i,t}$ and $E_{i,t}$; $C(K)$ is the annual cost function associated with the total reservoir capacity K.

The constraints to the problem basically involve the following:

a. Mass balance

$$ST_{t+1} = ST_t + QF_t + PP_t - EV_t - R_t, \quad \text{for all } t \quad (7.5.1b)$$

b. Relations between total release (R_t) and releases for various purposes

$$R_t = R_{ir,t} + R_{ws,t} + R_{in,t}, \quad \text{for all } t \quad (7.5.1c)$$

$$R_{hp,t} = R_{in,t}, \quad \text{for all } t \quad (7.5.1d)$$

where $R_{ir,t}$, $R_{ws,t}$, $R_{in,t}$ and $R_{hp,t}$ are, respectively, releases for irrigation, water supply, instream flow requirement, and hydropower generation during period t. The relations are problem specific.

TABLE 7.5.1
Typical constraints in reservoir system modeling (modified from Yeh, 1982)

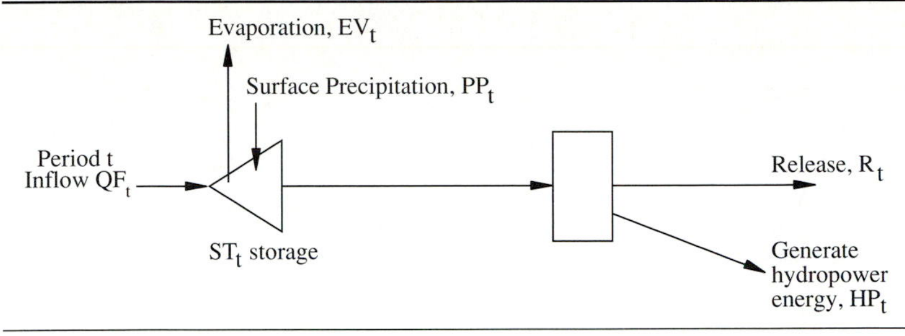

Constraint	Typical relationship	Comments
1. Conservation of mass	$ST_{t+1} = ST_t + QF_t + PP_t - EV_t - R_t$	PP_t and EV_t can be approximated as functions of $\overline{ST}_t = (ST_t + ST_{t+1})/2$.
2. Minimum and maximum	$ST_{\min,t} \leq ST_t \leq ST_{\max,t}$	Generally, $ST_{\min,t}$ and $ST_{\max,t}$ vary with t.
3. Hydropower requirements	$HP_t \geq HP_{t,\text{req'd}}$	$HP_{t,\text{req'd}}$ = required hydropower energy; HP_t is a nonlinear function of ST_t and R_t.
4. Water requirements	$R_t \geq W_{t,\text{req'd}}$	$W_{t,\text{req'd}}$ = required water in period t.
5. Minimum and maximum releases	$R_{\min,t} \leq R_t \leq R_{\max,t}$	$R_{\min,t}$ and $R_{\max,t}$ may vary with t.
6. Generator limitations	$HP_t \leq P_{\max} \cdot \Delta t$	$P_{\max}$ = maximum power output of hydropower plant, Δt = duration of period t.

c. Reservoir capacity and per period storage relation

$$ST_t \leq K - K_d, \quad \text{for all } t \tag{7.5.1e}$$

where K_d is the dead storage.

d. Irrigation release and target allocation

$$R_{ir,t} + D_{ir,t} - E_{ir,t} = T_{ir,t}, \quad \text{for all } t \tag{7.5.1f}$$

in which $D_{ir,t}$ and $E_{ir,t}$ are deficit and excess irrigation releases during period t with respect to the specified irrigation target allocation $T_{ir,t}$. The values of deficit and excess release cannot both be positive simultaneously. That is, they satisfy the condition of $D_{ir,t} \cdot E_{ir,t} = 0$.

e. Water supply release and target allocation

$$R_{ws,t} + D_{ws,t} - E_{ws,t} = T_{ws,t}, \quad \text{for all } t \tag{7.5.1g}$$

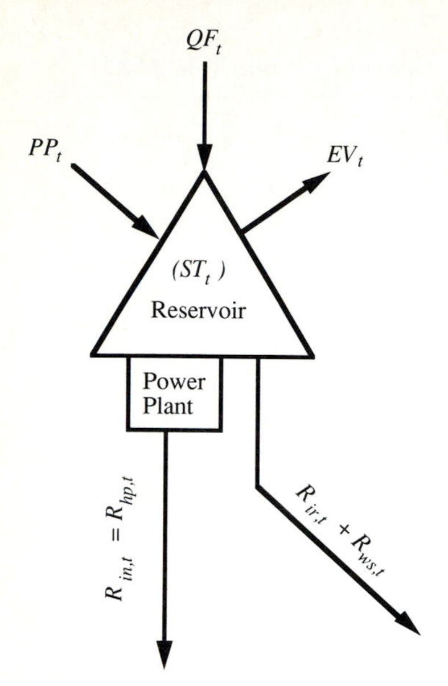

FIGURE 7.5.1
A single multiple-purpose reservoir.

in which $D_{ws,t}$ and $E_{ws,t}$ are deficit and excess water supply releases during period t with respect to the specified target allocation $T_{ws,t}$.

f. Instream flow release and target allocation

$$R_{in,t} + D_{in,t} - E_{in,t} = T_{in,t}, \quad \text{for all } t \tag{7.5.1h}$$

in which $D_{in,t}$ and $E_{in,t}$ are deficit and excess instream flow releases during period t with respect to the specified target allocation $T_{in,t}$.

g. Power supply and target allocation

$$e_p k R_{hp,t} h(K_d + ST_t, K_d + ST_{t+1}) + D_{hp,t} - E_{hp,t} = T_{hp,t}, \quad \text{for all } t \tag{7.5.1i}$$

in which e_p is the plant efficiency, k is the conversion factor, $D_{hp,t}$ and $E_{hp,t}$ are deficit and excess hydroelectric energy generated during period t with respect to the specified hydropower target allocation $T_{hp,t}$.

h. Recreation and target allocation

$$ST_t + K_d + D_{rec,t} - E_{rec,t} = T_{rec,t}, \quad \text{for all } t \tag{7.5.1j}$$

where $D_{rec,t}$ and $E_{rec,t}$ are deficit and excess active storage volume for recreation purposes during period t with respect to the specified target allocation $T_{rec,t}$.

The model is nonlinear due to the nonlinear storage-surface area-elevation relationship at a reservoir site (See Fig. 7.5.2). Furthermore, nonlinearity occurs in benefit and cost terms in the objective function Eq. (7.5.1a) and the power supply constraint Eq. (7.5.1i). Some simplified typical benefit and cost function relationships in water resource project development are shown in Fig. 7.5.3.

The direct solution approach is to apply appropriate constrained nonlinear optimization algorithms such as the generalized reduced gradient (GRG) method described

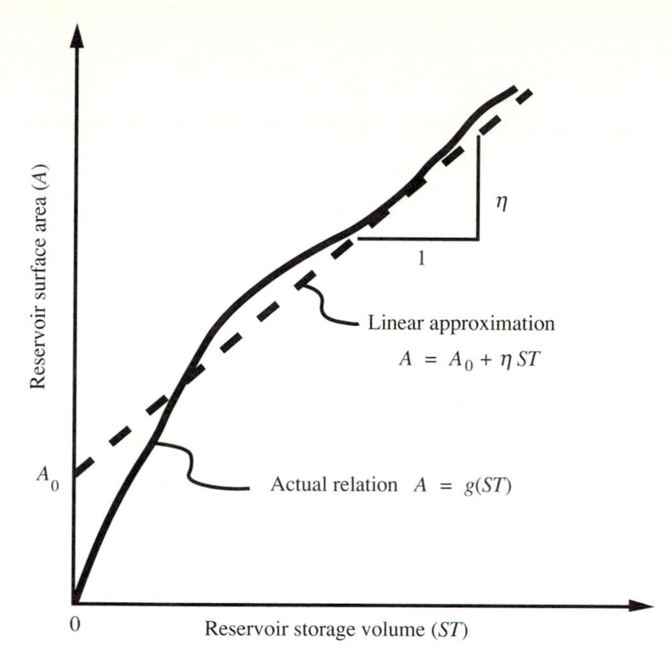

FIGURE 7.5.2
Reservoir storage-area relationship.

in Section 4.6. An alternative linear programming model is possible using a linearization of the storage-area relationship as shown in Fig. 7.5.2.

DYNAMIC PROGRAMMING (DP) APPROACH. Assuming the reservoir storage capacity, K, for the above single multiple-purpose reservoir example is known, DP can be used to solve the problem. In a DP framework the decision variables are the releases for water supply, irrigation, power generation, and instream flow requirement in each time period. For a given set of releases the amount of total benefit for each period could be calculated based on the benefit function. The decision on the releases for each period should be limited by the demands and the reservoir storage (the state variable) available. The DP sequential representation for an optimal multiple-purpose reservoir operation model is schematically depicted in Fig. 7.5.4 showing the stages (time period), state variable (storage) and decision variable (releases). The transition equation that links reservoir storage volume, hydrologic inputs and extracts, and releases from period to period is defined by the mass-balance equation, (7.5.1*b*).

Specifically, the elements of a DP model for determining the optimal reservoir releases over T periods can be expressed mathematically by the following equation.

a. Stage return function

$$r_t(R_t) = r_{ir}(R_{ir,t}, D_{ir,t}, E_{ir,t}|T_{ir,t}) + r_{ws}(R_{ws,t}, D_{ws,t}, E_{ws,t}|T_{ws,t})$$
$$+ r_{in}(R_{in,t}, D_{in,t}, E_{in,t}|T_{in,t})$$

$$+ r_{hp}(R_{hp,t}, D_{hp,t}, E_{hp,t}|T_{hp,t})$$

$$+ r_{rec}(R_{rec,t}, D_{rec,t}, E_{rec,t}|T_{rec,t}),$$

$$t = 1, 2, \ldots, T \tag{7.5.2}$$

where $r_t(R_t)$ is the total economic return for the period t associated with the total release R_t. Relations between total release and individual releases in each period are stated in constraint Eqs. (7.5.1c) and (7.5.1d).

b. Stage transition function

$$ST_{t+1} = ST_t + QF_t + PP_t - EV_t - R_t, \quad t = 1, 2, \ldots, T \tag{7.5.3}$$

subject to

$$K_d \leq ST_t \leq K, \quad t = 1, 2, \ldots, T \tag{7.5.4}$$

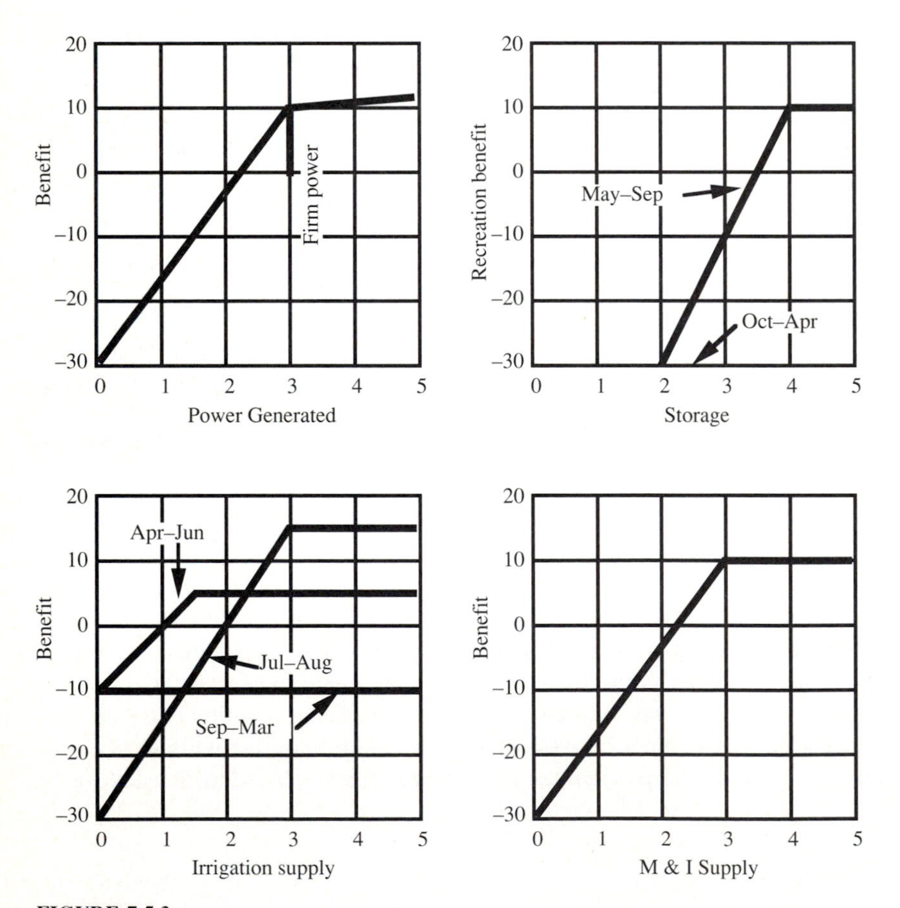

FIGURE 7.5.3
Typical simplified benefit functions (U.S. Army Corps of Engineers, 1977).

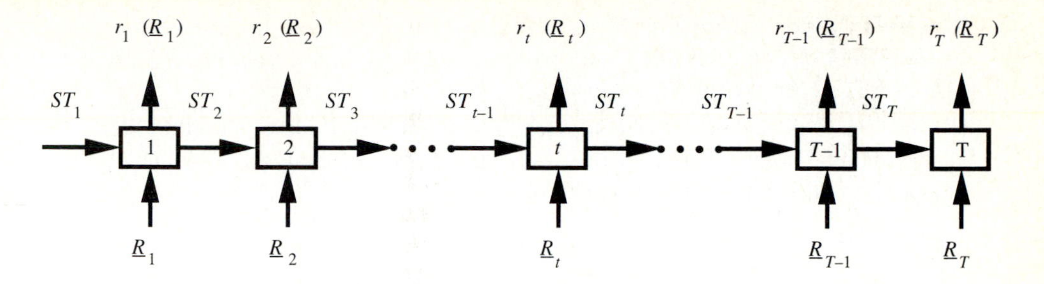

FIGURE 7.5.4
Representation of reservoir operation model by dynamic programming showing stages (time periods), state variables (ST) and decision (R).

c. Forward recursive formula

$$r_{t+1}(ST_{t+1}) = \underset{R_t}{\text{Max}}[r_t(R_t,\ ST_{t+1}) + r_t(ST_t)], \quad \text{for } t = 2, \ldots, T \qquad (7.5.5a)$$

$$r_{t+1}(ST_{t+1}) = \underset{R_t}{\text{Max}}[r_t(R_t,\ ST_t)], \quad \text{for } t = 1 \qquad (7.5.5b)$$

The computational procedure to implement the above DP algorithm for determining the optimal release policy over T periods involves the following basic steps;

1. Specify the initial storage volume ST_1 with $K_d \leq ST_1 \leq K$.
2. Discretize the storage space for each period.
3. Starting from period $t = 1$, determine the optimal releases during period $t = 1$ for all J_2 feasible storage levels $ST_{2,j},\ j = 1, 2, \ldots, J_2$ at the beginning of period $t = 2$; that is, perform the optimization

$$\begin{aligned} \text{Max } r_2(ST_{2,j}) = & \ r_{ir}(R_{ir,1}, D_{ir,1}, E_{ir,1}; T_{ir,1}) \\ & + r_{ws}(R_{ws,1}, D_{ws,1}, E_{ws,1}; T_{ws,1}) \\ & + r_{in}(R_{in,1}, D_{in,1}, E_{in,1}; T_{in,1}) \\ & + r_{hp}(R_{hp,1}, D_{hp,1}, E_{hp,1}; T_{hp,1}) \\ & + r_{rec}(R_{rec,1}, D_{rec,1}, E_{rec,1}; T_{rec,1}) \end{aligned} \qquad (7.5.6)$$

subject to

$$R_{ir,1} + R_{ws,1} + R_{in,1} + R_{hp,1} + R_{rec,1} = ST_1 + QF_1 + PP_1 - EV_1 - ST_{2,j} \qquad (7.5.7)$$

4. Store the optimal releases and the return for the period $t = 1$ associated with each feasible storage space at the beginning of period $t = 2$.

5. Repeat the computations in Steps (3)–(4) for $t = 2, 3, \ldots, T$ using the recursive formula, Eq. (7.5.5*a*). Only the optimal storage transition yielding the highest return is stored for further analysis.
6. Once the final stage $t = T$ is computed, a trace-back procedure is used to identify the optimal storage trajectory over the entire period of analysis, from which the optimal releases in each period can be found.

7.6 OPTIMAL SIZING AND OPERATION OF MULTIPLE-PURPOSE RESERVOIR SYSTEMS

The system considered in this section involves several multiple-purpose reservoirs which may be arranged in series, in parallel, or in a combination of both. Although the configuration of a multiple-reservoir system is more complicated than a single-reservoir system, the formulation of the optimization model is a straightforward extension of the single reservoir case as described in Section 7.5. The additional feature that should be incorporated in a multiple reservoir system model is the interrelation between reservoirs.

To demonstrate the essential features of the model without overcomplicating it, consider the reservoir system in Fig. 7.6.1 in which all reservoirs are multipurpose. The primary purposes to be considered in the example reservoir system include hydroelectric generation, municipal/industrial water supply, irrigation, recreation, and instream flow maintenance. Flood control is not considered.

Similar to Section 7.5 the objective is to determine the optimal storage capacity and release policy for each reservoir such that the total net benefit of the system operated over T periods is maximized. A diversion point is located downstream of reservoir 3 where water is withdrawn to supply municipal/industrial and irrigation needs. Furthermore, all hydropower generated by the three reservoirs is combined before it is distributed to the various users. The optimization model for this multiple-reservoir system can be formulated as the following:

$$
\begin{aligned}
\text{Maximize} & \sum_t B_{hp,t}\{R_{hp,s,t}, ST_{s,t}, T_{hp,t}, D_{hp,t}, E_{hp,t}\} \\
& + \sum_t B_{ws,t}\{R_{ws,3,t}, T_{ws,t}, D_{ws,t}, E_{ws,t}\} \\
& + \sum_t B_{ir,t}\{R_{ir,3,t}, T_{ir,t}, D_{ir,t}, E_{ir,t}\} \\
& + \sum_s \sum_t B_{in,s,t}\{R_{s,t}, T_{in,s,t}, D_{in,s,t}, E_{in,s,t}\} \\
& + \sum_s \sum_t B_{rec,s,t}\{ST_{s,t}, T_{rec,s,t}, D_{rec,s,t}, E_{rec,s,t}\} \\
& - \sum_s C(K_s)
\end{aligned}
\qquad (7.6.1a)
$$

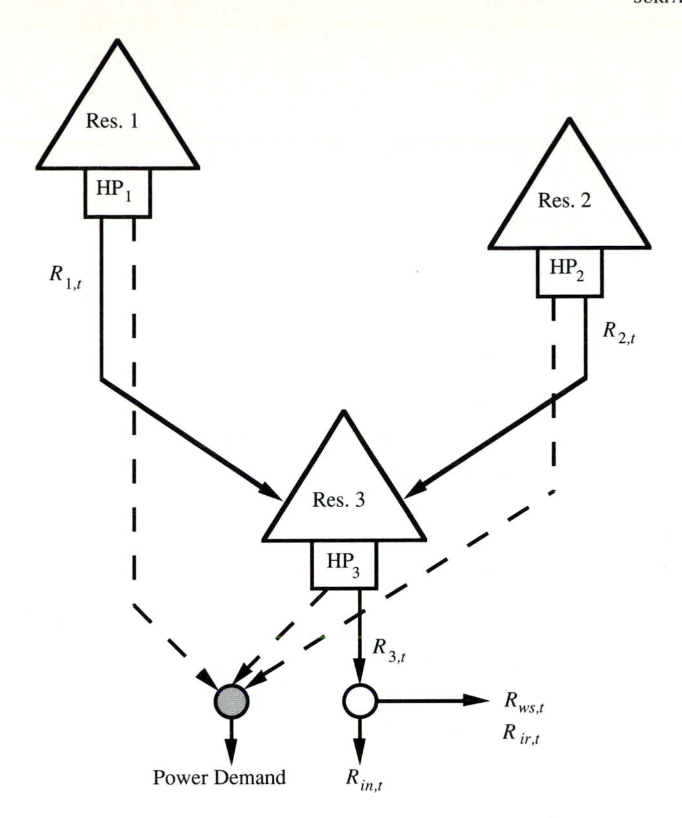

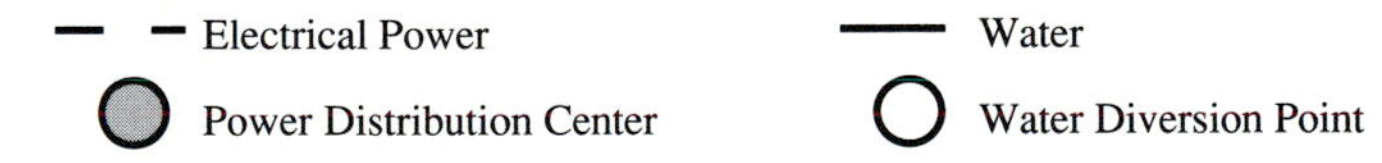

FIGURE 7.6.1
Schematic diagram of a multiple-purpose reservoir system.

in which the subscript s represents the reservoir site in the system, subject to

a. Mass balance for each reservoir:

$$ST_{s,t} - ST_{s,t+1} + PP_{s,t} - EV_{s,t} - R_{s,t} = -QF_{s,t}$$

$$\text{for } s = 1, 2 \text{ and } t = 1, \ldots, T \qquad (7.6.1b)$$

$$ST_{3,t} - ST_{3,t+1} + PP_{3,t} - EV_{3,t} + R_{1,t} + R_{2,t} - R_{3,t} = -QF_{3,t},$$

$$\text{for } t = 1, 2, \ldots, T \qquad (7.6.1c)$$

b. Hydropower generation:

$$\sum_s e_s k_s R_{s,t} h_{s,t}(K_{d,s} + ST_{s,t}, K_{d,s} + ST_{s,t+1}) + D_{hp,t} - E_{hp,t} = T_{hp,t},$$

$$\text{for } t = 1, 2, \ldots, T \qquad (7.6.1d)$$

in which $K_{d,s}$ is the dead storage at reservoir site s.

c. Water supply and irrigation:

$$R_{ws,3,t} + D_{ws,t} - E_{ws,t} = T_{ws,t} \quad \text{for } t = 1, 2, \ldots, T \qquad (7.6.1e)$$

$$R_{ir,3,t} + D_{ir,t} - E_{ir,t} = T_{ir,t}, \quad \text{for } t = 1,\ 2, \ldots, T \qquad (7.6.1f)$$

d. Instream flow requirement in each stream section:

$$R_{s,t} + D_{in,s,t} - E_{in,s,t} = T_{in,s,t}, \quad \text{for } s = 1,\ 2 \text{ and } t = 1, 2, \ldots, T \qquad (7.6.1g)$$

$$R_{3,t} - (R_{ws,3,t} + R_{ir,3,t}) + D_{in,3,t} - E_{in,3,t} = T_{in,3,t}$$
$$\text{for } t = 1, 2, \ldots, T \qquad (7.6.1h)$$

e. Recreation:

$$ST_{s,t} + K_{d,s} + D_{\text{rec},s,t} - E_{\text{rec},s,t} = T_{\text{rec},s,t},$$
$$\text{for } s = 1, 2, 3, \text{ and } t = 1, 2, \ldots, T \qquad (7.6.1i)$$

f. Constraints on reservoir storage:

$$ST_{s,t} + K_{d,s} - K_s \leq 0, \quad \text{for } s = 1, 2, 3, \text{ and } t = 1, 2, \ldots, T \qquad (7.6.1j)$$

MODEL SOLUTION. The model represented by Eqs. (7.6.1*a–j*) has $(23T+3)$ decision variables and $15T$ constraints. The model must be solved by nonlinear programming (NLP) techniques because of the nonlinearities in the objective function, in the constraints on power generation, and in the storage-area relationships in the mass-balance constraints. To solve such a model by linear programming, linearization of the nonlinear functions is required. A third approach is to employ dynamic programming (DP). When solving the multiple-reservoir system problems by a DP approach, each reservoir introduces a state variable. Because of the multidimensional nature of the problem, one way to solve a multiple-reservoir system is the use of discrete differential dynamic programming (DDDP) (Section 4.2). Fig. 7.6.2 illustrates the convergence of storage volume for reservoir 3 of the example system. Fig. 7.6.3 shows the initial and the optimal storage for all three reservoirs at the end of each time period.

7.7 RESERVOIR SIZING AND OPERATION UNDER HYDROLOGIC UNCERTAINTY: LP MODELS

In the design and operation of a reservoir system one must be aware of hydrologic, hydraulic, economic, environmental, legal and political uncertainties. Hydrologic uncertainty is attributed mainly due to the natural randomness of hydrologic events causing the amount of inflow, precipitation, evaporation, and seepage to be uncertain. Because these quantities change with respect to time, the analysis and incorporation of hydrologic uncertainties into reservoir sizing and operation is important.

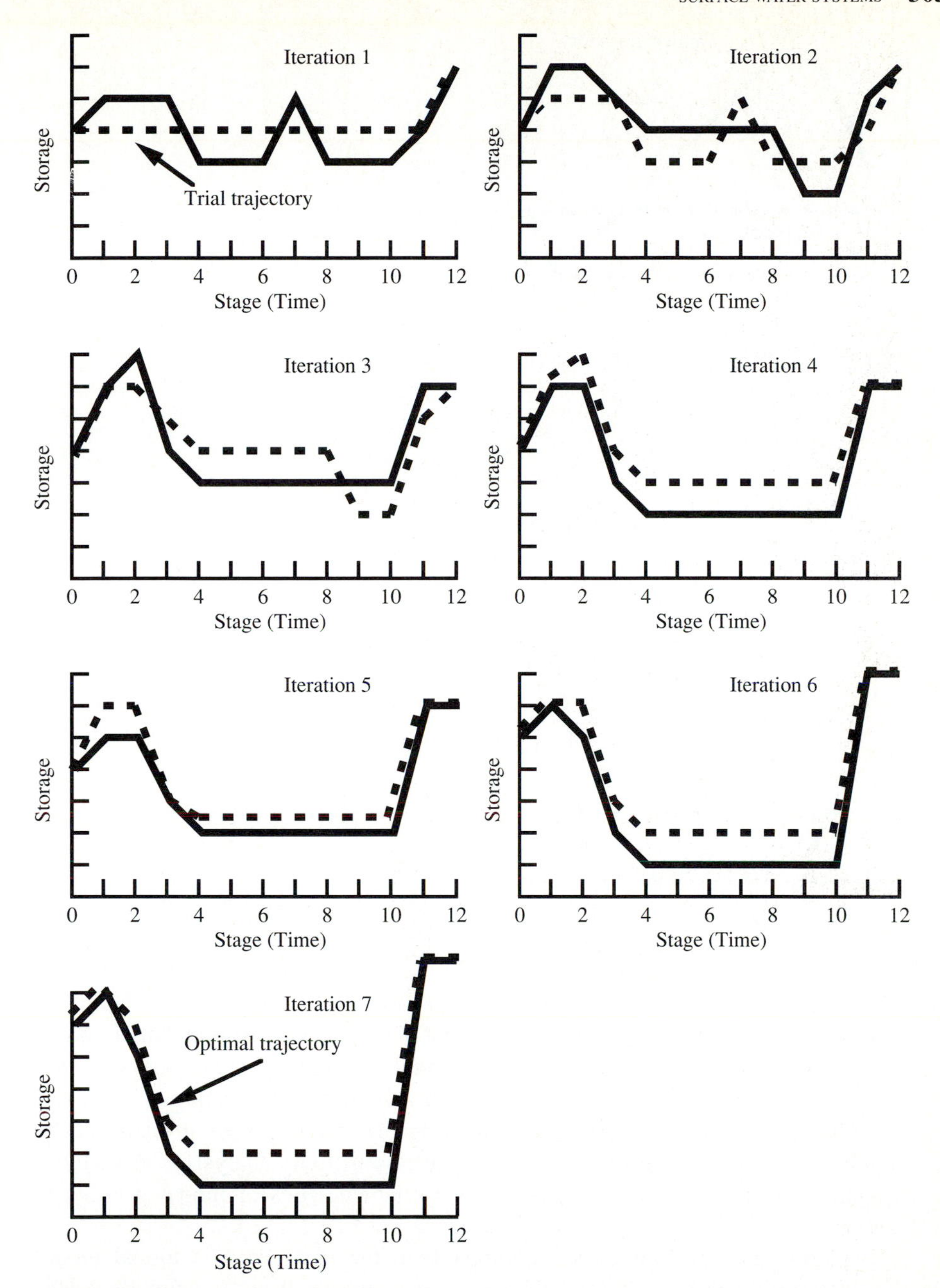

FIGURE 7.6.2
Convergence to optimal trajectory from trial trajectory for reservoir 3 in Fig. 7.6.1 (after Heidari et al., 1971).

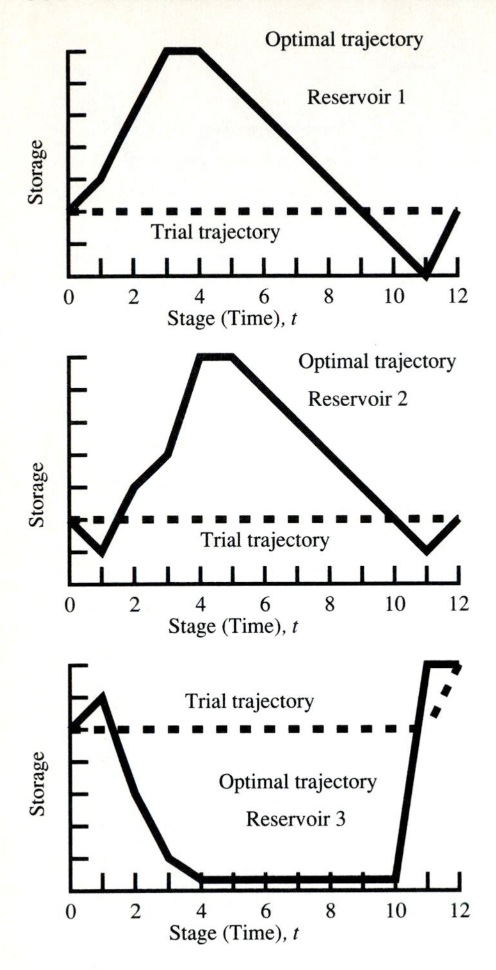

FIGURE 7.6.3
The optimal trajectory of three reservoir system in Fig. 7.6.1 (after Heidari et al., 1971).

In Sections 7.5 and 7.6 the models formulated are deterministic in that all model inputs and parameters are assumed known without uncertainty. The most common practice using a deterministic model to describe a system is to use averaged parameter values. The effect of parameter uncertainty on model results can be investigated by performing a sensitivity analysis. A disadvantage of sensitivity analysis is its inability to explicitly take into account the probabilistic characteristics of model parameters that are uncertain.

The firm yield or firm energy computed on the basis of a historical record can be "firm" only if the recorded flow sequence repeats itself indefinitely. This of course does not occur. The probability of a sequence of observations such as a streamflow sequence repeating itself is zero. There is a probability that the future flow is less than the firm yield determined from a flow-duration analysis. The conventional notion of firm yield should be adjusted in the context of a stochastic or uncertain environment.

7.7.1 Chance-Constrained Models Using Linear Decision Rules

In chance-constrained models for reservoir sizing and operation, deterministic constraints involving hydrologic parameters subject to uncertainty are replaced by probabilistic statements. Chance-constrained models can be used in a preliminary study for determining the cost-effective reservoir size and operation that satisfy the requirements on releases and storages for pre-specified levels of reliability. Because probabilistic statements are not mathematically operational (see Section 5.8), the solution of chance-constrained models, therefore, requires that probabilistic constraints are transformed into deterministic equivalents.

The chance-constrained reservoir optimization model for a single multiple-purpose conservation reservoir consists of constraints on the reliability of release and storages for an operation period such as a year. Without considering the target releases and target storages as discussed in Section 7.6.2, the primary constraints are storage constraints and release constraints given, respectively, as

$$P[ST_t \leq K_a] \geq \alpha_{ST_t} \quad t = 1, 2, \ldots, T \tag{7.7.1}$$

and

$$P[R_{\min,t} \leq R_t \leq R_{\max,t}] \geq \alpha_{R_t} \quad t = 1, 2, \ldots, T \tag{7.7.2}$$

in which $R_{\min,t}$ and $R_{\max,t}$ are the minimum and maximum specified releases during period t, and α_{ST_t} and α_{R_t} are the specified reliabilities associated with the storage and release constraints.

LINEAR DECISION RULES. Linear decision rules (LDR) in reservoir sizing and operation models are rules that relate release from a reservoir to the storage, inflows, and decision parameters. Consider the following LDR (Loucks and Dorfman, 1975)

$$R_t = ST_t + QF_t - EV_t - b_t \tag{7.7.3}$$

in which b_t is an unknown decision parameter for period t. In general, the decision variables b_t can be unrestricted-in-sign; with their sign depending on their relation to the physical parameters and the operation of the reservoir. Assume that the direct precipitation, PP_t, on the reservoir surface is negligible, then the continuity equation ($ST_{t+1} = ST_t + QF_t - EV_t - R_t$) along with Eq. (7.7.3) can be used to derive the decision rule used in the chance-constrained reservoir planning and operation model. Substituting ($ST_t = R_t - QF_t + EV_t + b_t$) from Eq. (7.7.3) into the continuity equation results in

$$\begin{aligned} ST_{t+1} &= ST_t + QF_t - EV_t - R_t \\ &= (R_t - QF_t + EV_t + b_t) + QF_t - EV_t - R_t \\ &= b_t \end{aligned} \tag{7.7.4}$$

Since the decision parameters $b_1, b_2, \ldots, b_{12}$ are equal to the end-of-period storage, they must be nonnegative. Equation (7.7.4) indicates that, using the LDR as described by Eq. (7.7.3), the decision parameter b_t for time period t equals the reservoir storage at the end of the time period. Since reservoir storage ST_t at the beginning of time period t is the end-of-period storage for time period $t-1$, hence, $ST_t = b_{t-1}$ which can be substituted in Eq. (7.7.3). The resulting decision rule is

$$R_t = QF_t - EV_t - b_t + b_{t-1} \tag{7.7.5}$$

Equations (7.7.4) and (7.7.5) are the decision rules that are used, respectively, in the storage chance-constraint Eq. (7.7.1) and the release chance-constraint Eq. (7.7.2).

Based on the LDR defined by Eq. (7.7.3), the resulting storage constraint $ST_t \leq K_a$ can be expressed in terms of the decision parameter as

$$b_{t-1} \leq K_a \tag{7.7.6}$$

Constraint Eq. (7.7.6) for storage is deterministic which does not involve random variables; therefore, the storage chance-constraint Eq. (7.7.1) can be replaced by Eq. (7.7.6). It should be noted that this results only from the use of the LDR stated by Eq. (7.7.3). Adoption of a different form of an LDR may result in the inclusion of random hydrologic components in which case the chance-constraint storage does not reduce to a deterministic constraint.

The release chance-constraint is developed by substituting Eq. (7.7.5) into Eq. (7.7.2), to obtain

$$P[R_{\min,t} \leq QF_t - EV_t - b_t + b_{t-1} \leq R_{\max,t}] \geq \alpha_{R_t} \tag{7.7.7}$$

The random variables in Eq. (7.7.7) are the inflow, QF_t, and the evaporation loss, EV_t, during time period t. The statistical properties of the stream inflow and evaporation rate (e_t) during time period t can be assessed by analyzing historical records. As discussed in Section 7.2.1, the amount of evaporation EV_t during time period t is a function of the reservoir surface area which is a nonlinear function of reservoir storage. Therefore, the release chance-constraint Eq. (7.7.7) is nonlinear.

For purposes of illustration, evaporation loss in time period t is considered as deterministic and can be estimated as

$$EV_t = e_t \cdot \frac{A_t(ST_t) + A_{t+1}(ST_{t+1})}{2} \tag{7.7.8}$$

Equation (7.7.8) can be simplified using the linear approximation shown in Fig. 7.5.2, as

$$EV_t = e_t \left[A_0 + \eta \left(\frac{ST_t + ST_{t+1}}{2} \right) \right] \tag{7.7.9a}$$

$$= e_t \left[A_0 + \eta \left(\frac{b_{t-1} + b_t}{2} \right) \right] \tag{7.7.9b}$$

Substituting Eq. (7.7.9*b*) into Eq. (7.7.7), and considering a deterministic evaporation rate (e_t) and a random inflow (QF_t), the release chance-constraint can be expressed as

$$P\left[R_{\min,t} \leq QF_t - e_t A_0 - \left(\frac{\eta e_t}{2} + 1 \right) b_t - \left(\frac{\eta e_t}{2} - 1 \right) b_{t-1} \leq R_{\max,t} \right] \geq \alpha_{R_t} \tag{7.7.10a}$$

or

$$P\left[R_{\text{min},t} \le \theta \le R_{\text{max},t}\right] \ge \alpha_{R_t} \tag{7.7.10b}$$

where

$$\theta = QF_t - e_t A_0 - \left(\frac{\eta e_t}{2} + 1\right) b_t - \left(\frac{\eta e_t}{2} - 1\right) b_{t-1}$$

The above release chance-constraint Eq. (7.7.10) is bounded on both sides, with QF_t being the only random variable. Utilizing the results presented in Section 5.8 for its deterministic equivalent, Eq. (7.7.10*b*) can be written as

$$P\left[\theta \le R_{\text{max},t}\right] - P\left[\theta \ge R_{\text{min},t}\right] \ge \alpha_{R_t} \tag{7.7.11}$$

Equation (7.7.11) can be decomposed into the following three equations

$$P\left[\theta \le R_{\text{max},t}\right] \ge \alpha_{R_{\text{max},t}} \tag{7.7.12a}$$

$$P\left[\theta \ge R_{\text{min},t}\right] \ge \alpha_{R_{\text{min},t}} \tag{7.7.12b}$$

and

$$\alpha_{R_{\text{min},t}} + \alpha_{R_{\text{max},t}} = 1 + \alpha_{R_t} \tag{7.7.12c}$$

where $\alpha_{R_{\text{min},t}}$ and $\alpha_{R_{\text{max},t}}$ are also unknown with $(1 - \alpha_{R_{\text{min},t}})$ and $(1 - \alpha_{R_{\text{max},t}})$, respectively, representing the left-and right-tail areas of the distribution as shown in Fig. 7.7.1.

The constraint Eq. (7.7.12*c*) is needed to satisfy the original release chance-constraint Eq. (7.7.2). Equation (7.7.12*c*) can be derived by referring to Fig. 7.7.1 in which the following identity must be satisfied.

$$1 = \left(1 - \alpha_{R_{\text{min},t}}\right) + \alpha_{R_t} + \left(1 - \alpha_{R_{\text{max},t}}\right) \tag{7.7.13}$$

A reasonable way to assume the values of the unknown $\alpha_{R_{\text{min},t}}$ and $\alpha_{R_{\text{max},t}}$ is to equally distribute the probability of noncompliance $(1 - \alpha_{R_t})$ on both ends of the distribution. In doing so, $\alpha_{R_{\text{min},t}} = \alpha_{R_{\text{max},t}} = (1 + \alpha_{R_t})/2$ and Eq. (7.7.12*c*) is satisfied.

Without considering the randomness of evaporation rate (e_t), Eqs. (7.7.12*a*) and (7.7.12*b*) can be written as

$$\begin{aligned} P\left[\theta \le R_{\text{max},t}\right] &= P\left[QF_t - e_t A_0 - \left(\frac{\eta e_t}{2} + 1\right) b_t - \left(\frac{\eta e_t}{2} - 1\right) b_{t-1} \le R_{\text{max},t}\right] \\ &= P\left[R_{\text{max},t} + e_t A_0 + \left(\frac{\eta e_t}{2} + 1\right) b_t + \left(\frac{\eta e_t}{2} - 1\right) b_{t-1} \ge QF_t\right] \\ &\ge \frac{1 + \alpha_{R_t}}{2} \end{aligned} \tag{7.7.14a}$$

and

$$\begin{aligned} P\left[\theta \ge R_{\text{min},t}\right] &= P\left[QF_t - e_t A_0 - \left(\frac{\eta e_t}{2} + 1\right) b_t - \left(\frac{\eta e_t}{2} - 1\right) b_{t-1} \ge R_{\text{min},t}\right] \\ &= P\left[R_{\text{min},t} + e_t A_0 + \left(\frac{\eta e_t}{2} + 1\right) b_t + \left(\frac{\eta e_t}{2} - 1\right) b_{t-1} \le QF_t\right] \\ &\ge \frac{1 + \alpha_{R_t}}{2} \end{aligned} \tag{7.7.14b}$$

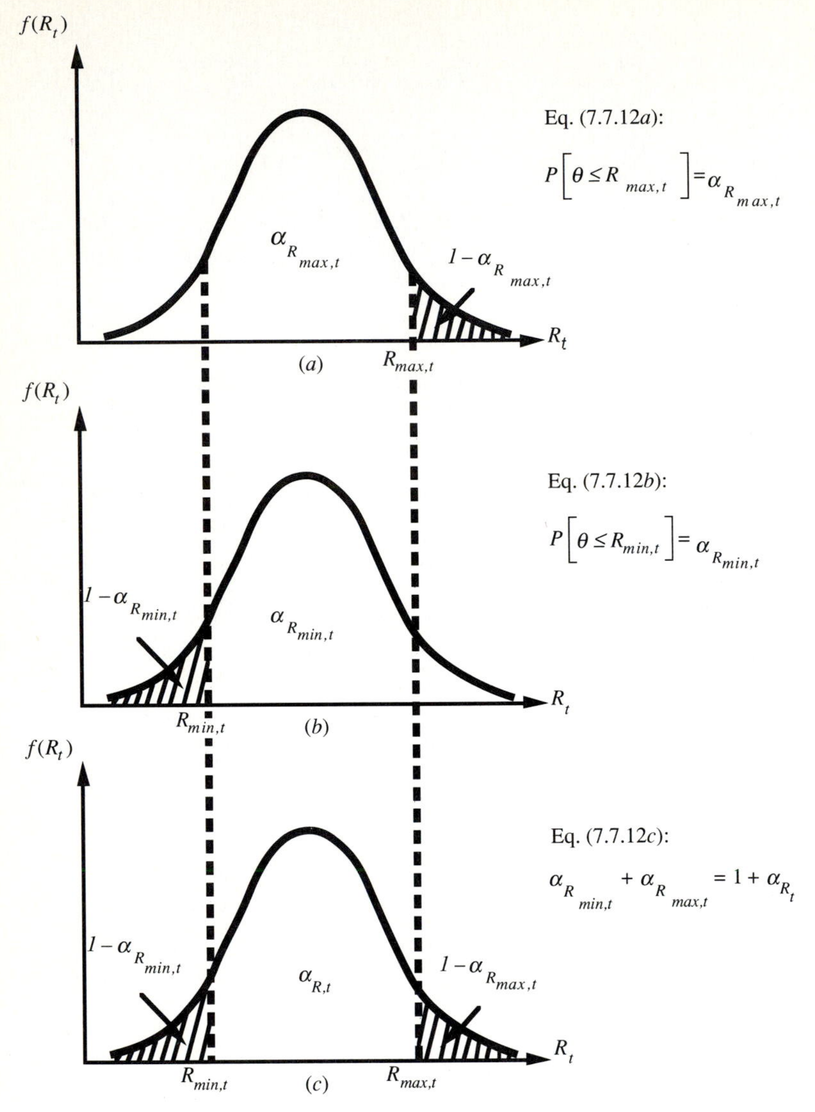

FIGURE 7.7.1
Probability levels for reservoir release chance constraints.

These chance-constraints are equivalent to the case in which only the RHS coefficient, QF_t, is random (see Section 5.8). Referring to Fig. 7.7.2*a* and assuming that QF_t is a normal random variable, the term on the RHS of the probabilistic expression in Eq. (7.7.14*a*) must satisfy the following inequality

$$R_{\max,t} + e_t A_0 + \left(\frac{\eta e_t}{2} + 1\right) b_t + \left(\frac{\eta e_t}{2} - 1\right) b_{t-1} \geq \mu_{QF_t} + \sigma_{QF_t} z_{(1+\alpha_{R_t})/2}$$

in which $\mu_{QF,t}$ and $\sigma_{QF,t}$ are the mean and standard deviation of the random inflow

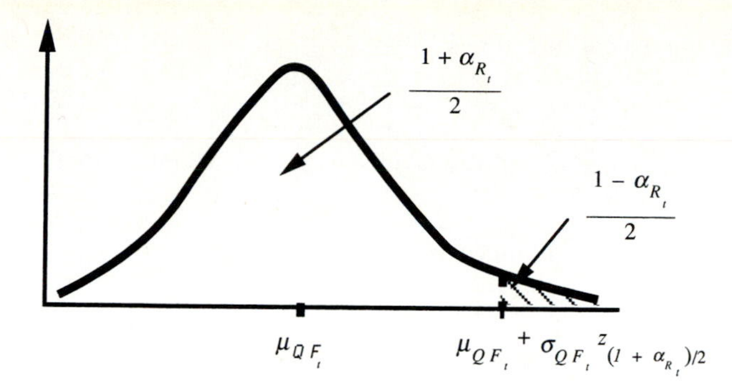

$$\text{(a) } P\left[QF_t \le \mu_{QF_t} + \sigma_{QF_t} z_{(1+\alpha_{R_t})/2}\right] = \frac{1 + \alpha_{R_t}}{2}$$

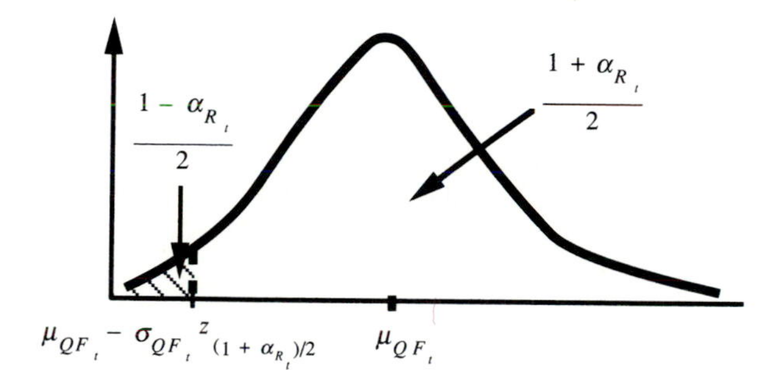

$$\text{(b) } P\left[QF_t \le \mu_{QF_t} - \sigma_{QF_t} z_{(1+\alpha_{R_t})/2}\right] = \frac{1 + \alpha_{R_t}}{2}$$

FIGURE 7.7.2
Probability correspondence for release chance-constraint.

during period t and $z_{(1+\alpha_{R_t})/2}$ is the value of standard normal variate with a cumulative probability of $\dfrac{1+\alpha_{R_t}}{2}$. The resulting deterministic equivalent of Eq. (7.7.14a) is then

$$\left(\frac{\eta e_t}{2} + 1\right) b_t + \left(\frac{\eta e_t}{2} - 1\right) b_{t-1} \ge -R_{\max,t} - e_t A_0 + \mu_{QF_t} + \sigma_{QF_t} z_{(1+\alpha_{R_t})/2} \quad (7.7.15a)$$

Similarly, referring to Fig. 7.7.2b, the term on the RHS of the probabilistic expression in Eq. (7.7.14b) must satisfy the following relation

$$R_{\min,t} + e_t A_0 + \left(\frac{\eta e_t}{2} + 1\right) b_t + \left(\frac{\eta e_t}{2} - 1\right) b_{t-1} \le \mu_{QF_t} - \sigma_{QF_t} z_{(1+\alpha_{R_t})/2}$$

Hence, the deterministic equivalent of Eq. (7.7.14b) is

$$\left(\frac{\eta e_t}{2}+1\right)b_t+\left(\frac{\eta e_t}{2}-1\right)b_{t-1}\leq -R_{\min,t}-e_tA_0+\mu_{QF_t}-\sigma_{QF_t}z_{(1+\alpha_{R_t})/2} \quad (7.7.15b)$$

In summary, the chance-constrained reservoir design and operation model described above can be stated as

$$\text{Minimize } K_a$$

subject to

1. storage constraints, Eq. (7.7.6)
2. release constraints: Eqs. (7.7.14a) and (7.7.14b).

in which K_a and b_t are the decision variables. This chance-constrained reservoir sizing and operation model is derived from the use of the LDR rule stated by Eq. (7.7.3). The above model could be expanded to consider randomness of evaporation. Also, it should be kept in mind that the above derivation assumes that inflows in each time period are normal random variables. This should be modified according to the distribution of inflows based on recorded data.

The above chance-constrained model may result in an infeasible solution for the specified range of releases and reliability requirement. This can be illustrated by referring to Eqs. (7.7.15a) and (7.7.15b). Since the LHS of Eqs. (7.7.15a) and (7.7.15b) are identical, the RHS of Eq. (7.7.15b) represents the upper bound that should satisfy the following relationship,

$$-R_{\max,t}+\sigma_{QF_t}z_{(1+\alpha_{R_t})/2}\leq -R_{\min,t}-\sigma_{QF_t}z_{(1+\alpha_{R_t})/2} \quad (7.7.16)$$

or simplified to

$$z_{(1+\alpha_{R_t})/2}\leq\frac{R_{\max,t}-R_{\min,t}}{2\sigma_{QF_t}} \quad (7.7.17)$$

Equation (7.7.17) shows that, under the condition of a tight range for releases and a high standard deviation of inflow, the RHS value could be smaller than the z-value for a specified reliability α. To achieve the same reliability for each time period t, the range of releases can be adjusted according to the variability of inflows during the period.

The use of an LDR in a chance-constrained model can lead to conservative results with reservoir capacity larger than needed to satisfy the specified reliability (Sniedovich, 1980; Stedinger et al., 1983). However, chance-constrained models theoretically have the distinct advantage of explicitly incorporating the stochastic nature of hydrologic inputs. The major limitations of chance-constrained models lie in their limited capability to handle multiobjective, multireservoir systems with interdependent benefits and costs (Yeh, 1982). The utility of chance-constrained models to a multiple-reservoir system in which inflows are cross-correlated is severely limited.

Example 7.7.1. Assume that the monthly stream flows listed in Table 7.2.1 for the Little Weiser River near Indian Valley, Idaho, can be described by a normal distribution.

Consider an 80 percent reliability requirement for the compliance of the storage and release constraints. Use the LDR Eq. (7.7.3) to develop a chance-constrained reservoir screening model for determining the reservoir size and operation at the site. Assume that direct precipitation on the reservoir surface is negligible and that monthly evaporation (from Example 7.2.1) is deterministic (nonrandom).

Solution. The objective function, Eq. (7.2.4*a*), of the model is to minimize the active storage capacity, Min K_a. The storage chance-constraints for each month, by Eq. (7.7.6), are

$$K_a - b_1 \geq 0 \quad \text{(for month 1)}$$

$$K_a - b_2 \geq 0 \quad \text{(for month 2)}$$

$$\vdots$$

$$K_a - b_{12} \geq 0 \quad \text{(for month 12)}$$

Without considering the random nature of monthly evaporation, the release chance-constraint, Eq. (7.7.7), can be expressed as

$$P\left(R_{\max,t} + EV_t + b_t - b_{t-1} \geq QF_t\right) \geq (1+\alpha)/2$$

and

$$P\left(R_{\min,t} + EV_t + b_t - b_{t-1} \leq QF_t\right) \geq (1+\alpha)/2$$

The deterministic equivalent of the above two release chance-constraints, referring to Eq. (7.7.15*a*–*b*), can be derived as

$$b_t - b_{t-1} \geq -R_{\max,t} - EV_t + \mu_{QF,t} + z_{(1+a)/2}\sigma_{QF,t}$$

and

$$b_t - b_{t-1} \leq -R_{\min,t} - EV_t + \mu_{QF,t} - z_{(1+a)/2}\sigma_{QF,t},$$

respectively. Therefore, the stated deterministic equivalent of release chance-constraints require the mean and standard deviation of the random inflows for each month, which are estimated from the historical flow data and are shown in the following table:

Month t	Mean μ_{QF_t}	Std. Dev. σ_{QF_t}	Month t	Mean μ_{QF_t}	Std. Dev. σ_{QF_t}
1	763	212	7	9892	7006
2	1337	961	8	20230	7548
3	1643	892	9	16200	10820
4	5370	4565	10	3152	2508
5	4136	2956	11	753	320
6	5270	1302	12	558	217

Since the desired reliability $\alpha = 0.80$ and all monthly streamflows are normal random variables, then $z_{(1+a)/2} = z_{0.90} = 1.28$ (from Table 5.2.1). Furthermore, suppose that the maximum and minimum monthly releases are 30000 AF and 100 AF, respectively.

The resulting release constraints for month 1 are

$$b_1 - b_{12} \geq -30000 - 270 + 763 + 1.28(212) = -29235.6$$

and

$$b_1 - b_{12} \leq -100 - 270 + 763 - 1.28(212) = 664.4$$

Similar constraints can be written for $t = 2, 3, \ldots, 12$. Once the values of the decision parameters $b_1, b_2, \ldots, b_{12}$ are determined by solving the LP model, they are used in Eq. (7.7.5) to determine the monthly reservoir release.

7.7.2 Yield Models

Yield models implicitly consider the reliability aspect of water yield in the determination of reservoir capacity and release policies. In a yield model, the total active capacity for a reservoir is decomposed into the **over-year capacity** and **within-year capacity**. The over-year capacity provides storage to accommodate fluctuation of annual flows. When shorter time scales are considered (monthly or weekly), the flow fluctuation will be greater. Therefore, within-year capacity is an additional storage provided to handle demand fluctuation with a shorter time period.

The reliability of annual yield of a stream can be assessed using annual total flow data in a flow-duration analysis. For example, the total annual inflows in Table 7.2.1 over the five-year period for the Little Weiser River near Indian Valley, Idaho are 41,418 AF, 63,119 AF, 51,109 AF, 93,153 AF, and 97,716 AF. The reliability of annual yield of 41,418 AF (the smallest quantity in the five-year record) is the probability that this quantity can be equalled or exceeded in each year. This probability is not equal to unity because the annual flow sequence will not repeat itself in the future. In general, the reliability of an annual yield which corresponds to the mth largest annual total flow in the record of n years can be estimated as $m/(n+1)$. Therefore, the reliability of having an annual yield of 41,418 AF ($m = 5$) for a five-year record ($n = 5$) in the future can be estimated as $5/6 = 83.3\%$. This method of estimating the annual yield reliability is identical to the plotting position approach in flood frequency analysis (see Chow et al., 1988).

Alternatively, the reliability of a specified annual yield or, conversely, the annual yield of a specified reliability at a given site without a reservoir can be assessed analytically using the following definition.

$$P\left(QF_y \geq Y_\alpha\right) = \alpha \tag{7.7.18}$$

in which QF_y is an annual (or yearly) total flow and Y_α is the annual yield with a reliability α under unregulated conditions.

Example 7.7.2. Based on the five annual total flows, given in Table 7.2.1, for the Little Weiser River near Indian Valley, Idaho, determine the annual yield with a reliability of 80 percent, 90 percent, and 95 percent under the unregulated condition. Assume that the annual total inflows are independent following a normal distribution.

Solution. The five annual total inflows are 41418 AF, 63119 AF, 51109 AF, 93153 AF, 97716 AF from which the mean $\overline{QF} = 69312$ AF and standard deviation $\sigma_{QF} =$

25128 AF can be calculated. Since the annual total inflows follow a normal distribution, the annual yield with a reliability α can be computed as

$$Y_\alpha = \overline{QF} - z_\alpha \sigma_{QF}$$

in which z_α is the standard normal deviate satisfying $P[Z \leq z_\alpha] = \alpha$. The annual yield, Y_α, with different reliability can be calculated in the following table.

α	z_α	Y_α (AF)
0.80	0.84	48204
0.90	1.28	37148
0.95	1.645	27976

It should be kept in mind that the above values for Y_α are annual yields at the site without being regulated by a reservoir. In other words, without a reservoir, the river basin can produce the above annual water yield with the corresponding reliability. From this table, it is observed that under the unregulated condition, the annual yield decreases as its reliability increases. To produce a higher annual release while maintaining the same level of reliability would require the construction of a reservoir to regulate the inflows.

Considering only surface inflows, the over-year active reservoir capacity required to produce annual release R_α with a reliability of $\alpha (R_\alpha = Y_\alpha)$ can be obtained by solving the following optimization model.

$$\text{Minimize } K_a^0 \tag{7.7.19a}$$

subject to

a. annual flow balance

$$ST_y - ST_{y+1} - \hat{R}_y = R_\alpha - QF_y, \quad y = 1, \ldots, Y \tag{7.7.19b}$$

b. reservoir capacity cannot be exceeded in any year

$$ST_y - K_a^0 \leq 0, \quad y = 1, 2, \ldots, Y \tag{7.7.19c}$$

where $\hat{R}_y$ is the release in excess of the annual release R_α in year y, K_a^0 is the over-year active capacity, and Y is the total number of years in the record. The decision variables in the above optimization model are K_a^0, ST_y, and $\hat{R}_y$. In fact, the above over-year model Eqs. (7.7.19*a*–*c*) is identical to the yield model Eqs. (7.2.4*a*–*c*) when the time interval is a year. The above model for determining the over-year capacity can be solved by the sequent-peak method or the golden-section method described in Section 7.2, using annual flow data. Evaporation can be added to the yield models.

When the distribution of within-year demand is different from the within-year flow distribution, additional storage capacity is required. To account for within-year flow variability, the total active storage required for a single reservoir can be determined by solving the following optimization model,

$$\text{Minimize } K_a \tag{7.7.20a}$$

subject to

a. flow balance for each time period t in each year y

$$ST_{t,y} - ST_{t+1,y} - \hat{R}_{t,y} = R_{\alpha,t} - QF_{t,y}, \quad \text{for all } t \text{ and } y \tag{7.7.20b}$$

b. reservoir capacity cannot be exceeded in any time period

$$ST_{t,y} - K_a \leq 0, \quad \text{for all } t \text{ and } y \tag{7.7.20c}$$

where K_a is the total active storage that takes into account seasonal variation of inflows through the constraints on flow balance, $R_{\alpha,t}$ is the release for the tth period in year y with a reliability of α which can be calculated in the same fashion as the annual value using Eq. (7.7.18) after substituting QF_t for QF_y, and $\hat{R}_{\alpha,t}$ is the release—in excess of $R_{\alpha,t}$. The decision variables are K_a, $ST_{t,y}$, and $\hat{R}_{t,y}$.

The model Eqs. (7.7.20*a*–*c*) which are identical to Eqs. (7.2.7*a*–*c*) form a complete yield model. For a stream flow record with Y years and M periods within each year, the complete yield model has $2MY$ constraints and $2MY+1$ decision variables. Therefore, the main disadvantage of the complete yield model is that the model size can become large in terms of the number of constraints and decision variables for a problem having a large number of time periods, especially for a multisite problem.

To reduce the size of the model, Loucks et al. (1981) pointed out that there generally exist a relatively short critical period within the entire record when the flow is low and releases from the storage reservoir is needed to satisfy the minimum required yield. If the critical period can be identified, only the flow continuity constraints for the critical period are needed. However, the critical period is a function of the unknown within-year releases $R_{\alpha,t}$, and the specified annual release, R_α.

Since the sum of within-year releases must be equal to the total annual release R_α, and the reservoir storage is expected to be depleted during the critical year, Loucks et al. (1981) point out that good results can be obtained by replacing the actual within-year inflows during the critical year by some appropriate fraction of the total annual release R_α. The resulting simpler yield model can be formulated as

$$\text{Minimize } K_a \tag{7.7.21a}$$

subject to

a. annual flow balance:

$$ST_y - ST_{y+1} - \hat{R}_y = R_\alpha - QF_y, \quad \text{for all years } y \tag{7.7.21b}$$

b. annual storage does not exceed over-year capacity:

$$ST_y - K_a^0 \leq 0, \quad \text{for all years } y \tag{7.7.21c}$$

c. flow balance for within-year periods:

$$ST_{y^*,t} - ST_{y^*,t+1} - R_{\alpha,t} = -\beta_t R_\alpha \quad \text{for all } t \text{ in the critical year } y^* \quad (7.7.21d)$$

d. within-year storage does not exceed within-year capacity $K_a - K_a^0$,

$$K_a - K_a^0 - ST_{y^*,t} \geq 0, \quad \text{for all } t \text{ in year } y^* \quad (7.7.21e)$$

in which β_t is the fraction of annual yield which satisfies $\Sigma_t \beta_t = 1$. A good choice for β_t is the ratio of inflow in the tth period of the driest year to the total annual inflow of that year. The decision variables in the above optimization model are K_a, ST_y, $\hat{R}_y$, K_a^0, $ST_{y^*,t}$, and $R_{\alpha,t}$.

If the within-year releases $R_{\alpha,t}$ are fixed and set to the predetermined fraction f_t of annual yield, constraint Eq. (7.7.21*d*) can be replaced by

$$ST_t - ST_{t+1} = -\beta_t R_\alpha + f_t R_\alpha \quad \text{for all within-year period } t \quad (7.7.21f)$$

It should be noted that the within-year constraints, Eqs. (7.7.21*d*) or (7.7.21*f*), do not contain the excess spill term as in Eq. (7.7.20*b*). The rationale is that, during the critical year, the reservoir storage volume should not exceed its active storage capacity, that is, the reservoir neither fills nor empties.

The simplified yield model as defined by Eqs. (7.7.21*a–e*), assuming that R_α is specified, has $2(M + Y)$ constraints and $2(M + Y) + 2$ decision variables, which is significantly less than the complete yield model, Eqs. (7.7.20*a–c*). The simplified yield model is only an approximation to the complete yield model, because it does not produce the exact solution to the active storage capacity of the reservoir. Studies (Loucks et al., 1981) show that the use of β_t based on the driest year of record provides as reasonable an estimate of active storage capacity as does the more complete and larger model.

The yield models described above do not consider evaporation and direct precipitation. Inclusion of such hydrologic elements in the complete yield model, Eqs. (7.7.20*a–c*) is straightforward as in Eqs. (7.2.8*a–c*). However, inclusion of evaporation and direct precipitation in the simplified yield model, Eqs. (7.7.21*a–e*), would be difficult because the exact storage volume at the beginning of each period in the critical year is not identified by the models. In such circumstances, the annual evaporation loss and within-year evaporation loss in each period t of the critical year can be approximated as

$$EV_y = \Sigma_t e_t A(ST_y, ST_t, ST_{t+1}) \quad (7.7.22)$$

$$EV_t = e_t A(ST_t, ST_{t+1}) \quad (7.7.23)$$

In a similar manner the direct precipitation amount can be estimated.

The yield models discussed above are for the sizing and operation of single reservoirs. Extension to the multiple-reservoir systems is generally straightforward as described in Section 7.6. To formulate a model for a multiple-reservoir system, interaction among reservoirs must be included. However, there is one special assumption that must be made to enable such extension, that is, the critical periods for all proposed reservoir sites must be identical. This could be a restrictive condition to satisfy in real-life problems, especially where the annual flows at different sites are not highly correlated.

There are other provisions that can be incorporated in yield models, such as allowing the possibility of failure to occur in certain years or allowing multiple releases with different levels of reliability. The objective of the yield model could be the maximization of net benefit which equals to the benefit from different levels of yield minus the cost of a reservoir.

Example 7.7.3. Use the monthly stream flow data given in Table 7.2.1 for the Little Weiser River near Indian Valley, Idaho, to develop a yield model that determines the minimum required active storage K_a which produces $R_{0.95}$ = 48,204 AF. Consider the critical year as the first year. This model should be able to determine the annual releases, $\hat{R}_1, \ldots, \hat{R}_5$ in excess of the specified release and the releases for each of the months during the critical year $R_{0.95,1} \ldots, R_{0.95,12}$. Assume that the randomness of the total annual inflow can be described by a normal distribution.

Solution. The objective function of the yield model is to minimize the total active storage (K_a) as stated in Eq. (7.2.4a). Note that the desired annual release $R_{0.95}$ = 48,204 AF is the same as the annual yield with an 80 percent reliability under the unregulated condition (see Example 7.7.21a). Hence, a reservoir capacity must be provided for enhancing the reliability of a given level of yield.

The over-year mass-balance constraint for each of the five years are formulated as

$$ST_1 - ST_2 - \hat{R}_1 = R_{0.95} - 41{,}418 = 6786 \quad \text{(for year 1)}$$

$$ST_2 - ST_3 - \hat{R}_2 = R_{0.95} - 63{,}119 = 14{,}915 \quad \text{(for year 2)}$$

$$ST_3 - ST_4 - \hat{R}_3 = R_{0.95} - 51{,}109 = 2905 \quad \text{(for year 3)}$$

$$ST_4 - ST_5 - \hat{R}_4 = R_{0.95} - 93{,}153 = 44{,}949 \quad \text{(for year 4)}$$

$$ST_5 - ST_1 - \hat{R}_5 = R_{0.95} - 97{,}764 = 49{,}512 \quad \text{(for year 5)}$$

The second set of constraints state that the over-year storage ST_y cannot exceed the over-year reservoir capacity (K_a^0) as

$$ST_y - K_a^0 \leq 0, \quad \text{for } y = 2, 3, 4, 5, 6.$$

The third constraint set consists of mass balance equations for each month in the critical year, that is, the first year. The constraint equations can be formulated as

$$ST_{1,1} - ST_{1,2} - R_{0.95,1} = -\beta_1 R_{0.95} \quad \text{(for month 1 in year 1)}$$

$$ST_{1,2} - ST_{1,3} - R_{0.95,2} = -\beta_2 R_{0.95} \quad \text{(for month 2 in year 1)}$$

$$\vdots$$

$$ST_{1,11} - ST_{1,12} - R_{0.95,11} = -\beta_{11} R_{0.95} \quad \text{(for month 11 in year 1)}$$

$$ST_{1,12} - ST_{1,13} - R_{0.95,12} = -\beta_{12} R_{0.95} \quad \text{(for month 12 in year 1)}$$

in which $ST_{y,t}$ represents the storage at the beginning of the period t in the critical year y, $ST_{1,13}$ should be made equal to $ST_{1,1}$, and $R_{0.95}$ = 48,204 AF/year. In the above constraint set, the parameter β_t for each month within the critical year can be estimated by the ratio of inflow in the tth month of the driest year to the total annual inflow. Since year 1 is the driest year in the record for this example, parameter β_t can be estimated as $\beta_t = QF_{1,t}/QF_{y=1}$ for $t = 1, 2, \ldots, 12$. The final set of constraints is to ensure that the storage in each month of the critical year does not exceed the within-year capacity $(K_a - K_a^0)$, that is, $K_a - K_a^0 - ST_{1,t} \geq 0$, for twelve months in critical year 1. The decision variables in this model are $\hat{R}_1, \ldots, \hat{R}_5$; $ST_2, \ldots, ST_6$; $R_{0.95,1}, \ldots, R_{0.95,12}$; $ST_{1,1}, \ldots, ST_{1,13}$; K_a and K_a^0.

7.8 RESERVOIR OPERATION UNDER HYDROLOGIC UNCERTAINTY: DP MODELS

Consider the random inflow QF_t in time period t which is discretized into a number of possible values indexed by $QF_{t,i}$ with $i = 1, 2, \ldots, I$. Because the inflows in different time periods are not entirely independent, the inflow sequence is frequently considered as a Markov process in stochastic DP modeling, in particular, a first-order Markov process. The **first-order Markov process** is a stochastic process in which the inflow in the current period t is solely dependent on the inflow of the previous time period $t - 1$. Inflows separated by more than one time period (or lag) are considered to be independent. The discretized inflows to a reservoir following a first-order Markov process are characterized by the so-called **transition probability** $p_{ij} = P(QF_{t,j}|QF_{t-1,i})$ representing the likelihood of observing an inflow of $QF_{t,j}$ during the current period t given that the inflow of the previous time period $t - 1$ is $QF_{t-1,i}$. These conditional probabilities can best be summarized in the **transition probability matrix** $P = [p_{ij}]$ shown in Fig. 7.8.1.

Consider the reservoir operation for a particular time period t in which the initial storage ST_t and the final storage ST_{t+1} are fixed. Then the release R_t, from the continuity equation, is dependent on the inflow QF_t and evaporation EV_t of the period. Due to the random nature of the inflows, the stochastic DP model must consider the releases and the associated effects under all possible inflow conditions. By discretizing the feasible storage space of the reservoir as the deterministic DP approach, the stage representation of a stochastic DP model for a single reservoir is illustrated in Fig. 7.8.2.

Referring to time period t in Fig. 7.8.2, the release $R_{t,k,i,m}$ during this period with initial storage $ST_{t,k}$, inflow $QF_{t,i}$, and final storage $ST_{t+1,m}$ must satisfy the following continuity equation, which is the transformation or transition function in a DP model

$$ST_{t+1,m} = ST_{t,k} + QF_{t,i} - EV_t - R_{t,k,i,m} \tag{7.8.1}$$

in which evaporation in period t is assumed known. Therefore, during the DP computation, the release $R_{t,k,i,m}$ in time period t (stage t) associated with an initial storage (input state k) $ST_{t,k}$, an inflow $QF_{t,i}$, and a final storage (output state m) $ST_{t+1,m}$ can be determined from Eq. (7.8.1). Without losing generality, the following discus-

Inflows in Time Period (t-1) \ Inflows in Time Period t	$QF_{t,1}$	$QF_{t,2}$	...	$QF_{t,j}$	...	$QF_{t,I}$
$QF_{t-1,1}$	p_{11}	p_{12}	...	p_{1j}	...	p_{1I}
$QF_{t-1,2}$	p_{21}	p_{22}	...	p_{2j}	...	p_{2I}
⋮	⋮	⋮	⋮	⋮	⋮	⋮
$QF_{t-1,i}$	p_{i1}	p_{i2}	...	p_{ij}	...	p_{iI}
⋮	⋮	⋮	⋮	⋮	⋮	⋮
$QF_{t-1,I}$	p_{I1}	p_{I2}		p_{Ij}		p_{II}

FIGURE 7.8.1
Transition probability matrix of the first-order Markov inflows.

sions of the stochastic DP algorithm consider evaporation as known or as a function of beginning and end of month storage.

To derive the recursive formula for the stochastic DP reservoir operation model, consider a problem involving T periods as shown in Fig. 7.8.2. It is assumed that the final storage at the end of last time period is fixed at a specified level, that is, $ST_{T+1} = ST^*$. Using a backward recursive algorithm, the optimal return associated with a given initial storage $ST_{T,m}$ and inflow $QF_{T,i}$ can be expressed as

$$f_T^*(ST_{T,m}, QF_{T,j}) = r_T(R_{T,m,j,*}), \quad \text{for all } m \text{ and } j \tag{7.8.2}$$

where $r_T(R_{T,m,j,*})$ is the return for time period T based on the release $R_{T,m,j,*} = -ST^* + ST_{T,m} + QF_{T,j} - EV_T$.

A DP backward algorithm moves backward one period in time to $(T - 1)$. Referring to Fig. 7.8.2, the release $R_{T-1,k,i,m}$ associated with the connection shown in this stage is determined by $R_{T-1,k,i,m} = -ST_{T,m} + ST_{T-1,k} + QF_{T-1,i} - EV_{T-1}$. Notice that, in period $(T - 1)$, the connection of the state is made from a given combination of initial storage and inflow, that is, $(ST_{T-1,k}, QF_{T-1,i})$ to only the final storage $ST_{T,m}$ for the period. Inflows in the next period $QF_{T,j}$ are of no concern in the state transition consideration as shown in Eq. (7.8.1). Because the inflow $QF_{T,j}$ in the next period T is not known, when period $T - 1$ is under consideration for a specified initial storage and final storage transition, all future possible inflows $QF_{T,j}$ and the corresponding returns must be considered. Since the future return cannot be predicted exactly, the expected return from future time periods is commonly used

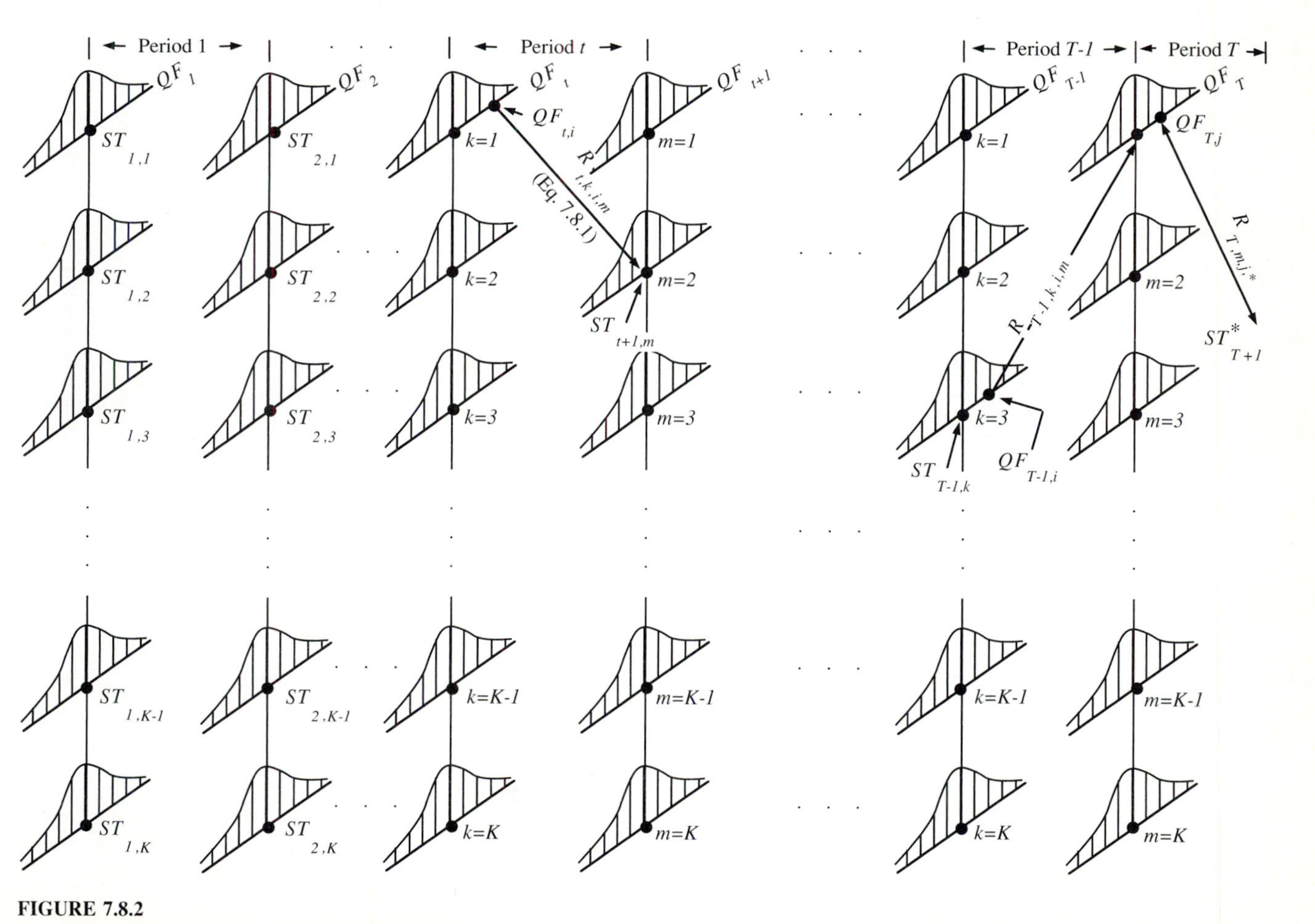

FIGURE 7.8.2
Stage-state representation of stochastic DP reservoir operation model.

in the recursive equation. The expected return from period T, given the release $R_{T-1,k,i,m}$ in period $(T-1)$ with initial storage $ST_{T-1,k}$ and inflow $QF_{T-1,i}$, is,

$$\begin{aligned}\bar{f}_T^*(ST_{T,m}) &= \sum_{i=1}^{I} P(Q_{T,j}|Q_{T-1,i}) f_T^*(ST_{T,m}, QF_{T,j}) \\ &= \sum_{i=1}^{I} p_{ij} f_T^*(ST_{T,m}, QF_{T,j})\end{aligned} \tag{7.8.3}$$

Then, the total return from the last two time periods with initial storage $ST_{T-1,k}$ and inflow $QF_{T-1,i}$ in period $T-1$ is the sum of the returns associated with the release $R_{T-1,k,i,m}$ and the expected return from period T, that is,

$$f_{T-1}(ST_{T-1,k}, QF_{T-1,i}) = r_{T-1}(R_{T-1,k,i,m}) + \bar{f}_T^*(ST_{T,m}) \tag{7.8.4}$$

from which the best end-of-period storage transition $ST_{T,m}$ is to be identified. The recursive optimization for a given combination of initial storage and inflow, that is, $(ST_{T-1,k}, QF_{T-1,i})$ in period $(T-1)$ can be expressed as

$$\begin{aligned}&f_{T-1}^*(ST_{T-1,k}, QF_{T-1,i}) \\ &\qquad = \max\left\{r(R_{T-1,k,i,m}) + \bar{f}_T^*(ST_{T,m})\right\} \text{ over } (ST_{T,m}) \text{ in period } T\end{aligned} \tag{7.8.5}$$

In general, the backward recursive formula for a stochastic DP reservoir operational model is expressed as

$$\begin{aligned}&f_t^*(ST_{t,k}, QF_{t,i}) \\ &\qquad = \begin{cases} r_t(R_{t,k,i,m}) & \text{for } t = T. \\ \max\limits_{ST_{t+1,m}} \left[r_t(R_{t,k,i,m}) + \bar{f}_{t+1}^*(ST_{t+1,m})\right] & \text{for } t = 1, 2, \ldots, T-1. \end{cases}\end{aligned} \tag{7.8.6}$$

REFERENCES

Chow, V. T., D. R. Maidment, and L. W. Mays: *Applied Hydrology*, McGraw-Hill, Inc., New York, 1988.

Goodman, A. S.: *Principles of Water Resources Planning*, Prentice-Hall, Englewood Cliffs, N.J., 1984.

Heidari, M., V. T. Chow, and D. D. Meredith: "Water Resources Systems Analysis by Discrete Differential Dynamic Programming," *Hydraulic Engineering Series*, No. 24, Department of Civil Engineering, University of Illinois, Urbana, 1971.

Loucks, D. P. and P. Dorfman: "An Evaluation of Some Linear Decision Rules in Chance-Constrained Models for Reservoir Planning and Operation," *Water Resources Research*, AGU, 11 (6): 777–782, 1975.

Loucks, D. P., J. R. Stedinger, and D. A. Haith: *Water Resource Systems Planning and Analysis*, Prentice-Hall, Englewood Cliffs, N.J., 1981.

ReVelle, C. S., E. Joeres, and W. Kirby: "The Linear Decision Rule in Reservoir Management and Design, 1; Development of the Stochastic Model," *Water Resources Research*, AGU, 5(4): 767–777, 1969.

Sniedovich, M.: "Analysis of a Chance-Constrained Reservoir Control Model," *Water Resources Research*, AGU, 16(5): 849–854, 1980.

Stedinger, J. R., B. F. Sule, and D. Pei: "Multiple Reservoir System Screening Models," *Water Resources Research*, AGU, 19(6): 1383–1393, 1983.

U.S. Army Corps of Engineers, Hydrologic Engineering Center: *HEC-5, Simulation of Flood Control and Conservation Systems Users Manual*, Davis, Calif., April 1982.

U.S. Army Corps of Engineers, Hydrologic Engineering Center: *Hydrologic Engineering Methods for Water Resources Development: Reservoir System Analysis for Conservation*, vol. 9, Davis, Calif., June 1977.

Yeh, W. W-G: State of the Art Review: Theories and Applications of Systems Analysis Techniques to the Optimal Management and Operation of a Reservoir System, UCLA-ENG-82-52, University of California, Los Angeles, June 1982.

PROBLEMS

7.2.1 Table 7.2.1 contains monthly runoff volumes for water years 1966 to 1970 for the Little Weiser River near Indian Valley, Idaho. Construct the flow-duration curve using this five years of data and determine the firm-yield for this site.

7.2.2 Using the monthly flow data in Table 7.2.1 for the Little Weiser River, construct the cumulative mass curve over a 5-year period and determine the required active storage capacity to produce a firm yield of 2000 AF/month.

7.2.3 Use Problem 7.2.2 to determine the active storage capacity required to produce 2000 AF/month firm yield by computing the cumulative difference between supply and demand.

7.2.4 Use GAMS to solve Example 7.2.2.

7.2.5 Using the water balance principle and the golden section method (Section 4.4.2) determine the firm yield for the Little Weiser River site based on the monthly streamflow data for several reservoir capacities. Write a computer program to solve this problem.

7.7.1 Solve Example 7.7.1 for the optimal active storage capacity and the corresponding decision parameters $b_1, b_2, \ldots, b_{12}$ using GAMS.

7.7.2 Solve Example 7.7.1 using GAMS for the optimal active storage capacity and the corresponding decision parameters $b_1, b_2, \ldots, b_{12}$ assuming that monthly inflows are lognormal random variables.

7.7.3 Referring to Example 7.7.1, determine the minimum achievable release reliability when the range of releases is between 200 AF and 20,000 AF.

7.7.4 Solve Problem 7.7.3 assuming that monthly inflows follow a lognormal distribution.

7.7.5 Formulate the chance-constrained reservoir planning and design model using the following LDR

$$R_t = ST_{t-1} - b_t$$

proposed by ReVelle, Joeres, and Kirby (1969). Assume that inflows are normal random variables and the minimum storage for each time period is $ST_{\min,t}$.

7.7.6 Using the data given in Example 7.7.1, solve the chance-constrained reservoir model developed in Problem 7.7.5, using GAMS with $\alpha = 0.80$ for both release and storage channel constraints and $ST_{\min,t} = 0$ AF. Furthermore, the monthly release cannot exceed the beginning-of-month storage.

7.7.7 Compare the numerical solutions for Problems 7.7.1 and 7.7.6 and discuss the reasons for the discrepancies in model solution.

7.7.8 Solve the yield model in Example 7.7.3 numerically for the optimal capacity and monthly releases.

7.7.9 Resolve Example 7.7.3 numerically for the optimal reservoir capacity and monthly release using $R_{0.95}$ = 37,148 AF and $R_{0.95}$ = 43,305 AF. Develop the release-storage curve for the 95 percent reliability.

7.7.10 Develop the release-storage curve using the yield model equation (7.7.21*a*–*e*) for 80 percent reliability based on the Little Weiser River data in Table 7.2.1.

CHAPTER 8

GROUNDWATER SYSTEMS

8.1 BASIC PRINCIPLES OF GROUNDWATER SYSTEMS

This chapter presents various tools that can be used in groundwater management. Using the definition of van der Heijde et al. (1985), **groundwater management** can be defined to include "planning, implementation, and adaptive control of policies and programs related to the exploration, inventory, development, and operation of water resources containing groundwater." The numerical modeling of groundwater is relatively new and was not extensively pursued until the mid-1960s. Since that time there has been significant progress in the development and application of numerical models for groundwater management.

Numerical-simulation models have been used extensively to evaluate groundwater resources and develop a better understanding of the flow characteristics of aquifers. Simulation models have been used to explore hydrogeological problems and to predict the impacts of various groundwater management alternatives such as the impacts of pumpage and recharge, groundwater-surface water interactions, the migration of chemical contaminants, saltwater intrusion, etc. Numerical-simulation models are typically used in a repetitive manner for groundwater management problems, looking at various scenarios to attempt to find the one that best achieves an objective. In contrast to simulation models, optimization models directly consider the objective of the management in addition to various types of constraints that are typically placed upon the management policies. For a more thorough review of various groundwater optimization models refer to Willis and Yeh (1987).

8.1.1 Groundwater Hydrology

Groundwater hydrology is "the science of the occurrence, distribution, and movement of water below the surface of the earth." The origins of groundwater are through infiltration, influent streams, seepage from reservoirs, artificial recharge, condensation, seepage from oceans, water trapped in sedimentary rock (connate water), and juvenile water. Any significant quantity of subsurface water is stored in subsurface formations called **aquifers**. These aquifers consist of unconsolidated rocks, mainly gravel and sand which are usually of large areal extent and are essentially underground storage reservoirs. Aquifers are classified as unconfined or confined depending upon the presence or absence of a water table (Fig. 8.1.1).

An **unconfined aquifer** is one in which a water table serves as the upper surface of the zone of saturation, also known as a **free, phreatic**, or **nonartesian aquifer**. Changes in the water table (rising or falling) correspond to changes in the volume of water in storage within an aquifer. A **confined aquifer** is one in which the groundwater is confined under pressure greater than atmospheric by overlying, relatively impermeable strata. Confined aquifers are also known as **artesian** or **pressure aquifers**. Water enters such aquifers in an area where the confining bed rises to the surface or ends underground and is known as a **recharge area** (see Fig. 8.1.1). Changes of the water levels in wells penetrating confined aquifers result primarily from changes in pressure rather than changes in storage volumes. A confined aquifer becomes an unconfined aquifer when the **piezometric surface** (hydrostatic pressure level or head) falls below the bottom of the upper confining bed.

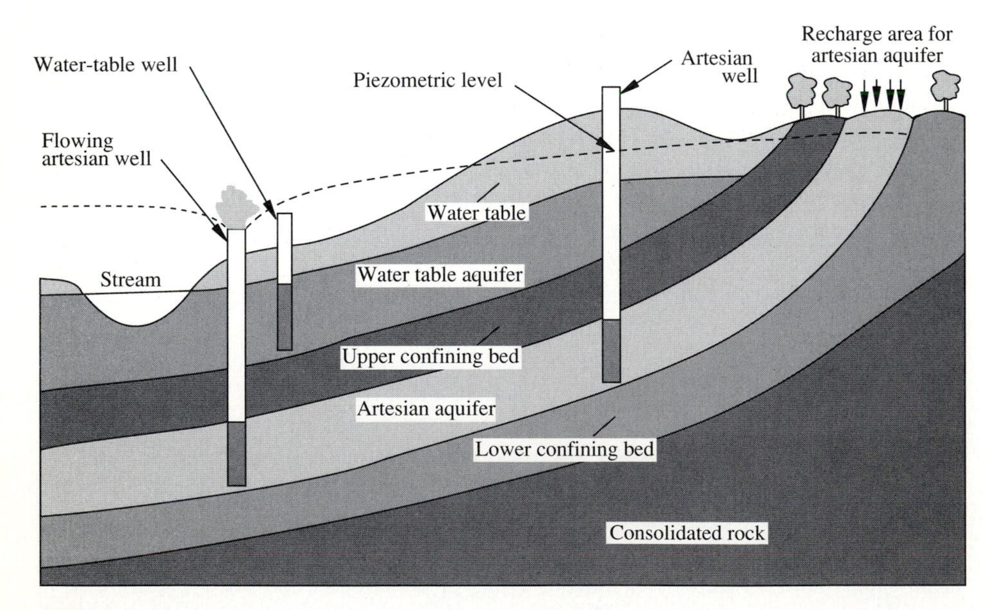

FIGURE 8.1.1
Subsurface distribution of water (Gehm and Bregman, 1976).

Aquifers perform two important functions—a storage function and a transmission function. In other words, aquifers *store* water, functioning as a reservoir, and also transmit water, functioning as a pipeline. An important property of an aquifer related to storage is the **porosity**, Φ. The porosity is a measure of the opening or pore volume divided by the total volume. Porosity represents the amount of water an aquifer can hold but does not indicate the amount of water a porous material will yield.

When water is drained from a saturated material under the influence of gravity, only a portion of the total saturated volume in the pores is released. Part of the water is retained in interstices due to the molecular attraction, adhesion, and cohesion. The **specific yield**, S_y, is the volume of water drained from a saturated sample of unit volume. **Specific retention**, S_r, is the quantity of water that is retained in the unit volume after gravity drainage. The sum of the specific yield and specific retention is the porosity Φ.

The **storage coefficient** S of an aquifer is the volume of water the aquifer releases from or takes into storage per unit surface area of the aquifer per unit decline or rise of head. Considering a vertical column of a unit cross-sectional area extending through a confined aquifer and an unconfined aquifer, in both cases the storage coefficient S equals the volume of water released from the aquifer when the piezometric surface or water table per unit area declines a unit length. The storage coefficient then has dimensions of L^3/L^3 or is dimensionless. In the case of unconfined aquifers, the storage coefficient corresponds to the specific yield. Confined aquifers typically have storage coefficients in the range $5 \times 10^{-5} \leq S \leq 5 \times 10^{-3}$ (Todd, 1980). These small values indicate that large pressure changes are required to produce substantial water yields. The storage coefficient is determined in the field by pump tests (Walton, 1970; Bouwer, 1978; Bear, 1979; Freeze and Cherry, 1979; Todd, 1980; Kashef, 1986; de Marsily, 1986).

Permeability is the property related to the transmission function of an aquifer, which is the measure of the ease of movement of groundwater through aquifers and can be thought of as the hydraulic conductivity. The **hydraulic conductivity** or **coefficient of permeability** K is the rate of flow of water through a unit cross-sectional area of the aquifer under a unit hydraulic gradient (per unit length of head loss).

Closely related to the hydraulic conductivity is the transmissivity which indicates the capacity of an aquifer to transmit water through its entire thickness. **Transmissivity** is the flow rate of water through a unit width vertical strip extending the saturated thickness of an aquifer under a unit hydraulic gradient. The transmissivity for a confined aquifer is equal to the hydraulic conductivity multiplied by the saturated thickness of the aquifer:

$$T = Kb \tag{8.1.1}$$

in which b is the saturated thickness of a confined aquifer. For an unconfined aquifer the saturated thickness is the head h, so that

$$T = Kh \tag{8.1.2}$$

Aquifers assumed to have the same hydraulic conductivity from one location to another are referred to as **isotropic**. **Anisotropic** conditions exist where the hydraulic conductivity varies with location.

8.1.2 Groundwater Movement

Groundwater in its natural state is invariably moving and this movement is governed by hydraulic principles. The flow through aquifers is expressed by **Darcy's law**. This law states that the flow rate through porous media is proportional to the headloss and inversely proportional to the length of the flow path. In general form, Darcy's law relates the **Darcy flux** to the rate of headloss per unit length of porous medium, $\partial h/\partial l$, as

$$v = -K\frac{\partial h}{\partial l} \tag{8.1.3}$$

where v is the Darcy flux or velocity or specific discharge, (L/T), and l is the distance along the average direction of flow. The negative sign is used so that v is positive in the direction of decreasing h. The total discharge through a cross-sectional area, A, of porous media is then

$$q = vA = -KA\frac{dh}{dl} \tag{8.1.4}$$

Pumping water from an aquifer removes water from storage surrounding the well causing the water table for unconfined aquifers, or the piezometric surface for confined aquifers, to lower. The amount the water table or piezometric surface is lowered is the **drawdown**, s. A drawdown curve such as shown in Fig. 8.1.2 for a confined aquifer and in Fig. 8.1.3 for an unconfined aquifer shows the variation of the drawdown with distance from the well. The drawdown curve for radial flow to a well actually describes a **cone of depression**, which is an **area of influence** (wherein the drawdown $s > 0$) of the well defines the outer limit of the cone of depression.

The governing equation for radial flow is the well-known diffusion equation

$$\frac{1}{r}\frac{\partial}{\partial r}\left(r\frac{\partial h}{\partial r}\right) = \frac{\partial^2 h}{\partial r^2} + \frac{1}{r}\frac{\partial h}{\partial r} = \frac{S}{T}\frac{\partial h}{\partial t} \tag{8.1.5}$$

where r is the radial distance from a pumped well, and t is the time since the beginning of pumping. For steady-state conditions, that is, $\partial h/\partial t = 0$, Eq. (8.1.5) reduces to

$$\frac{1}{r}\frac{\partial}{\partial r}\left(r\frac{\partial h}{\partial r}\right) = 0 \tag{8.1.6}$$

Table 8.1.1 lists the various equations used for steady and unsteady radial flow to wells in confined and unconfined aquifers.

8.1.3 Types of Groundwater Quantity Management Models

Aquifer simulation models have been used to examine the effects of various groundwater management strategies. Use has been primarily of the "case study" or "what if" type. The analyst specifies certain quantities and the model predicts the techni-

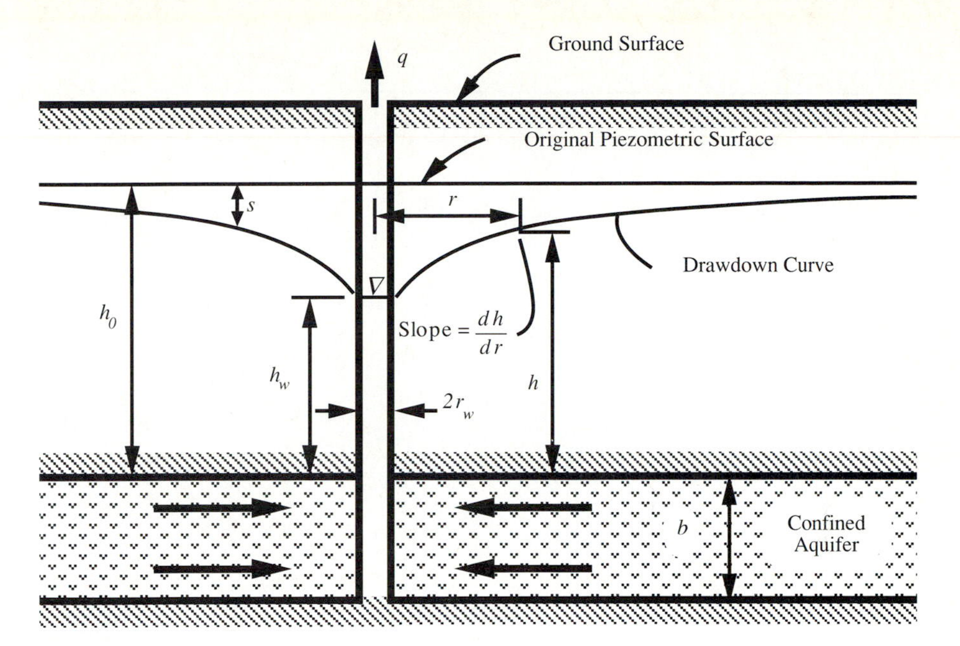

FIGURE 8.1.2
Confined aquifer.

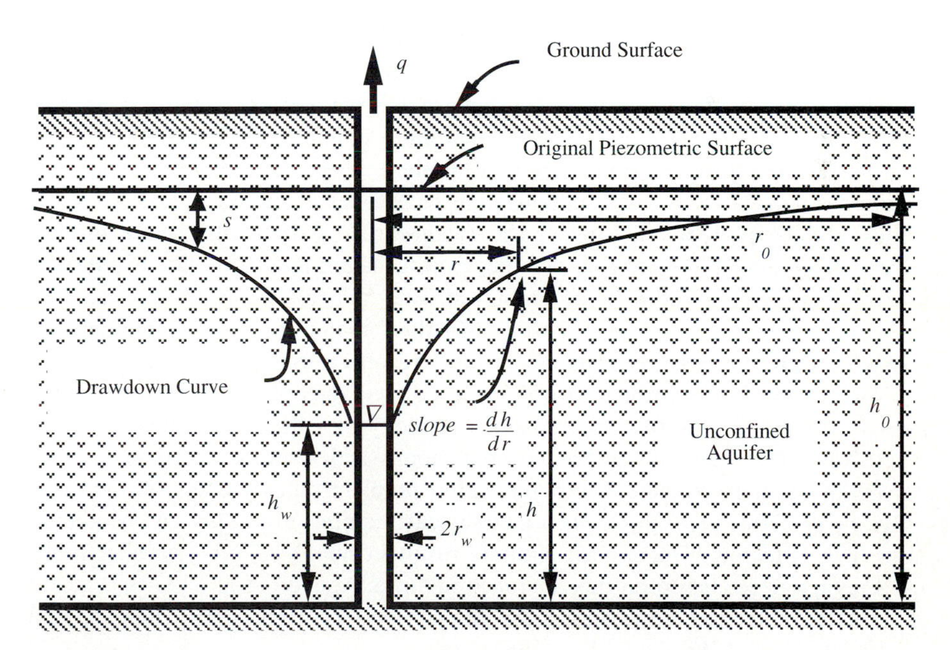

FIGURE 8.1.3
Unconfined aquifer.

TABLE 8.1.1
Radial flow to wells

	Governing equation	Well discharge	
Steady-state			
Confined	$q = -2\pi r K b \frac{dh}{dr}$	$q = 2\pi K b \frac{(h_2 - h_1)}{\ln(r_2/r_1)}$	Thiem equation
Unconfined	$q = -2\pi r K h \frac{dh}{dr}$	$q = \pi K \frac{(h_2^2 - h_1^2)}{\ln\left(\frac{r_2}{r_1}\right)}$	
Unsteady-state			
Confined	$\frac{\partial^2 h}{\partial r^2} + \frac{1}{r}\frac{\partial h}{\partial r} = \frac{S}{T}\frac{\partial h}{\partial t}$	$q = \frac{4\pi s T}{W(u)}$ where $W(u) = -0.5772 - \ln(u) + u - \frac{u^2}{2 \cdot 2!} + \frac{u^3}{3 \cdot 3!} - \frac{u^4}{4 \cdot 4!} + \cdots$ and $u = \frac{r^2 S}{4Tt}$	Theis equation
Confined		$q = \frac{4\pi T s}{\left[-0.5772 - \ln\left(\frac{r^2 S}{4Tt}\right)\right]}$	Cooper-Jacob Approximation

cal and perhaps economic consequences of this choice. The analyst evaluates these consequences and uses judgement and intuition to specify the next case.

Optimization methods have been used in groundwater management for more than a decade. Most uses focused on explicitly combining simulation and optimization, resulting in so-called **simulation-management models**. A classification of groundwater management models based upon optimization techniques is presented in Fig. 8.1.4. Gorelick (1983) also considers two basic categories: (a) **hydraulic management models** that are aimed at managing pumping and recharge; and (b) **policy evaluation models** that also can consider the economics of water allocations. Hydraulic management models have been developed based upon three major approaches: **embedding approach**, **optimal control approach**, and **unit response matrix approach**.

The **embedding approach** incorporates the equations of the simulation model (represented as a set of difference equations) directly into the optimization problem to be solved. This method has limited applications and is mostly used in groundwater hydraulic management. The optimization problem quickly becomes too large to solve by available algorithms when a large-scale aquifer, especially an unconfined aquifer, is considered. Unconfined aquifers result in nonlinear programming problems. Previous work based on this approach includes Aguado et al. (1974); Aguado et al. (1977); Willis and Newman (1977); Aguado and Remson (1980); Remson and Gorelick (1980); and Willis and Liu (1984).

The **optimal control approach** is based upon concepts from optimal control theory with the basic methodology being to couple optimization techniques with a groundwater simulator to implicitly solve the governing equation of groundwater flow

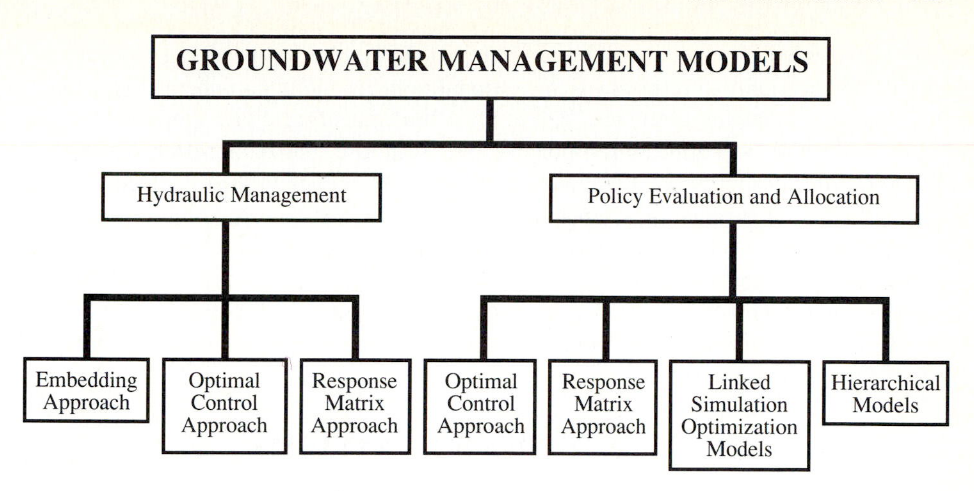

FIGURE 8.1.4
Classification of optimization models for groundwater management.

for each iteration of the optimization procedure. This methodology could be thought of as a variation of the embedding approach with the governing equations being solved implicitly. The state variables that represent the heads and the control variables that represent the pumpages are implicitly related through the simulator. The simulator equations are used to express the state variables in terms of the control variables yielding a much smaller reduced optimization problem that must be solved many times. Wanakule, Mays and Lasdon (1986) presented a general groundwater management model, based upon nonlinear programming and a groundwater simulation model, that can be used to solve both hydraulic management problems and groundwater policy evaluation (allocation) problems.

The **response matrix approach** generates a unit response matrix by solving the simulation model several times, each with unit pumpage at a single pumping node. Superposition is used to determine total drawdowns. This yields a smaller optimization problem, but the method has two major limitations. It is exact only for a confined aquifer but has good accuracy for an unconfined aquifer with relatively small drawdowns compared to the aquifer thickness. A drawdown correction method may be used to improve accuracy for an unconfined aquifer with larger drawdowns, but acceptable accuracy cannot be guaranteed (Heidari, 1982). In addition, the response matrix must be recomputed when exogeneous factors such as aquifer boundary conditions or potential well locations change. An alternative is to treat these factors as decision variables, but then more variables and constraints are included in the optimization problem. Work stemming from this approach includes that by Maddock (1972, 1974), Maddock and Haimes (1975), Morel-Seytoux and Daly (1975), Morel-Seytoux et al. (1980), Heidari (1982), Illangasekare and Morel-Seytoux (1982), and Willis (1984).

Groundwater policy evaluation and allocation models are used for water allocation purposes involving economic management objectives subject to institutional policies as constraints in addition to the hydraulic management constraints. Applications of these types of models have been to transient aquifer problems that consider the agricultural economy in response to institutional policies and to conjunctive use of surface water-groundwater problems. Figure 8.1.4 illustrates four types of approaches for policy evaluation and allocation problems. The response matrix approach is used for problems considering hydraulic-economic response (Gorelick, 1983). Linked simulation optimization models use the results of an external groundwater simulation model as input to a series of subarea economic optimization models (Gorelick, 1983). Examples of the linked simulation-optimization models include: Young and Bredehoeft (1972), Daubert and Young (1982), Bredehoeft and Young (1983). Hierarchical models use subarea decomposition and a response matrix approach (Haimes and Dreizen, 1977; Bisschop et al., 1982).

8.2 SIMULATION OF GROUNDWATER SYSTEMS

8.2.1 Development of Governing Equations

Darcy's law relates the Darcy flux v with dimension (L/T) to the rate of headloss per unit length of porous medium $\partial h/\partial l$, in Eq. (8.1.3). The negative sign indicates that the total head is decreasing in the direction of flow because of friction. This law applies to a cross section of porous medium which is large compared with the cross section of individual pores and grains of the medium. At this scale, Darcy's law describes a steady uniform flow of constant velocity, in which the net force on any fluid element is zero. For unconfined saturated flow, the two forces are gravity and friction. Darcy's law can also be expressed in terms of the transmissivity Eq. (8.1.1) or (8.1.2) for confined conditions as

$$v = -\frac{T}{b}\frac{\partial h}{\partial l} \tag{8.2.1}$$

or for unconfined conditions as

$$v = -\frac{T}{h}\frac{\partial h}{\partial l} \tag{8.2.2}$$

Considering two-dimensional (horizontal) flow, a general flow equation can be derived by considering flow through a rectangular element (control volume) shown in Fig. 8.2.1. The flow components ($q = Av$) for the four sides of the element are expressed using Darcy's law where $A = \Delta x \cdot h$ for unconfined conditions and $A = \Delta x \cdot b$ for confined conditions so that

$$q_1 = -T_{x_{i-1,j}}\Delta y_j \left(\frac{\partial h}{\partial x}\right)_1 \tag{8.2.3a}$$

$$q_2 = -T_{x_{i,j}}\Delta y_j \left(\frac{\partial h}{\partial x}\right)_2 \tag{8.2.3b}$$

$$q_3 = -T_{y_{i,j+1}}\Delta x_i \left(\frac{\partial h}{\partial y}\right)_3 \tag{8.2.3c}$$

$$q_4 = -T_{y_{i,j}}\Delta x_i \left(\frac{\partial h}{\partial y}\right)_4 \tag{8.2.3d}$$

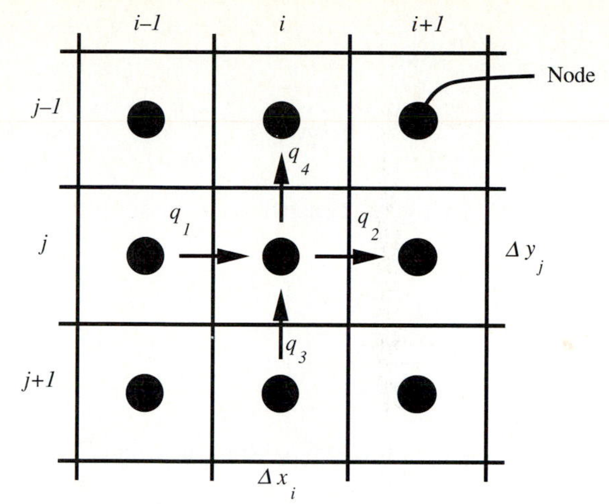

Cell to Cell Water Transfers

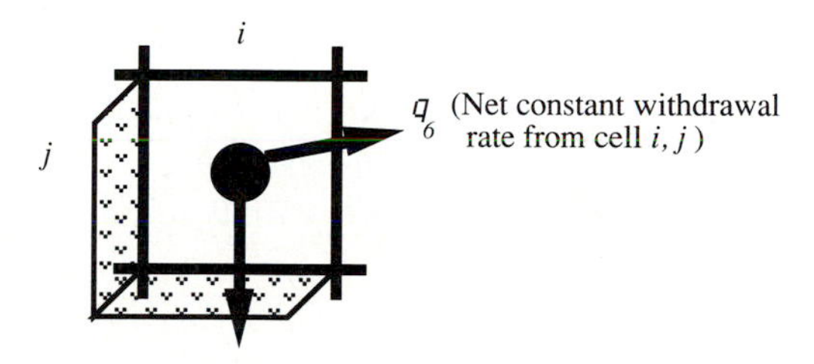

FIGURE 8.2.1
Finite difference grid.

where $T_{x_{i,j}}$ is the transmissivity in the x flow direction from element (i, j) to element $(i+1, j)$. The terms $(\partial h/\partial x)_1$, $(\partial h/\partial x)_2, \ldots$, define the hydraulic gradients at the element sides $1, 2, \ldots$.

The rate at which water is stored or released in the element over time is

$$q_5 = S_{i,j} \Delta x_i \Delta y_i \frac{\partial h}{\partial t} \tag{8.2.4}$$

in which $S_{i,j}$ is the storage coefficient for element (i, j). In addition, the flow rate q_6 for constant net withdrawal or recharge from the element over time interval Δt is considered

$$q_6 = q_{i,j,t} \tag{8.2.5}$$

in which $q_{i,j,t}$ has a positive value for pumping whereas it has a negative value for recharge.

By continuity the flow into and out of a grid or cell is

$$q_1 - q_2 + q_3 - q_4 = q_5 + q_6 \tag{8.2.6}$$

Substituting in Eqs. (8.2.3)–(8.2.5) gives

$$-T_{x_{i-1,j}}\Delta y_j\left(\frac{\partial h}{\partial x}\right)_1+T_{x_{i,j}}\Delta y_j\left(\frac{\partial h}{\partial x}\right)_2-T_{y_{i,j+1}}\Delta x_i\left(\frac{\partial h}{\partial y}\right)_3$$
$$+T_{y_{i,j}}\Delta x_i\left(\frac{\partial h}{\partial y}\right)_4=S_{i,j}\Delta x_i\Delta y_j\frac{\partial h}{\partial t}+q_{i,j,t} \quad (8.2.7)$$

Dividing Eq. (8.2.7) by $\Delta x_i\Delta y_j$ and simplifying for constant transmissivities in the x and y directions yields

$$-T_x\left[\frac{\left(\frac{\partial h}{\partial x}\right)_1-\left(\frac{\partial h}{\partial x}\right)_2}{\Delta x_i}\right]-T_y\left[\frac{\left(\frac{\partial h}{\partial y}\right)_3-\left(\frac{\partial h}{\partial y}\right)_4}{\Delta y_j}\right]=S_{i,j}\frac{\partial h}{\partial t}+\frac{q_{i,j,t}}{\Delta x_i\Delta y_j} \quad (8.2.8)$$

For Δx and Δy infinitesimally small the terms in brackets [ ] become second derivatives of h, then Eq. (8.2.8) reduces to

$$T_x\frac{\partial^2 h}{\partial x^2}+T_y\frac{\partial^2 h}{\partial y^2}=S\frac{\partial h}{\partial t}+W \quad (8.2.9)$$

which is the general partial differential equation for unsteady flow in the horizontal direction in which $W=q_{i,j,t}/\Delta x_i\Delta y_j$ is a sink term with dimensions (L/T).

In the more general case for unsteady, two-dimensional heterogeneous anisotropic case, Eq. (8.2.9) is expressed as

$$\frac{\partial}{\partial x}\left(T_x\frac{\partial h}{\partial x}\right)+\frac{\partial}{\partial y}\left(T_y\frac{\partial h}{\partial y}\right)=S\frac{\partial h}{\partial t}+W \quad (8.2.10a)$$

or more simply

$$\frac{\partial}{\partial x_i}\left(T_{i,j}\frac{\partial h}{\partial x_j}\right)=S\frac{\partial h}{\partial t}+W \qquad i,j=1,2 \quad (8.2.10b)$$

8.2.2 Finite Difference Equations

The partial derivative expressions for Darcy's law, Eqs. (8.2.3 *a–d*), can be expressed in finite difference form for time t in Eq. (8.2.7) using

$$\left(\frac{\partial h}{\partial x}\right)_1=\left(\frac{h_{i-1,j,t}-h_{i,j,t}}{\Delta x_i}\right) \quad (8.2.11a)$$

$$\left(\frac{\partial h}{\partial x}\right)_2=\left(\frac{h_{i,j,t}-h_{i+1,j,t}}{\Delta x_i}\right) \quad (8.2.11b)$$

$$\left(\frac{\partial h}{\partial y}\right)_3=\left(\frac{h_{i,j+1,t}-h_{i,j,t}}{\Delta y_j}\right) \quad (8.2.11c)$$

$$\left(\frac{\partial h}{\partial y}\right)_4=\left(\frac{h_{i,j,t}-h_{i,j-1,t}}{\Delta y_i}\right) \quad (8.2.11d)$$

and the time derivative in Eq. (8.2.7) is

$$\frac{\partial h}{\partial t} = \left(\frac{h_{i,j,t} - h_{i,j,t-1}}{\Delta t} \right) \tag{8.2.12}$$

Substituting Eqs. (8.2.11) and (8.2.12) into Eq. (8.2.7) yields

$$\begin{aligned} &- T_{x_{i-1,j}} \Delta y_j \left(\frac{h_{i-1,j,t} - h_{i,j,t}}{\Delta x_i} \right) + T_{x_{i,j}} \Delta y_j \left(\frac{h_{i,j,t} - h_{i+1,j,t}}{\Delta x_i} \right) \\ &- T_{y_{i,j+1}} \Delta x_i \left(\frac{h_{i,j+1,t} - h_{i,j,t}}{\Delta y_j} \right) + T_{y_{i,j}} \Delta x_i \left(\frac{h_{i,j} - h_{i,j-1,t}}{\Delta y_j} \right) \\ &- S_{i,j} \Delta x_i \Delta y_j \left(\frac{h_{i,j,t} - h_{i,j,t-1}}{\Delta t} \right) - q_{i,j,t} = 0 \end{aligned} \tag{8.2.13}$$

which can be further simplified to

$$A_{i,j} h_{i,j,t} + B_{i,j} h_{i-1,j,t} + C_{i,j} h_{i+1,j,t} + D_{i,j} h_{i,j+1,t} + E_{i,j} h_{i,j-1,t} + F_{i,j,t} = 0 \tag{8.2.14}$$

where

$$A_{i,j} = \left[T_{x_{i-1,j}} \frac{\Delta y_j}{\Delta x_i} + T_{x_{i,j}} \frac{\Delta y_j}{\Delta x_i} + T_{y_{i,j+1}} \frac{\Delta x_i}{\Delta y_j} + T_{y_{i,j}} \frac{\Delta x_i}{\Delta y_j} - S_{i,j} \frac{\Delta x_i \Delta y_j}{\Delta t} \right] \tag{8.2.15a}$$

$$B_{i,j} = -T_{x_{i-1,j}} \frac{\Delta y_j}{\Delta x_i} \tag{8.2.15b}$$

$$C_{i,j} = -T_{x_{i,j}} \frac{\Delta y_i}{\Delta x_i} \tag{8.2.15c}$$

$$D_{i,j} = -T_{y_{i,j+1}} \frac{\Delta x_i}{\Delta y_j} \tag{8.2.15d}$$

$$E_{i,j} = -T_{y_{i,j}} \frac{\Delta x_i}{\Delta y_j} \tag{8.2.15e}$$

$$F_{i,j,t} = S_{i,j} \frac{\Delta x_i \Delta y_j}{\Delta t} - q_{i,j,t} \tag{8.2.15f}$$

The coefficients $A_{i,j}$, $B_{i,j}$, $C_{i,j}$, and $D_{i,j}$ are linear functions of the thickness of cell (i, j) and the thickness of one of the adjacent cells. For artesian conditions, this thickness is a known constant, so if cell (i, j) and its neighbors are artesian, Eq. (8.2.14) is linear for all t. For unconfined (water table) conditions, the thickness of cell (i, j) is $h_{i,j,t} - BOT_{i,j}$, where $BOT_{i,j}$ is the average elevation of the bottom of the aquifer for cell (i, j). Then for unconfined conditions, Eq. (8.2.14) involves products of heads and is nonlinear in terms of the heads.

An **iterative alternating direction implicit** (IADI) **procedure** can be used to solve the set of equations. The IADI procedure involves reducing a large set of equations to several smaller sets of equations. One such smaller set of equations is generated by writing Eq. (8.2.14) for each cell or element in a column but assuming that the head for the nodes on the adjacent columns are known. The unknowns in this set of equations are the heads for the nodes along the column. The head for the nodes along adjoining columns are not considered unknowns. This set of equations is solved by Gauss elimination and the process is repeated until each column is treated. The next step is to develop a set of equations along each row, assuming the head for the nodes along adjoining rows are known. The set of equations for each row is solved and the process is repeated for each row in the finite difference grid.

Once the sets of equations for the columns and the sets of equations for the rows have been solved, one "iteration" has been completed. The iteration process is repeated until the procedure converges. Once convergence is accomplished, the terms $h_{i,j}$ represent the heads at the end of the time step. These heads are used as the beginning heads for the following time step. For a more detailed discussion of the iterative alternating direction implicit (IADI) procedure, see Peaceman and Rachford (1955), Prickett and Lonnquist (1971), or Wang and Anderson (1982). Two widely used two-dimensional finite-difference models for groundwater flow are by Prickett and Lonnquist (1971) and by Trescott et al. (1976).

An example of the application of a two-dimensional finite-difference groundwater model is the Edwards (Balcones Fault Zone) aquifer shown in Fig. 1.2.12. This aquifer has been modeled using the GWSIM groundwater simulation model developed by the Texas Water Development Board (1974). GWSIM is a finite difference simulation model which uses the IADI method similar to the model by Prickett and Lonnquist (1971). The finite difference grid for the Edwards aquifer is shown in Fig. 8.2.2, which has 856 active cells to describe the aquifer.

8.3 HYDRAULIC MANAGEMENT MODELS: EMBEDDING APPROACH

8.3.1 Steady-State One-Dimensional Problems for Confined Aquifers

Considering a confined aquifer with flow in one-dimension and fixed-head boundaries as shown in Fig. 8.3.1 with pumping wells that are fully penetrating, the governing equation for steady-state flow can be derived from Eq. (8.2.9) as

$$\frac{\partial^2 h}{\partial x^2} = \frac{W}{T_x} \tag{8.3.1}$$

where $\partial h/\partial t = 0$.

Using a central differencing scheme, Eq. (8.3.1) can be written in finite difference form as

$$\frac{h_{i+1} - 2h_i + h_{i-1}}{(\Delta x)^2} = \frac{W_i}{T_x} \tag{8.3.2}$$

Aguado et al. (1974) formulated the following type of linear programming model for determining the optimal steady-state pumpage from a one-dimensional confined aquifer with fixed head boundaries. The optimization problem can be stated as

$$\text{Maximize } Z = \sum_{i \in I} h_i \tag{8.3.3}$$

subject to Eq. (8.3.2) for each well.

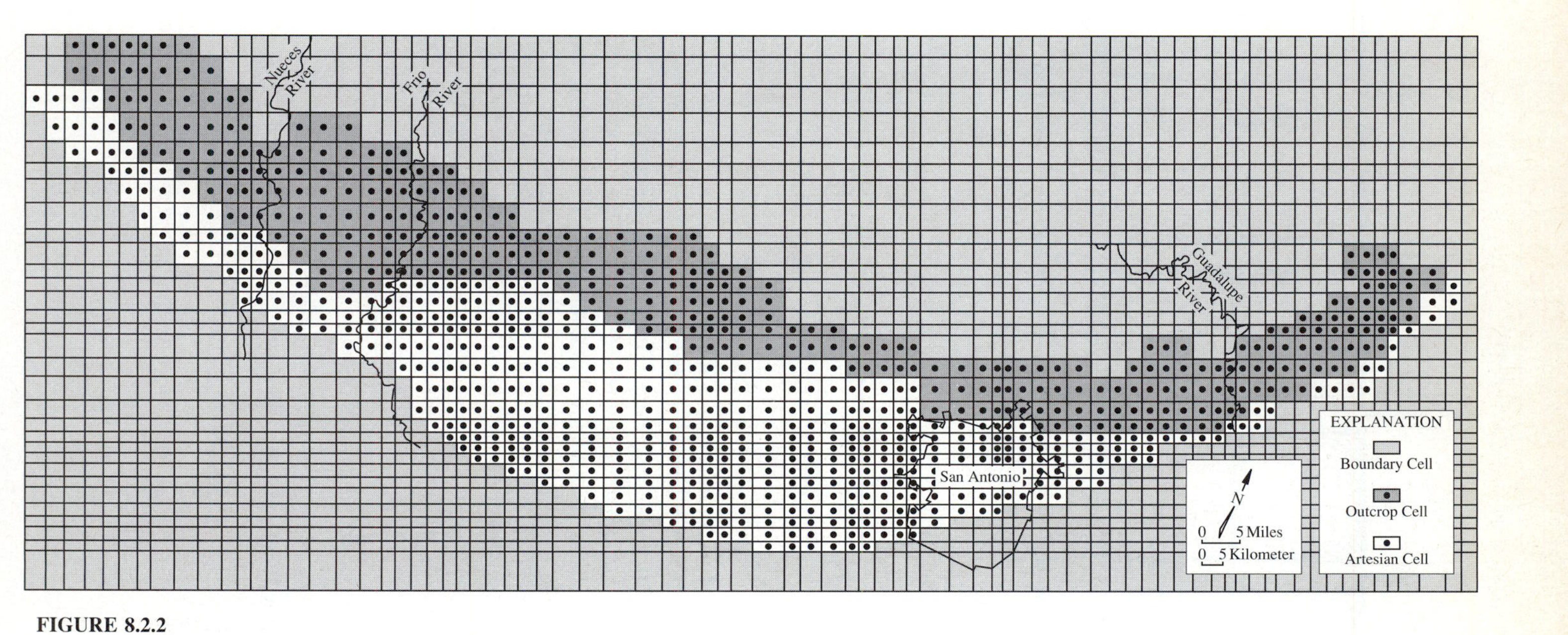

FIGURE 8.2.2
Cell map used for the digital computer model of the Edwards (Balcones fault zone) aquifer (after Klemt et al., 1979).

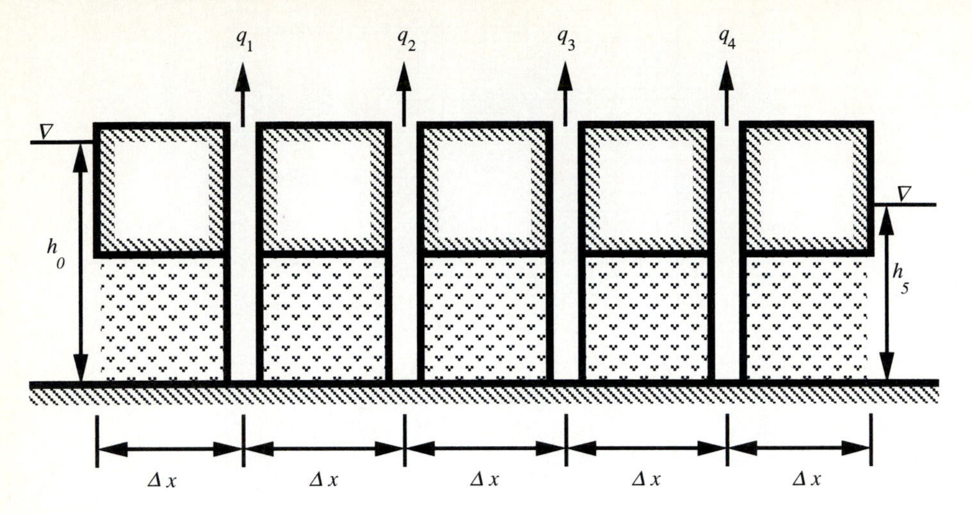

FIGURE 8.3.1
Confined one-dimensional aquifer.

$$\sum_{i \in I} W_i \geq W_{\min} \tag{8.3.4}$$

$$h_i \geq 0 \qquad i \in I \tag{8.3.5a}$$

$$W_i \geq 0 \qquad i \in I \tag{8.3.5b}$$

where I is the set of wells and $W_{\min}$ is the minimum total production rate for the wells. The unknowns in this problem are h and W. Once the model is solved the pumpage can be determined from $W = q_i/\Delta x_i^2$. The head maintenance objective of Eq. (8.3.3) is practical for managing some aquifers; however, other types of objective functions could be used, such as minimizing pumping costs. The above model formulation considers negligible well diameters and negligible well losses.

Example 8.3.1. Develop an LP model for determining the optimal (maximize heads) steady-state pumpage of the one-dimensional confined aquifer shown in Fig. 8.3.1. The wells are equally spaced at a distance of Δx apart with constant head boundaries h_0 and h_5.

Solution. The objective function is simply

$$\text{Maximize } Z = h_1 + h_2 + h_3 + h_4$$

subject to the following finite difference equations

$$-2h_1 + h_2 - \frac{(\Delta x)^2}{T} W_1 = -h_0$$

$$h_1 - 2h_2 + h_3 - \frac{(\Delta x)^2}{T} W_2 = 0$$

$$h_2 - 2h_3 + h_4 - \frac{(\Delta x)^2}{T} W_3 = 0$$

$$h_3 - 2h_4 - \frac{(\Delta x)^2}{T} W_4 = -h_5$$

and the constraint on the production rate

$$W_1 + W_2 + W_3 + W_4 \geq W_{\min}$$

$$h_i \geq 0 \qquad i = 1, \ldots, 4$$

$$W_i \geq 0 \qquad i = 1, \ldots, 4$$

The unknowns in this LP model are $h_1, \ldots, h_4$ and $W_1, \ldots, W_4$. Additional constraints can be used to force the heads to be decreasing in the direction of flow, which are

$$h_4 \geq h_5; \qquad h_3 - h_4 \geq 0; \qquad h_2 - h_3 \geq 0; \qquad h_1 - h_2 \geq 0; \qquad \text{and } h_1 \leq h_0$$

UNCONFINED AQUIFERS. A one-dimensional unconfined aquifer is shown in Fig. 8.3.2 with constant head boundaries and fully penetrating wells that are equally spaced. The governing equation for steady-state flow can be derived from Eq. (8.2.10)

$$\frac{\partial}{\partial x}\left(T_x \frac{\partial h}{\partial x}\right) = W \tag{8.3.6}$$

where $T_x = Kh$ so that

$$\frac{d^2h^2}{dx^2} = \frac{2W}{K} \tag{8.3.7}$$

In order to simplify the notation, the substitution $w = h^2$ can be made to linearize the problem so that the finite difference expression can be written as

$$\frac{d^2w}{dx^2} = \frac{w_{i+1} - 2w_i + w_{i-1}}{(\Delta x)^2} = \frac{2W_i}{K} \tag{8.3.8}$$

assuming that the hydraulic conductivity K is constant throughout the aquifer. The governing Eq. (8.3.8) for each well is now linear. Aguado et al. (1974) formulated

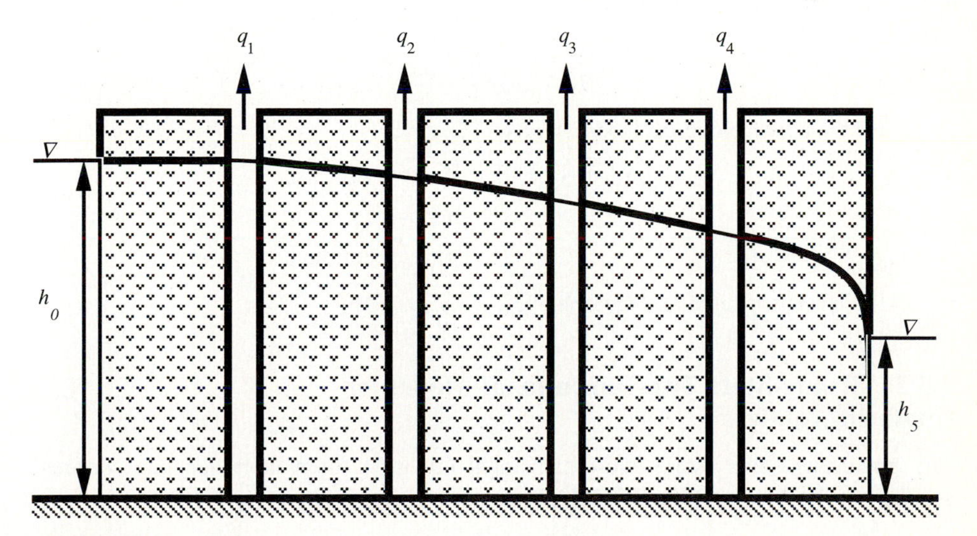

FIGURE 8.3.2
Unconfined one-dimensional aquifer.

the following linear programming model for determining the optimal steady-state pumpage from a one-dimensional unconfined aquifer

$$\text{Maximize } Z = \sum_{i \in I} w_i \tag{8.3.9}$$

subject to Eq. (8.3.8) for each well and

$$\sum_{i \in I} W_i \geq W_{\min} \tag{8.3.10}$$

$$w_i \geq 0 \qquad i \in I \tag{8.3.11a}$$

$$W_i \geq 0 \qquad i \in I \tag{8.3.11b}$$

The unknowns in the LP model are w_i and W_i. The heads h_i can be determined from $h_i = \sqrt{w_i}$ once the LP model has been solved for the unknowns.

Example 8.3.2. Develop an LP model for determining the optimal steady-state pumpage of the one-dimensional unconfined aquifer in Fig. 8.3.2 to maximize heads. The wells are equally spaced at a distance of Δx apart with constant head boundaries h_0 and h_5.

Solution. The objective function is simply

$$\text{Maximize } Z = w_1 + w_2 + w_3 + w_4$$

subject to the following finite difference equations

$$-2w_1 + w_2 - \frac{2(\Delta x)^2}{K} W_1 = -w_0$$

$$w_1 - 2w_2 + w_3 - \frac{2(\Delta x)^2}{K} W_2 = 0$$

$$w_2 - 2w_3 + w_4 - \frac{2(\Delta x)^2}{K} W_3 = 0$$

$$w_3 - 2w_4 - \frac{2(\Delta x)^2}{K} W_4 = -w_5$$

and the constraint on the production rate.

$$W_1 + W_2 + W_3 + W_4 \geq W_{\min}$$

$$w_i \geq 0 \qquad i = 1, \ldots, 4$$

$$W_i \geq 0 \qquad i = 1, \ldots, 4$$

The unknowns in this LP model are $w_1, \ldots, w_4$ and $W_1, \ldots, W_4$.

8.3.2 Steady-State Two-Dimensional Model for Confined Aquifers

The governing steady-state two-dimensional equation for a homogeneous confined aquifer can be derived from Eq. (8.2.9) as

$$\frac{\partial^2 h}{\partial x^2} + \frac{\partial^2 h}{\partial y^2} = \frac{W}{T} \tag{8.3.12}$$

for which $\partial h/\partial t = 0$ and $T_x = T_y = T$. Using central differences, Eq. (8.3.12) can be expressed in finite difference form as

$$\frac{h_{i+1,j} - 2h_{i,j} + h_{i-1,j}}{(\Delta x)^2} + \frac{h_{i,j+1} - 2h_{i,j} + h_{i,j-1}}{(\Delta y)^2} = \frac{W_{i,j}}{T} \tag{8.3.13}$$

which can be reduced for $\Delta x = \Delta y$ to

$$h_{i+1,j} - 4h_{i,j} + h_{i-1,j} + h_{i,j+1} + h_{i,j-1} = \frac{(\Delta x)^2 W_{i,j}}{T} \tag{8.3.14}$$

An LP model for the optimal steady-state pumpage from a two-dimensional confined aquifer can be formulated as

$$\text{Maximize } Z = \sum_{i,j \in I} h_{i,j} \tag{8.3.15}$$

subject to Eq. (8.3.14) for each cell and

$$\sum_{i,j \in I} W_{i,j} \geq W_{\min} \tag{8.3.16}$$

$$h_{i,j} \geq 0 \tag{8.3.17a}$$

$$W_{i,j} \geq 0 \tag{8.3.17b}$$

where I represents the set of pumping wells. The unknowns in the LP model are $h_{i,j}$ for all the cells and $W_{i,j}$ for the pumping cells.

Example 8.3.3. Develop an LP model for determining the optimal steady-state pumping from the two-dimensional confined aquifer shown in Fig. 8.3.3. This aquifer has constant (fixed) heads along the aquifer boundaries. This aquifer has three pumping cells (2,2), (3,2) and (3,3), as shown in the Fig. 8.3.3.

Solution. The objective function is simply

$$\text{Maximize } Z = h_{2,2} + h_{3,2} + h_{3,3}$$

subject to the finite difference Eq. (8.3.14) written for each cell in the aquifer. The finite difference equations for cell (1,1) is

$$h_{2,1} - 4h_{1,1} + h_{0,1} + h_{1,2} + h_{1,0} = 0$$

$W_{1,1} = 0$ because there is no pumping and heads $h_{0,1}$ and $h_{1,0}$ are known constant heads so this constraint can be written as

$$h_{2,1} - 4h_{1,1} + h_{1,2} = -h_{0,1} - h_{1,0}$$

with the known values on the RHS. The finite difference equations for pumping cell (2,2) is

$$h_{3,2} - 4h_{2,2} + h_{1,2} + h_{2,3} + h_{2,1} - \frac{(\Delta x)^2}{T} W_{2,2} = 0$$

The finite difference equation can be written for each of the remaining cells in the aquifer. The pumpage constraint Eq. (8.3.16) is simply

$$W_{2,2} + W_{3,2} + W_{3,3} \geq W_{\min}$$

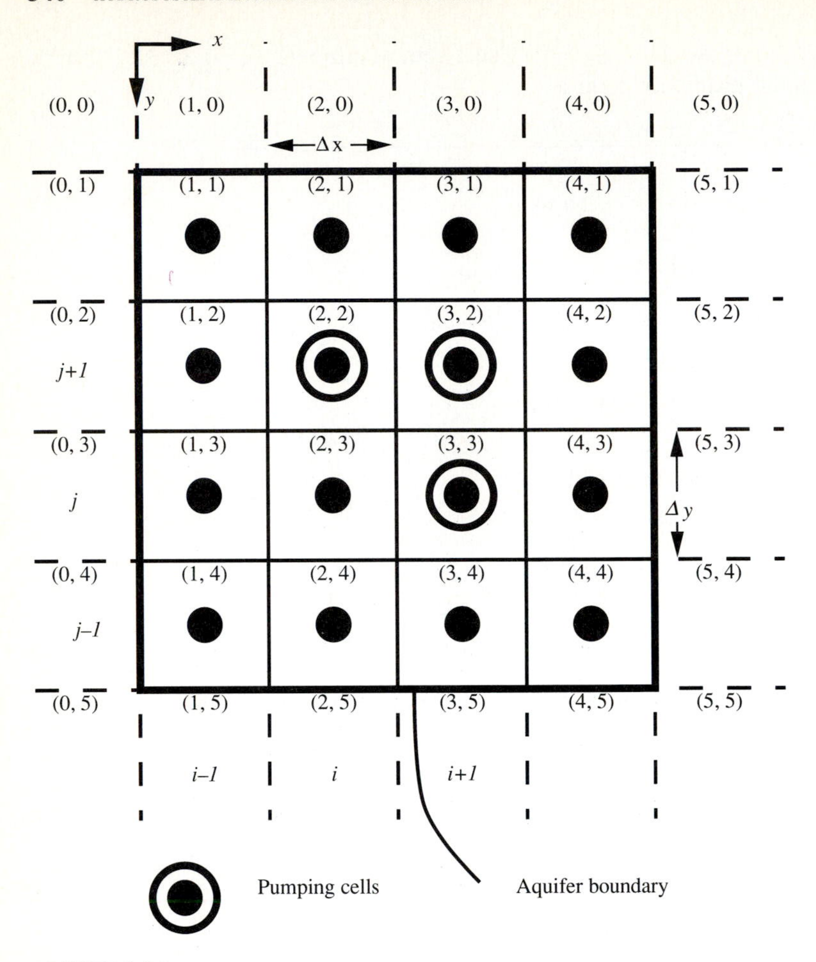

FIGURE 8.3.3
Confined two-dimensional aquifer (plan view).

8.3.3 Transient, One-Dimensional Problem for Confined Aquifers

The governing equation for transient, one-dimensional confined problems has the following form derived from Eq. (8.2.9)

$$T\frac{\partial^2 h}{\partial x^2} = S\frac{\partial h}{\partial t} + W \tag{8.3.18}$$

Using the Crank-Nicholson scheme (Remson et al., 1971) the finite difference approximation of the second-order derivative in Eq. (8.3.18) can be expressed as

$$\frac{\partial^2 h}{\partial x^2} = \frac{1}{2}\left[\frac{h_{i+1,t} - 2h_{i,t} + h_{i-1,t}}{(\Delta x)^2} + \frac{h_{i+1,t-1} - 2h_{i,t-1} + h_{i-1,t-1}}{(\Delta x)^2}\right] \tag{8.3.19}$$

The finite difference equation for Eq. (8.3.18) is determined using Eq. (8.3.19) and the finite difference approximation Eq. (8.2.12) for $\partial h/\partial t$,

$$T\left[\frac{h_{i+1,t}-2h_{i,t}+h_{i-1,t}}{2(\Delta x)^2}+\frac{h_{i+1,t-1}-2h_{i,t-1}+h_{i-1,t-1}}{2(\Delta x)^2}\right] - S\left[\frac{h_{i,t}-h_{i,t-1}}{\Delta t}\right]-\frac{W_{i,t}+W_{i,t-1}}{2}=0 \quad (8.3.20)$$

which can be simplified to

$$h_{i+1,t}-2h_{i,t}+h_{i-1,t}+h_{i+1,t-1}-2h_{i,t-1}+h_{i-1,t-1} - \frac{2(\Delta x)^2}{T(\Delta t)}S[h_{i,t}-h_{i,t-1}]-\frac{(\Delta x)^2}{T}(W_{i,t}+W_{i,t-1})=0 \quad (8.3.21)$$

Aguado et al. (1974) proposed an LP model for transient one-dimensional flow with an objective of maximizing the sum of heads in the last time step, τ, that is,

$$\text{Maximize } Z=\sum_i h_{i,\tau}$$

The constraints are Eqs. (8.3.21) written for each cell along with

$$\sum_i W_{i,t}\geq W_{\min,t} \qquad t=1,\ldots,\tau \quad (8.3.23)$$

Example 8.3.4. Develop an LP model for the transient, one-dimensional confined aquifer in Fig. 8.3.1 considering two time periods. The head boundaries at h_0 and h_5 are known for time $t=0$, $t=1$, and $t=2$.

Solution. The objective function is simply

$$\text{Maximize } Z=h_{1,2}+h_{2,2}+h_{3,2}+h_{4,2}$$

and the constraints for well $i=1$ and time period $t=1$ is

$$h_{2,1}-2h_{1,1}+h_{0,1}+h_{2,0}-2h_{1,0}+h_{0,0} - 2\frac{(\Delta x)^2}{T(\Delta t)}S(h_{1,1}-h_{1,0})-\frac{(\Delta x)^2}{T}(W_{1,1}+W_{1,0})=0$$

in which $h_{0,1}$ is the constant boundary and $W_{1,0}$, $h_{2,0}$, $h_{1,0}$, and $h_{0,0}$ represent the known initial conditions. This constraint can be rearranged with only the unknowns on the LHS

$$h_{2,1}-2h_{1,1}-\frac{2(\Delta x)^2}{T(\Delta t)}Sh_{1,1}-\frac{(\Delta x)^2}{T}W_{1,1}= -h_{0,1}-h_{2,0}+2h_{1,0}-h_{0,0}-\frac{2(\Delta x)^2}{T(\Delta t)}Sh_{1,0}+\frac{(\Delta x)^2}{T}W_{1,0}$$

The constraint for well $i=2$ and time period $t=2$ is

$$h_{3,2}-2h_{2,2}+h_{1,2}+h_{3,1}-2h_{2,1}+h_{1,1} - \frac{2(\Delta x)^2}{T(\Delta t)}S(h_{2,2}-h_{2,1})-\frac{(\Delta x)^2}{T}(W_{2,2}+W_{2,1})=0$$

The finite difference equations are written for each well at each time period.

The pumpage constraints are

$$W_{1,1} + W_{2,1} + W_{3,1} + W_{4,1} \geq W_{\min,1}$$

and

$$W_{1,2} + W_{2,2} + W_{3,2} + W_{4,2} \geq W_{\min,2}$$

8.3.4 Steady-State Two-Dimensional Problem for Unconfined Aquifers

This type of problem may be typical for dewatering of a construction or mining site (see Fig. 8.3.4). Aguado et al. (1974) presented an LP model for solving this problem. The governing Eq. (8.2.10) can be expressed for a homogeneous and isotropic aquifer as

$$\frac{\partial^2 h^2}{\partial x^2} + \frac{\partial^2 h^2}{\partial y^2} = \frac{2W}{K} \tag{8.3.24}$$

or

$$\frac{\partial^2 w}{\partial x^2} + \frac{\partial^2 w}{\partial y^2} = \frac{2W}{K} \tag{8.3.25}$$

where $w = h^2$. Using the central differencing scheme of Eq. (8.3.2), the finite difference representation of Eq. (8.3.25) is

$$\frac{w_{i+1,j} - 2w_{i,j} + w_{i-1,j}}{(\Delta x)^2} + \frac{w_{i,j+1} - 2w_{i,j} + w_{i,j-1}}{(\Delta y)^2} = \frac{2W_{i,j}}{K} \tag{8.3.26}$$

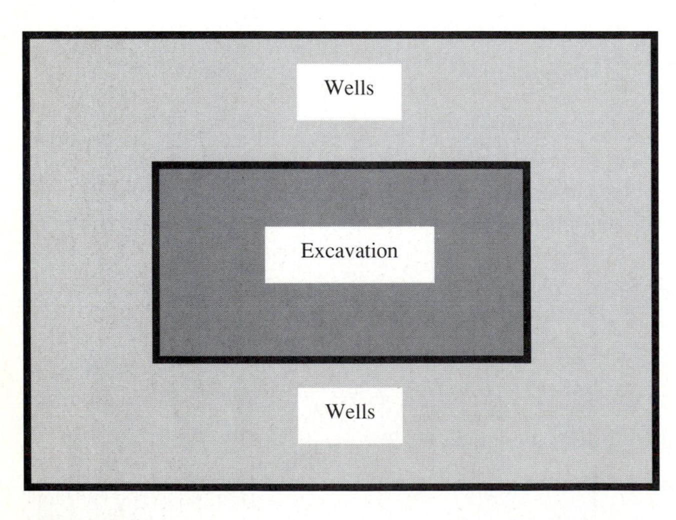

FIGURE 8.3.4
Excavation site for two-dimensional unconfined aquifer.

For the purposes of this modeling effort, both the heads and the pumpages are to be determined so that the unknowns are w and W, then Eq. (8.3.26) is

$$\left[\frac{1}{(\Delta x)^2}\right] w_{i+1,j} + \left[\frac{1}{(\Delta x)^2}\right] w_{i-1,j} - \left[\frac{2}{(\Delta x)^2} + \frac{2}{(\Delta y)^2}\right] w_{i,j} + \left[\frac{1}{(\Delta y)^2}\right] w_{i,j+1} + \left[\frac{1}{(\Delta y)^2}\right] w_{i,j-1} - \frac{2}{K} W_{i,j} = 0 \quad (8.3.27)$$

An LP model for the dewatering problem could be stated to minimize the amount of steady-state pumpage required to maintain head levels below specified levels. The LP model statement is

$$\text{Minimize } Z = \sum_{i,j \in I} W_{i,j} \quad (8.3.28)$$

subject to Eq. (8.3.27) written for each cell and the maximum allowable head requirement in the excavation site

$$w_{i,j} \leq w_r \quad (8.3.29)$$

and nonnegativity constraints

$$w_{i,j} \geq 0 \quad (8.3.30a)$$

$$W_{i,j} \geq 0 \qquad i, j \in I \quad (8.3.30b)$$

The number of wells and their locations are affected by the well spacing in the two-dimensional grid.

8.4 POLICY EVALUATION AND ALLOCATION MODELS: RESPONSE MATRIX APPROACH

The response matrix approach uses a groundwater simulation model external to the optimization model to develop unit responses. A **unit response** describes the influence of a pulse stimulus (such as a unit pumpage or injection over a time period) at a selected location (well) or cell upon the hydraulic heads for the other locations (wells) or cells throughout an aquifer. The **response matrix** consists of all the unit responses. Maddock (1972) derived a function relating the drawdown in a confined aquifer to pumpages through the use of a unit response function, β. This function, also referred to as an **algebraic technological function**, defines the drawdown $s_{k,n}$ in the kth cell at the end of the nth time period,

$$s_{k,n} = \sum_{p=1}^{n} \sum_{j=1}^{J} \beta_{k,j,n,p} q_{j,p} \quad (8.4.1)$$

where the unit response function $\beta_{k,j,n,p}$ is the change in drawdown (unit drawdown) in the kth cell at the end of the nth time period due to a unit pumpage from the jth cell (j may equal k) during the pth time period; $q_{j,p}$ is the quantity pumped from j during the pth time period; and J is the number of cells.

Equation (8.4.1) is based on the assumption that (1) flow is horizontal only, (2) wells are fully penetrating, (3) the transmissivity and storage coefficient may be nonhomogeneous, and (4) the pumpage is constant over a time period but can vary for different time periods. The unit response function, β, can be calculated analytically or numerically. To numerically determine β, a separate groundwater simulation is required for each pumped well or cell. For each simulation a unit pumpage over a time period is assumed and the simulation determines the response or drawdown at the other well locations or cells, for this unit pulse. Maddock (1974) extended the idea of the algebraic technological function to unconfined aquifers.

A groundwater management model by Heidari (1982) was formulated based upon the response function approach as follows:

$$\text{Maximize } Z = \sum_{j=1}^{J} \sum_{n=1}^{N} q_{j,n} \tag{8.4.2}$$

subject to:

a. satisfying the governing equation of flow through the response matrix

$$s_{k,n} = \sum_{p=1}^{n} \sum_{j=1}^{J} \beta_{k,j,n,p} q_{j,p} \qquad k = 1, \ldots, J \qquad n = 1, \ldots, N \tag{8.4.3}$$

b. pumpage cannot exceed $\bar{q}_{j,n}$, which is the smaller of the appropriated right or its capacity.

$$0 \leq q_{j,n} \leq \overline{q}_{j,n} \qquad j = 1, \ldots, J \qquad n = 1, \ldots, N \tag{8.4.4}$$

c. drawdown at each well or cell cannot exceed an upper limit, $\overline{s}_{k,n}$

$$0 \leq s_{k,n} \leq \overline{s}_{k,n} \qquad k = 1, \ldots, J \qquad n = 1, \ldots, N \tag{8.4.5}$$

d. demand, Q_n, for each time period n should be satisfied

$$\sum_{j=1}^{J} q_{j,n} \geq Q_n \qquad n = 1, \ldots, N \tag{8.4.6}$$

The upper limit on drawdown $\overline{s}_{k,n}$ can be defined as a fraction (γ) of the saturated thickness b_k of the aquifer at the kth well (or cell), so that

$$\overline{s}_{k,n} = \gamma \cdot b_k \tag{8.4.7}$$

Actually, constraint Eqs. (8.4.3) and (8.4.5) can be combined. Heidari (1982) applied the above LP model to the Pawnee Valley in Kansas to determine optimal pumpage for policies with and without net appropriations as constraints.

Example 8.4.1. Develop a policy evaluation model using LP for a simple confined aquifer defined by four cells with a pumping well in each cell. A total of four time periods are to be considered. The unit response function $\beta_{k,j,n,p}$ is known and the objective is to maximize pumpage.

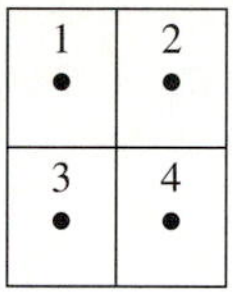

1 •	2 •
3 •	4 •

Solution

$$\text{Maximize } Z = \sum_{j=1}^{4} \sum_{n=1}^{4} q_{j,n}$$
$$= q_{1,1} + q_{1,2} + q_{1,3} + q_{1,4} + \cdots + q_{4,1} + q_{4,2} + q_{4,3} + q_{4,4}$$

subject to:

a. satisfying governing equations of flow through the response matrix

WELL 1 ($k = 1$)

($n = 1$)

$$s_{1,1} = \sum_{p=1}^{n=1} \sum_{j=1}^{4} \beta_{k,j,n,p}\, q_{j,p}$$
$$= \beta_{1,1,1,1}\, q_{1,1} + \beta_{1,2,1,1}\, q_{2,1} + \beta_{1,3,1,1}\, q_{3,1} + \beta_{1,4,1,1}\, q_{4,1}$$

($n = 2$)

$$s_{1,2} = \sum_{p=1}^{n=2} \sum_{j=1}^{4} \beta_{k,j,n,p}\, q_{j,p}$$
$$= \beta_{1,1,2,1}\, q_{1,1} + \beta_{1,2,2,1}\, q_{2,1} + \beta_{1,3,2,1}\, q_{3,1} + \beta_{1,4,2,1}\, q_{4,1}$$
$$+ \beta_{1,1,2,2}\, q_{1,2} + \beta_{1,2,2,2}\, q_{2,2} + \beta_{1,3,2,2}\, q_{3,2} + \beta_{1,4,2,2}\, q_{4,2}$$

($n = 3$)

$$s_{1,3} = \sum_{p=1}^{n=3} \sum_{j=1}^{4} \beta_{k,j,n,p}\, q_{j,p}$$
$$= \beta_{1,1,3,1}\, q_{1,1} + \beta_{1,2,3,1}\, q_{2,1} + \beta_{1,3,3,1}\, q_{3,1} + \beta_{1,4,3,1}\, q_{4,1}$$
$$+ \beta_{1,1,3,2}\, q_{1,2} + \beta_{1,2,3,2}\, q_{2,2} + \beta_{1,3,3,2}\, q_{3,2} + \beta_{1,4,3,2}\, q_{4,2}$$
$$+ \beta_{1,1,3,3}\, q_{1,3} + \beta_{1,2,3,3}\, q_{2,3} + \beta_{1,3,3,3}\, q_{3,3} + \beta_{1,4,3,3}\, q_{4,3}$$

($n = 4$)

$$s_{1,4} = \sum_{p=1}^{n=4} \sum_{j=1}^{4} \beta_{k,j,n,p}\, q_{j,p}$$
$$= \beta_{1,1,4,1}\, q_{1,1} + \beta_{1,2,4,1}\, q_{2,1} + \beta_{1,3,4,1}\, q_{3,1} + \beta_{1,4,4,1}\, q_{4,1}$$
$$+ \beta_{1,1,4,1}\, q_{1,2} + \beta_{1,2,4,2}\, q_{2,2} + \beta_{1,3,4,2}\, q_{3,2} + \beta_{1,4,4,2}\, q_{4,2}$$
$$+ \beta_{1,1,4,3}\, q_{1,3} + \beta_{1,2,4,3}\, q_{2,3} + \beta_{1,3,4,3}\, q_{3,3} + \beta_{1,4,4,3}\, q_{4,3}$$
$$+ \beta_{1,1,4,4}\, q_{1,4} + \beta_{1,2,4,4}\, q_{2,4} + \beta_{1,3,4,4}\, q_{3,4} + \beta_{1,4,4,4}\, q_{4,4}$$

Well 2 ($k = 2$)

($n = 1$)

$$s_{2,1} = \sum_{p=1}^{n=1} \sum_{j=1}^{4} \beta_{k,j,n,p}\, q_{j,p}$$
$$= \beta_{2,1,1,1}\, q_{1,1} + \beta_{2,2,1,1}\, q_{2,1} + \beta_{2,3,1,1}\, q_{3,1} + \beta_{2,4,1,1}\, q_{4,1}$$

($n = 2$)

$$s_{2,2} = \sum_{p=1}^{n=2} \sum_{j=1}^{4} \beta_{k,j,n,p}\, q_{j,p}$$
$$= \beta_{2,1,2,1}\, q_{1,1} + \beta_{2,2,2,1}\, q_{2,1} + \beta_{2,3,2,1}\, q_{3,1} + \beta_{2,4,2,1}\, q_{4,1}$$
$$+ \beta_{2,1,2,2}\, q_{1,2} + \beta_{2,2,2,2}\, q_{2,2} + \beta_{2,3,2,2}\, q_{3,2} + \beta_{2,4,2,2}\, q_{4,2}$$

The remaining governing constraints are written in a similar manner.

b. pumpage cannot be greater than the appropriated right

$$\begin{gathered} 0 \le q_{1,1} \le \overline{q}_{1,1} \\ 0 \le q_{1,2} \le \overline{q}_{1,2} \\ \vdots \\ 0 \le q_{2,1} \le \overline{q}_{2,1} \\ \vdots \\ 0 \le q_{3,1} \le \overline{q}_{3,1} \\ \vdots \\ 0 \le q_{4,1} \le \overline{q}_{4,1} \end{gathered}$$

c. drawdown at each well cannot exceed an upper limit

$$\begin{gathered} 0 \le s_{1,1} \le \overline{s}_{1,1} \\ 0 \le s_{1,2} \le \overline{s}_{1,2} \\ \vdots \\ 0 \le s_{2,1} \le \overline{s}_{2,1} \\ \vdots \\ 0 \le s_{3,1} \le \overline{s}_{3,1} \\ \vdots \\ 0 \le s_{4,1} \le \overline{s}_{4,1} \end{gathered}$$

d. demand requirement

$$\begin{gathered} q_{1,1} + q_{2,1} + q_{3,1} + q_{4,1} \ge Q_1 \\ q_{1,2} + q_{2,2} + q_{3,2} + q_{4,2} \ge Q_2 \\ q_{1,3} + q_{2,3} + q_{3,3} + q_{4,3} \ge Q_3 \\ q_{1,4} + q_{2,4} + q_{3,4} + q_{4,4} \ge Q_4 \end{gathered}$$

8.5 GROUNDWATER MANAGEMENT MODEL: OPTIMAL CONTROL APPROACH

The general groundwater management problem (GGMP) can be expressed mathematically as follows:

GGMP.

$$\text{Optimize } Z = f(\mathbf{h}, \mathbf{q}) \tag{8.5.1}$$

subject to

a. the general groundwater flow constraints Eq. (8.2.10) or (8.2.14);

$$\mathbf{g}(\mathbf{h}, \mathbf{q}) = \mathbf{0} \tag{8.5.2}$$

b. the simple bounds;

$$\underline{\mathbf{q}} \leq \mathbf{q} \leq \overline{\mathbf{q}} \tag{8.5.3}$$

$$\underline{\mathbf{h}} \leq \mathbf{h} \leq \overline{\mathbf{h}} \tag{8.5.4}$$

c. other constraint sets, such as demand constraints;

$$\mathbf{w}(\mathbf{h}, \mathbf{q}) \leq \mathbf{0} \tag{8.5.5}$$

where ($^-$) represents an upper bound and (_) represents a lower bound.

Both head **h** and pumpage or recharge **q** are vectors of decision variables which have maximum dimensions equal to the product of the number of active nodes within the aquifer boundary and time steps. Fixed pumpages or recharges are considered to be constants. By convention, available pumpages have a positive value and the elements of **q** have a negative value where there is available recharge. Usually the number of variable pumpages and/or recharges (hereafter the terms pumpages that refer to **q** will imply both pumpages and/or recharges) is small and results in a much smaller dimension of **q** than **h**.

The objective function Eq. (8.5.1), which may be either maximization (e.g., sum of heads) or minimization (e.g., minimize pumpage), can be a linear or nonlinear function. Also, it may be nonseparable or contain only terms of pumpages or heads, but has to be differentiable. Constraint Eq. (8.5.2) represents a system of equations governing ground-water flow which are finite difference or simulator equations when **q** is unknown. The upper ($\overline{\mathbf{q}}$) and lower ($\underline{\mathbf{q}}$) bounds on pumpages physically may or may not exist. Unlike pumpage, the lower bound on heads ($\underline{\mathbf{h}}$) can be viewed as the bottom elevation of the aquifer while the upper head bound ($\overline{\mathbf{h}}$) can be regarded as ground surface elevations for the unconfined cells. In addition to constraint Eqs. (8.5.2)–(8.5.4), constraint Eq. (8.5.5) may be included to impose restrictions such as water demands, operating rules, budgetary limitations, etc.

The above general optimization model, Eqs. (8.5.1)–(8.5.5), was solved by Wanakule, Mays and Lasdon (1986) using an optimal control framework. A generalized reduced gradient method (Section 4.6) makes up the overall optimization framework along with a simulation model to perform function evaluations (solution of the general groundwater flow constraints) at each iteration of the optimization. The solution procedure reduces the problem by expressing dependent or state variables (heads, **h**) in terms of the independent variables called control variables (pumpage on recharge, **q**) by implicitly using a simulator to solve the flow Eq. (8.5.2) for heads given the pumpages and recharges. Because the simulator solves only the general flow constraints Eq. (8.5.2), the bound constraint on heads, Eq. (8.5.4), may not be satisfied. The head bound constraints are satisfied by incorporating them into the objective function as an augmented Lagrangian penalty function (Section 4.7). The reduced problem that is solved by the optimizer is to optimize (maximize or minimize) the augmented Lagrangian objective subject to constraint Eqs. (8.5.3) and (8.5.5). Wanakule, Mays, and Lasdon (1986) used GRG2, the generalized reduced gradient model by Lasdon, et al. (1978) and GWSIM, the groundwater simulation model developed by the Texas Water Development Board (1974). GWSIM is a finite difference simulation model based upon the IADI method.

REFERENCES

Aguado, E., and I. Remson: "Ground-water Management with Fixed Charges," *J. Water Resour. Plann. Manage. Div., Am. Soc. Civ. Eng.*, **106**, 375–382, 1980.

Aguado, E., I. Remson, M. F. Pikul, and W. A. Thomas: "Optimum Pumping for Aquifer Dewatering," *J. Hydraul. Div., Am. Soc. Civ. Eng.*, **100**, 860–877, 1974.

Aguado, E., N. Sitar, and I. Remson: "Sensitivity Analysis in Aquifer Studies," *Water Resour. Res.*, **13**, 733–737, 1977.

Bear, J.: *Hydraulics of Groundwater*, McGraw-Hill, Inc., New York, 1979.

Bisschop, J. W., J. H. Candler, and G. T. O'Mara: "The Indian Basin model: A Special Application of Two-level Linear Programming," *Math Program*, **20**, 30–38, 1982.

Bouwer, H.: *Groundwater Hydrology*, McGraw-Hill, Inc., New York, 1978.

Bredehoeft, J. D. and R. A. Young: "Conjunctive Use of Groundwater and Surface Water for Irrigated Agriculture: Risk Aversion," *Water Resource Res.*, AGU, **19**(5):1111–1121, 1983.

Daubert, J. T., and R. A. Young: "Groundwater Development in Western River Basins: Large Economic Gains with Unseen Costs," *Ground Water*, **20**(1):80–85, 1982.

de Marsily, G.: *Quantitative Hydrogeology*, Academic, Orlando, Florida, 1986.

Fletcher, R.: *Practical Methods of Optimization, Vol. 1*, Wiley, New York, 1981.

Freeze, R. A. and J. A. Cherry: *Groundwater*, Prentice-Hall, Englewood Cliffs, N.J., 1979.

Gehm, H. W. and J. I. Bregman: *Handbook of Water Resources and Pollution Control*, Van Nostrand Reinhold Company, New York, 1976.

Gorelick, S. M.: "A Review of Distributed Parameter Groundwater Management Modeling Methods," *Water Resour. Res.*, **19**, 305–319, 1983.

Haimes, Y. Y. and Y. C. Dreizen: "Management of Groundwater and Surface Water Via Decomposition." *Water Resources Res.* AGU, **13**(1):69–77, 1977.

Heidari, M.: "Application of Linear Systems Theory and Linear Programming to Groundwater Management in Kansas," *Water Resour. Bulletin*, **18**, 1003–1012, 1982.

Illangasekare, T. and H. J. Morel-Seytoux: "Stream-aquifer Influence Coefficients as Tools for Simulation and Management," *Water Resour. Res.*, **18**, 168–176, 1982.

Kashef, A.-A. I.: *Groundwater Engineering*, McGraw-Hill, Inc., New York, 1986.

Klemt, W. B., T. R. Knowles, G. R. Elder, and T. W. Sieh: "Groundwater Resources and Model Applications for the Edwards (Balcones Fault Zone) Aquifer in the San Antonio Region, Texas," Report 239, Texas Department of Water Resources, Austin, Oct., 1979.

Lasdon, L. S., A. D. Warren, A. Jain, and M. Ratner: "Design and Testing of a Generalized Reduced Gradient Code for Nonlinear Programming," *Assoc. Comput. Mach. Trans. Math. Software*, **4**, 34–50, 1978.

Luenberger, D. G.: *Linear and Nonlinear Programming*, Addison-Wesley, Reading, Mass., 1984.

Maddock, T., III: "Algebraic Technological Function for a Simulation Model," *Water Resour. Res.*, **8**, 129–134, 1972.

Maddock, T., III: "Nonlinear Technological Functions for Aquifers Whose Transmissivities Vary with Drawdown," *Water Resour. Res.*, **10**, 877– 881, 1974.

Maddock, T., III, and Y. Y. Haimes: "A Tax System for Groundwater Management," *Water Resour. Res.*, **11**, 7–14, 1975.

Mantell, J. and L. S. Lasdon: "A GRG Algorithm for Econometric Control Problems," *Ann. Econ. Soc. Manage.*, **6**, 581–597, 1978.

Morel-Seytoux, H. J., and C. J. Daly: "A Discrete Kernel Generator for Stream-Aquifer Studies," *Water Resour. Res.*, **11**, 253–260, 1975.

Morel-Seytoux, H. J., G. Peters, R. Young, and T. Illangasekare: "Groundwater Modeling for Management," Paper presented at the International Symposium on Water Resource Systems, Water Resour. Dev. and Training Cent., Univ. of Roorkee, Roorkee, India, 1980.

Norman, A. L., L. S. Lasdon, and J. K. Hsin: "A Comparison of Methods for Solving and Optimizing a Large Nonlinear Econometric Model, Discussion Paper," Cent. for Econ. Res., Univ. of Tex., Austin, 1982.

Peaceman, D. W. and H. H. Rachford, Jr.: "The Numerical Solution of Parabolic and Elliptic Differential Equations," *J. Soc. Ind. and App. Math.*, vol. 3, 28–41, 1955.

Prickett, T. A., and C. G. Lonnquist: "Selected Digital Computer Techniques for Groundwater Resource Evaluation," *Bull. Ill. State Water Surv.*, 55, 1971.

Remson, I., G. M. Hornberger, and F. J. Molz: *Numerical Methods in Subsurface Hydrology*, Wiley-Interscience, New York, 1971.

Remson, I., and S. M. Gorelick: "Management Models Incorporating Groundwater Variables," in *Operation Research in Agriculture and Water Resources*, D. Yaron and C. S. Tapiero, eds., North-Holland, Amsterdam, 1980.

Rockafellar, R. T.: "A Dual Approach to Solving Nonlinear Programming Problems by Unconstrained Optimization," *Math. Programming*, **5**, 354–373, 1973.

Texas Water Development Board: "GWSIM—Groundwater Simulation Program," *Program Document and User's Manual*, UM S7405, Austin, Texas, 1974.

Todd, D. K.: *Ground Water Hydrology*, 2d ed., Wiley, New York, 1980.

Trescott, P. C., G. F. Pinder, and S. P. Larson: "Finite-difference Model for Aquifer Simulation in Two-Dimensions with Results of Numerical Experiments," in *U.S. Geological Survey Techniques of Water Resources Investigations*, Book 7, C1, U.S. Geological Survey, Reston, Va., 1976.

van de Heide, P., Y. Bachmat, J. Bredehoeft, B. Andrews, D. Holtz, and S. Sebastian: *Groundwater Management: The Use of Numerical Models*, Water Resources Monograph Series, vol. 5, 2d. ed., American Geophysical Union, Washington, D.C., 1985.

Walton, W. C.: *Groundwater Resource Evaluation*, McGraw-Hill, Inc., New York, 1970.

Wanakule, N., L. W. Mays and L. S. Lasdon: "Optimal Management of Large-scale Aquifers: Methodology and Application," *Water Resources Research*, vol. 22, no. 4, 447–465, April 1986.

Wang, H. F., and M. P. Anderson: *Introduction to Groundwater Modeling: Finite Difference and Finite Element Models*, W. H. Freeman, San Francisco, 1982.

Willis, R.: "A Unified Approach to Regional Groundwater Management," in *Groundwater Hydraulics, Water Resour. Monogr. 9*, by J. S. Rosenshein and G. D. Bennett, eds., AGU, Washington, D.C., 1984.

Willis, R., and P. Liu: "Optimization Model for Ground-Water Planning," *J. Water Resour. Plann. Manage. Div., Am. Soc. Civ. Eng.*, **110**, 333–347, 1984.

Willis, R., and B. A. Newman: "Management Model for Groundwater Development," *J. Water Resour. Plann. Manage. Div., Am,. Soc. Civ. Eng.*, **13**, 159–171, 1977.

Willis, R., and W. W.-G. Yeh: *Groundwater Systems Planning and Management*, Prentice-Hall, Englewood Cliffs, N.J., 1987.

Young, R. A. and Bredehoeft, J. D.: "Digital Computer Simulation for Solving Management Problems of Conjunctive Groundwater and Surface Water Systems," *Wat. Resour. Res.*, **8**(3):533–556, 1972.

PROBLEMS

8.1.1 Using the first-order analysis of uncertainty, determine an expression for the coefficient of variation of the well drawdown using the Cooper-Jacob approximation for unsteady radial flow in a confined aquifer. Consider T and S to be random random variables.

8.1.2 Using the expression developed in Problem 8.1.1 for the coefficient of variation of well drawdown Ω_s, determine Ω_s for a discharge of Q = 1000 m^3/day at a time of 1 day since pumping started.

Parameter	Mean	Coefficient of variation
T	1000 m^3/day	0.05
S	0.0001	0.001
r	200 m	0.0

8.1.3 A pumping test is a frequently used field technique to determine the properties of aquifers. It is done by pumping a well at a constant rate and observing the drawdown of the piezometric surface or water table in an observation well at some distance from the pump well. Then an appropriate analytical solution for groundwater flow is used to determine the transmissivity and storage coefficient. Therefore, aquifer properties so determined will be valid in the vicinity of the pumping well when flow satisfies the conditions used in developing the analytical solution. Utilize the Cooper-Jacob equation to determine the optimal transmissivity (T) and storage coefficient (S) that minimizes the sum of square of errors between the observed and calculated drawdowns at two observation wells over time.

Time (days)	Drawdown (m) Observation wells	
	No. 1 (r = 100m)	No. 2 (r = 200m)
0.001	0.087	0.015
0.005	0.200	0.100
0.01	0.252	0.147
0.05	0.370	0.270
0.1	0.435	0.320
0.5	0.555	0.450
1	0.610	0.500
5	0.745	0.630
10	0.805	0.690

8.1.4 Resolve Problem 8.1.3 to minimize the sum of absolute error between the observed and calculated drawdowns at two observation wells.

8.3.1 Set up and solve the linear programming model to determine the optimal pumpage for the one-dimensional confined aquifer in Fig. 8.3.1 where Δx = 80 m, T = 1000 m^2/day, $W_{\min}$ = 500 m/day, h_0 = 40 m and h_5 = 35 m. Solve this model for $h_1, \ldots, h_4$ and $W_1, \ldots, W_4$.

8.3.2 Set up the LP model to determine the optimal pumpage for the one-dimensional confined aquifer in Fig. 8.3.1 where Δx = 10 ft, T = 10,000 ft^2/day, h_0 = 100 ft, h_4 = 110 ft and $W_{\min}$ = 100 ft/day.

8.3.3 Solve the LP model set up in Problem 8.3.2.

8.3.4 Set up and solve the LP model to determine the optimal pumpage for the one-dimensional unconfined aquifer in Fig. 8.3.2 where Δx = 80 m, K = 150 m/day, $W_{\min}$ = 500 m/day, h_0 = 40 m and h_5 = 35 m.

8.3.5 Set up the complete LP model for the two-dimensional confined aquifer described in Example 8.3.3.

8.3.6 Set up and solve the LP model for the two-dimensional confined aquifer shown in Fig. 8.3.3 to determine the optimal pumpage from cells (2,2), (3,2) and (3,3); Δx = 100 ft, T = 10,000 ft^2/day, $W_{\min}$ = 2 ft/day. The boundary condition is a constant head of 20 ft.

8.3.7 Set up the complete LP model for the transient, one-dimensional confined aquifer problem in Example 8.3.4.

8.3.8 Solve the model for the transient, one-dimensional confined aquifer developed in Problem 8.3.7 for Δx = 100 ft, Δt = 1 day, T = 10,000 ft^2/day, S = 0.001, h_0 = 120 ft, h_5 = 105 ft and $W_{\min}$ = 200 ft/day. At t = 0, h_1 = 118.5 ft, h_2 = 116 ft, h_3 = 113 ft, h_4 = 103.5 ft, and $W_{1,0} = W_{2,0} = W_{3,0} = W_{4,0} = 0$.

8.4.1 Develop a groundwater allocation model for an irrigation district to determine the optimal cropping pattern and groundwater allocation that maximizes the net return from agricultural production. The irrigation district has the responsibility for determining the optimal cropping pattern (acreage devoted to each crop) with the districts authority. Water is pumped from an extensive confined aquifer system for which steady state model is considered two-dimensional response of the aquifer. The profit per acre for a crop is (the unit revenue of a crop times the yield per acre) minus (the unit cost of groundwater times the amount of groundwater applied per acre). The yield would be a nonlinear function defined by a production function (Chapter 2). Define the objective function, constraints, parameters, and decision variables. What would be the solution methodology?

8.4.2 Develop a conjunctive groundwater and surface water planning model that can optimally distribute, over time, the water resources for various water demands. The decisions to be made are the groundwater and surface water allocation in each planning period. The objective function is to maximize the net discounted benefits from operating the system over the planning horizon, including costs for capital, operation and maintenance. Constraints would include: (a) conservation or balance equations for the surface water system; (b) capacity limitations of the surface water system; (c) groundwater system flow equations to define aquifer response; (d) bound constraints on groundwater levels. Others may be required such as for artificial recharge, pumping schedules. Use GW_{ij}^t and SW_{kj}^t to represent, respectively, groundwater allocated from groundwater source i to demand j in time period t and surface water allocated from source k to demand j in time period t. Define the objective function, constraints, parameters, and decision variables. What would be the solution methodology?

8.4.3 In the planning stage of groundwater management, applications of analytical solutions of groundwater flow such as those given in Table 8.1.1 can be made. Those analytical equations can be used to calculate the unit response matrix in groundwater management models. Now, consider an undeveloped, homogenous, confined groundwater aquifer (see Fig. 8.P.1) in which there are three potential pump wells and five control points at which the drawdowns are to be observed. For steady-state management, the Thiem equation

$$s_{kl} = \frac{\ln(r_{01}/r_{kl})}{2\pi T} q_l$$

can be used in which s_{kl} = drawdown at control point k resulting from a pumpage of q_l at the well location l; r_{ol} = radius of influence of potential production well; r_{kl} = distance between control point k and well location l; and T = aquifer transmissivity. Assume that the radius of influence of all pump wells is 700 feet (213 meters) and the aquifer transmissivity is 5000 gallon/day/ft (0.0007187 m^2/sec). Based on the information about the maximum allowable drawdown and pump capacities given in Fig. 8.P.1, formulate an LP model to determine the optimal pumpage for each well.

8.4.4 Refer to the hypothetical groundwater system as shown in Fig. 8.P.1. Consider now that the management is to be made for three periods of 50 days each. This is a management problem for transient groundwater systems which can be approximately described by the Theis equation or Cooper-Jacob equation as given in Table 8.1.1.

(*a*) Using the Cooper-Jacob equation define the unit response function.

(*b*) Assume that the storage coefficient of the aquifer is 0.002 and transmissivity is 5000 ft^2/day (465 m^2/day). Further, the maximum allowable drawdowns (in ft) at each control point are given as

	Control point				
Period	**1**	**2**	**3**	**4**	**5**
1	5	5	8	5	5
2	8	8	10	8	8
3	10	10	15	10	10

Develop the LP model and solve it for maximizing total pumpage for all three 50-day periods.

8.4.5 Assume that the transmissivity of the hypothetical aquifer (Fig. 8.P.1) is a lognormal random variable with mean 5000 gallon/day/ft and coefficient of variation of 0.3. Convert the LP model developed in Problem 8.4.3 into a chance-constrained model such that the specified drawdown at the five control points will not be exceeded with 95 percent reliability. Also, solve the model.

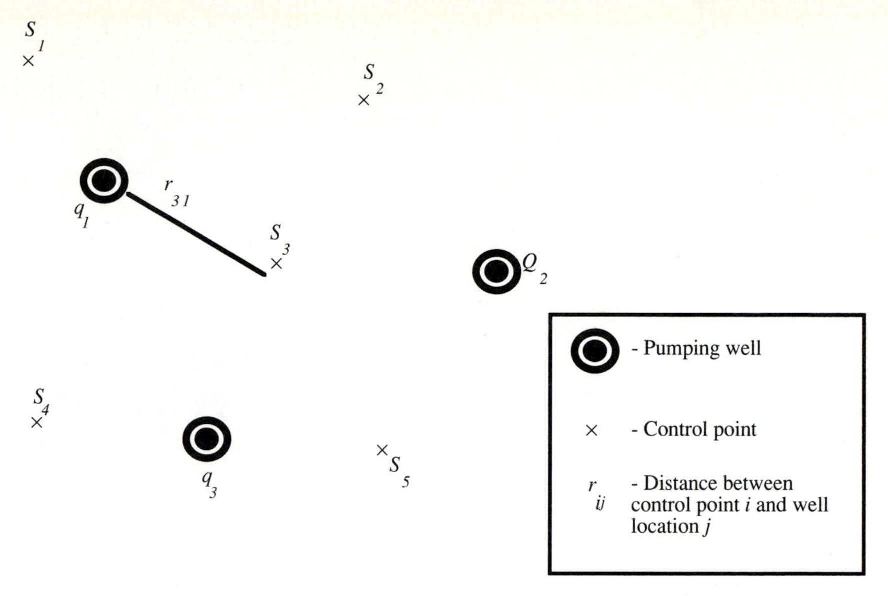

Distance (ft) between potential pump wells and control points

Pumping Well	Control Points					Pump Capacity (gpd)
	1	2	3	4	5	
1	158 ft	381	158	255	430	200,000
2	515	255	292	474	158	200,000
3	447	447	200	200	200	200,000
Max. Allowable Drawdown	7 ft	7	15	7	7	

FIGURE 8.P.1
Location of pumping wells and control points for a hypothetical groundwater basin (Problems 8.4.3–8.4.5).

CHAPTER

9

WATER DISTRIBUTION SYSTEMS

9.1 DESCRIPTION AND PURPOSE OF WATER DISTRIBUTION SYSTEMS

Water utilities are implemented to construct, operate, and maintain water supply systems. The basic function of these water utilities is to obtain water from a source, treat the water to an acceptable quality, and deliver the desired quantity of water to the appropriate place at the required time. The analysis of a water utility is usually to evaluate one or more of the major functional components of the utility: source development; raw-water transmission; raw-water storage, treatment, finished-water storage; and finished-water distribution as well as associated subcomponents. Because of their interaction, finished-water storage is usually evaluated in conjunction with finished-water distribution and raw-water storage is usually evaluated in conjunction with the source. Figure 9.1.1 illustrates the functional components of a water utility.

The purpose of a water-distribution network is to supply the system's users the amount of water demanded and to supply this water with adequate pressure under various loading conditions. A **loading condition** is defined as a pattern of nodal demands. A system may be subject to a number of different loading conditions: fire demands at different nodes; peak daily demands; a series of patterns varying throughout a day; or a critical load when one or more pipes are broken. In order to insure that a design is adequate, a number of loading conditions including critical

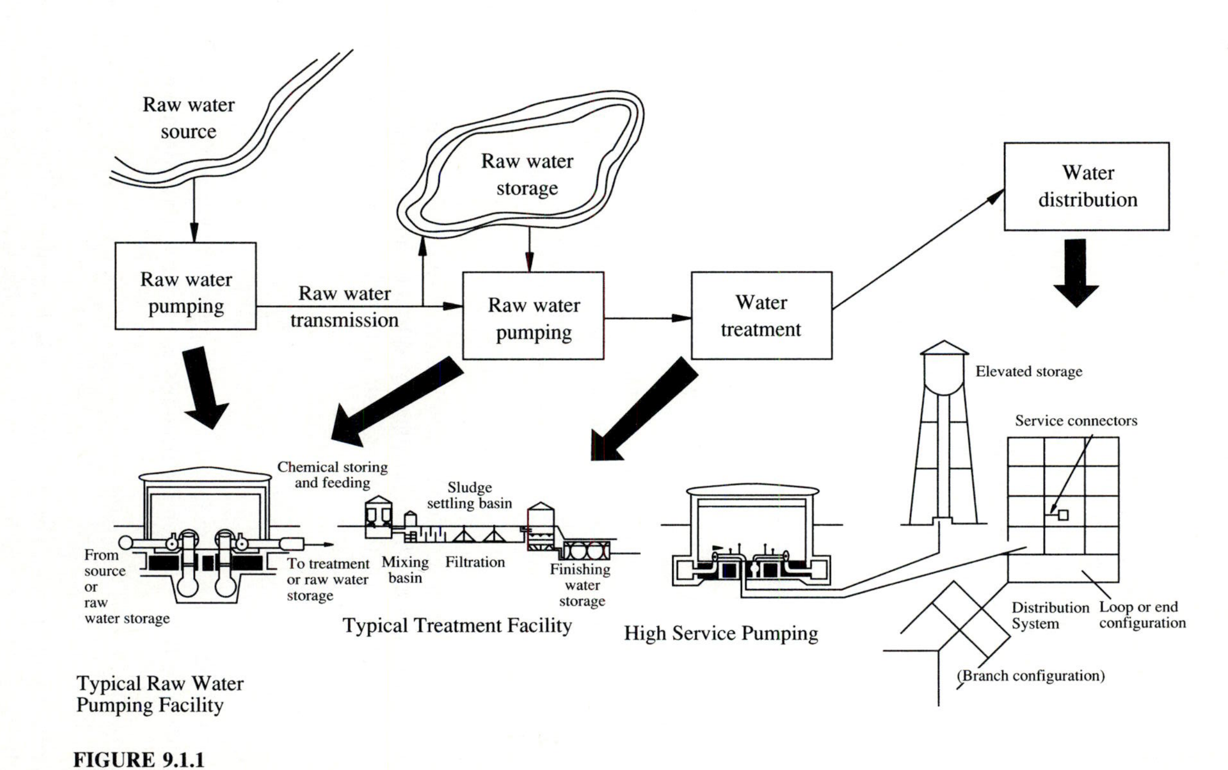

FIGURE 9.1.1
Functional components of a water utility (Cullinane, 1989).

conditions must be considered. The ability to operate under a variety of load patterns is required to have a reliable network.

Water distribution systems are composed of three major components: pumping stations; distribution storage; and distribution piping. These components may be further divided into subcomponents which in turn can be divided into sub-subcomponents. For example, the pumping station component consists of structural, electrical, piping, and pumping unit subcomponents. The pumping unit can be divided further into sub-subcomponents: pump, driver, controls, power transmission, and piping and valves. The exact definition of components, subcomponents and sub-subcomponents depends on the level of detail of the required analysis and to a somewhat greater extent the level of detail of available data. In fact, the concept of component-subcomponent-sub-subcomponent merely defines a hierarchy of building blocks used to construct the water distribution system. The purpose of this chapter is to present the mathematics and methodologies required to understand both simulation and optimization models for the design and analysis of water distribution networks. In addition, the methodologies for performing reliability analysis are explained.

9.2 WATER DISTRIBUTION SYSTEM COMPONENTS

The various components of water distribution systems and their purposes are described in this section. The principal elements in the system are pipe sections or links which are of constant diameter and may contain fittings such as bends and valves. Pipes are manufactured using a variety of materials, for example, cast iron, welded steel, concrete, or wood staves. Typically for water distribution systems, the relationship used to describe flow or velocity in pipes is the Hazen-Williams equation:

$$V = 1.318 C_{HW} R^{0.63} S_f^{0.54} \tag{9.2.1}$$

where V is the average flow velocity in ft/s, C_{HW} is the Hazen-Williams roughness coefficient as listed in Table 9.2.1 for pipes of different materials and ages, R is the hydraulic radius in ft and S_f is the friction slope ft/ft. This equation can be expressed in terms of headloss (ft) as

$$h_L = 3.02 L D^{-1.167} \left(\frac{V}{C_{HW}} \right)^{1.852} \tag{9.2.2}$$

where L is the length of pipe in ft and D is the pipe diameter in ft.

Another pipe flow equation for headloss is the Darcy-Weisbach equation

$$h_L = f \frac{L}{D} \frac{V^2}{2g} \tag{9.2.3}$$

where f is the friction factor. The friction factor is a function of the Reynold's number and the relative roughness which is the absolute roughness of the interior pipe surface divided by the pipe diameter. The friction factor can be determined from a Moody diagram. Equations (9.2.2) and (9.2.3) are for computing headloss due to friction.

TABLE 9.2.1

Hazen-Williams roughness coefficient for selected pipe materials and ages (from Wood, 1980)

Type of pipe	Condition		Hazen-Williams Coeff. C_{HW}
Cast iron	New	All sizes	130
	5 years old	12″ and over	120
		8″	119
		4″	118
	10 years old	24″ and over	113
		12″	111
		4″	107
	20 years old	24″ and over	100
		12″	96
		4″	89
	30 years old	30″ and over	90
		16″	87
		4″	75
	50 years old	40″ and over	77
		24″	74
		4″	55
Welded steel	Values of C_{HW} are the same for cast iron pipes, 5 years older		
Riveted steel	Values of C_{HW} are the same as for cast iron pipes, 10 years older		
Wood stave	Average values, regardless of age		120
Concrete or concrete lined	Large size, good workmanship, steel forms		140
	Large size, good workmanship, wooden forms		120
	Centrifugally spun		135
Plastic or drawn tubing			150

In addition to headloss due to friction, energy loss along a pipe link is increased by a number of so called **minor loss components** such as bends and valves. Minor loss components may produce substantial head loss in a pipe section. The minor losses are proportional to the velocity head and are a function of the type of fitting and in the case of valves, its percent open,

$$h_{Lm} = M\frac{V^2}{2g} \qquad (9.2.4)$$

where h_{Lm} is the minor head loss, V is the flow velocity, and g is acceleration due to gravity. Table 9.2.2 lists the values of M for the most common fittings.

Nodes are classified in two categories, junction nodes and fixed-grade nodes. **Junction nodes** are connections of two or more pipes or where flow is removed from or input to the system. Changes in pipe diameter are typically modeled as a junction node. A **fixed-grade node** (FGN) is a node where the pressure and elevation are fixed. Reservoirs, tanks, and large constant pressure mains are examples of fixed grade nodes. Valves can be adjusted to vary the head loss across them and may even be totally closed to stop flow. This flexibility may be useful when operating a system to force flow in certain directions or to close off sections of a system to allow for maintenance or repairs of water mains. **Check valves** allow flow only in one direction and if conditions exist for flow reversal, a check valve closes to stop flow through the valve. Check valves, may be installed at the discharge end of a pump to prevent backflow.

Another type of valve is a **pressure-regulating valve**, also called a **pressure-reducing valve** (PRV), which are used to maintain a constant specified pressure at the downstream side of the valve for all flows which are lower than the upstream head. When connecting high- and low-pressure water systems, the PRV permits flow from the high pressure system if the pressure on the low side is not excessive. The head

TABLE 9.2.2
Loss coefficient for common fittings (from Wood, 1980)

Fitting	M
Glove valve, fully open	10.0
Angle valve, fully open	5.0
Swing check valve, fully open	2.5
Gate valve, fully open	0.2
Gate valve, 3/4 open	1.0
Gate valve, 1/2 open	5.6
Gate valve, 1/4 open	24.0
Short-radius elbow	0.9
Medium-radius elbow	0.8
Long-radius elbow	0.6
45 elbow	0.4
Closed return bend	2.2
Tee, through side outlet	1.8
Tee, straight run	0.3
Coupling	0.3
45 Wye, through side outlet	0.8
45 Wye, straight run	0.3
Entrance	
square	0.5
bell mouth	0.1
re-entrant	0.9
Exit	1.0

loss through the valve varies depending upon the downstream pressure, not on the flow in the pipe.

Tanks are used to store water through the day to allow pumps to operate closer to their maximum efficiency and minimize energy requirements. In a simple system with one pump, it is most economical from the pumping cost standpoint to operate this pump at its peak efficiency but the demands vary through time so this is not possible. Adding a tank to the system, which would act as a buffer to store water during lower loads and release it to the system during high demands, would allow the pump to operate near the average demand. Assuming the average demand is the pump's rated capacity, this will be the most efficient system with respect to pumping cost. The cost of building the storage tank and rest of the system, however, must be added to the energy cost to determine the overall cost of the design.

9.3 PUMPS AND PUMPING HYDRAULICS

The purpose of valves is to reduce the head in the system while pumps are used to increase the energy. Many different types of pumps exist but **centrifugal pumps** are most frequently used in water distribution systems. Centrifugal pumps impart energy to the water through a rotating element called an impeller and may be classified in two types, centrifugal and axial-flow, depending upon the direction the water is forced. The number and angles of the blades on the impeller and the speed of the pump motor affect the operating characteristics of centrifugal pumps.

A **pump head-characteristic curve** is a graphical representation of the total dynamic head versus the discharge that a pump can supply. These curves which are determined from pump tests are supplied by the pump manufacturer. When two or more pumps are operated, the pump station losses, which are the headlosses associated with the piping into and out of the pump, should be subtracted from the manufacturer's pump curve to derive the **modified head-characteristic curve**, as shown in Fig. 9.3.1.

Two points of interest on the pump curve are the shutoff head, and the normal discharge or rated capacity. The **shutoff head** is the head output by the pump at zero discharge while the **normal discharge** (or head) or **rated capacity** is the discharge (or head) where the pump is operating at its most efficient level. Variable speed motors can drive pumps at a series of rotative speeds which would result in a set of pump

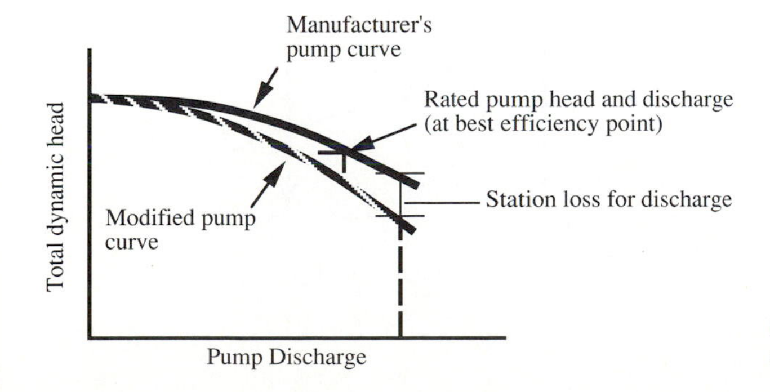

FIGURE 9.3.1
Modified pump curve.

curves for the single pump. Typically, to supply a given flow and head, a set of pumps are provided to operate in series or parallel and the number of pumps working depends on the flow requirements. This makes it possible to operate the pumps near their peak efficiency.

Pump manufacturers also provide pump characteristic curves for various speeds and for various impeller sizes, as shown in Fig. 9.3.2. Multiple-pump operation for one or more pumps in parallel or in series requires addition of the modified head-characteristic curves. For pumps operating in parallel, the modified head-characteristic curves are added horizontally with the respective heads remaining the same (Fig. 9.3.3). For pumps operating in series, the modified head-characteristic curves are added vertically with the respective discharges remaining the same.

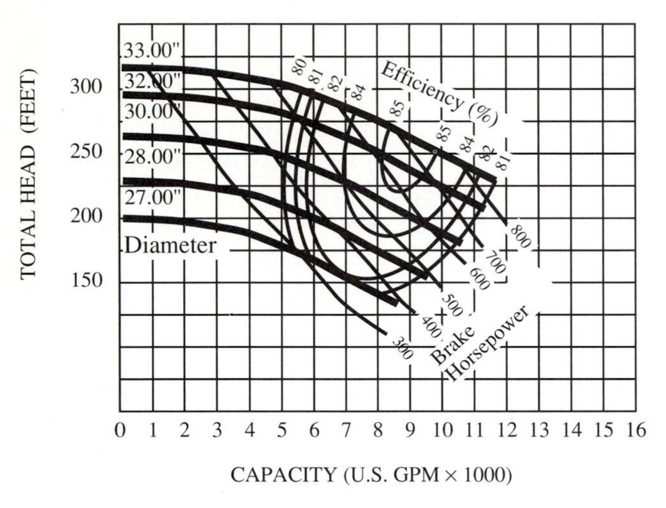

FIGURE 9.3.2
Manufacturer's pump performance curves.

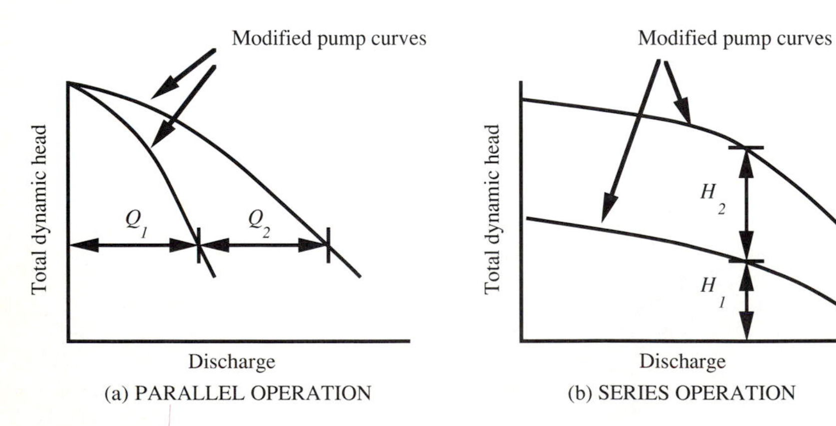

FIGURE 9.3.3
Pumps operating in series and parallel.

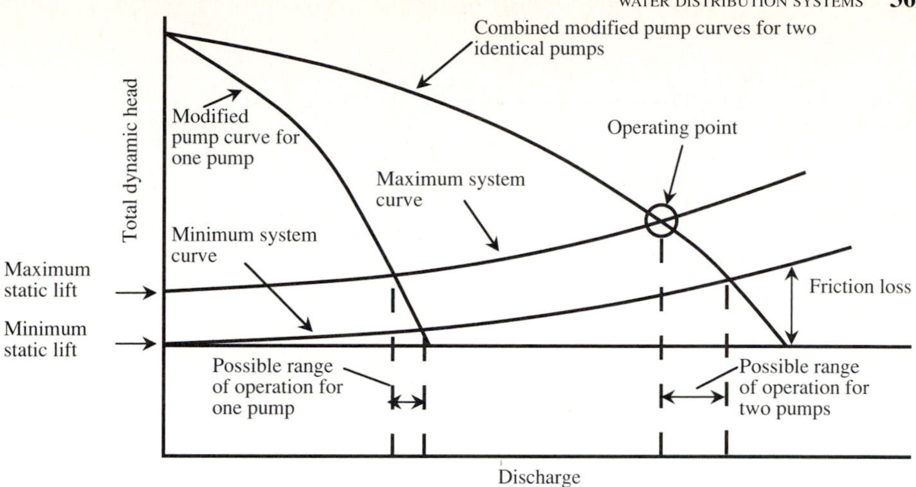

FIGURE 9.3.4
Operating point for pumps.

A **system head curve** is a graphical representation of the **total dynamic head** (TDH), defined as the static head plus the headloss, versus discharge. The headlosses are a function of the flow rate, size and length of pipe, and size, number and type of fitting. Figure 9.3.4 illustrates a system head curve for a minimum and maximum static lift along with the modified pump head characteristic curves. Note that the operation points are where the system head curve and the modified pump curves intersect.

Pump manufacturers also provide curves relating the **brake horse-power** (required by pump) to the pump discharge (see Fig. 9.3.2). The brake horsepower, E_p, is calculated using

$$E_p = \frac{QH\gamma}{550e} \tag{9.3.1}$$

where Q is the pump discharge in cfs, H is the total dynamic head in ft, γ is the specific weight of water in lb/ft^3 and e is the pump efficiency.

The **pump efficiency** is the power delivered by the pump to the water (water horsepower) divided by the power delivered to the pump by the motor (brake horsepower). **Efficiency curves**, as shown in Fig. 9.3.2 define how well the pump is transmitting energy to water.

9.4 NETWORK SIMULATION

9.4.1 Conservation Laws

The distribution of flows through a network under a certain loading pattern must satisfy the conservation of mass and the conservation of energy. Figure 9.4.1 shows a simple example network consisting of 19 pipes. Assuming water is an incompressible fluid, by the conservation of mass, flow at each of the junction nodes must be conserved, that is,

$$\sum Q_{\text{in}} - \sum Q_{\text{out}} = Q_{\text{ext}} \tag{9.4.1}$$

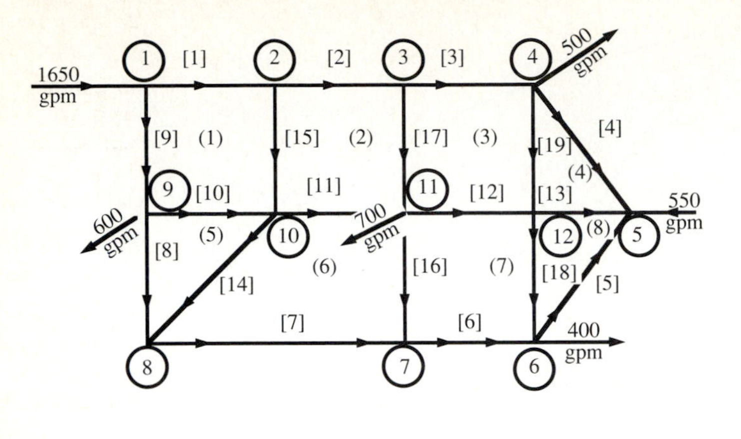

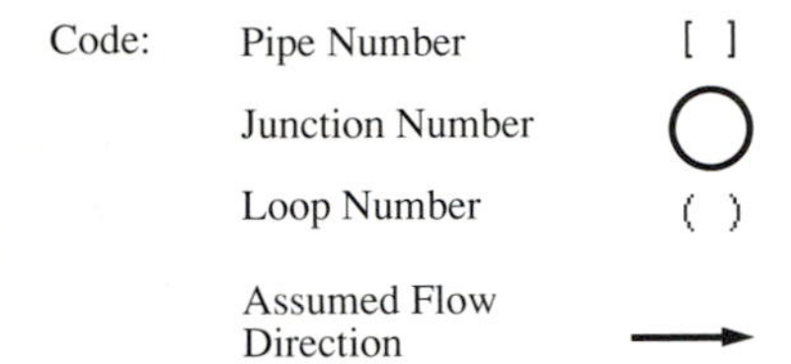

FIGURE 9.4.1
Example network (Wood and Charles, 1972).

where Q_{in} and Q_{out} are the pipe flows into and out of the node, respectively, and Q_{ext} is the external demand or supply at the node.

For each **primary loop**, which is an independent closed path, the conservation of energy must hold; that is, the sum of energy or head losses, h_L, minus the energy gains due to pumps, H_{pump}, around the loop must be equal to zero,

$$\sum_{i,j \in I_p} h_{L_{i,j}} - \sum_{k \in J_p} H_{\text{pump},k} = 0 \tag{9.4.2}$$

where $h_{Li,j}$ refers to the headloss in the pipe connecting nodes i and j; I_p is the set of pipes in the loop p; k refers to pumps; J_p is the set of pumps in loop p; and $H_{\text{pump},k}$ is the energy added by pump k contained in the loop and summed over the number of pumps. Equation (9.4.2) must be written for all independent loops.

Energy must be conserved between fixed-grade nodes which are points of known constant grade (elevation plus pressure head). If there are N_F such nodes then there are $N_F - 1$ independent equations of the form;

$$\Delta E_{\text{FGN}} = \sum_{i,j \in I_p} h_{L_{i,j}} - \sum_{k \in J_p} H_{\text{pump},k} \tag{9.4.3}$$

where ΔE_{FGN} is the difference in total grade between the two FGN's. The total number of equations, $N_J + N_L + (N_F - 1)$, also defines the number of pipes in the network in which N_J is the number of junction nodes and N_L is the total number of independent loops.

The change in head which occurs across each component is related to the flow through the component. By substituting the appropriate relationships for each component into the continuity and energy equations, it is possible to set up a system of nonlinear equations with the same number of unknowns. This set of equations can be solved by iterative techniques for the unknowns. Several computer programs have been written to automate these procedures. These models, called **network solvers** or **simulation models**, are now widely accepted and applied. This section presents the equations used to describe the relationships between head loss and flow and then discusses how each component is represented in a network simulation model.

The energy loss for water flow in a pipe is typically described by the Hazen-Williams Eq. (9.2.1) which can be expressed in terms of flowrate Q as

$$h_L = \frac{KLQ^{1.852}}{C_{HW}^{1.852} D^{4.87}} = K_p Q^{1.852} \tag{9.4.4}$$

where K is a coefficient, L and D are the pipe length and diameter, respectively, and K_p is the product of the constant values. The energy loss using the Darcy-Weisbach equation is

$$h_L = f\frac{L}{D}\frac{V^2}{2g} = \frac{8fL}{\pi^2 g D^5} Q^2 = K_p Q^2 \tag{9.4.5}$$

Similarly, energy losses that are minor losses in valves, expansions, contractions, etc. are given by

$$h_{Lm} = M\frac{V^2}{2g} = K_m Q^2 \tag{9.4.6}$$

in which K_m is a combined coefficient including M, g and the pipe diameter.

The relationship between the added head, H_{pump}, and discharge, Q, is typically a concave curve with H_{pump} increasing as Q decreases as shown in Fig. 9.3.2. For the normal operating range, this curve is usually well approximated by a quadratic or exponential equation, that is,

$$H_{\text{pump}} = AQ^2 + BQ + H_c \tag{9.4.7}$$

or

$$H_{\text{pump}} = H_c - CQ^n \tag{9.4.8}$$

with A, B, and n being coefficients and H_c is the cutoff head or maximum head. Also associated with a pump is an efficiency curve which defines the relationship of energy consumption and pump output (Fig. 9.3.2). Efficiency, e, is a function of Q and appears in Eq. (9.3.1) as a function of power, E_p, that is,

$$e = \frac{\gamma Q H_{\text{pump}}}{550 E_p} \tag{9.4.9}$$

A pump achieves maximum efficiency at the design or rated discharge. Depending upon the simulation model a pump may also be described by a curve of constant power, E_p.

As noted in the previous section, the limiting constraints in the design problem are usually the pressure restrictions at the nodes. Since the head losses in the system increase almost quadratically with the flowrates as seen in the Hazen-Williams equation, less head is required for patterns with lower total demand and as the demand level increases the head needed increases but faster than linearly. This relationship is a system curve from which the least cost operation of pumps can be determined (Fig. 9.3.4).

9.4.2 Network Equations

The governing conservation equations can be written in terms of the unknown nodal heads or the pipe flows using loop equations, head or nodal equations, or ΔQ equations. The loop or flow equations consist of the junction relationships written with respect to the N_p unknown flowrates. The component equations with pipe flows are substituted for h_L in the energy equations to form an additional $N_L + (N_F - 1)$ equations. This results in N_p equations written with respect to the N_p unknown flowrates.

The head or node equations use only flow continuity and consider the nodal heads as unknown rather than the pipe flows. In this case, additional equations are required for each pump and valve increasing the total number of equations. For a link the difference in head between the connected nodes i and j is equal to $h_{L,i,j}$.

$$h_{L,i,j} = H_i - H_j \tag{9.4.10}$$

This relationship can be substituted into the Hazen-Williams equation which in turn is rewritten and substituted for Q in the continuity equations.

The following **nodal equation** results for the node shown in Fig. 9.4.2 with assumed flow directions defined by the arrows (flow from a junction is negative)

$$-\left(\frac{H_i - H_j}{K_{p,i,j}}\right)^{0.54} - \left(\frac{H_i - H_{j+1}}{K_{p,i,j+1}}\right)^{0.54} + \left(\frac{H_{j+2} - H_i}{K_{p,j+2,i}}\right)^{0.54} - Q_{ext,i} = 0 \tag{9.4.11}$$

where $K_{p,i,j}$ is the coefficient defined in Eqs. (9.4.4) and (9.4.5) for the pipe connecting nodes i and j. These nodal equations can be written for each junction and component node resulting in a system of nonlinear equations with the same number of unknowns which is the total number of nodal heads. Similarly, the equations for the other components can be rewritten with respect to the nodal heads. The nodal equations can be linearized in an iterative solution technique.

The ΔQ equations directly use the **loop equations** and implicitly ensure that the node equations are satisfied. In this formulation the energy equation for each loop is written in terms of the flows,

$$\sum_{(i,j)\in I_p} K_{p,i,j}(Q_{i,j} + \Delta Q_{i,j})^n = 0 \tag{9.4.12}$$

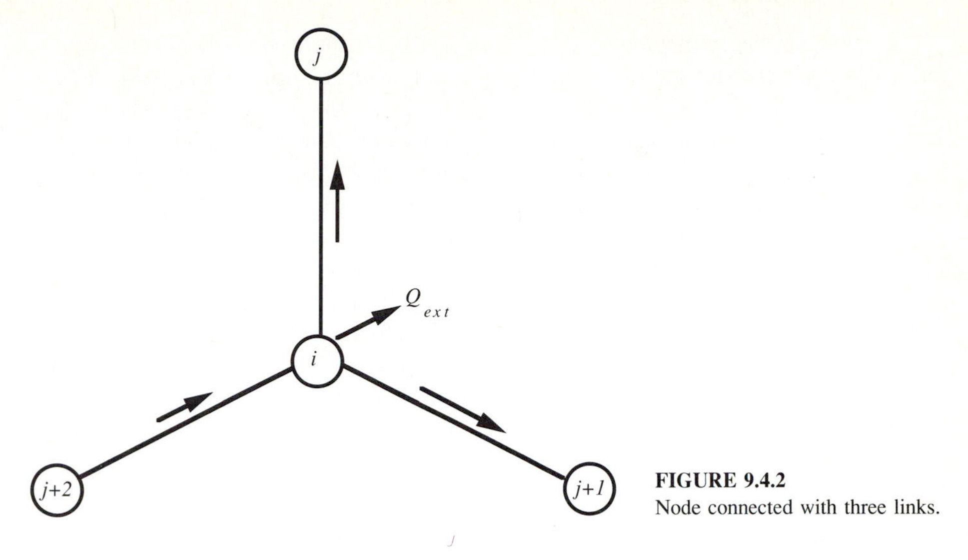

FIGURE 9.4.2
Node connected with three links.

where I_p is the set of pipes in loop p. An initial solution which satisfies the flow continuity is supplied at the beginning of the algorithm. The loop corrective factors ΔQ are computed to achieve the equalities and are defined such that continuity is preserved. An iterative approach is used to update and converge to the proper solution.

EXTENDED PERIOD SIMULATION. The equations described above are the relationships between flow and head for the main components in the network and can be solved for a single demand pattern operating in a steady state. An **extended period simulation** (EPS) analyzes a series of demand patterns in sequence. The purpose of an EPS is to determine the variation in tank levels and their effect on the pressures in the system. The water surface elevation in a tank varies depending upon the pressure distribution at the node where the tank is connected to the system.

Unlike a single-period analysis where a tank level is considered as fixed, in a EPS the tank levels change with progressive simulations to account for inflow and outflow. In a steady state simulator, flows are assumed to be constant throughout a subperiod. Tank levels, which are modeled as FGNs, are adjusted using simple continuity at the end of the subperiod and these new levels are then used as the fixed grades for the next subperiod. The accuracy of the simulation is dependent upon the length of the subperiods and the magnitude of flows to and from the tank.

9.4.3 Network Simulation Algorithms

Several iterative solution approaches have been applied to solve the sets of equations described in the previous section; including the **linear theory method**, the **Newton-Raphson method**, and the **Hardy-Cross technique**. Due to the nature of the equations the linear theory method for solving the flow equations and the Newton-Raphson technique (see Chap. 4) for solving the node equations are considered most efficient. This section summarizes the linear theory method.

The linear theory method was presented by Wood and Charles (1972) for simple networks and later extended to include pumps and other appurtenances (Wood, 1980). Martin and Peters (1963) published an algorithm using the Newton-Raphson method for a pipe system. Shamir and Howard (1968) showed that pumps and valves could be incorporated as well as being able to solve for unknowns besides the nodal heads. Other works using the Newton-Raphson method have been published that are based on exploiting the matrix structure (Epp and Fowler 1970; Lemieux, 1972; and Gessler and Walski, 1985) or using permutations of mixed solution techniques (Liu, 1969). The third algorithm, the Hardy-Cross method (Linsley and Franzini, 1979), is typically associated with the ΔQ equations. The method developed in 1936 (by Hardy-Cross) is attractive for hand calculations and easily coded, however, it is basically the Newton-Raphson method applied to one loop at a time. It requires more computation time than the other two methods and for large complex networks it has been found to have slow convergence.

Comparing the other approaches, the Newton-Raphson method may converge more quickly than the linear method for small systems whereas it may converge very slowly for large networks compared to the linear method (Wood and Charles, 1972). The linear theory method, however, has the capability to analyze all components, with more flexibility in the representation of pumps and better convergence properties. The University of Kentucky model, KYPIPE, by Wood (1980) is a widely used and accepted program based on the linear theory method.

The linear gradient method solves for the discharge Q using the path (energy) equations,

$$\Delta E = \sum h_L + \sum h_{Lm} - \sum H_{\text{pump}} \tag{9.4.13}$$

and using Eqs. (9.4.4) or (9.4.5), (9.4.6) and (9.4.7)

$$\Delta E = \sum (K_p Q^n + K_m Q^2) - \sum (AQ^2 + BQ + H_c) \tag{9.4.14}$$

where $n = 1.852$ for the Hazen-Williams equation and $n = 2$ for the Darcy-Weisbach equation for fully turbulent flow.

The pressure head (grade) difference in a pipe section with a pump for $Q = Q_r$ can be expressed as

$$f(Q_r) = K_p Q_r^n + K_m Q_r^2 - (AQ_r^2 + BQ_r + H_c) \tag{9.4.15}$$

where r represents the rth iteration. The gradient, $\partial f / \partial Q$ evaluated at Q_r, is

$$G_r = \left[\frac{\partial f}{\partial Q}\right]_{Q_r} = nK_p Q_r^{n-1} + 2K_m Q_r - (2AQ_r + B) \tag{9.4.16}$$

The nonlinear energy equations are linearized in terms of the unknown flowrate Q_{r+1} in each pipe using

$$\begin{aligned} f(Q_{r+1}) &= f(Q_r) + \left[\frac{\partial f}{\partial Q}\right]_{Q_r} (Q_{r+1} - Q_r) \\ &= f(Q_r) + G_r(Q_{r+1} - Q_r) \end{aligned} \tag{9.4.17}$$

The **path equations** (either from one fixed grade to another one or around a loop) can be written as

$$\Delta E = \sum f(Q_{r+1}) = \sum f(Q_r) + \sum G_r(Q_{r+1} - Q_r) \tag{9.4.18}$$

where the $\sum$ refers to summing over each pipe and ΔE is a known head difference. For a loop $\Delta E = 0$, so that

$$\sum G_r Q_{r+1} = \sum (G_r Q_r - f(Q_r)) \tag{9.4.19}$$

For a path between two fixed grade nodes, ΔE is a constant, then by Eq. (9.4.18)

$$\sum G_r Q_{r+1} = \sum (G_r Q_r - f(Q_r)) + \Delta E \tag{9.4.20}$$

Equations (9.4.19) and/or (9.4.20) are used to formulate $N_L + (N_F - 1)$ equations and are combined with the N_J continuity Eq. (9.4.1) to form a set of $N_p = N_L + (N_F - 1) + N_J$ linear equations (number of pipes) in terms of the unknown flowrate Q_{r+1} in each pipe. Using a set of initial flowrates Q_r in each pipe, the system of linear equations is solved for Q_{r+1} using a matrix procedure. This new set of flowrates Q_{r+1} is used as the known values to obtain a second solution of the linear equations. This procedure continues until the change in flowrates $|Q_{r+1} - Q_r|$ is insignificant and meets some convergence criteria.

Example 9.4.1. Develop the system of equations to solve for the pipe flows of the 19 pipe water distribution network shown in Fig. 9.4.1. The equations are to be based upon the linear theory method using loop equations.

Solution. Let $Q_1, Q_2, \ldots$ represent the flow in pipe 1, pipe 2, ...

CONSERVATION OF FLOW AT EACH NODE:

Node 1:	$Q_1 + Q_9 = 1{,}650$	Node 7:	$Q_7 + Q_{16} - Q_6 = 0$
Node 2:	$Q_1 - Q_2 - Q_{15} = 0$	Node 8:	$Q_8 + Q_{14} - Q_7 = 0$
Node 3:	$Q_2 - Q_3 - Q_{17} = 0$	Node 9:	$Q_9 - Q_{10} - Q_8 = 600$
Node 4:	$Q_3 - Q_4 - Q_{19} = 500$	Node 10:	$Q_{10} + Q_{15} - Q_{11} - Q_{14} = 0$
Node 5:	$Q_4 + Q_5 + Q_{13} = -550$	Node 11:	$Q_{11} + Q_{17} - Q_{12} - Q_{16} = 700$
Node 6:	$Q_{18} + Q_6 - Q_5 = 400$	Node 12:	$Q_{12} + Q_{19} - Q_{13} - Q_{18} = 0$

Including all 12 of the conservation of flow constraints results in one redundant equation so that only 11 of the above constraints are needed.

CONSERVATION OF ENERGY (LOOP EQUATIONS):

Loop 1: $K_{p,1}Q_1^n + K_{p,15}Q_{15}^n - K_{p,10}Q_{10}^n - K_{p,9}Q_9^n = 0$
Loop 2: $K_{p,2}Q_2^n + K_{p,17}Q_{17}^n - K_{p,11}Q_{11}^n - K_{p,15}Q_{15}^n = 0$
Loop 3: $K_{p,3}Q_3^n + K_{p,19}Q_{19}^n - K_{p,12}Q_{12}^n - K_{p,17}Q_{17}^n = 0$
Loop 4: $K_{p,4}Q_4^n - K_{p,13}Q_{13}^n - K_{p,19}Q_{19}^n = 0$
Loop 5: $K_{p,10}Q_{10}^n + K_{p,14}Q_{14}^n - K_{p,8}Q_8^n = 0$

Loop 6: $K_{p,11}Q_{11}^n + K_{p,16}Q_{16}^n - K_{p,7}Q_7^n - K_{p,14}Q_{14}^n = 0$
Loop 7: $K_{p,12}Q_{12}^n + K_{p,18}Q_{18}^n - K_{p,6}Q_6^n - K_{p,16}Q_{16}^n = 0$
Loop 8: $K_{p,13}Q_{13}^n - K_{p,5}Q_5^n - K_{p,18}Q_{18}^n = 0$

The above eight conservation of energy equations are linearized using $k = K_pQ^{n-1}$

Loop 1: $k_1Q_1 + k_{15}Q_{15} - k_{10}Q_{10} - k_9Q_9 = 0$
Loop 2: $k_2Q_2 + k_{17}Q_{17} - k_{11}Q_{11} - k_{15}Q_{15} = 0$
Loop 3: $k_3Q_3 + k_{19}Q_{19} - k_{12}Q_{12} - k_{17}Q_{17} = 0$
Loop 4: $k_4Q_4 - k_{13}Q_{13} - k_{19}Q_{19} = 0$
Loop 5: $k_{10}Q_{10} + k_{14}Q_{14} - k_8Q_8 = 0$
Loop 6: $k_{11}Q_{11} + k_{16}Q_{16} - k_7Q_7 - k_{14}Q_{14} = 0$
Loop 7: $k_{12}Q_{12} + k_{18}Q_{18} - k_6Q_6 - k_{16}Q_{16} = 0$
Loop 8: $k_{13}Q_{23} - k_5Q_5 - k_{18}Q_{18} = 0$

This system of 19 equations (11 conservation of flow equations and 8 energy equations) can be solved for the 19 unknown discharges.

9.5 OPTIMIZATION MODELS FOR DESIGN OF BRANCHED SYSTEMS

Hydraulic simulation models provide a very powerful tool for determining the hydraulics of a water distribution system. These models can be used in a trial-and-error fashion to determine the hydraulic characteristics (pressure heads, pump operation, tank levels, etc.) for a particular network design. However, these models have no ability to determine the optimal or minimum cost system. This section presents a description of modeling branched pipe systems.

The purpose of a water distribution system is to supply the water demanded by the user at an adequate pressure. A designer's problem is to determine the minimum cost system while satisfying the demands at the required pressures. The cost of the system includes the initial investment for the components, such as pipes, tanks, valves and pumps, and the energy cost for pumping the water throughout the system. The design or optimization problem can be stated as;

Minimize: Capital Investment Cost + Energy Cost

Subject to:

1. Hydraulic Constraints
2. Satisfy Water Demands
3. Meet Pressure Requirements

The design of a branched water distribution system such as an irrigation system can be formulated as an LP problem (Gupta, 1969; Calhoun, 1971; and Gupta et al., 1972). An example system is illustrated in Fig. 9.5.1 for which the objective is to

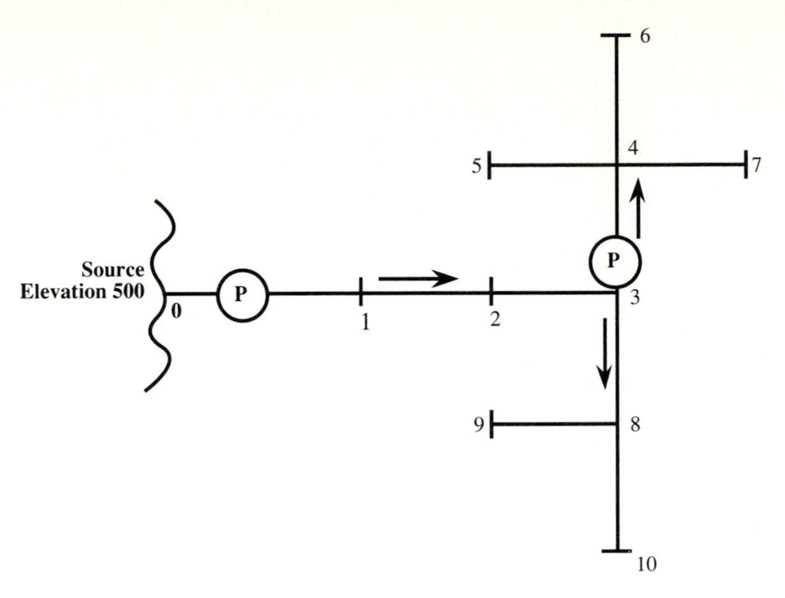

FIGURE 9.5.1
Network for Example 9.5.1

determine the length of pipe segment denoted as $X_{i,j,m}$ of the mth diameter in the pipe reach between nodes i and j. A branching network can be supplied from one or more sources and designed for a single loading condition.

The LP model can be stated as

$$\text{Minimize } Z = \sum_{(i,j)\in I} \sum_{m\in M_{i,j}} c_{i,j,m} X_{i,j,m} \tag{9.5.1}$$

subject to

a. Length constraints for each link to force the sum of the lengths of each diameter to equal the total reach length.

$$\sum_{m\in M_{i,j}} X_{i,j,m} = L_{i,j} \qquad (i,j) \in I \tag{9.5.2}$$

b. Conservation of energy constraints written from the source (fixed-grade node) with known elevation in H_s to the delivery points.

$$H_{\min,n} \le H_s + E_p - \sum_{(i,j)\in I_n} \sum_{m\in M_{i,j}} J_{i,j,m} X_{i,j,m} \le H_{\max,n} \qquad n = 1, \ldots, N \tag{9.5.3}$$

c. Nonnegativity

$$X_{i,j,m} \ge 0 \tag{9.5.4}$$

where $M_{i,j}$ = the set of candidate pipe diameters for the pipe connecting nodes i and j

$c_{i,j,m}$ = the cost per unit length of the mth diameter for the link connecting nodes i and j

I = the set of pipe links that define the network

I_n = the set of pipes that defines the path to node n (delivery point n)

$L_{i,j}$ = the length of the link connecting nodes i and j

$J_{i,j,m}$ = hydraulic gradient of the pipe of diameter m connecting nodes i and j

H_s = known elevation of water source which is a fixed grade node

E_p = known energy head added to the system

$H_{\min,n}$ = minimum allowable head requirement at delivery point n

$H_{\max,n}$ = maximum allowable head requirement at delivery point n

N = total number of delivery points

This formulation can be expanded to consider the additional pumping head XP required as a decision variable,

$$\text{Minimize } Z = \sum_{(i,j)\in I} \sum_{m\in M_{i,j}} c_{i,j,m} X_{i,j,m} + \sum_k CP_k XP_k \tag{9.5.5}$$

subject to

a. Equation (9.5.2)—length constraints

b.

$$H_{\min,n} \le H_s + \sum_k XP_k - \sum_{(i,j)\in I_n} \sum_{m\in M_{ij}} J_{i,j,m} X_{i,j,m} \le H_{\max,n} \qquad n = 1, \ldots, N \tag{9.5.6}$$

c.

$$XP_k \ge 0 \tag{9.5.7a}$$

$$X_{i,j,m} \ge 0 \tag{9.5.7b}$$

where CP_k is the unit cost of pumping head at location k and XP is the pumping head at location k.

Example 9.5.1. Develop an LP model to determine the minimum cost pipe diameter and pumpage for the network shown in Fig. 9.5.1. Pumps are located on the downstream sides of nodes 0 and 3. The LP model must be able to solve for unknown lengths of various pipe sizes for each reach. The pipes to be considered for the distribution network have a Darcy-Weisbach friction factor of f = 0.02 and each reach is

3000 ft long. Consider three pipe diameters for each reach. Write out the constraints and objective function. The piping system has a constant elevation of 500 ft. The demand discharges and minimum required pressure head at the supply nodes are listed in Table 9.5.1. The cost of pumping head is $500.00 per foot and the pipe costs are:

Diameter (in)	8	10	12	15	18	21	24	27	30	36	42
Cost ($/ft)	8	10	12	15	18	21	24	27	30	36	42

Solution. The objective is to minimize the cost

$$\text{Minimize } Z = CP_1XP_1 + CP_2XP_2 + c_{0,1,1}X_{0,1,1} + c_{0,1,2}X_{0,1,2} + c_{0,1,3}X_{0,1,3} + \cdots + c_{8,10,1}X_{8,10,1} + c_{8,10,2}X_{8,10,2} + c_{8,10,3}X_{8,10,3}$$

where CP_1 and CP_2 are the unit costs of pump head at locations 1 and 2. The constraints are length constraints:

$$X_{0,1,1} + X_{0,1,2} + X_{0,1,3} = 3000 \text{ (for reach 0–1)}$$

$$X_{1,2,1} + X_{1,2,2} + X_{1,2,3} = 3000 \text{ (for reach 1–2)}$$

$$X_{2,3,1} + X_{2,3,2} + X_{2,3,3} = 3000 \text{ (for reach 2–3)}$$

$$\vdots \qquad\qquad \vdots$$

The hydraulic constraint to delivery points 5 is:

$$500 + XP_1 - J_{0,1,1}X_{0,1,1} - J_{0,1,2}X_{0,1,2} - J_{0,1,3}X_{0,1,3} - J_{1,2,1}X_{1,2,1} - J_{1,2,2}X_{1,2,2} - J_{1,2,3}X_{1,2,3} - J_{2,3,1}X_{2,3,1} - J_{2,3,2}X_{2,3,2} - J_{2,3,3}X_{2,3,3} + XP_2 - J_{3,4,1}X_{3,4,1} - J_{3,4,2}X_{3,4,2} - J_{3,4,3}X_{3,4,3} - J_{4,5,1}X_{4,5,1} - J_{4,5,2}X_{4,5,2} - J_{4,5,3}X_{4,5,3} \geq 550$$

where Q is in cfs and D is in feet. The hydraulic gradient (headloss per unit pipe length) is

TABLE 9.5.1
Demand discharges and minimum required pressure heads for Example 9.5.1

Nodes	Demand discharge Q (cfs)	Minimum required pressure head elevation (ft)
5	4	550
6	4	550
7	4	550
9	6	550
10	6	550

$$J = \frac{8fQ^2}{\pi^2 g D^5}$$

For the pipe reach connecting nodes 0 and 1 considering a diameter of 18 inches (1.5 ft), with a flow of 24 cfs, the headloss per unit length of pipe is

$$J_{0,1,1} = \frac{(8)(0.02)(24)^2}{\pi^2(32.2)(1.5)^5} = 0.0382\frac{ft}{ft}$$

9.6 OPTIMIZATION MODELS FOR DESIGN OF LOOPED SYSTEMS

9.6.1 General Problems

The general problem is to determine the minimum cost design including layout of a water distribution network subject to meeting the constraints. Thus, given a distribution of demands the model should select which of the candidate components are necessary and determine their optimal sizes in the final network. This section is concerned only with the design and analysis of the water distribution systems assuming that external demands and pressure requirements are predefined. If a system is to be designed considering demands at present and at some time in the future, the change in system performance due to, for example, changing pipe roughness, must also be externally defined.

To ensure that the constraints are satisfied, the equations that define the pressure and flow distributions in the system must be satisfied. The flow and loop or nodal equations which define the flow within a pipe as well as those representing pumps, valves, storage tanks, and other network components are nonlinear functions. In addition, the cost equations of the various components are also typically nonlinear. This high degree of nonlinearity causes great difficulty in determining an optimal (minimum cost) design of the network. The problem is further complicated by the fact that the present manner in which redundancy is introduced is to analyze the system under more than one set of demands (**multiple loading conditions**). Thus, instead of a single set of n nonlinear equations being considered, the number of equations are n times the number of loads. The technology to determine optimal solutions to large highly nonlinear mathematical programming problems has just become available in the recent past by the introduction of such models as GRG2, MINOS, and GAMS-MINOS (see Section 4.9)

The overall optimization problem for water distribution network design can be stated mathematically in terms of the nodal pressure heads, **H**, and the various design parameters, **D** as follows:

$$\text{Objective: Minimize Cost } = f(\mathbf{D}, \mathbf{H}) \tag{9.6.1}$$

Subject to:

a. Conservation of flow and energy constraints

$$\mathbf{G}(\mathbf{H}, \mathbf{D}) = \mathbf{0} \tag{9.6.2}$$

b. Head bounds $\underline{\mathbf{H}} \leq \mathbf{H} \leq \overline{\mathbf{H}}$ (9.6.3)

c. Design constraints $\underline{\mathbf{u}} \leq \mathbf{u}(\mathbf{D}) \leq \overline{\mathbf{u}}$ (9.6.4)

d. General constraints $\underline{\mathbf{w}} \leq \mathbf{w}(\mathbf{H}, \mathbf{D}) \leq \overline{\mathbf{w}}$ (9.6.5)

where the decision variables $\mathbf{D}$ define the dimensions for each component in the system such as diameter of the pipes, pump size, valve setting, and tank volume or elevation. The objective function can be linear or nonlinear allowing for various types of components to be designed by the model. Each component to be designed has a term associated with it in the objective; therefore, the formulation allows for variation of the cost equations to account for site specific costs and/or construction staging. This gives the model the capability to design expansions of existing systems or to design new networks.

The relationships, $\mathbf{G}(\mathbf{H}, \mathbf{D}) = \mathbf{0}$, are the set of nonlinear Eqs. (9.4.1), (9.4.2), and (9.4.3) that define the pressure and flow distribution in the system and make up the majority of constraints in the problem. The nodal demands are parameters in this set of equations, and if the equalities are satisfied in the model, the user's demands are met. The equations are written in a general form that allows all types of systems; pipelines, branched or looped, to be analyzed with all levels of complexity. This formulation does not restrict the number of equations in the set $\mathbf{G}$, so that one or more demand patterns can be considered by the model while designing the system. The vector $\mathbf{H}$ is the pressure head at specified nodes in the system with $\underline{\mathbf{H}}$ and $\overline{\mathbf{H}}$ being the lower and upper bounds. The design constraints $\mathbf{u}$ are usually simple bounds but are shown as functions for the general formulation and are typically set by physical limitations or the availability of the components. The general constraint set $\mathbf{w}$ includes limits on terms which are functions of both the nodal pressures and the design variables. A limitation of velocity in a pipe is one example of such a constraint.

9.6.2 A Linear Programming Model

Different variations of models have been developed in the literature to linearize the optimization model described by Eqs. (9.6.1)–(9.6.5). These include the models by Alperovits and Shamir (1977), Shamir (1979), Quindry et al. (1981), and Morgan and Goulter (1985).

Morgan and Goulter (1985) present a heuristic LP-based procedure for the least-cost layout and design of water-distribution networks. This model links an LP procedure with a Hardy-Cross network solver. The Hardy-Cross solver determines the flow and pressure head distributions for the designs determined in the LP model. The procedure is iterative in that repeated solutions of the LP and Hardy-Cross are used. The LP model considers decision variables as replacement sizes for pipes determined in the previous LP solution. A mathematical model statement is

$$\text{Minimize } Z = \sum_{i,j} [(c_{i,j,m+1} - c_{i,j,m})X_{i,j,m+1} + (c_{i,j,m-1} - c_{i,j,m})X_{i,j,m-1}] \quad (9.6.6)$$

subject to

a. Pressure head constraints to ensure minimum pressure heads at demand point n

$$\sum_{i,j \in I_n} [(J_{i,j,m+1} - J_{i,j,m})X_{i,j,m+1} + (J_{i,j,m-1} - J_{i,j,m})X_{i,j,m-1}] \leq \underline{H}_n - H_n \tag{9.6.7}$$

$X_{i,j,m+1}$ and $X_{i,j,m-1}$ are the length of pipe of the mth diameter in link i, j replaced by a pipe of the $m + 1$ or $m - 1$ diameter, respectively; I_n is the set of pipes that defines the path to node n.

b. Length constraints to ensure that no more than the existing length of pipe in link (i, j) is replaced.

$$X_{i,j,m+1} \leq L_{i,j} \tag{9.6.8}$$

$$X_{i,j,m-1} \leq L_{i,j} \tag{9.6.9}$$

c. Nonnegativity constraints

$$X_{i,j,m+1} \geq 0 \tag{9.6.10}$$

$$X_{i,j,m-1} \geq 0 \tag{9.6.11}$$

where $c_{i,j,m-1}$, $c_{i,j,m}$, $c_{i,j,m+1}$ are the unit costs of pipe of the $(m - 1)$th, mth, or $(m + 1)$th diameter in link (i, j).

The solution procedure begins with an initial flow pattern and pipe network design. The LP model is then solved for the decision variables $X_{i,j,m+1}$ and $X_{i,j,m-1}$ which redefine the pipe sizes in each link for the given flow distribution and pressure head distribution. Once the LP model is solved then a Hardy-Cross network solver is used to redefine the flow distribution and pressure head distribution in the new network design. This process is repeated iteratively until the best solution is found.

9.7 WATER DISTRIBUTION SYSTEM DESIGN MODEL

The LP model discussed in Section 9.6.2 is limited to the design of pipe sizes. This section presents a methodology to solve the optimization problem (9.6.1)–(9.6.5) in which the solution technique is based upon the concepts of **optimal control theory**. The generalized reduced gradient method (see Section 4.6) makes up the overall optimization framework along with a simulation model used to perform function evaluations (solving the conservation of flow and energy constraints) at each iteration of the optimization. In other words, this optimization problem, which is highly nonlinear, is solved directly, but reduced in complexity by incorporating a network simulator to solve the flow conservation and energy (loop) constraints. Such a methodology allows a detailed analysis of the system components under various loading conditions and reduces the constraint size so that large water distribution systems with a large number of components can be designed as new systems or can be analyzed in existing systems.

The set of constraints **G** in Eq. (9.6.2) is a system of nonlinear (in terms of diameter) equations which define the pressure and flow distribution in the network for each of the demand patterns. These may include independent critical loads and a series of demand patterns linked through time by the tank elevations. Each subset of constraints in **G** is a system of n nonlinear equations with n unknowns. If multiple loads or a large system is analyzed, the problem size may and typically does exceed that which can be solved by existing nonlinear programming (NLP) codes. The solution methodology exploits these equality relationships by employing an approach similar to that used for discrete time optimal control problems (Lasdon and Mantell, 1978; Norman et al., 1982). The technique reduces the problem by expressing dependent variables in terms of independent variables, called **control variables**, via the equality constraints (9.6.2). This step results in a reduced problem to be solved with a new objective and a significantly smaller number of constraints, many of which are simple bounds. The reduced problem can now be solved by existing NLP codes. In this problem the pressure heads **H** are defined as the **state** or **basic variables** and written with respect to the design parameters **D** called the **control** or **nonbasic variables**. Water distribution simulation models such as KYPIPE can be used to solve the network Eq. (9.6.2) for the nodal heads given a set of design parameters such as pipe, pump, and storage tank sizes.

The solution technique can be applied to the original problem, Eq. (9.6.2), under assumptions for the various components as described later. The set of network Eqs. (9.6.2) can be solved (given a set **D**) to determine the nodal heads for a sequence of demand patterns. Assuming the gradient of **H** with respect to **D** is nonsingular in this range for all demands, the problem defined by Eqs. (9.6.1)–(9.6.5) can be transformed into a reduced form. Since **H** can be written in terms of **D** as **H**(**D**), the objective can be transformed to a reduced objective as a function of **D** only and the set of Eqs. (9.6.2) can be removed from the constraint set to be solved implicitly by the network simulator each time they are required to be evaluated in the optimization procedure. The new reduced problem is then:

$$\text{Min. Cost } f(\mathbf{D}, \mathbf{H}(\mathbf{D})) = F(\mathbf{D}) \tag{9.7.1}$$

subject to

$$\underline{\mathbf{H}} \leq \mathbf{H}(\mathbf{D}) \leq \overline{\mathbf{H}} \tag{9.7.2}$$

$$\underline{\mathbf{u}} \leq \mathbf{u}(\mathbf{D}) \leq \overline{\mathbf{u}} \tag{9.7.3}$$

$$\underline{\mathbf{w}} \leq \mathbf{w}(\mathbf{H}(\mathbf{D}), \mathbf{D}) \leq \overline{\mathbf{w}} \tag{9.7.4}$$

The set of equations **G** are satisfied by the network simulator that calculates the implicit function **H**(**D**) when required by the optimization procedure. The known nodal heads **H**(**D**) are substituted into Eqs. (9.7.1), (9.7.2), (9.7.3), and (9.7.4) to compute the values of the objective function and constraints. Figure 9.7.1 shows the linkage between the optimization and the simulation models.

To solve the reduced problem Eqs. (9.7.1)–(9.7.4), generalized reduced gradient algorithms (Section 4.6) require the gradients of the objective function, called **reduced gradients**, with respect to the design parameters. The functions $f(\mathbf{D}, \mathbf{H}(\mathbf{D}))$ and **H**(**D**) are differentiable implicit functions not known in a closed form; therefore, the gradients cannot be directly calculated. By using the two-step procedure of Lasdon

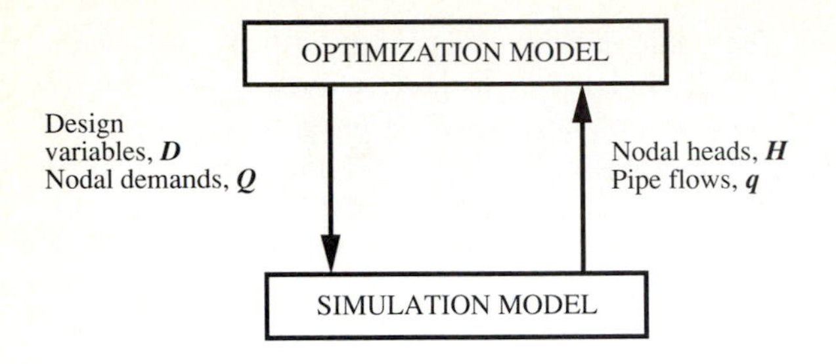

FIGURE 9.7.1
Optimization-simulation model linkage.

and Mantell (1978), the reduced gradients can be computed efficiently by solving a system of linear equations for each loading condition or sequence of demand patterns (Lansey and Mays, 1989).

Typically in NLP algorithms, the control variable bounds are met by restricting the step size; however, since the state variables **H** are implicit functions of the component sizes **D** they are not considered when determining the step size. If the state variables are violated, more iterations would be required to return to feasibility. The technique used to resolve this problem is the use of an augmented Lagrangian penalty function method (see Section 4.7) where the basic variables are included in the objective function through a penalty term. This penalty term is attached to the original problem objective before the variable reduction is made. The new augmented objective which results from incorporating the head bound Eq. (9.7.2) is of the form:

$$\text{Min } AL(\mathbf{H}, \mathbf{D}, \boldsymbol{\lambda}, \boldsymbol{\psi}) = f(\mathbf{H}, \mathbf{D}) + \frac{1}{2}\sum_i \psi_i \text{ Min } \left[0, c_i - \frac{\lambda_i}{\psi_i}\right]^2 + \frac{1}{2}\sum_i \frac{\lambda_i^2}{\psi_i} \tag{9.7.5}$$

where i is the index for each bound constraint, and ψ_i and λ_i are the penalty weights and Lagrange multipliers for the ith bound, respectively. Also, c_i is the violation of the bounds either above the upper or below the minimum and defined as

$$c_i = \min\left(\underline{c}_i, \overline{c}_i\right) \qquad \text{for all } i \tag{9.7.6}$$

where $\underline{c}_i = H_i - \underline{H}_i$ and $\overline{c}_i = \overline{H}_i - H_i$. By this definition $\underline{c}_i$ and $\overline{c}_i$ cannot both be negative since only one bound may be violated at one time. Thus, the head bounds are included directly in the objective and are considered when determining the change in the design parameter values.

Applying the reduction technique to the original problem, Eqs. (9.6.1)–(9.6.5), the new objective function in the reduced optimization problem is:

$$\text{Min } RAL(\mathbf{D}, \boldsymbol{\lambda}, \boldsymbol{\psi}) = AL(\mathbf{H}(\mathbf{D}), \mathbf{D}, \boldsymbol{\lambda}, \boldsymbol{\psi}] \tag{9.7.7}$$

subject to the constraints, Eqs. (9.7.3) and (9.7.4).

The reduced gradients can be calculated again using the reduced-gradient procedure described and the gradient terms are determined for the augmented Lagrangian function as previously described. This method has been shown to converge given the correct weights and multipliers, which are determined by an updating procedure (Lansey and Mays, 1989). The general constraints, that is, Eq. (9.7.4), which are also implicit functions, can be incorporated in a similar manner or considered directly as constraints in the model.

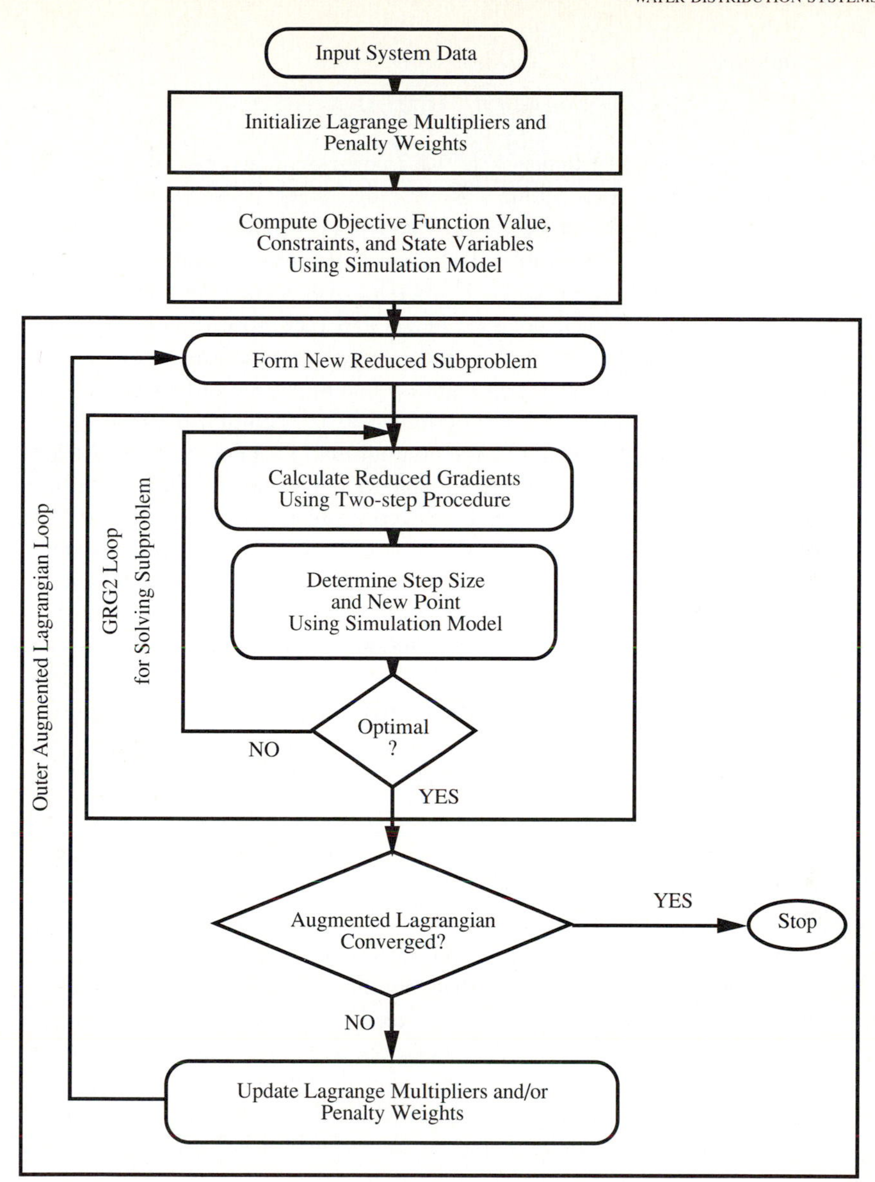

FIGURE 9.7.2
Flow chart of optimal control algorithm.

The reduced problem Eq. (9.7.7) subject to Eqs. (9.7.3) and (9.7.4) is solved using GRG2, a general NLP code by Lasdon and Waren (1982), based upon the generalized reduced gradient method. Figure 9.7.2 presents a flow chart of the algorithm. The model attempts to converge to the optimal solution and the correct Lagrange multipliers $\boldsymbol{\lambda}$ and penalty weights ψ through a two-step optimization procedure using the objective function,

$$\text{Min } Z = \min_{\boldsymbol{\lambda},\boldsymbol{\psi}}\{\min_{\mathbf{D}} AL[\mathbf{H}(\mathbf{D}), \mathbf{D}, \boldsymbol{\lambda}, \boldsymbol{\psi}]\} \qquad (9.7.8)$$

The flow chart of the algorithm in Fig. 9.7.2 consists of two nested loops. The inner loop contains the operations within GRG2 which determine the change of the design variables. It is an iterative process which converges to the optimal solution for the given objective function. Within this inner loop is where the simulation and optimization models are linked. KYPIPE, the University of Kentucky simulation model can be used to determine the nodal pressure heads and pipe flows given the design parameters and nodal demand (Fig. 9.7.1). Once the optimum of the subproblem is found, convergence of the augmented Lagrangian problem is tested by examining the infeasibilities and execution is halted if the optimal feasible solution is found. When a solution of this subproblem is found which is not optimal, the penalty term variables are adjusted and a new subproblem is formed for GRG2. This new problem continues to have the same constraints as the original reduced problem as Eqs. (9.7.3)–(9.7.4) but has a new augmented Lagrangian objective function Eq. (9.7.5) with the updated weights and multipliers. The outer loop continues until an iteration limit is reached or the optimum is determined. A computer code was developed by Lansey and Mays (1989) to implement the algorithm and solve the reduced problem. It consists of three main components; the nonlinear optimization model GRG2; the hydraulic simulation model KYPIPE; and the routines linking these models.

9.8 WATER DISTRIBUTION SYSTEM RELIABILITY

The operator of a water utility has an interest in two types of reliability, *mechanical* reliability and *hydraulic* reliability. **Mechanical reliability** relates to the operability of equipment. **Hydraulic reliability** refers to consistently providing the specified quantity of water at the required pressures, and at the time and location desired.

Mechanical reliability is the ability of distribution system components to provide continuing and long-term operation without the need for frequent repairs, modifications, or replacement of components or subcomponents. Mechanical reliability is usually defined as the probability that a component or subcomponent performs its mission within specified limits for a given period of time in a specified environment. When quantified, mechanical reliability is merely an expression of the probability that a piece of equipment is operational at any given time.

Hydraulic reliability, as it relates to water distribution system design, can be defined as the ability of the system to provide service with an acceptable level of interruption in spite of abnormal conditions. The evaluation of hydraulic reliability relates directly to the basic function of the water distribution system, that is, delivery of the specified quantity of water to the appropriate place at the required time under the desired pressure. It is important to note that mechanical reliability should be explicitly considered in the determination of hydraulic reliability. Thus, the evaluation of the reliability of a water distribution system must consider the quantity of water delivered, the residual pressure at which the water is delivered, the time at which the water must be delivered, and the location within the system to which the water is delivered. Thus, what is referred to as reliability can be measured in terms of availability, where **availability** (see Section 5.5) is the percentage of time that the demand can be supplied at or above the required residual pressure.

Implicit in the evaluation of availability is the necessity to define failure. Whereas this is fairly straightforward when defining the mechanical reliability of operating equipment, failure is a somewhat more ambiguous term when discussing hydraulic performance. When does hydraulic failure occur? The best approach may be to base the definition of failure on a performance criteria. Thus, **hydraulic failure** occurs when the system cannot supply the specified amount of water to the specified location, at the specified time, and at the specified pressure.

9.8.1 Component Reliability

Mathematically, the reliability $\alpha(t)$ of a component (see Section 5.5) is defined as the probability that the component experiences no failures during the time interval from time zero to time t, given that it is new or repaired at time zero.

$$\alpha(t) = \int_t^{\infty} f(t)\, dt \tag{9.8.1}$$

where $f(t)$ is the probability density function (PDF) of the time-to-failure of the component. The PDF is either assumed or developed from equipment failure data, using various statistical data (se Section 5.5).

The concept of reliability is suitable for the evaluation of nonrepairable components; however, for repairable components such as those most often found in water distribution systems, it is much more appropriate to use the concept of availability. Whereas the reliability is the probability that the component experiences no failures during the interval from time zero to time t, the availability of a component is the probability that the component is in operational condition at time t, given that the component was as good as new at time zero. The reliability generally differs from the availability because reliability requires the continuation of the operational state over the whole time interval.

In order to discuss availability the concept of maintenance of water distribution systems needs to be defined. There are two basic categories of maintenance events: *corrective* maintenance and *preventive* maintenance.

Corrective maintenance is defined as the activity of repair after a breakdown has taken place or as unscheduled maintenance due to equipment failure. Corrective maintenance activities have four time periods that contribute to the unavailability of the component: (1) time between the failure and the recognition of failure (initial response time); (2) time awaiting repair materials; (3) time awaiting manpower; and (4) active repair time. The initial response time, time awaiting materials, and time awaiting manpower are utility and maintenance event specific and are often neglected in the evaluation of reliability. Active repair time is the time required to disassemble, correct the deficiency, reassemble, and return the failed equipment to an operational state. The **corrective maintenance time** (CMT) is the time (usually expressed in hours/year) that a component is nonoperational because of corrective maintenance activities.

Preventive maintenance has a variety of potential meanings. In its most limited form, preventive maintenance is merely the inspection of equipment to prevent breakdowns before they occur. A broader definition includes activities such as repetitive servicing, upkeep, and overhaul (e.g. lubrication, painting, and cleaning). Preventive maintenance may also be called planned or routine maintenance. Since preventive maintenance is scheduled, only the active repair time contributes to the unavailability of the component. It should be noted that not all preventive maintenance activities result in component unavailability. The **preventive maintenance time** (PMT) is the time (usually expressed as hours/year) that a component is nonoperational because of preventive maintenance activities.

Component availability can be expressed mathematically as the fraction of clock time that the component is operational, that is, available for service. On an annual basis, this can be calculated using the following equation

$$A = \frac{(8760 - \text{CMT} - \text{PMT})}{8760} \tag{9.8.2}$$

where A is the availability; CMT is the corrective maintenance time, hr/yr; PMT is the preventive maintenance time, hr/yr; and there are 8760 hours in a year. Typically, mean values are used for the corrective and preventive maintenance times.

Accurate calculation of the mechanical reliability requires knowledge of the precise reliability of the basic components and the impact on mission accomplishment caused by the set of all possible failures. Thus, for a large system with many interactive components, such as a water distribution system, it is extremely difficult to analytically compute the mechanical reliability. There is no comprehensive data base of failure and repair information for components and subcomponents for water distribution systems.

9.8.2 System Reliability

The reliability of series-parallel systems is generally straightforward as discussed in Chapter 5. Water distribution systems have a nonseries-parallel configuration and the evaluation is much more difficult. Several techniques have been developed for sys-

TABLE 9.8.1
Typical failure and repair data for components other than pipes

Subcomponent	Mean-time-between-failure (hrs $\times 10^6$)	Mean-time-to-repair (hrs)	Preventive maintenance (hrs/yr)	Availability
Pumps	.032066	9.6	2.	.99116
Power transmission	.035620	2.3	7.	.99898
Motors	.066700	6.9	14.	.99816
Valves	.014440	11.6	41.	.96446
Controls	.083580	3.7	9.	.99870

tem reliability evaluation; however, their application in practice to water distribution systems has been minimal. Several recent developments on determining water distribution component reliability and system reliability are described in Mays (1989) and Mays et al. (1989).

One useful method for evaluating system reliability is the **minimum cut set method**. A **cut set** is defined as a set of elements which, if it fails, causes the system to fail regardless of the condition of the other elements in the system. A **minimal cut** is one in which there is no proper subset of elements whose failure alone will cause the system to fail. In other words, a minimal cut is such that if any component is removed from the set, the remaining elements collectively are no longer a cut set. The minimal cut sets are denoted as C_i, $i = 1, \ldots, m$ and C_i' denotes the complement of C_i, that is, the failure of all elements of the cut C_i. The system reliability is

$$\alpha_s = 1 - P\left[\bigcup_{i=1}^{m} C_i\right] = P\left[\bigcap_{i=1}^{m} C_1'\right] \tag{9.8.3}$$

Example 9.8.1. Determine the system reliability for the five pipe water distribution network in Fig. 9.8.1 using the minimum cut set method. Node 1 is the source node and nodes 3, 4, and 5 are demand nodes. The components subject to possible failure are the five pipes, each of which has a 5 percent failure probability due to breakage or other causes that require it to be removed from service. The system reliability is defined as the probability that water can reach all three demand nodes from the source node. The states of serviceability of each pipe are independent.

Solution. Based on the system reliability as defined, the minimum cut sets for the example network are

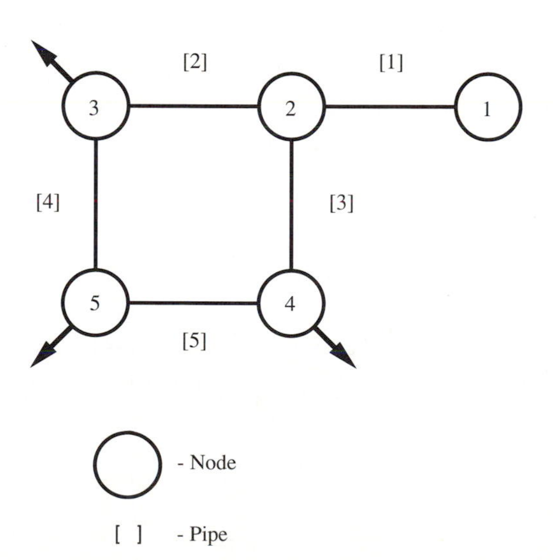

FIGURE 9.8.1
Example water distribution network.

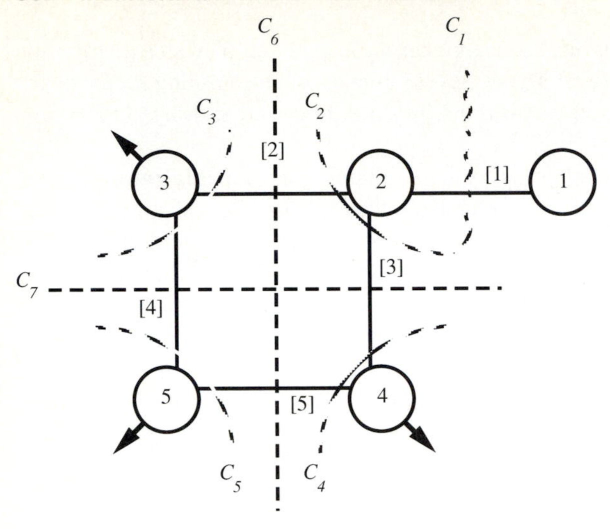

C_i = the ith outset
[ ] = pipe

FIGURE 9.8.2
Cut sets for the example water distribution network.

$$C_1 = \{F_1\}, \qquad C_2 = \{F_2 \bigcap F_3\},$$

$$C_3 = \{F_2 \bigcap F_4\}, \qquad C_4 = \{F_3 \bigcap F_4\},$$

$$C_5 = \{F_4 \bigcap F_5\}, \qquad C_6 = \{F_2 \bigcap F_5\},$$

$$C_7 = \{F_3 \bigcap F_4\},$$

where C_i = the ith cut set and F_k = the failure state of pipe link k. The above seven cut sets for the example network is shown in Fig. 9.8.2. The system unreliability α'_s is the probability of occurrence of the union of the cut set, that is,

$$\alpha'_s = P\left[\bigcup_{i=1}^{7} C_i\right]$$

The system reliability can be obtained by subtracting α'_s from 1. However, the computation, in general, will be very cumbersome for finding the probability of the union of large numbers of events, even if they are independent. In this circumstance, it is computationally easier to compute the system reliability, by Eq. (9.8.3) as

$$\alpha_s = P\left[\overline{\bigcup_{i=1}^{7} C_i}\right] = P\left[\bigcap_{i=1}^{7} C'_1\right]$$

where the overbar "−" represents the complement of the event. Since all cut sets behave independently, all their complements also behave independently. The probability of the intersection of a number of independent events, as described in Section 5.1, is

$$\alpha_s = \prod_{i=1}^{7} P(C'_i)$$

where

$$P(C_1') = 0.95,$$

$$P(C_2') = P(C_3') = \cdots = P(C_7') = 1. - (.05)(.05) = 0.9975$$

Hence, the system reliability of the example network is

$$\alpha_s = (0.95)(0.9975)^6 = 0.9360$$

REFERENCES

Alperovits, E. and U. Shamir: "Design of Optimal Water Distribution Systems," *Water Resources Research*, AGU, vol. 13, no. 6, pp. 885–900, 1977.

Calhoun, C.: "Optimization of Pipe Systems by Linear Programming," *Control of Flow in Closed Conduits*, J. P. Tullis, ed., Colorado State University, Ft. Collins, pp. 175–192, 1971.

Cullinane, Jr., M. J.: "Methodologies for the Evaluation of Water Distribution System Reliability/Availability," Ph.D. dissertation. University of Texas at Austin, May 1989.

Epp, R. and A. Fowler: "Efficient Code for Steady-State Flows in Networks," *Journal of the Hydraulics Division*, ASCE, vol. 96, no. HY1, pp. 43–56, January 1970.

Gessler, J. and T. Walski: "Water Distribution System Optimization," Technical Report EL-85-11, U.S. Army Engineer Waterways Experiment Station, Vicksburg, Miss., 1985.

Gupta, I.: "Linear Programming Analysis of a Water Supply System," *AIIE Trans.* **1**(1), pp. 56–61, 1969.

Gupta, I., M. S. Hassan and J. Cook: "Linear Programming Analysis of a Water Supply System with Multiple Supply Points," *AIIE Transactions* **4**(3), pp. 200–204, 1972.

Lansey, K. E. and L. W. Mays: "Water Distribution System Design for Multiple Loading," *Journal of Hydraulic Engineering*, ASCE, vol. 115, no. 10, October 1989.

Lasdon, L. S. and J. Mantell: "A GRG Algorithm for Econometric Control Problems," *Annuals of Economic and Social Management*, vol. 6, no. 5, 1978.

Lasdon, L. S. and A. D. Waren: GRG2 User's Guide, Department of General Business, University of Texas at Austin, 1982.

Lemieux, P., "Efficient Algorithm for Distribution Networks," *Journal of the Hydraulics Division*, ASCE, vol. 98, no. HY11, pp. 1911–1920, November 1972.

Linsley, R. K. and J. B. Franzini: *Water Resources Engineering*, 3d edition, McGraw-Hill, Inc., 1979.

Liu, K. T. H.: "The Numerical Analysis of Water Supply Network by Digital Computer," Proceedings of the Thirteenth Congress, International Association of Hydraulic Research, vol. 1, pp. 35–42, 1969.

Martin, D. W. and G. Peters: "The Application of Newton's Method to Network Analysis by Digital Computer," *Journal of the Institute of Water Engineers*, vol. 17, pp. 115–129, 1963.

Mays, L. W., ed.: *Reliability Analysis of Water Distribution Systems*, American Society of Civil Engineers, New York, 1989.

Mays, L. W., Y. Bao, L. Brion, M. J. Cullinane, Jr., N. Duan, K. Lansey, Y.-C. Su, and J. Woodburn: "New Methodologies for the Reliability-Based Analysis and Design of Water Distribution Systems," Technical Report CRWR 227, Center for Research in Water Resources, University of Texas at Austin, July 1989.

Morgan, D. R. and I. Goulter: "Optimal Urban Water Distribution Design," *Water Resources Research*, AGU, vol. 21, no. 5, pp. 642–652, 1985.

Norman, A. L., L. S. Lasdon, and J. K. Hsin: "A Comparison of Methods for Solving and Optimizing a Large Nonlinear Econometric Model," Discussion Paper, Center for Economic Research, University of Texas, Austin, 1982.

Quindry, G. E., E. D. Brill, Jr., and J. C. Liebman: "Optimization of Looped Water Distribution Systems," *Journal of Environmental Engineering*, ASCE, vol. 107, no. EE4, pp. 665–679, 1981.

Shamir, U., and C. D. Howard: "Water Distribution System Analysis," *Journal of Hydraulic Division*, ASCE, vol. 94, no. HY1, pp. 219–234, 1968.

Shamir, U.: "Optimization in Water Distribution Systems Engineering," *Mathematical Programming*, no. 11, pp. 65–75, 1979.

Wood, D. and C. Charles: "Hydraulic Network Analysis Using Linear Theory," *Journal of Hydraulics Division*, ASCE, vol. 98, no. HY7, pp. 1157–1170, 1972.

Wood, D.: "Computer Analysis of Flow in Pipe Networks Including Extended Period Simulation–User's Manual," Office of Engineering, Continuing Education and Extension, University of Kentucky, 1980.

PROBLEMS

9.5.1 Complete all the constraints for the LP model in Example 9.5.1.

9.5.2 Solve Example 9.5.1 using GAMS or any other available LP code.

9.5.3 Using the branching network in Fig. 9.5.1 for Example 9.5.1 resolve this problem using GAMS for the following demands

Node	5	6	7	9	10
Demand (cfs)	6	6	6	10	10

9.5.4 Develop the LP model to determine the minimum cost pipe diameters and pumpage for a system similar to the one in Fig. 9.5.1 except that no pump exists at node 3 and the pipes connecting nodes 3 to 8, 8 to 9, and 8 to 10 do not exist. All other information in Example 9.5.1 is valid.

9.5.5 Solve the LP model for Problem 9.5.4 using GAMS or any other available LP code.

9.6.1 Describe in detail the model developed by Alperovits and Shamir (1977) and Shamir (1979) in the minimum cost design of water distribution systems. Also discuss the solution procedure.

9.6.2 Describe in detail the model developed by Quindry, Brill, and Liebman (1981) for the minimum cost design of water distribution systems.

9.6.3 Set up the LP model described in Section 9.6.2 by Morgan and Goulter (1985) for the water distribution system in Fig. 9.P.1. Use the pipe cost data presented in Example 9.5.1.

9.6.4 Develop a nonlinear programming model (i.e. define the objective function and constraints) for the pipe network in Fig. 9.P.1. This network has two loops and eight pipes each 3,280 feet in length. All nodes are at the same elevation and the pressure head at the source, node 1, is 196.8 feet. The minimum nodal pressure head requirement at each node is 100 feet, and the mean Hazen-Williams roughness coefficient is 100 for each pipe. The cost for each pipe is

$$\text{Cost} = 0.331 L D^{1.51}$$

where D is the pipe diameter in inches and L is the pipe length in feet.

9.6.5 Solve the NLP model formulated in Problem 9.6.4 using GAMS-MINOS.

9.8.1 Use the minimum cut set method to determine the system reliability of the water distribution system network in Fig. 9.P.1 with a 3 percent failure probability for each pipe.

9.8.2 Studying the pipe break record of a water distribution system reveals that the number of pipe breaks can be expressed as

$$N(t) = N(t_0)e^{A(t-t_0)}$$

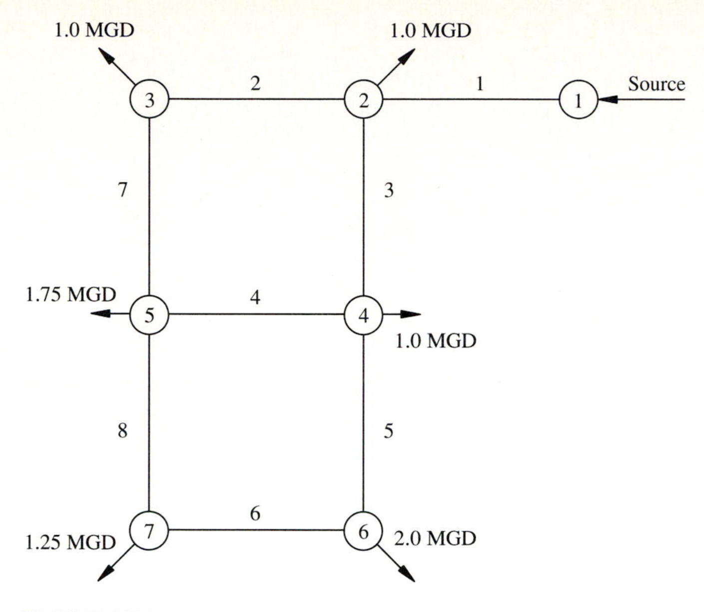

FIGURE 9.P.1

in which $N(t)$ = number of breaks per 1000-ft long pipe in year t; t = time (year); t_0 = base year when pipe was installed; and A = growth rate coefficient. The operational cost of the system with regard to a pipe consists of two items: (1) maintenance cost for repairing pipe breaks; and (2) cost of replacing the deteriorated pipe section. The unit cost of repairing breaks for a 1000-ft long pipe is C_b whereas the unit cost of replacing a 1000-ft long pipe is C_r. Consider that in the time frame t is continuous, allowing fractions of a year. The single payment present worth factor is $e^{-r(t-t_p)}$ in which t_p is the year to which the present value is to be computed and r is the nominal interest rate which is related to annual interest rate i as $r = \ln(1 + i)$. Therefore, the present value (PV_{t_p}) in year t_p for the cashflow $C(t)$, $t \geq t_p$ can be calculated using

$$PV_{t_p} = \int_{t_p}^{t} e^{-r(t-t_p)} C(t)\, dt$$

Use the above information to formulate an optimization model and solve the model to determine the optimal replacement time (t_r) that minimizes the present value of total cost at year t_p.

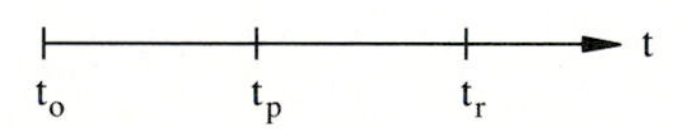

PART III

WATER EXCESS ENGINEERING AND MANAGEMENT

CHAPTER 10

HYDROLOGY AND HYDRAULICS FOR WATER EXCESS MANAGMENT

This chapter presents some of the basic hydrologic and hydraulic methodologies needed for water excess management, serving as a survey of the more widely used methods. For greater detail on the principles of these methodologies the reader is referred to the books (Bedient and Huber, 1988; Chow et al., 1988; Viessman et al., 1989; and Bras, 1990). The various types of methodologies include both stochastic or deterministic models and lumped or distributed models. Definitions for these model classifications follow those of Chow et al. (1988). **Deterministic** models do not consider randomness whereas **stochastic** models have outputs that are at least partially random. **Lumped** models spatially average a system, regarding the system as a single point in space without dimensions. **Distributed** models consider various points in space and define model variables as functions of space dimensions. The major topics discussed herein are rainfall-runoff analysis (deterministic, lumped), river and reservoir hydrologic routing (deterministic, lumped), flood flow frequency analysis (stochastic, lumped), water surface profile analysis (deterministic, distributed), and hydraulic routing (deterministic, distributed) for flood forecasting.

10.1 FLOODPLAIN HYDROLOGIC AND HYDRAULIC ANALYSIS

The hydrologic and hydraulic analysis of floods is required for the planning, design, and management of many types of facilities including hydrosystems within a floodplain or watershed. These analyses are needed for determining potential flood elevations and depths, areas of inundation, sizing of channels, levee heights, right of way limits, design of highway crossings and culverts, and many others. The typical requirements include (Hoggan, 1989):

1. *Floodplain information studies.* Development of information on specific flood events such as the 10-, 100-, and 500-year frequency events.
2. *Evaluations of future land-use alternatives.* Analysis of a range of flood events (different frequencies) for existing and future land uses to determine flood hazard potential, flood damage, and environmental impact.
3. *Evaluation of flood loss reduction measures.* Analysis of a range of flood events (different frequencies) to determine flood damage reduction associated with specific design flows.
4. *Design studies.* Analysis of specific flood events for sizing facilities to assure their safety against failure.
5. *Operation studies.* Evaluation of a system to determine if the demands placed upon it by specific flood events can be met.

The methods used in hydrologic and hydraulic analysis are determined by the purpose and scope of the project and the data availability. Figure 10.1.1 is a schematic of hydrologic and hydraulic analysis for floodplain studies. The types of hydrologic analysis are either to perform a rainfall-runoff analysis or a flood flow frequency analysis. If an adequate number of historical annual instantaneous peak discharges (**annual maximum series**) are available then a flood flow frequency analysis can be performed to determine peak discharges for various return periods. Otherwise a rainfall-runoff analysis must be performed using a historical storm or a design storm for a particular return period to develop a storm runoff hydrograph.

Determination of water surface elevations can be performed using a steady-state water surface profile analysis if only peak discharges are known or one can select the peak discharges from generated storm runoff hydrographs. For a more detailed and comprehensive analysis, an unsteady flow analysis based upon a hydraulic routing model and requiring the storm runoff hydrograph can be used to more accurately define maximum water surface elevations. The unsteady flow analysis also provides more detailed information such as the routed discharge hydrographs at various locations throughout a river reach.

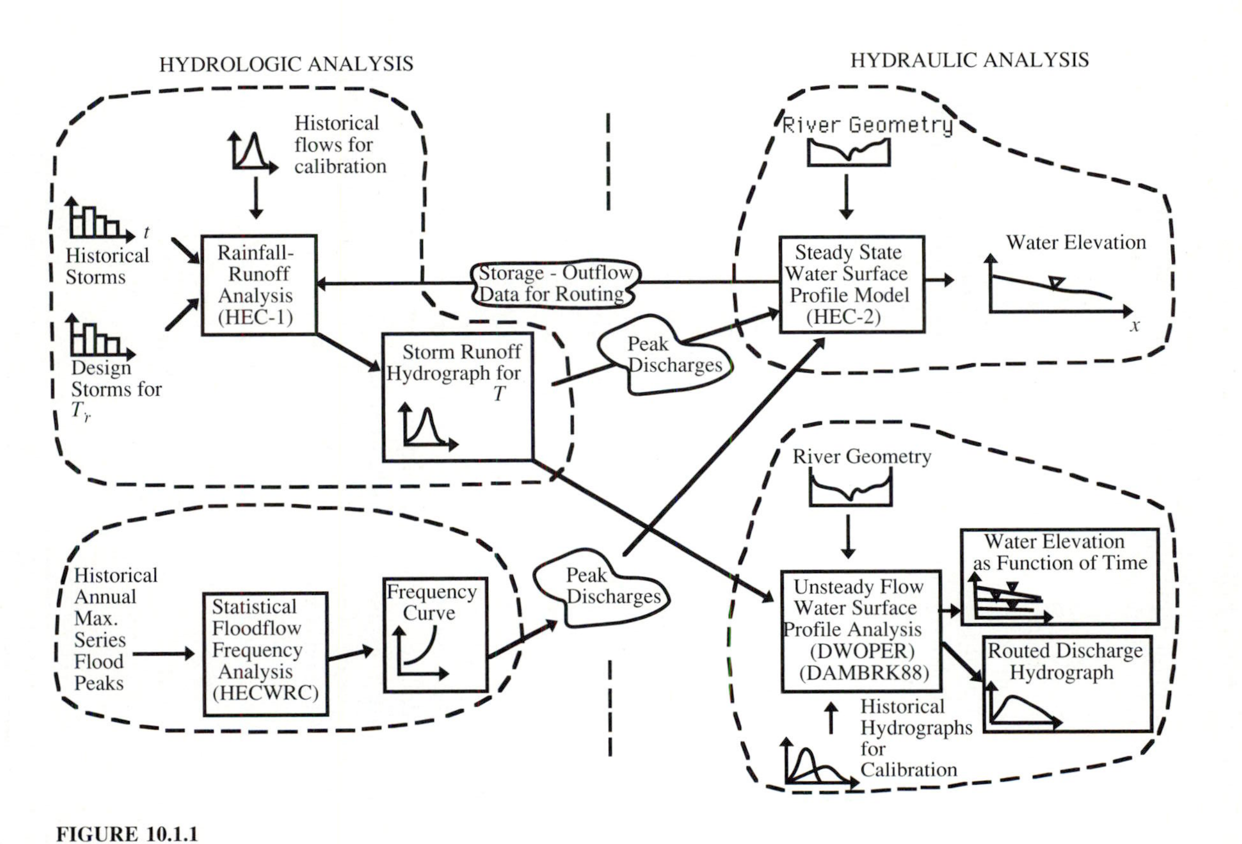

FIGURE 10.1.1
Components of a hydrologic-hydraulic floodplain analysis.

10.2 STORM HYDROGRAPH DETERMINATION: RAINFALL-RUNOFF ANALYSIS

A **streamflow** or **discharge hydrograph** is a graph or table showing the flow rate as a function of time at a given location on the stream (Chow et al., 1988). In effect, the hydrograph is "an integral expression of the physiographic and climatic characteristics that govern the relations between rainfall and runoff of a particular drainage basin" (Chow, 1964).

The objective of rainfall-runoff analysis is illustrated in Fig. 10.2.1 where the system is a watershed or river catchment, the input is the rainfall hyetograph, and the output is the runoff or discharge hydrograph. Figure 10.2.2 defines the variable processes (or steps) used to determine the total runoff hydrograph from the rainfall input. This section describes the unit hydrograph approach and its application to the determination of the storm runoff hydrograph.

10.2.1 Hydrologic Losses

Rainfall excess, or **effective rainfall**, is that rainfall which is neither retained on the land surface nor infiltrated into the soil. After flowing across the watershed surface, rainfall excess becomes **direct runoff** at the watershed outlet. The graph of rainfall excess vs. time, or the **rainfall excess hyetograph**, is a key component in the study of rainfall-runoff relationships. The difference between the observed total rainfall hyetograph and the rainfall excess hyetograph is the **abstractions**, or **losses**. Losses are primarily water absorbed by infiltration with some allowance for interception and surface storage.

Infiltration is the process of water penetrating from the ground surface into the soil. There are many factors that influence the infiltration rate, including the condition of the soil surface and its vegetative cover, the properties of the soil, such as its porosity and hydraulic conductivity, and the current moisture content of the soil. The **infiltration rate** f expressed in inches per hour or centimeters per hour, is the rate at which water enters the soil at the surface. Under conditions of ponded water on the surface, infiltration occurs at the **potential infiltration rate**. Most infiltration equations describe the potential rate. The **cumulative infiltration** F is the accumulated depth of water infiltrated during a specified time interval and is the integral of the infiltration rate over that interval of time

$$F_t = F(t) = \int_0^t f(\tau)\, d\tau \qquad (10.2.1)$$

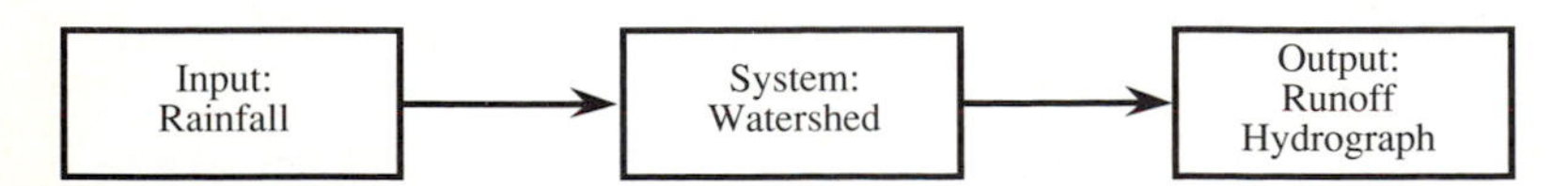

FIGURE 10.2.1
Rainfall-runoff modeling.

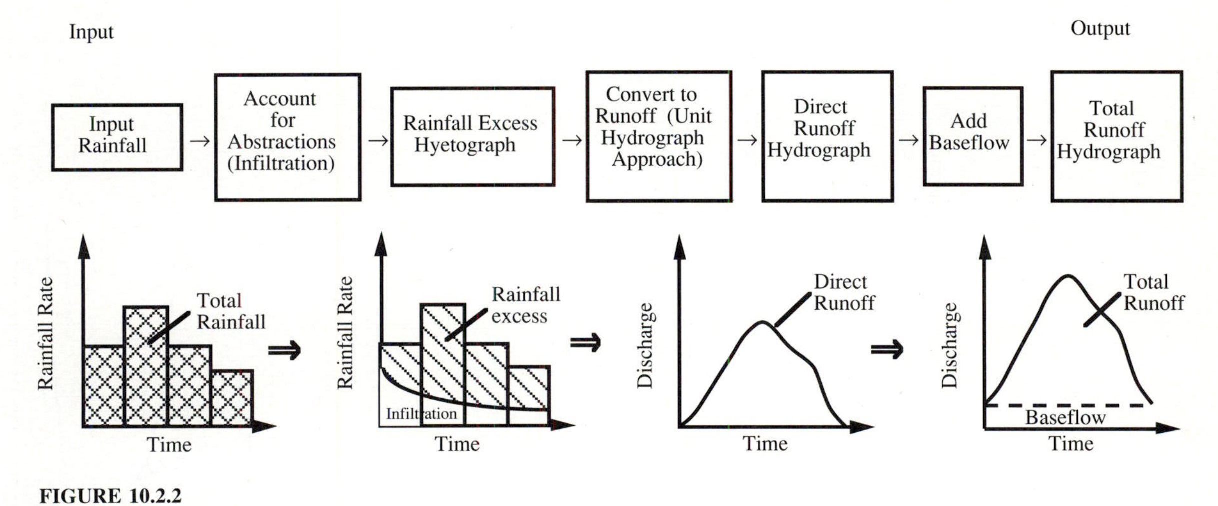

FIGURE 10.2.2
Steps to define storm runoff.

The infiltration rate is the time derivative of the cumulative infiltration. Three widely used and accepted infiltration equations (Green–Ampt, Horton, and SCS) are presented in Table 10.2.1. The interrelationships of rainfall, infiltration rate, and cumulative infiltration are shown in Fig. 10.2.3.

TABLE 10.2.1
Infiltration equations

Cumulative infiltration (F_t)	Infiltration rate (f_t)	Comments
Green-Ampt equation $F_t - \psi\Delta\theta \ln\left(1 + \frac{F_t}{\psi\Delta\theta}\right) = Kt$	$f_t = K\left(\frac{\psi\Delta\theta}{F_t} + 1\right)$	• Hydraulic conductivity (K) • Wetting front soil suction head (ψ) • Change in moisture content ($\Delta\theta$) $\Delta\theta = \eta - \theta_i$ • Porosity η • Initial moisture content (θ_i)
Horton's equation $F_t = f_c t + (f_0 - f_c)(1 - e^{-kt}$	$f_t = f_c + (f_0 - f_c)e^{-kt}$	• Constant infiltration rate (f_c) • Decay constant (k)
SCS method $F_t = \frac{S(P_t + I_a)}{P_t - I_a + S}$	$f_t = \frac{S^2 \frac{dP_t}{dt}}{(P_t - I_a + S)^2}$	• Potential maximum retention (S) $S = \frac{1000}{CN} - 10$ • Dimensionless curve number (CN) $0 \leq CN \leq 100$ • Initial abstraction I_a $I_a = 0.2S$ • Total rainfall to time t, P_t

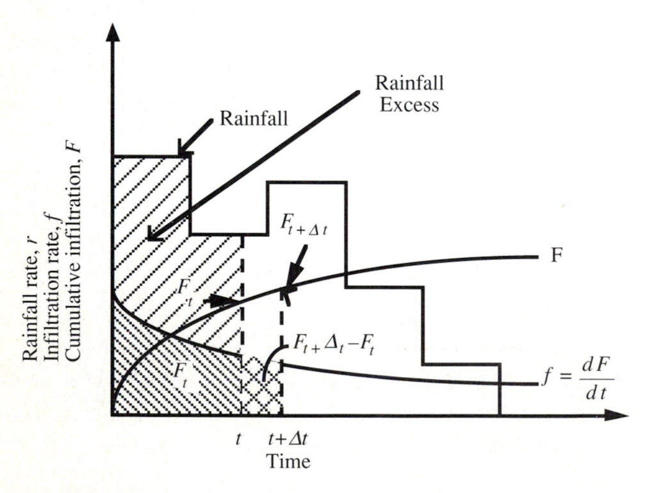

FIGURE 10.2.3
Rainfall, infiltration rate, and cumulative infiltration.

10.2.2 Unit Hydrograph Approach

A **unit hydrograph** is the direct runoff hydrograph resulting from 1 in (or 1 cm in SI units) of rainfall excess generated uniformly over a drainage area at a constant rate for an effective duration. Essentially the unit hydrograph is the unit pulse response function of a linear hydrologic system. The unit hydrograph is a simple linear model that can be used to derive the hydrograph resulting from any amount of rainfall excess. The following basic assumptions are inherent in the unit hydrograph approach:

1. The rainfall excess has a constant intensity within the effective duration.
2. The rainfall excess is uniformly distributed throughout the entire drainage area.
3. The base time of the direct runoff hydrograph resulting from a rainfall excess of given duration is constant.
4. The ordinates of all direct runoff hydrographs of a common base time are directly proportional to the total amount of direct runoff represented by each hydrograph.
5. For a given watershed, the hydrograph resulting from a given rainfall excess reflects the unchanging characteristics of the watershed.

The following discrete **convolution equation** is used to compute direct runoff Q_n given rainfall excess P_m and the unit hydrograph U_{n-m+1} (Chow et al., 1988)

$$Q_n = \sum_{m=1}^{n \le M} P_m U_{n-m+1} \tag{10.2.2}$$

in which n represents the time. The reverse process called **deconvolution**, is used to derive a unit hydrograph given P_m and Q_n. Suppose that there are M pulses of excess rainfall and N pulses of direct runoff in the storm considered; then N equations can be written for $Q_n, n = 1, 2, \ldots, N$, in terms of $N - M + 1$ unknown unit hydrograph ordinates.

When observed rainfall-runoff data are not available, a synthetic unit hydrograph must be used. The most commonly used method is Snyder's method. This method relates the time from the centroid of the rainfall to the peak of the unit hydrograph to physiographical characteristics of the watershed.

Once the unit hydrograph has been determined, it may be applied to find the direct runoff and streamflow hydrographs. A rainfall hyetograph is selected from which the abstractions are subtracted to define the rainfall excess hyetograph. The time interval used in defining the rainfall excess hyetograph ordinates must be the same as that for which the unit hydrograph was specified.

10.2.3 U.S. Army Corps of Engineers Hydrologic Engineering Center, HEC-1

The HEC-1 computer program was developed by the U.S. Army Corps of Engineers (1990) Hydrologic Engineering Center (HEC) for simulating the rainfall-runoff process for watersheds ranging in size and complexity from small urban catchments to large multibasin river systems. This model can be used to determine runoff from synthetic as well as historical events. A watershed or basin is represented as an interconnected system of components (see Fig. 10.2.4), each of which models an aspect of the rainfall-runoff process within a subbasin. The components are the land surface runoff component, the streamflow routing component, the reservoir component, the diversion component, and a pump component.

The unit hydrograph or kinematic wave procedure can be used in the land surface runoff component to determine the direct runoff hydrograph. The land surface runoff components for the example basin in Fig. 10.2.4 are 10, 20, 30, 40, 50, and 60. The stream routing components, 1020, 3040, 2050, 5060, and 6070 in Fig. 10.2.4, are used to represent the flood movement in the channel using hydrologic routing. The input is the upstream hydrograph which is routed to a downstream point using the Muskingum method, the level pool routing method, or the kinematic wave routing method. A reservoir component, such as 70 in Fig. 10.2.4, is similar to the streamflow routing component using a level pool routing procedure. The diversion component can be used to represent channel diversions, stream bifurcation, or any transfer of flow from one point in a watershed to another point in or out of a particular watershed. The pump component can be used in simulating pumping plants that lift runoff out of low ponding areas such as behind levees.

10.2.4 Continuous Simulation Models

The HEC-1 model, which uses the unit hydrograph approach, is an **event model** which is used to simulate individual rainfall-runoff events. Event models emphasize

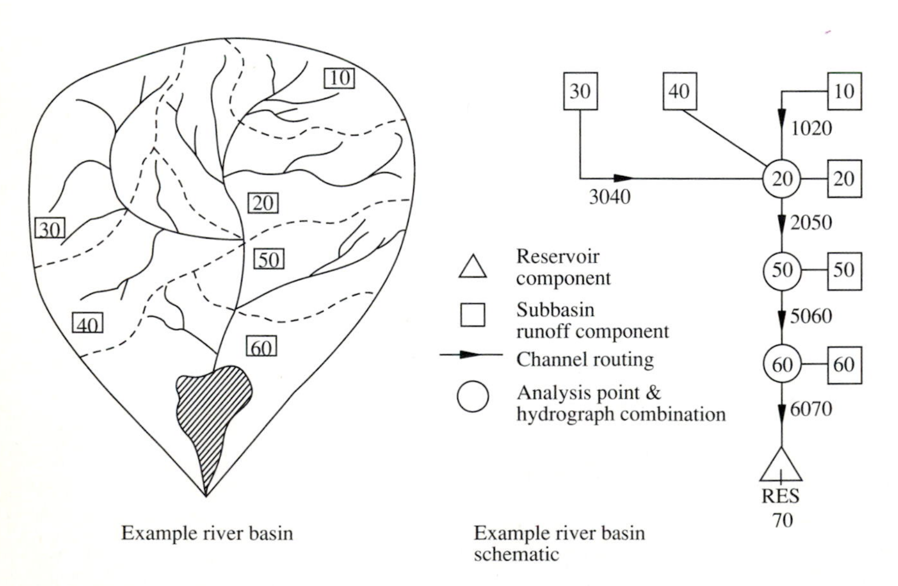

FIGURE 10.2.4
Rainfall-runoff modeling using HEC-1 (U.S. Army Corps of Engineers, 1989).

infiltration and surface runoff with the objective to determine direct runoff. These models are used to calculate flood flows where the major contributor to runoff is direct runoff. Event models do not consider moisture accounting between storm events. **Continuous models** explicitly account for all runoff components including both surface flow and indirect runoff (interflow and groundwater flow). These models account for the overall moisture balance of a watershed on a long term basis and therefore are suited for long term runoff-volume forecasting. Continuous models consider evapotranspiration and long term abstractions that define the rate of moisture recovery during periods without precipitation.

Three continuous models used in the U.S. are: (1) the Streamflow Synthesis and Reservoir Regulation (SSARR) model developed by the U.S. Army Corps of Engineer North Pacific Division (1986); (2) the Stanford Watershed Model (SWM) developed at Stanford University (Crawford and Linsley, 1966); and (3) the Sacramento Model developed by the Joint Federal—State River Forecast Center, the U.S. National Weather Service, and the State of California Department of Water Resources (Burnach, Ferral, and McGuire, 1973). A modified version of the Sacramento model was incorporated into the U.S. National Weather Service River Forecast System (NWSRFS) (Peck, 1976). The Sacramento model is discussed in more detail in the following paragraphs. Because the calculation of runoff from rainfall is based upon soil moisture accounting, the Sacramento model is also referred to as the Sacramento soil-moisture accounting model.

The Sacramento model is deterministic and has lumped input and lumped parameters within a soil moisture accounting area. This model, which can be used to model headwaters, divides the soil vertically into two main soil moisture accounting zones. The **upper zone** accounts for interception storage and the upper soil layer; the **lower zone** accounts for the bulk of the soil moisture and the ground water storage capacity. A conceptual diagram of the Sacramento model is shown in Fig. 10.2.5.

The upper and lower zones store **tension water** and free water (see Fig. 10.2.5). Tension water is tightly bound to the soil particles and therefore is not readily available for movement. Free water is not bound to the soil particles and can move both horizontally and vertically through the soil profile. Tension water is depleted only by evapotranspiration, whereas free water is transferred by percolation, interflow, evapotranspiration, and tension water replenishment. In the upper zone, tension water requirements (**upper zone tension water storage**, UZTWS) must be met before water is transferred to free water storage. In the lower zone, part of the incoming water can become free water without fulfilling tension water requirements.

Movement of water from the upper to the lower zone is determined by a percolation function that relates capacities and contents of both zones and free water depletion coefficients. Two types of evapotranspiration information are used: (1) seasonal evapotranspiration demand curve consisting of average monthly values, or (2) actual potential evapotranspiration data with monthly adjustment factors to account for seasonal changes in the vegetative cover and ground condition.

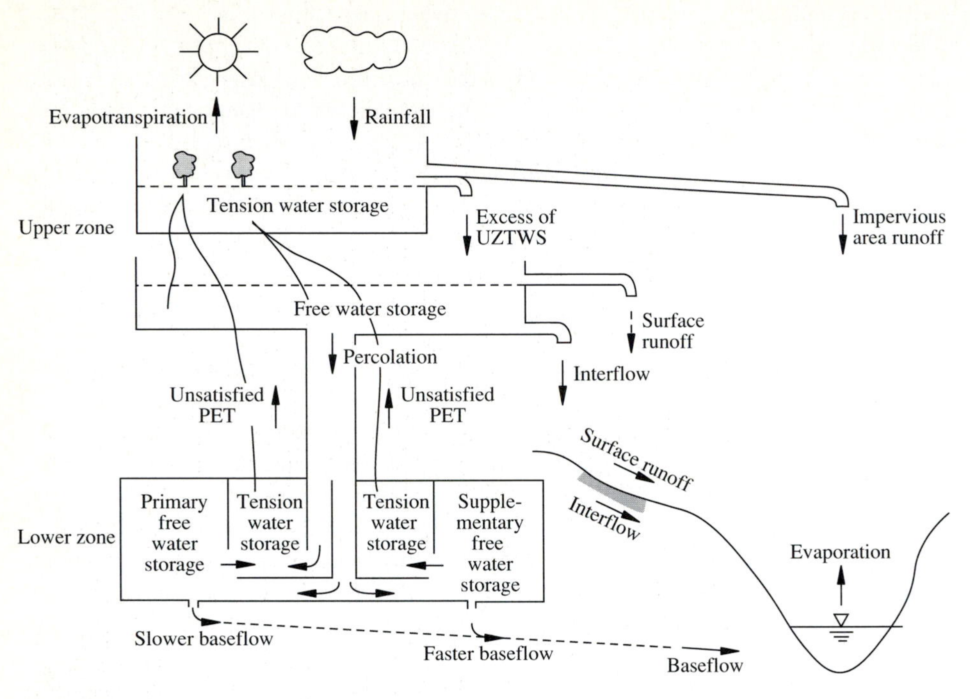

FIGURE 10.2.5
Conceptual diagram of Sacramento model (Brazil and Hudlow, 1981).

The NWS version of the Sacramento model uses 6-hour computational time intervals for calibration simulations and for operational forecasts (Brazil and Hudlow, 1981). Both pervious and impervious areas are considered. Impervious areas directly connected to a stream contribute directly to stream flow without traveling through the soil. Saturated soils near a stream also act as impervious areas and this area can change in size depending upon the soil moisture. The Sacramento model generates five components of channel flow (Brazil and Hudlow, 1981):

1. Direct runoff resulting from precipitation applied to the impervious and temporary impervious areas.
2. Surface runoff resulting form precipitation input which is at a greater rate than the upper zone intake.
3. Interflow which is lateral drainage from the upper zone free water storage.
4. Supplemental base flow which is drainage from the lower zone free water supplemental storage.
5. Primary base flow which is drainage from the lower free water primary storage.

The channel inflow for each time interval is the sum of the above flow components.

10.3 HYDROLOGIC ROUTING: RESERVOIRS AND RIVERS

Flow routing is a procedure to determine the time and magnitude of flow (i.e., the flow hydrograph) at a point on a watercourse from known or assumed hydrographs at one or more points upstream. If the flow is a flood, the procedure is specifically known as flood routing. Routing by lumped system methods is called **hydrologic routing**, and routing by distributed systems methods is called **hydraulic routing**. Flow routing by distributed-system methods is described in Section 10.6.

For hydrologic routing, input $I(t)$, output $Q(t)$, and storage $ST(t)$ are related by the continuity equation

$$\frac{dST(t)}{dt} = I(t) - Q(t) \qquad (10.3.1)$$

If an inflow hydrograph, $I(t)$, is known, Eq. (10.3.1) cannot be solved directly to obtain the outflow hydrograph, $Q(t)$, because both Q and ST are unknown. A second relationship, or storage function, is required to relate ST, I, and Q; coupling the storage function with the continuity equations provides a solvable combination of two equations and two unknowns.

The specific form of the storage function depends on the nature of the system being analyzed. In reservoir routing by the level pool method, storage is a nonlinear function of Q, $ST = f(Q)$ and the function $f(Q)$ is determined by relating reservoir storage and outflow to reservoir water level. In the Muskingum method for channel flow routing, storage is linearly related to I and Q.

The effect of storage is to redistribute the hydrograph by shifting the centroid of the inflow hydrograph to the position of that of the outflow hydrograph in a time of redistribution. In very long channels, the entire flood wave also travels a considerable distance and the centroid of its hydrograph may then be shifted by a time period longer than the time of redistribution. This additional time may be considered as the time of translation. The total time of flood movement between the centroids of the inflow and outflow hydrographs is equal to the sum of the time of redistribution and the time of translation. The process of redistribution modifies the shape of the hydrograph, while translation changes its position.

10.3.1 Hydrologic Reservoir Routing

Level pool routing is a procedure for calculating the outflow hydrograph from a reservoir assuming a horizontal water surface, given its inflow hydrograph and storage-outflow characteristics. The change in storage over a time interval Δt, $ST_{j+1} - ST_j$, can be expressed as

$$ST_{j+1} - ST_j = \left(\frac{I_j + I_{j+1}}{2}\right)\Delta t - \left(\frac{Q_j + Q_{j+1}}{2}\right)\Delta t \qquad (10.3.2)$$

from Eq. (10.3.1). The inflow values at the beginning and end of the jth time interval are I_j and I_{j+1}, respectively, and the corresponding values of the outflow are Q_j and Q_{j+1}. The values of I_j and I_{j+1} are known from the inflow hydrograph. The values of Q_j and ST_j are known at the beginning of the jth time interval from calculations for the previous time interval. Hence, Eq. (10.3.2) contains two unknowns, Q_{j+1} and ST_{j+1}, which are isolated by multiplying Eq. (10.3.2) through by $2/\Delta t$, and rearranging the result to produce:

$$\left(\frac{2ST_{j+1}}{\Delta t} + Q_{j+1}\right) = (I_j + I_{j+1}) + \left(\frac{2ST_j}{\Delta t} - Q_j\right) \tag{10.3.3}$$

In order to calculate the outflow, Q_{j+1}, a storage-outflow function relating $2ST/\Delta t + Q$ and Q is needed, which can be developed using elevation-storage and elevation-outflow relationships. The relationship between water surface elevation and reservoir storage can be derived from topographic maps. The elevation-discharge relation is derived from equations relating head and discharge for various types of spillways and outlet works. The value of Δt is taken as the time interval of the inflow hydrograph. For a given value of water surface elevation, the values of storage ST and discharge Q are determined, then the value of $2ST/\Delta t + Q$ is calculated and plotted on the horizontal axis of a graph with the value of the outflow Q on the vertical axis.

In routing the flow through time interval j, all terms on the RHS of Eq. (10.3.3) are known, and so the value of $2ST_{j+1}/\Delta t + Q_{j+1}$ can be computed. The corresponding value of Q_{j+1} can be determined from the storage-outflow function $2ST/\Delta t + Q$ versus Q, either graphically or by linear interpolation of tabular values. To set up the data required for the next time interval, the value of $(2ST_{j+1}/\Delta t - Q_{j+1})$ is calculated using

$$\left(\frac{2ST_{j+1}}{\Delta t} - Q_{j+1}\right) = \left(\frac{2ST_{j+1}}{\Delta t} + Q_{j+1}\right) - 2Q_{j+1} \tag{10.3.4}$$

The computation is then repeated for subsequent routing periods.

10.3.2 Hydrologic River Routing

The **Muskingum method** is a commonly used hydrologic river routing method that is based upon a variable discharge-storage relationship. This method models the storage volume of flooding in a river channel by a combination of wedge and prism storages. During the advance of a flood wave, inflow exceeds outflow, producing a wedge of storage. During the recession, outflow exceeds inflow, also resulting in a wedge. In addition, there is a prism of storage which is formed by a volume of constant cross section along the length of a prismatic channel.

Assuming that the cross-sectional area of the flood flow is directly proportional to the discharge at the section, the **volume of prism storage** is equal to KQ where K is a proportionality coefficient, and the **volume of wedge storage** is equal to $KX(I - Q)$, where X is a weighting factor having the range $0 \leq X \leq 0.5$. The total storage is defined as the sum of two components,

$$ST = KQ + KX(I - Q) \tag{10.3.5}$$

which can be rearranged to give the storage function for the Muskingum method

$$ST = K[XI + (1 - X)Q] \tag{10.3.6}$$

and represents a linear model for routing flow in streams.

The value of X depends on the shape of the modeled wedge storage. The value of X ranges from 0 for reservoir-type storage to 0.5 for a full wedge. When $X = 0$, there is no wedge and hence no backwater; this is the case for a level-pool reservoir. In natural streams, X is between 0 and 0.3 with a mean value near 0.2. Great accuracy in determining X may not be necessary because the results of the method are relatively insensitive to the value of this parameter. The parameter K is the time of travel of the flood wave through the channel reach. For hydrologic routing, the values of K and X are assumed to be constants throughout the range of flow.

The values of storage at times j and $j + 1$ can be written, respectively, as

$$ST_j = K[XI_j + (1 - X)Q_j] \tag{10.3.7}$$

and

$$ST_{j+1} = K[XI_{j+1} + (1 - X)Q_{j+1}] \tag{10.3.8}$$

Using Eqs. (10.3.7) and (10.3.8), the change in storage over time interval Δt is

$$ST_{j+1} - ST_j = K\{[XI_{j+1} + (1 - X)Q_{j+1}] - [XI_j + (1 - X)Q_j]\} \tag{10.3.9}$$

The change in storage can also be expressed, using Eq. (10.3.2). Combining Eqs. (10.3.9) and (10.3.2) and simplifying gives

$$Q_{j+1} = C_1 I_{j+1} + C_2 I_j + C_3 Q_j \tag{10.3.10}$$

which is the routing equation for the Muskingum method where

$$C_1 = \frac{\Delta t - 2KX}{2K(1 - X) + \Delta t} \tag{10.3.11}$$

$$C_2 = \frac{\Delta t + 2KX}{2K(1 - X) + \Delta t} \tag{10.3.12}$$

$$C_3 = \frac{2K(1 - X) - \Delta t}{2K(1 - X) + \Delta t} \tag{10.3.13}$$

Note that $C_1 + C_2 + C_3 = 1$.

If observed inflow and outflow hydrographs are available for a river reach, the values of K and X can be determined. Assuming various values of X and using known values of the inflow and outflow, successive values of the numerator and denominator of the following expression for K, derived from Eqs. (10.3.9) and (10.3.2), can be computed using

$$K = \frac{0.5\Delta t[(I_{j+1} + I_j) - (Q_{j+1} + Q_j)]}{X(I_{j+1} - I_j) + (1 - X)(Q_{j+1} - Q_j)} \tag{10.3.14}$$

The computed values of the numerator and denominator are plotted for each time interval, with the numerator on the vertical axis and the denominator on the horizontal axis. This usually produces a graph in the form of a loop. The value of X that produces a loop closest to a single line is taken to be the correct value for the reach, and K, according to Eq. (10.3.14), is equal to the slope of the line. Since K is the time required for the flood wave to traverse the reach, its value may also be estimated as the observed time of travel of peak flow through the reach.

10.4 HYDROLOGIC FREQUENCY ANALYSIS FOR FLOODPLAIN DETERMINATION

10.4.1 Flood Flow Frequency Analysis

The primary objective of the frequency analysis of hydrologic data is to determine the recurrence interval of a hydrologic event of a given magnitude. The **recurrence interval** is defined as the average interval of time within which the magnitude of a hydrologic event will be equaled or exceeded once, on the average. The term **frequency** is often used interchangeably with recurrence interval; however, it should not be construed to indicate a regular or stated interval of occurrence or recurrence. **Hydrologic frequency analysis** is the approach of using probability and statistical analysis to estimate future frequencies (probabilities of hydrologic events occurring) based upon information contained in hydrologic records. Because of the range of uncertainty and diversity of methods in determining flood flow estimates and the varying results that can be obtained using the various methods, the U.S. Water Resources Council (WRC) (1981) attempted to promote a uniform or consistent approach to flood flow frequency studies.

FREQUENCY FACTOR METHOD. Computation of the magnitudes of extreme events, such as flood flows, requires that the probability distribution function be invertible, that is, given a value for return period T or $[F(x_T) = T/(T-1)]$, the corresponding value of x_T can be determined. The magnitude x_T of a hydrologic event may be represented as the mean $\overline{x}$ plus a departure of the variate from the mean. This departure is equal to the product of the standard deviation s_x and a **frequency factor** K_T. The departure Δx_T and the frequency factor K_T are functions of the return period and the type of probability distribution to be used in the analysis, as Chow (1951) proposed the following frequency factor equation

$$x_T = \overline{x} + K_T s_x \tag{10.4.1}$$

When the variable analyzed is $y = \log(x)$ or $y = \ln(x)$, then the same method is applied to the statistics for the logarithms of the data, using

$$y_T = \overline{y} + K_T s_y \tag{10.4.2}$$

and the required value of x_T is found by taking the antilog of y_T. For a given distribution, a $K-T$ relationship can be determined between the frequency factor and the corresponding return period. Table 10.4.1 lists values of the frequency factor for the Pearson Type III (and log-Pearson Type III) distribution for various values of the return period and coefficient of skewness.

10.4.2 U.S. Water Resources Council Guidelines

The U.S. Water Resources Council (WRC) has recommended that the log-Pearson III be used as a base method for flood flow frequency studies (U.S. Water Resources Council, 1981). This was an attempt to promote a consistent, uniform approach to flood flow frequency determination for use in all federal planning involving water and related land resources. This choice of the log-Pearson Type III is, however, to some extent arbitrary, in that no rigorous statistical criteria exist on which a comparison of distributions can be based.

The skew coefficient is very sensitive to the size of the sample; thus, it is difficult to obtain an accurate estimate from small samples. Because of this, the U.S. Water Resources Council (1981) recommended using a generalized estimate of the skew coefficient when estimating the skew for short records. As the length of record increases, the skew is usually more reliable. The guidelines recommend the use of a weighted skew, G_w, based upon the equation

$$G_w = WG_s + (1 - W)G_m \tag{10.4.3}$$

where W is a weight, G_s is the skew coefficient computed using the sample data, and G_m is a map skew, values of which are found in Fig. 10.4.1. The weighted skew is derived as a weighted average between skew coefficients computed from sample data (sample skew) and regional or map skew coefficients (referred to as a generalized skew in U.S. Water Resources Council, 1981). The weight that minimizes the variance or mean square error of the weighted skew can be determined by

$$W = \frac{\text{Var}(G_m)}{\text{Var}(G_s) + \text{Var}(G_m)} \tag{10.4.4}$$

where $\text{Var}(G_s)$ is the variance of the sample skew and $\text{Var}(G_m)$ is the variance of the map skew. Determination of W using Eq. (10.4.4) requires the values of $\text{Var}(G_m)$ and $\text{Var}(G_s)$. The value of $\text{Var}(G_m)$, estimated for the map skew provided by the WRC, is 0.3025. Alternatively, $\text{Var}(G_m)$ could be derived from a regression study relating the skew to physiographical and meteorological characteristics of the basins and determining $\text{Var}(G_m)$ as the square of the standard error of the regression equation (Tung and Mays, 1981).

The weighted skew G_w can be determined by substituting Eq. (10.4.4) into Eq. (10.4.3), resulting in

$$G_w = \frac{\text{Var}(G_m) \cdot G_s + \text{Var}(G_s) \cdot G_m}{\text{Var}(G_m) + \text{Var}(G_s)} \tag{10.4.5}$$

The variance (mean square error) of the station skew for log-Pearson Type III random variables can be obtained from the results of work by Wallis et al. (1974). Their results showed that the $\text{Var}(G_s)$ is a function of record length and population skew. For use in calculating G_w, this function $\text{Var}(G_s)$ can be approximated with sufficient accuracy using

$$\text{Var}(G_s) = 10^{A-B[\log(N/10)]} \tag{10.4.6}$$

where

TABLE 10.4.1
K_T values for Pearson Type III distribution

	Recurrence interval in years										
	1.0101	**1.0526**	**1.1111**	**1.2500**	**2**	**5**	**10**	**25**	**50**	**100**	**200**
	Exceedance probability										
Skew coeff.	**.99**	**.95**	**.90**	**.80**	**.50**	**.20**	**.10**	**.04**	**.02**	**.01**	**.005**
	Positive Skew										
3.0	−0.667	−0.665	−0.660	−0.636	−0.396	0.420	1.180	2.278	3.152	4.051	4.970
2.9	−0.690	−0.688	−0.681	−0.651	−0.390	0.440	1.195	2.277	3.134	4.013	4.909
2.8	−0.714	−0.711	−0.702	−0.666	−0.384	-.460	1.210	2.275	3.114	3.973	4.847
2.7	−0.740	−0.736	−0.724	−0.681	−0.376	0.479	1.224	2.272	3.093	3.932	4.783
2.6	−0.769	−0.762	−0.747	−0.696	−0.368	0.499	1.238	2.267	3.071	3.889	4.718
2.5	−0.799	−0.790	−0.771	−0.711	−0.360	0.518	1.250	2.262	3.048	3.845	4.652
2.4	−0.832	−0.819	−0.795	−0.725	−0.351	0.537	1.262	2.256	3.023	3.800	4.484
2.3	−0.867	−0.850	−0.819	−0.739	−0.341	0.555	1.274	2.248	2.997	3.753	4.515
2.2	−0.905	−0.882	−0.844	−0.752	−0.330	0.574	1.284	2.240	2.970	3.705	4.444
2.1	−0.946	−0.914	−0.869	−0.765	−0.319	0.592	1.294	2.230	2.942	3.656	4.372
2.0	−0.990	−0.949	−0.895	−0.777	−0.307	0.609	1.302	2.219	2.912	3.605	4.298
1.9	−1.037	−0.984	−0.920	−0.788	−0.294	0.627	1.310	2.207	2.881	3.553	4.223
1.8	−1.087	−1.020	−0.945	−0.799	−0.282	0.643	1.318	2.193	2.848	3.499	4.147
1.7	−1.140	−1.056	0.970	−0.808	−0.268	0.660	1.324	2.179	2.815	3.444	4.069
1.6	−1.197	−1.093	−0.994	−0.817	−0.254	0.675	1.329	2.136	2.780	3.388	3.990
1.5	−1.256	−1.131	−1.018	−0.825	−0.240	0.690	1.333	2.146	2.743	3.330	3.910
1.4	−1.318	−1.168	−1.041	−0.832	−0.225	0.705	1.337	2.128	2.706	3.271	3.838
1.3	−1.383	−1.260	−1.064	−0.838	−0.210	0.719	1.339	2.108	2.666	3.211	3.745
1.2	−1.449	−1.243	−1.086	−0.844	−0.195	0.732	1.340	2.087	2.626	3.149	3.661
1.1	−1.518	−1.280	−1.107	−0.848	−0.180	0.745	1.341	2.066	2.585	3.087	3.575
1.0	−1.588	−1.317	−1.128	−0.852	−0.164	0.758	1.340	2.043	2.542	3.022	3.489
.9	−1.660	−1.353	−1.147	−0.854	−0.148	0.769	1.339	2.018	2.498	2.957	3.401
.8	−1.733	−1.388	−1.166	−0.856	−0.132	0.780	1.336	1.993	2.453	2.891	3.312
.7	−1.806	−1.423	−1.183	−0.857	−0.116	0.790	1.333	1.967	2.407	2.824	3.223
.6	−1.880	−1.458	−1.200	−0.857	−0.099	0.800	1.328	1.939	2.359	2.755	3.132
.5	−1.955	−1.491	−1.216	−0.856	−0.083	0.808	1.323	1.910	2.311	2.686	3.041
.4	−2.029	−1.524	−1.231	−0.855	−0.066	0.816	1.317	1.880	2.261	2.615	2.949
.3	−2.104	−1.555	−1.245	−0.853	−0.050	0.824	1.309	1.849	2.211	2.544	2.856
.2	−2.178	−1.586	−1.258	−0.850	−0.033	0.830	1.301	1.818	2.159	2.472	2.763
.1	−2.252	−1.616	−1.270	−0.846	−0.017	0.836	1.292	1.785	2.107	2.400	2.670
.0	−2.326	−1.645	−1.282	−0.842	0	0.842	1.282	1.751	2.054	2.326	2.576

$$A = -0.33 + 0.08|G_s| \quad \text{if} \quad |G_s| \leq 0.90 \tag{10.4.7a}$$

$$-0.52 + 0.30|G_s| \quad \text{if} \quad |G_s| > 0.90 \tag{10.4.7b}$$

$$B = 0.94 - 0.26|G_s| \quad \text{if} \quad |G_s| < 1.50 \tag{10.4.7c}$$

$$0.55 \quad \text{if} \quad |G_s| > 1.50 \tag{10.4.7d}$$

in which $|G_s|$ is the absolute value of the sample skew for the station record (used as an estimate of population skew) and N is the record length in years.

TABLE 10.4.1
K_T values for Pearson Type III distribution (*Continued*)

	Recurrence interval in years										
	1.0101	**1.0526**	**1.1111**	**1.2500**	**2**	**5**	**10**	**25**	**50**	**100**	**200**
	Exceedance probability										
Skew coeff.	**.99**	**.95**	**.90**	**.80**	**.50**	**.20**	**.10**	**.04**	**.02**	**.01**	**.005**
	Negative Skew										
−.1	−2.400	−1.673	−1.292	−0.836	0.017	0.846	1.270	1.716	1.000	2.252	2.482
−.2	−2.472	−1.700	−1.301	−0.830	0.033	0.850	1.258	1.680	1.945	2.178	2.388
−.3	−2.544	−1.726	−1.309	−0.824	0.050	0.853	1.245	1.643	1.890	2.104	2.294
−.4	−2.615	−1.750	−1.317	−0.816	0.066	0.855	1.231	1.606	1.834	2.029	2.201
−.5	−2.686	−1.774	−1.323	−0.808	0.083	0.856	1.216	1.567	1.777	1.955	2.108
−.6	−2.755	−1.797	−1.328	−0.800	0.099	0.857	1.200	1.528	1.720	1.880	2.016
−.7	−2.824	−1.819	−1.333	−0.790	0.116	0.857	1.183	1.488	1.663	1.806	1.929
−.8	−2.891	−1.839	−1.336	−0.780	0.132	0.856	1.166	1.448	1.606	1.733	1.837
−.9	−2.957	−1.858	−1.339	−0.769	0.148	0.854	1.147	1.407	1.549	1.660	1.749
−1.0	−3.022	−1.877	−1.340	−0.758	0.164	0.852	1.128	1.366	1.492	1.588	1.664
−1.1	−3.087	−1.894	−1.341	−0.745	0.180	0.848	1.107	1.324	1.435	1.518	1.581
−1.2	−3.149	−1.910	−1.340	−0.732	0.195	0.844	1.086	1.282	1.379	1.449	1.501
−1.3	−3.211	−1.925	−1.339	−0.719	0.210	0.838	1.064	1.240	1.324	1.383	1.424
−1.4	−3.271	−1.938	−1.337	−0.705	0.225	0.832	1.041	1.198	1.270	1.318	1.351
−1.5	−3.330	−1.951	−1.333	−0.690	0.240	0.825	1.018	1.157	1.217	1.256	1.282
−1.6	−3.388	−1.962	−1.329	−0.675	0.254	0.817	0.994	1.116	1.166	1.197	1.216
−1.7	−3.444	−1.972	−1.324	−0.660	0.268	0.808	0.970	1.075	1.116	1.140	1.155
−1.8	−3.499	−1.981	−1.318	−0.643	0.282	0.799	0.945	1.035	1.069	1.087	1.097
−1.9	−3.553	−1.989	−1.310	−0.627	0.294	0.788	0.920	0.996	1.023	1.037	1.044
−2.0	−3.605	−1.996	−1.302	−0.609	0.307	0.777	0.895	0.959	0.980	0.990	0.995
−2.1	−3.656	−2.001	−1.294	−0.592	0.319	0.765	0.869	0.923	0.939	0.946	0.949
−2.2	−3.705	−2.006	−1.284	−0.574	0.330	0.752	0.844	0.888	0.900	0.905	0.907
−2.3	−3.753	−2.009	−1.274	−0.555	0.341	0.739	0.819	0.855	0.864	0.867	0.869
−2.4	−3.800	−2.011	−1.262	−0.537	0.351	0.725	0.795	0.823	0.830	0.832	0.833
−2.5	−3.845	−2.012	−1.250	−0.518	0.360	0.711	0.771	0.793	0.798	0.799	0.800
−2.6	−3.889	−2.013	−1.238	−0.499	0.368	0.696	0.747	0.764	0.768	0.769	0.769
−2.7	−3.932	−2.012	−1.224	−0.479	0.376	0.681	0.724	0.738	0.740	0.740	0.741
−2.8	−3.973	−2.010	−1.210	−0.460	0.384	0.666	0.702	0.712	0.714	0.714	0.714
−2.9	−4.013	−2.007	−1.195	−0.440	0.390	0.651	0.681	0.683	0.689	0.690	0.690
−3.0	−4.051	−2.003	−1.180	−0.420	0.396	0.636	0.660	0.666	0.666	0.667	0.667

COMPUTER PROGRAM HECWRC. The computer program HECWRC (U.S. Army Corps of Engineers, 1982) can be used to perform flood flow frequency analysis of annual maximum flood series, following the procedures in the U.S. Water Resources Council Bulletin 17B (1981). This program is available from the U.S. Army Corps of Engineers Hydrologic Engineering Center in Davis, California, in both a mainframe computer version and a microcomputer version.

Example 10.4.1. Use the U.S. Water Resources Council method to determine the 2-, 10-, 25-, 50-, and 100-year peak discharges for the station record of the San Gabriel

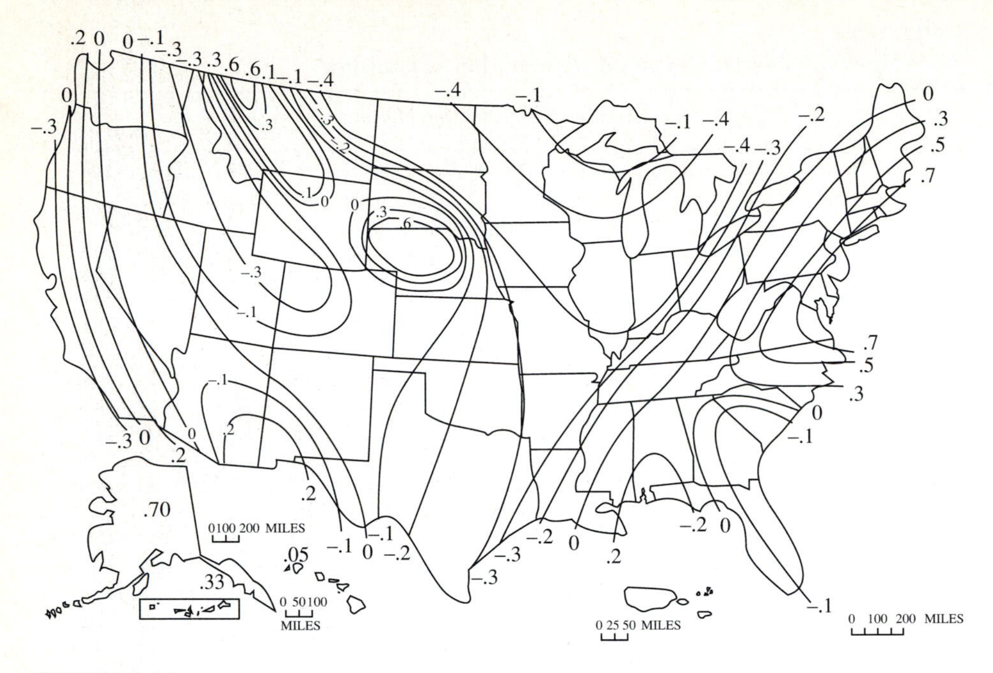

FIGURE 10.4.1
Generalized skew coefficients of annual maximum streamflow. (*Source*: Guidelines for determining flood flow frequency, Bulleting 17B. Hydrology Subcommittee, Interagency Advisory Committee on Water Data, U.S. Geological Survey, Reston, Va. Revised with corrections March 1982.)

River at Georgetown, Texas. The map skew is −0.3. The 39-year record for 1935–1973 is given in Table 10.4.2.

Solution. *Step 1.* The first step is to transform the sample data to their logarithmic values, $y_i = \log x_i$, $i = 1, \ldots, N$ where $N = 39$.

Step 2. Compute the sample statistics. The sample mean of the log-transformed values is

$$\overline{y} = \frac{1}{N} \sum_{i=1}^{N} y_i = 4.0838$$

The sample standard deviation is

$$s_y = \left[\frac{1}{N-1} \sum_{i=1}^{N} (y_i - \overline{y})^2 \right]^{1/2} = 0.4605$$

The skew coefficient is

$$G_s = \frac{N \sum_{i=1}^{N} (y_i - \overline{y})^3}{(N-1)(N-2)s_y^3} = -0.480$$

TABLE 10.4.2
Annual peak discharges for San Gabriel River at Georgetown, Texas

Year	Discharge (cfs)	Year	Discharge (cfs)	Year	Discharge (cfs)
1935	25,100	1948	14,000	1961	22,800
1936	32,400	1949	6,600	1962	4,040
1937	16,300	1950	5,080	1963	858
1938	24,800	1951	5,350	1964	13,800
1939	903	1952	11,000	1965	26,700
1940	34,500	1953	14,300	1966	5,480
1941	30,000	1954	24,200	1967	1,900
1942	18,600	1955	12,400	1968	21,800
1943	7,800	1956	5,660	1969	20,700
1944	37,500	1957	155,000	1970	11,200
1945	10,300	1958	21,800	1971	9,640
1946	8,000	1959	3,080	1972	4,790
1947	21,000	1960	71,500	1973	18,100

Step 3. Compute the weighted skew using the map skew of −0.3. The variance of the station skew is computed using Eq. (10.4.6). From Eq. (10.4.7*a*), for $|G_s| \leq 0.90$ and Eq. (10.4.7*c*) for $|G_s| \leq 1.50$, then coefficients A and B are, respectively,

$$A = -0.33 + 0.08 \times 0.480 = -0.292$$

and

$$B = 0.94 - 0.26 \times 0.480 = 0.815$$

Then, using Eq. (10.4.6)

$$\begin{aligned}\text{Var}(G_s) &= 10^{A-B\log(N/10)} \\ &= 10^{-0.292-0.815\log(39/10)} \\ &= 0.168\end{aligned}$$

The weight is now computed using Eq. (10.4.4)

$$\begin{aligned}W &= \text{Var}(G_m)/[\text{Var}(G_s) + \text{Var}(G_m)] \\ &= 0.325/(0.168 + 0.3025) \\ &= 0.642\end{aligned}$$

Hence, $1 - W = 0.358$ so the weighted skew coefficient is computed using Eq. (10.4.3)

$$\begin{aligned}G_w &= WG_s + (1 - W)G_m \\ &= (0.642)(-0.48) + (0.358)(-0.3) \\ &= -0.416\end{aligned}$$

For purposes of computation, simply use $G_w = -0.4$.

Step 4. Compute the discharges for the specified return periods. The frequency factors are found in Table 10.4.1. As an example, the frequency factor for the 100-year return period corresponding to $G_w = -0.4$ is $K_{100} = 2.02$. Using Eq. (10.4.2), the value of y_{100} is computed as

$$y_T = \overline{y} + K_T s_y = 4.0838 + 2.020(0.4605) = 5.014$$

so the estimated 100-year discharge is $Q_{100} = (10)^{5.014} = 103{,}279$ cfs. The computed discharges for the other return periods are listed in Table 10.4.3.

10.5 FLOODPLAIN ELEVATIONS: WATER SURFACE PROFILE DETERMINATION

STANDARD STEP METHOD. Once the discharge has been determined, the next step (see Fig. 10.1.1) in a floodplain analysis is to compute the water surface profile for this discharge for determining the **floodways** and **floodway fringes**, as defined in Fig. 10.5.1. This section presents the standard step method of water surface profile analysis for steady, gradually-varied, nonuniform, one-dimensional flow which has a total head H at a particular cross-section location (see Fig. 10.5.2)

$$H = z + y + \alpha \frac{V^2}{2g} \tag{10.5.1}$$

where z is the elevation of the channel bottom above a datum, y is the depth, V is the mean velocity at a cross-section, α is the energy correction factor for nonuniform velocities, $V^2/2g$ is the velocity head, and g is the acceleration due to gravity. Refer to Problem 10.5.1 to derive α.

The change in head with respect to distance L along the channel is

$$\frac{dH}{dL} = \frac{dz}{dL} + \frac{dy}{dL} + \frac{d}{dL}\left(\alpha \frac{V^2}{2g}\right) \tag{10.5.2}$$

The total energy loss term is $\frac{dH}{dL} = -S_f - S_e$ where S_f is the friction slope, S_e is the contraction-expansion loss and the channel bottom slope is $S_0 = -\frac{dz}{dL}$; so Eq. (10.5.1) is

TABLE 10.4.3
Results of flood flow frequency analysis for San Gabriel River discharges

Return period T (yr)	Frequency factor K_T	$\log Q_T$	Discharges Q_T (cfs)
2	0.069	4.116	13048
5	0.854	4.477	30009
10	1.228	4.649	44614
25	1.600	4.821	66177
50	1.826	4.925	84088
100	2.020	5.014	103276

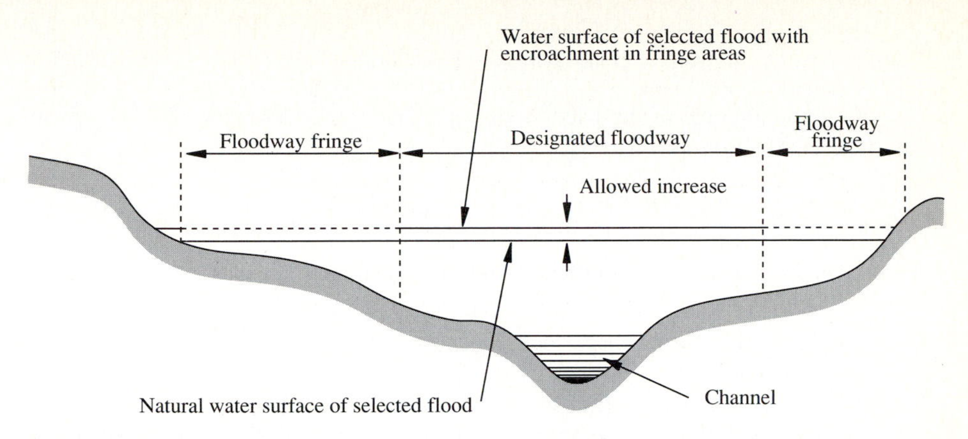

FIGURE 10.5.1
Definition of floodway fringe. The floodway fringe is the area between the designated floodway limit and the limit of the selected flood. The floodway limit is defined so that encroachment limited to the floodway fringe will not significantly increase flood elevation. The 100-year flood is commonly used and a one-foot allowable increase is standard in the United States. (Chow, Maidment, and Mays, 1988)

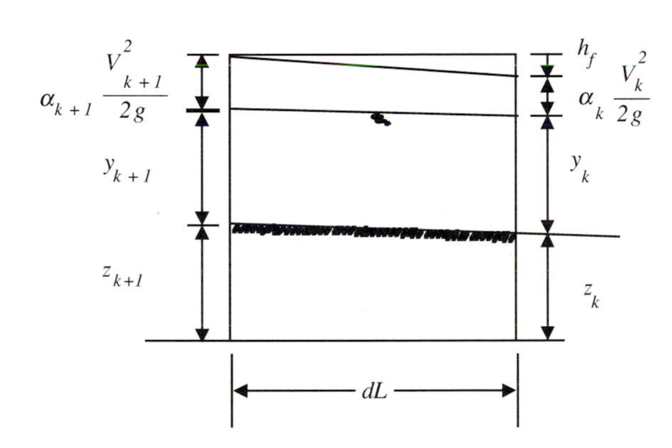

FIGURE 10.5.2
Channel reach.

$$-S_f - S_e = -S_0 + \frac{dy}{dL} + \frac{d}{dL}\left(\alpha\frac{V^2}{2g}\right) \tag{10.5.3}$$

which can be rearranged to

$$S_0 dL = dy + d\left(\alpha\frac{V^2}{2g}\right) + (S_f + S_e)dL \tag{10.5.4}$$

The differentials dy and $d\left(\alpha\frac{V^2}{2g}\right)$ are defined over the channel reach in finite difference form $dy = y_k - y_{k+1}$ and $d\left(\alpha\frac{V^2}{2g}\right) = \alpha_k\frac{V_k^2}{2g} - \alpha_{k+1}\frac{V_{k+1}^2}{2g}$; so Eq. (10.5.4) is expressed as

$$S_0 dL + y_{k+1} + \alpha_{k+1}\frac{V_{k+1}^2}{2g} = y_k + \alpha_k\frac{V_k^2}{2g} + S_f dL + S_e dL \tag{10.5.5}$$

The total loss is the friction loss $h_f = S_f dL$ plus the contraction-expansion losses, $h_0 = S_e dL$

$$h_f + h_0 = S_f dL + h_0 = \frac{1}{2}\left(S_{f_{k+1}} + S_{f_k}\right) dL + h_0 \tag{10.5.6}$$

where S_f is the average of the friction slopes at the two end sections of the channel reach. The contraction loss is

$$h_0 = C_c \left[\alpha_k \frac{V_k^2}{2g} - \alpha_{k+1} \frac{V_{k+1}^2}{2g}\right] \qquad \text{for } d\left(\alpha \frac{V^2}{2g}\right) < 0 \tag{10.5.7}$$

where C_c is the constraction coefficient and the expansion loss is

$$h_0 = C_e \left[\alpha_k \frac{V_k^2}{2g} - \alpha_{k+1} \frac{V_{k+1}^2}{2g}\right] \qquad \text{for } d\left(\alpha \frac{V^2}{2g}\right) > 0 \tag{10.5.8}$$

where C_e is the expansion coefficient.

The most widely used method to evaluate the friction slope has been to use Manning's equation

$$Q = \frac{1.486}{n} A R^{2/3} S_f^{1/2} \tag{10.5.9}$$

where n is the Manning's roughness factor and R is the hydraulic radius. Using the conveyance, $K = \frac{1.486}{n} A R^{2/3}$, the friction slope can be expressed as

$$S_f = \frac{Q^2}{K^2} = \frac{Q^2}{2}\left(\frac{1}{K_k^2} + \frac{1}{K_{k+1}^2}\right) \tag{10.5.10}$$

The slope term can be expressed as $S_0 dx = z_{k+1} - z_k$. Equation (10.5.5) becomes

$$z_{k+1} + y_{k+1} + \alpha_{k+1} \frac{V_{k+1}^2}{2g} = z_k + y_k + \alpha_k \frac{V_k^2}{2g} + \frac{Q^2}{2}\left(\frac{1}{K_k^2} + \frac{1}{K_{k+1}^2}\right) dL + h_0 \tag{10.5.11}$$

Typically, in water surface profile computations, water surface elevations are more important than depths. Denoting water surface elevations as $h = z + y$, then Eq. (10.5.11) is simplified to

$$h_{k+1} + \alpha_{k+1} \frac{V_{k+1}^2}{2g} = h_k + \alpha_k \frac{V_k^2}{2g} + \frac{Q^2}{2}\left(\frac{1}{K_k^2} + \frac{1}{K_{k+1}^2}\right) dL + h_0 \tag{10.5.12}$$

The **standard step procedure** is described in the following steps:

a. Start at a point in the channel where the depth is known or assumed. This is a **downstream boundary condition** for subcritical flow and an **upstream boundary condition** for supercritical flow. Computation proceeds upstream for subcritical flow and downstream for supercritical flow.

b. Choose a water surface elevation h'_{k+1} at the upstream end of the reach for subcritical flow or h'_k at the downstream end of the reach for supercritical flow. This water surface elevation will be slightly lower or higher depending upon the type of profile.

c. Next compute the conveyance, corresponding friction slope and expansion-contraction loss terms in Eq. (10.5.12) using the assumed water surface elevation.

d. Solve Eq. (10.5.12) for h_{k+1} (subcritical flow) or h_k (for supercritical flow).

e. Compare the calculated water surface elevation h with the assumed water surface elevation h'. If the calculated and assumed elevations do not agree within an acceptable tolerance (e.g., 0.01 ft) then set $h'_{k+1} = h_{k+1}$ (for subcritical flow) and $h'_k = h_k$ (for supercritical flow) and return to step **c**.

U.S. ARMY CORPS OF ENGINEERS HEC-2. The HEC-2 computer program was developed by the U.S. Army Corps of Engineers (1990) for calculating water surface profiles for steady, gradually varied flow in natural or manmade channels. This computer program is based upon the standard step method to compute changes in water surface elevation between adjacent cross-sections. Both subcritical and/or supercritical flow can be modeled. The effects of obstructions to flow such as bridges, culverts, weirs, and buildings located in the floodplain are included. The methodology is based upon the following assumptions: one dimensional steady gradually varied flow; average friction slope is assumed constant between adjacent cross-sections; and rigid boundary condition (no scour or sedimentation is considered).

10.6 HYDRAULICS OF FLOOD FORECASTING: DISTRIBUED ROUTING

Hydraulic (distributed) flow routing procedures are becoming popular for purposes of flood routing because these methodologies allow computation of the flowrate and water level to be computed as functions of both space and time. These methods are also commonly referred to as unsteady flow routing and are based upon the one-dimensional flow equations referred to as the Saint-Venant equations. As a comparison, the hydrologic routing procedures discussed in Section 10.3 are lumped procedures and compute flowrate as a function of time alone. The water surface profile determination using the standard step method in Section 10.5 is a distributed model but is for a fixed or steady-state discharge.

The **conservation form** of the Saint-Venant equations is used to describe unsteady flow because this form provides the versatility required to simulate a wide range of flows from gradual long-duration flood waves in rivers to abrupt waves similar to those caused by a dam failure. The computer models for hydraulic routing presented in Section 10.7 are based upon numerical solutions of the Saint-Venant equations. These equations, given below, are derived in detail by Chow et al. (1988). The continuity equation is

$$\frac{\partial Q}{\partial x} + \frac{\partial (A + A_0)}{\partial t} - q = 0 \qquad (10.6.1)$$

and the momentum equation is

$$\frac{1}{A}\frac{\partial Q}{\partial t} + \frac{1}{A}\frac{\partial(\beta Q^2/A)}{\partial x} + g\left(\frac{\partial h}{\partial x} + S_f + S_e\right) - \beta q v_x + W_f B = 0 \qquad (10.6.2)$$

where

x = longitudinal distance along the channel or river

t = time

A = cross-sectional area of flow

A_0 = cross-sectional area of off-channel dead storage (contributes to continuity, but not momentum)

q = lateral inflow per unit length along the channel

h = water surface elevation

v_x = velocity of lateral flow in the direction of channel flow

S_f = friction slope

S_e = eddy loss slope

B = width of the channel at the water surface

W_f = wind shear force

β = momentum correction factor

g = acceleration due to gravity.

The following assumptions apply to the Saint-Venant equations:

1. The flow is one-dimensional; depth and velocity vary only in the longitudinal direction of the channel. This implies that the velocity is constant and the water surface is horizontal across any section perpendicular to the longitudinal axis.
2. Flow is assumed to vary gradually along the channel so that hydrostatic pressure prevails and vertical accelerations can be neglected (Chow, 1959).
3. The longitudinal axis of the channel is approximated as a straight line.
4. The bottom slope of the channel is small and the channel bed is fixed; that is, the effects of scour and deposition are negligible.
5. Resistance coefficients for steady uniform turbulent flow are applicable so that relationships such as Manning's equation can be used to describe resistance effects.
6. The fluid is incompressible and of constant density throughout the flow.

The momentum equation consists of terms for the physical processes that govern the flow momentum. The term defining the change in water surface elevation can be expressed as $g\frac{\partial h}{\partial x} = g\left(\frac{\partial y}{\partial x} - S_0\right)$ where y is the depth and S_0 is the channel bottom slope. The terms in the momentum equation are: the **local acceleration** term $\frac{1}{A}\frac{\partial Q}{\partial t}$ which describes the change in momentum due to the change in velocity over time; the **convective acceleration** term $\frac{1}{A}\frac{\partial(\beta Q^2/A)}{\partial x}$ which describes the change in momentum due to change in velocity along the channel; the **pressure force** term $g\frac{\partial h}{\partial x}$ is proportional to the change in the water depth along the channel; the **gravity force** term, gS_0, is proportional to the bed slope S_0; and the **friction force** term, gS_f, is proportional to the friction slope S_f. The local and convective acceleration terms represent the effect of inertial forces on the flow.

When the water level or flow rate is changed at a particular point in a channel with a subcritical flow, the effects of these changes propagate back upstream. These **backwater effects** can be incorporated into distributed routing methods through the local acceleration, convective acceleration, and pressure terms. Hydrologic (lumped) routing methods may not perform well in simulating the flow conditions when backwater effects are significant and the river slope is mild, because these methods have no hydraulic mechanisms to describe upstream propagation of changes in flow momentum.

10.7 U.S. NATIONAL WEATHER SERVICE MODELS FOR RIVER ROUTING

DAMBRK MODEL. Forecasting downstream flash floods due to dam failures is an application of flood routing that has received considerable attention. The most widely used dam-breach model is the U.S. National Weather Service DAMBRK model by Fread (1977, 1980, 1981.) This model consists of three functional parts: (1) temporal and geometric description of the dam breach; (2) computation of the breach outflow hydrograph; and (3) routing the breach outflow hydrograph downstream.

Breach formation is the growth of the opening in the dam as it fails. The shape of a breach (triangular, rectangular, or trapezoidal) is specified by the slope z and the terminal width B_w of the bottom of the breach. The DAMBRK model assumes the breach bottom width starts at a point and enlarges at a linear rate until the terminal width is attained at the end of the failure time interval T. The breach begins when the reservoir water surface elevation h exceeds a specified value h_{cr} allowing for overtopping failure or piping failure.

The DAMBRK model uses hydrologic storage routing or the dynamic wave model to compute the reservoir outflow where a single reservoir is involved or in the first upstream reservoir in a series of reservoirs. The reservoir outflow hydrograph is then routed downstream using the full dynamic wave model described in Section 10.6; alternatively a dynamic wave model for flood routing in meandering rivers with flood plains can be used. The DAMBRK model can simulate several reservoirs located sequentially along a valley with a combination of reservoirs breaching. Highway and railroad bridges with embankments can be treated as internal boundary conditions. **Internal boundary conditions** are used to describe the flow at locations along a waterway where the Saint-Venant equations are not applicable. In other words, these are locations such as spillways, breaches, waterfalls, bridge openings, highway embankments, and so on, where the flow is rapidly rather than gradually varied.

FLDWAV MODEL. The FLDWAV model (Fread, 1985) is a synthesis of the U.S. National Weather Service DWOPER and DAMBRK models, with additional modeling capabilities not available in either of the other models. FLDWAV is a generalized dynamic wave model for one-dimensional unsteady flows in a single or branched waterway that can consider a variety of river mechanics capabilities (see Fig. 10.7.1).

The following special features and capacities are included in FLDWAV: variable Δt and Δx computational intervals; irregular cross-sectional geometry; off-channel storage; roughness coefficients that vary with discharge or water surface elevation, and with distance along the waterway; capability to generate linearly interpolated cross sections and roughness coefficients between input cross sections; automatic computation of initial steady flow and water elevations at all cross sections along the waterway; external boundaries of discharge or water surface elevation time series (hydrographs), a single-valued or looped depth-discharge relation (tabular or computed);

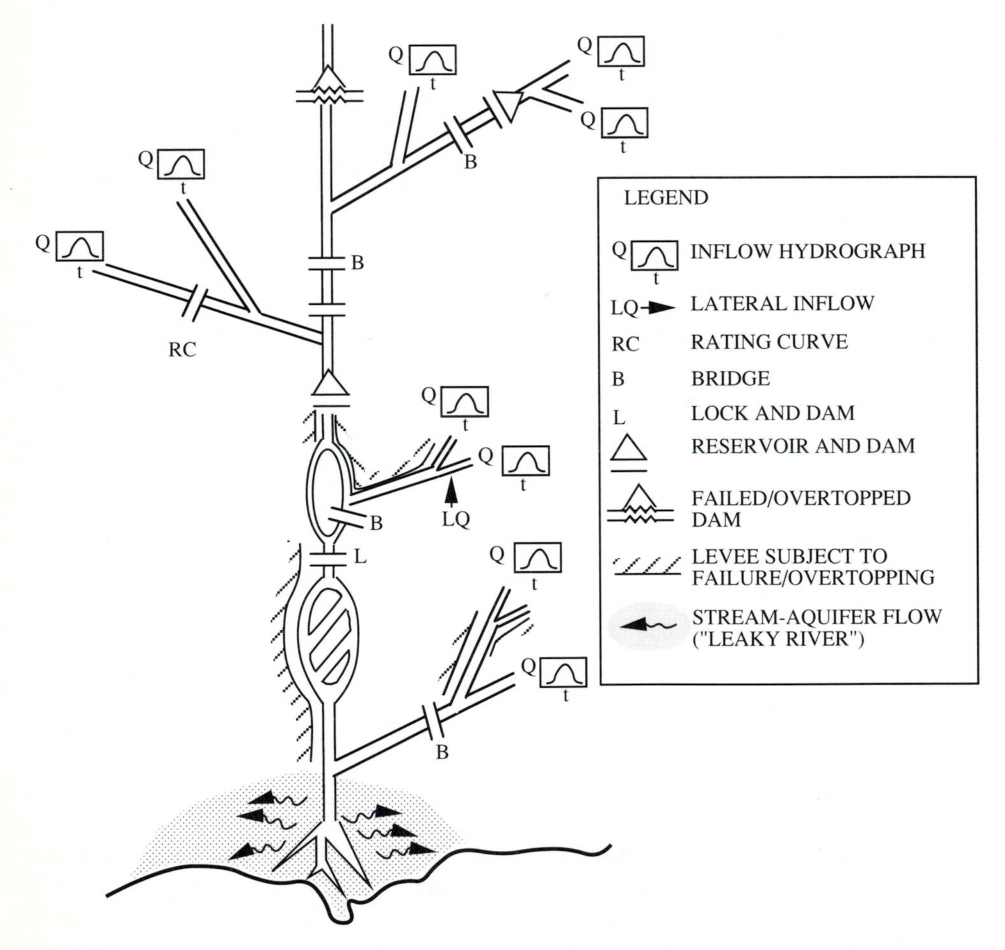

FIGURE 10.7.1
Schematic of a complex river system illustrating some of the hydraulic features that FLDWAV is capable of handling. (Fread, 1990)

time-dependent lateral inflows (or outflows); internal boundaries enable the treatment of time-dependent dam failures, spillway flows, gate controls, or bridge flows, or bridge-embankment overtopping flow; short-circuiting of flood-plain flow in a valley with a meandering river; levee failure and/or overtopping; a special computational technique to provide numerical stability when treating flows that change from supercritical to subcritical, or conversely, with time and distance along the waterway; and an automatic calibration technique for determining the variable roughness coefficient by using observed hydrographs along the waterway.

REFERENCES

Bedient, P. B. and W. C. Huber: *Hydrology and Floodplain Analysis*, Addison Wesley, Reading, Mass., 1988.

Bras, R. L.: *Hydrology: An Introduction to Hydrologic Science*, Addison Wesley, Reading, Mass., 1990.

Brazil, L. F. and M. D. Hudlow: Calibration Procedures Used with the National Weather Service River Forecast System, IFAC Symposium or Water and Related Land Resource System (Cleveland, Ohio, 1980), Yacov Haimes, Pergamon Press, Oxford and New York, 1981.

Burnach, J. C., R. L. Ferral, and R. A. McGuire: A Generalized Streamflow Simulation System, Conceptual Modeling for Digital Computers, Joint Federal-State River Forecast Center, U.S. Department of Commerce, NOAA National Weather Service, and State of California Dept. of Water Resources, March 1973.

Chow, V. T.: A General Formula for Hydrologic Frequency Analysis, *Trans. Am. Geophysical Union*, vol. 32, no. 2, pp. 231–237, 1951.

Chow, V. T.: *Open-Channel Hydraulics*, McGraw-Hill, New York, 1959.

Chow, V. T., ed.: *Handbook of Applied Hydrology*, McGraw-Hill Inc., New York, 1964.

Chow, V. T., D. R. Maidment, and L. W. Mays: *Applied Hydrology*, McGraw-Hill Inc., New York, 1988.

Crawford, N. H. and R. K. Linsley: Digital Simulation in Hydrology: Stanford Watershed Model IV, Technical Report No. 39, Department of Civil Engineering, Stanford University, Stanford, Calif. July 1966.

Fread, D. L.: The Development and Testing of Dam-Break Flood Forecasting Model: Proceedings, Dam-break Flood Modeling Workshop, U.S. Water Resources Council, Washington, D.C., pp. 164–197, 1977.

Fread, D. L.: Capabilities of NWS Model to Forecast Flash Floods Caused by Dam Failures. Preprint Volume, Second Conference on Flash Floods, March 18–20, Am. Meteorol. Soc., Boston, pp. 171–178, 1980.

Fread, D. L.: Some Limitations of Dam-Breach Flood Routing Models. Preprint, A. Soc. Civ. Eng. Fall Convention, St. Louis, Mo., October 1981.

Fread, D. L.: Channel routing, Chap. 14 in *Hydrological Forecasting*, M. G. Anderson and T. P. Burt, eds., Wiley, New York, pp. 437–503, 1985.

Fread, D. L.: Personal Communication, Hydrologic Research Laboratory, U.S. National Weather Service, Silver Spring, Md., 1990.

Hoggan, D. H.: *Computer Assisted Floodplain Hydrology and Hydraulics*, McGraw-Hill, Inc., New York, 1989.

Peck, E. L.: Catchment Modeling and Initial Parameter Estimation for the National Weather Service River Forecast System, NOAA Technical Memorandum NWS HYDRO-31, U. Department of Commerce, Silver Spring, Md., 1976.

Singh, K. P.: Unit Hydrograph—A Comparative Study, *Water Resources Bulletin*, AWRA, **12**(2): 381–392, 1976.

Soil Conservation Service: *National Engineering Handbook*, Section 4, Hydrology, U.S. Dept. of Agriculture, Washington, D.C., 1972.

Tung, Y. K. and L. W. Mays: Reducing Hydrologic Parameter Uncertainty, *Journal of the Water Resources Planning and Managment Division*, ASCE, vol. 107, no. WR1, pp. 245–262, March 1981.

U.S. Army Corps of Engineers, Hydrologic Engineering Center: Flood Flow Frequency Analysis, Computer Program 723-X6-L7750 Users Manual, Davis, Calif., February, 1982.

U.S. Army Corps of Engineers, Hydrologic Engineering Center: HEC-1, Flood Hydrograph Package, Users Manual, Davis, Calif., 1990.

U.S. Army Corps of Engineers, Hydrologic Engineering Center: HEC-2, Water Surface Profiles, Users Manual, Davis, Calif., 1990.

U.S. Army Corps of Engineers, North Pacific Division: Program Description and User Manual for SSARR Model, Stream Flow Synthesis and Reservoir Regulation, Portland, Ore., 1986.

U.S. Water Resources Council (now called the Interagency Advisory Committee on Water Data), Guidelines in Determining Flood Flow Frequency, Bulletin, 17B, available from Office of Water Data Coordination, U.S. Geological Survey, Reston, VA 22092, 1981.

Viessman, Jr., W., G. L. Lewis, and J. W. Knapp: *Introduction to Hydrology*, Harper and Row, New York, 1989.

Wallis, J. R., N. C. Matalas, and J. R. Slack: Just a Moment, *Water Resources Research*, vol. 10, no. 2, pp. 211–219, April 1974.

PROBLEMS

10.2.1 Derive the equation for cumulative infiltration using the Green-Ampt equation found in Table 10.2.1.

10.2.2 Derive the equation for cumulative infiltration using Horton's equation found in Table 10.2.1.

10.2.3 Derive the equations for infiltration rate and cumulative infiltration using the SCS equation found in Table 10.2.1.

10.2.4 Use the Green-Ampt method to compute the infiltration rate and cumulative infiltration for a silty clay soil ($\eta = 0.479$, $\psi = 29.22$ cm, $K = 0.05$ cm/hr) at 0.25 hour increments up to 4 hours from the beginning of infiltration. Assume an initial effective saturation of 30 percent and continuous ponding.

10.2.5 The parameters for Horton's equation are $f_0 = 3.0$ in/h, $f_c = 0.5$ in/h, and $k = 4.0h^{-1}$. Determine the infiltration rate and cumulative infiltration at 0.25 hour increments up to 4 hours from the beginning of infiltration. Assume continuous ponding.

10.2.6 The one-hour unit hydrograph for a watershed is given below. Determine the runoff from this watershed for the storm pattern given. The abstractions have a constant rate of 0.3 in/hr.

Time (hr)	1	2	3	4	5	6
Precipitation (in)	0.5	1.0	1.5	0.5		
Unit hydrograph (cfs)	10	100	200	150	100	50

10.2.7 Use the rainfall pattern in Problem 10.2.6 and compute the infiltration rate and cumulative infiltration for each hour of the storm by the SCS method. Also what is the rainfall excess for each hour? Use an SCS curve number of $CN = 80$.

10.2.8 A 6-hour rainfall pattern for Maricopa County, Arizona is

Time (hr)	0.0	0.25	0.50	0.75	1.0	1.25	1.50	1.75	
% depth	0.	0.5	0.9	1.4	2.2	3.0	3.8	4.7	
Time (hr)	2.0	2.25	2.50	2.75	3.0	3.25	3.50	3.75	
% depth	5.4	6.2	7.5	8.8	10.7	12.7	20.5	36.6	
Time (hr)	4.0	4.25	4.50	4.75	5.0	5.25	5.50	5.75	6.00
% depth	82.3	90.0	92.0	93.9	95.2	96.5	97.7	98.8	100.0

Determine the cumulative abstractions for a 100-year 6-hour rainfall of 3.0 inches with a curve number of 85.

10.2.9 Compute the one hour synthetic unit hydrograph for a 50 mi^2 watershed with a main stream length of 4.0 mi and a main channel length from the watershed outlet to the point opposite the center of gravity of the watershed of 1.5 mi. Use C_t = 2.0 and C_p = 0.6.

10.2.10 The March 16–18, 1936, storm on a watershed of the North Branch Potomac River near Cumberland, Maryland, resulted in the following rainfall-runoff values (from Singh, 1976)

Time (hr)	4	8	12	16	20	24	28
Effect Rainfall (in)	0.12	0.88	0.80	1.00	0.24		
Direct Runoff (in/hr)	0.003	0.011	0.036	0.090	0.140	0.140	0.110
Time (hr)	32	36	40	44	48	52	56
Direct Runoff (in/hr)	0.082	0.057	0.037	0.025	0.014	0.008	0.004
Time (hr)	60	64					
Direct Runoff (in/hr)	0.002	0.001					

The drainage area of the watershed is 897 mi^2. Determine the 4-hr unit hydrograph using linear programming to minimize the sum of absolute deviations.

10.2.11 Referring to Problem 10.2.10, determine the 4-hr unit hydrograph using linear programming to minimize the largest absolute deviation.

10.2.12 Referring to Problem 10.2.10, determine the 4-hr unit hydrograph by ordinary least squares method described in Section 6.3. Check the resulting unit hydrograph to see if it is a reasonable one.

10.2.13 Referring to the unit hydrograph derived in Problem 10.2.12 by the ordinary least squares method, propose an optimization model to eliminate the drawback as identified in Problem 10.2.12. Solve the model for the optimal unit hydrograph using the data given in Problem 10.2.10.

10.3.1 Consider a 2-acre stormwater detention basin with vertical walls. The triangular inflow hydrograph increases linearly from zero to a peak of 60 cfs at 60 min and then decreases linearly to a zero discharge at 180 min. Route the inflow hydrograph through the detention basin using the following head- discharge relationship for the 5-ft pipe spillway.

Head (ft)	0.0	0.5	1.0	1.5	2.0	2.5	3.0	3.5	4.0	4.5	5.0	5.5	6.0	6.5	7.0	7.5	8.0	8.5	9.0	9.5	10.0
Discharge (cfs)	0	3	8	17	30	43	60	78	97	117	137	156	173	190	205	218	231	242	253	264	275

The pipe is located at the bottom of the basin. Assuming the basin is initially empty, use the level pool routing procedure with a 10-minute time interval to determine the maximum depth in the detention basin. (from Chow, et al., 1988).

10.3.2 Route the following upstream inflow hydrograph through a downstream flood control channel reach using the Muskingum method. The channel reach has a $K = 2.5$ hours and $X = 0.2$. Use a routing interval of 1 hour.

Time (hr)	1	2	3	4	5	6	7
Inflow (cfs)	90	140	208	320	440	550	640
Time (hr)	8	9	10	11	12	13	14
Inflow (cfs)	680	690	670	630	570	470	390
Time (hr)	15	16	17	18	19	20	
Inflow (cfs)	330	250	180	130	100	90	

10.3.3 Use the U.S. Army Corps of Engineers HEC-1 computer program to solve Problem 10.3.2.

10.3.4 Using Eq. (10.3.10) and the following inflow-outflow hydrographs determine the optimal K and X for the Muskingum model by linear programming.

Time (hr)	0	6	12	18	24	30	36	42
Inflow (cms)	22	23	35	71	103	111	109	100
Outflow (cms)	22	21	21	26	34	44	65	66
Time (hr)	48	54	60	66	72	78	84	90
Inflow (cms)	86	71	59	47	39	32	28	24
Outflow (cms)	75	82	85	84	80	73	64	54
Time (hr)	96	102	108	114	120	126		
Inflow (cms)	22	21	20	19	19	18		
Outflow (cms)	44	36	30	25	22	19		

10.3.5 Resolve Problem 10.3.1 by the ordinary least squares method. (Do not forget the constraint that $C_1 + C_2 + C_3 = 1$).

10.4.1 Use the U.S. Army Corps of Engineers HECWRC computer program to solve Example 10.4.1.

10.4.2 Considering the annual peak for the San Gabriel River (Example 10.4.1) for water years 1940–1973 (Table 10.4.3), determine the frequency discharge relationship following the U.S. Water Resources Council guidlines. Compare the discharges with those found in Example 10.4.1.

10.5.1 Derive the energy correction factor $\alpha = \frac{(A_t)^2 \sum K_i/A_i}{(K_t)^3}$ where A_t and K_t refer to the total cross-sectional area of flow and conveyance, and A_i and K_i refer to the ith portion of the total cross-sectional area of flow and conveyance.

CHAPTER
11

URBAN STORMWATER MANAGEMENT SYSTEMS

11.1 URBAN STORMWATER MANAGEMENT SYSTEM

Urban stormwater management systems include all appurtenances that guide, control, and modify the quantity and quality of urban runoff. A typical urban drainage system consists of various subsystems (Fig. 11.1.1) which convey rainfall from where it impacts to the receiving waters. Five basic subsystems characterize these systems: (1) overland flow or surface runoff subsystem; (2) storm sewer or conduit transport subsystem; (3) detention or storage subsystem; (4) main channel or open channel transport subsystem; and (5) receiving water subsystem which includes rivers, lakes, or oceans.

The overland flow subsystem transforms rainfall input into surface water runoff. Output or runoff hydrographs from the overland flow subsystem is input into the storm sewer subsystem which transports runoff to a detention subsystem, a main channel subsystem, or a receiving water subsystem. Output or discharge releases from a detention or storage subsystem can be input into a main channel subsystem or a receiving water subsystem. Output from a main channel subsystem could be inflow or input to a detention storage or receiving water subsystem.

Systems concepts are increasingly being used as an aid in understanding and developing solutions to complex urban water problems. The problems encountered in

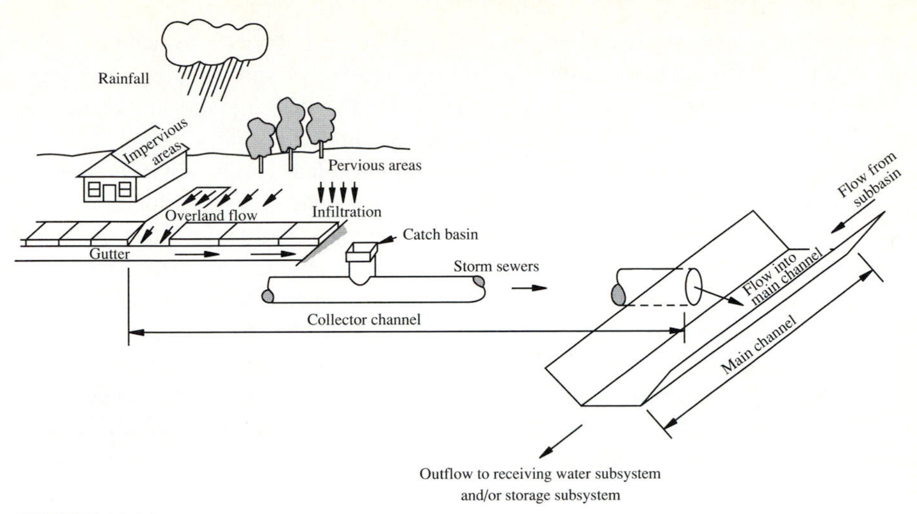

FIGURE 11.1.1
Typical urban drainage pattern (after U.S. Army Corps of Engineers, 1979).

urban water systems, which are inherently distributed systems, must be analyzed to account for both spatial and temporal variations. Urban watersheds vary in space since the ground-surface slope and cover and the soil type change, at different locations in the watershed. They are temporarily varied because the hydrological characteristics change with respect to time along with the process of urbanization. The mathematical formulation of urban water systems distributed in both time and space is a complicated task. Consequently, the spatial variation is sometimes ignored and the system is treated as a lumped system. Some spatial variation can be taken care of by dividing the whole watershed system into several subsystems that are considered to be lumped, then linking these lumped system models together to produce a model of the entire system.

Models can be used as tools for planning and management. Numerous investigations and research on the effect of urbanization on runoff volume have been done. In particular, several watershed simulation models have been developed. In urban stormwater management the determination of runoff yield and the optimal design of sewer networks are very important. Runoff prediction models range from the well-known rational formula to advanced simulation models such as the Storm Water Management Model (SWMM), (Huber et al., 1975). In urban water resources management storm water runoff alleviation is one of the major tasks. Storm sewer systems play an important role in this work. To design an adequate storm sewer system a rather accurate estimate of stormwater runoff yield is required.

11.2 STORM SEWER DESIGN

11.2.1 Design Philosophy

Most storm sewer systems are converging-branch or simply tree-type systems. A storm sewer system may consist of a large number of sewers, junctions, manholes and inlets in addition to other regulating or operational devices such as gates, valves, weirs, overflows, regulators, and pumping stations. These devices do have an effect upon the system, hydraulically dividing it into a number of subsystems. The factors involved in the design of storm sewer systems are the determination of diameters, slopes, and crown or invert elevations for each pipe in the system.

From an engineering viewpoint the drainage problem can be divided into two aspects: runoff prediction and system design. In the recent years, considerable effort has been devoted to runoff prediction in urban areas. The second aspect of the drainage problem, design methodology, has received less attention. The basic types of design models are **hydraulic design models** and **optimization design models** (Section 11.4). The hydraulic design models determine the sewer sizes using only hydraulic considerations. The sewer system layout is predetermined and the sewer slope is generally assumed to follow the ground slope. The basic concept is to determine the minimum sewer size that has a capacity to carry the design discharge under full pipe gravity flow conditions. Many of the so called "sewer design methods" are actually flow simulation or prediction methods to provide the design hydrographs.

The following constraints and assumptions are commonly used in storm sewer design practice:

a. Free-surface flow exists for the design discharges, that is, the sewer system is designed for "gravity flow" so that pumping stations and pressurized sewers are not considered.

b. The sewers are commercially available circular pipes no smaller than 8 in. diameter.

c. The design diameter is the smallest commercially available pipe that has flow capacity equal to or greater than the design discharge and satisfies all the appropriate constraints.

d. Storm sewers must be placed at a depth that will not be susceptible to frost, drain basements, and allow sufficient cushioning to prevent breakage due to ground surface loading. Therefore, minimum cover depths must be specified.

e. The sewers are joined at junctions such that the crown elevation of the upstream sewer is no lower than that of the downstream sewer.

f. To prevent or reduce excessive deposition of solid material in the sewers, a minimum permissible flow velocity at design discharge or at barely full-pipe gravity flow is specified (e.g., 2.5 ft/s).

g. To prevent the occurrence of scour and other undesirable effects of high-velocity flow, a maximum permissible flow velocity is also specified. Maximum velocities in sewers are important mainly because of the possibilities of excessive erosion on the sewer inverts.

h. At any junction or manhole the downstream sewer cannot be smaller than any of the upstream sewers at that junction.

i. The sewer system is a dendritic network converging towards downstream without closed loops.

11.2.2 Rational Method

The rational method, which can be traced back to the mid-19th century, is still probably the most popular method used for the design of storm sewers. Although criticisms have been raised on the adequacy of the method, and several other more advanced methods have been proposed, the rational method, because of its simplicity, is still in continued use for sewer designs when high accuracy of runoff rate is not required.

Using the rational method, the storm runoff peak is estimated by using the rational formula

$$Q = CiA \tag{11.2.1}$$

where the peak runoff rate Q is in cfs; C is a runoff coefficient (Table 11.2.1), i is the average rainfall intensity in inches per hour, and A is the drainage area in acres. In urban areas, the drainage area usually consists of subareas or subcatchments of substantially different surface characteristics. As a result, a composite analysis is required that must account for the various surface characteristics. The areas of the subcatchments are denoted by A_j and the runoff coefficients of each subcatchment are denoted by C_j. Then the peak runoff is computed using the following form of the rational formula.

$$Q = i \sum_{j=1}^{m} C_j A_j \tag{11.2.2}$$

in which m is the number of subcatchments drained by a sewer.

The **rainfall intensity** (i) is the average rainfall rate in inches per hour which is considered for a particular drainage basin or sub-basin. The intensity is selected on the basis of design rainfall duration and design frequency of occurrence. The design duration is equal to the time of concentration for the drainage area under consideration. The frequency of occurrence is a statistical variable which is established by design standards or chosen by the engineer as a design parameter.

The **time of concentration** (t_c) used in the rational method is the time associated with the peak runoff from the watershed to the point of interest. Runoff from a watershed usually reaches a peak at the time when the entire watershed is contributing, in which case, the time of concentration is the time for a drop of water to flow from the remotest point in the watershed to the point of interest. Runoff may reach a peak prior to the time the entire watershed is contributing. A trial-and-error procedure can be used to determine the critical time of concentration. The time of concentration to any point in a storm drainage system is the sum of the inlet time (t_0) and the flow time (t_f) in the upstream sewers connected to the catchment, that is,

$$t_c = t_0 + t_f \tag{11.2.3}$$

where the flow time is

$$t_f = \sum \frac{L_j}{V_j} \tag{11.2.4}$$

TABLE 11.2.1
Runoff coefficients for use in the rational method (Chow, Maidment, and Mays, 1988)

	Return period (years)						
Character of surface	**2**	**5**	**10**	**25**	**50**	**100**	**500**
Developed							
Asphaltic	0.73	0.77	0.81	0.86	0.90	0.95	1.00
Concrete/roof	0.75	0.80	0.83	0.88	0.92	0.97	1.00
Grass areas (lawns, parks, etc.)							
Poor condition (grass cover less than 50% of the area)							
Flat, 0–2%	0.32	0.34	0.37	0.40	0.44	0.47	0.58
Average, 2–7%	0.37	0.40	0.43	0.46	0.49	0.53	0.61
Steep, over 7%	0.40	0.43	0.45	0.49	0.52	0.55	0.62
Fair condition (grass cover on 50% to 75% of the area)							
Flat, 0–2%	0.25	0.28	0.30	0.34	0.37	0.41	0.53
Average, 2–7%	0.33	0.36	0.38	0.42	0.45	0.49	0.58
Steep, over 7%	0.37	0.40	0.42	0.46	0.49	0.53	0.60
Good condition (grass cover larger than 75% of the area)							
Flat, 0–2%	0.21	0.23	0.25	0.29	0.32	0.36	0.49
Average, 2–7%	0.29	0.32	0.35	0.39	0.42	0.46	0.56
Steep, over 7%	0.34	0.37	0.40	0.44	0.47	0.51	0.58
Undeveloped							
Cultivated Land							
Flat, 0–2%	0.31	0.34	0.36	0.40	0.43	0.47	0.57
Average, 2–7%	0.35	0.38	0.41	0.44	0.48	0.51	0.60
Steep, over 7%	0.39	0.42	0.44	0.48	0.51	0.54	0.61
Pasture/Range							
Flat, 0–2%	0.25	0.28	0.30	0.34	0.37	0.41	0.53
Average, 2–7%	0.33	0.36	0.38	0.42	0.45	0.49	0.58
Steep, over 7%	0.37	0.40	0.42	0.46	0.49	0.53	0.60
Forest/Woodlands							
Flat, 0–2%	0.22	0.25	0.28	0.31	0.35	0.39	0.48
Average, 2–7%	0.31	0.34	0.36	0.40	0.43	0.47	0.56
Steep, over 7%	0.35	0.39	0.41	0.45	0.48	0.52	0.58

in which L_j is the length of the jth pipe along the flow path, and V_j is the average flow velocity in the pipe. The inlet time t_0 is the longest time of overland flow of water in a catchment to reach the storm sewer inlet draining the catchment.

Example 11.2.1. Determine the diameters for pipes 5.1, 5.2, 5.3, 6.1, and 7.1 in the Goodwin Avenue drainage basin in Urbana, Illinois (Fig. 11.2.1). The catchment characteristics are listed in Table 11.2.2. The capacities of sewers in the system should handle surface runoff generated from a 2-year storm. The rainfall-intensity-duration relationship for a 2-year return period is also given in Table 11.2.2. A 14.65 ac drainage area with $\Sigma CA = 10.52$ and a time of concentration $t_c = 15.2$ min flows into the upstream end of pipe 5.1 through a pipe that is not shown.

Solution. The inlet times for subcatchments in the watershed are given in Table 11.2.2. Shown in Table 11.2.3 are the computations for the design of the sewer pipes. The Manning's roughness coefficient n is 0.014 for all the sewers.

This example demonstrates that in the rational method each sewer is designed individually and independently (except for the computation of sewer flow time) and the corresponding rainfall intensity i is computed repeatedly for the area drained by the sewer. For a given sewer, all the different areas drained by this sewer have the same i. Thus, as the design progresses towards the downstream sewers, the drainage area increases and usually the time of concentration increases accordingly. This increasing t_c in turn gives a decreasing i which should be applied to the entire area drained by the sewer.

TABLE 11.2.2
Catchment characteristics of Goodwin Avenue drainage basin and rainfall intensity-duration-frequency relationship (Yen, 1978)

Catchment draining to manhole	Ground elevation at manhole (ft)	Area A (acres)	Runoff coefficient C	Inlet time (min)	Length of outflow pipe from the manhole (ft)
51	720.12	1.25	0.70	10.3	230
52	721.23	0.70	0.65	11.8	70
53	720.26	1.50	0.55	17.6	130
61	719.48	0.60	0.75	9.0	160
71	715.39	2.30	0.70	12.0	240
81	715.10				

Rainfall intensity duration frequency relationship

Duration t_D (min)	5	10	15	20
Rainfall intensity i (in/hr)	5.4	4.18	3.51	3.1

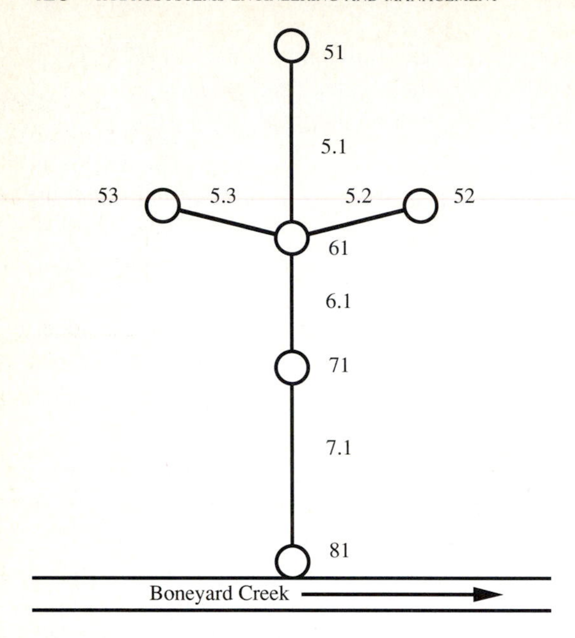

FIGURE 11.2.1
Goodwin Avenue drainage basin and storm sewer system in Urbana, Illinois. (Yen, 1978)

11.3 HYDROGRAPH DESIGN METHOD

Hydrograph design methods consider design hydrographs as input to the upstream end of sewers and use some form of routing to propagate the inflow hydrograph to the downstream end of the sewer. The routed hydrograph is added to the surface runoff hydrograph to the manhole at the downstream junction, the routed hydrographs for each sewer is added also. The combined hydrographs for all upstream connecting pipes plus the hydrograph for the surface runoff represents the design inflow hydrograph to the next (adjacent) downstream sewer pipe. The pipe size and sewer slope are selected based upon solving for the commercial size pipe that can handle the peak discharge of the inflow hydrograph and maintain a gravity flow.

A simple hydrograph design method that is a rather effective method is the **hydrograph time lag method** (Yen, 1978). The hydrograph time lag method is a hydrologic (lumped) routing method. The inflow hydrograph of a sewer is shifted without distortion by the sewer flow time t_f to produce the sewer outflow hydrograph. The outflow hydrographs of the upstream sewers at a manhole are added, at the corresponding times, to the direct manhole inflow hydrograph to produce the inflow hydrograph for the downstream sewer in accordance with the continuity relationship

$$\sum Q_{ij} + Q_j - Q_o = \frac{dS}{dt} \tag{11.3.1}$$

in which Q_{ij} is the inflow from the ith upstream sewer into the junction j, Q_o is the outflow from the junction into the downstream sewer, Q_j is the direct inflow into the manhole or junction, and S is the water stored in the junction structure or manhole. For point type junctions where there is no storage, dS/dt is 0.

TABLE 11.2.3

Rational method computation for Example 11.2.1 (Goodwin Ave.)

(1)	(2)	(3)	(4)	(5)	(6)	(7)	(8)	(9)	(10)	(11)	(12)	(13)	(14)	(15)	(16)	(17)	(18)	(19)
				Increment														
Sewer	Length L (ft)	Slope S_0	Total area drained (ac)	Catchment	Area (ac)	C	CA	ΣCA	Inlet time (min)	Upstream sewer flow time (min)	t_c (min)	t_D (min)	i (in./hr.)	Design discharge Q_p (cfs)	Computed diameter d_r (ft)	Pipe size used d (ft)	Flow velocity (fps)	Sewer flow time (min)
5.1	230	0.0028		5.1	1.25	0.70	0.88		10.3	—	10.3							
											15.2							
			14.65					10.52				15.2	3.50	36.8	3.13	3.50	3.8	1.00
5.2	70	0.0250	0.70	5.2	0.70	0.65	0.46	0.46	11.8	—	11.8	11.8	3.90	1.79	0.67	0.67	5.1	0.23
5.3	130	0.0060	1.70	5.3	1.70	0.55	0.94	0.94	17.6	—	17.6	17.6	3.30	3.10	1.07	1.25	2.5	0.86
6.1	160	0.017		6.1	0.6	0.75	0.45		9.0	—	9.0							
				5.3					17.6	0.86	18.5							
			17.65					12.37				18.5	3.2	39.58	3.77	4.0	12.6	0.21
7.1	240	0.0012		7.1	2.3	0.70	1.61											
			19.95					13.98				18.7	3.2	44.7	3.94	4.0		

Col. (3) Slopes chosen to be greater than or equal to ground slopes.

Col. (4) Total area drained by sewer is equal to sum of areas of subcatchments drained by sewer.

Col. (9) Summation of CA for all areas drained by sewer, equal to sum of contributing areas in (9) plus areas in (8) for that sewer.

Col. (10) Inlet time is sum of overland flow inlet time.

Col. (11) Sewer flow time of immediate upstream sewer.

Col. (12) Time of concentration t_c for each of the possible critical flow paths plus the sewer flow time for each flow path.

Col. (13) Rainfall duration t_D is assumed equal to the largest of the different times of concentration of different flow paths to arrive at the entrance of the sewer considered.

Col. (14) Rainfall intensity (i) for duration in Col. (13).

Col. (15) Design discharge computed using Eq. (11.2.2), product of Cols. (9) and (14).

Col. (16) Required sewer diameter computed using $d_r = \left[0.032\left(\frac{Q}{\sqrt{S_0}}\right)\right]^{3/8}$ for $n = 0.014$.

Col. (17) Next larger commercial size pipe to that computed in (16).

Col. (18) Flow velocity from $V = 4Q/(\pi d^2)$.

Col. (19) Sewer flow time is L/V Col. (9) divided by Col. (18).

The sewer flow time t_f that is used to shift the hydrograph is estimated by

$$t_f = L/V \tag{11.3.2}$$

in which L is the length of the sewer and V is a sewer flow velocity. The velocity could be computed, assuming a full-pipe flow, using

$$V = \frac{4Q_p}{\pi d^2} \tag{11.3.3}$$

where Q_p is the peak discharge and d is the pipe diameter. Also steady uniform flow equations such as Manning's equation or the Darcy-Weisbach equation could be used to compute the velocity.

In this method, the continuity relationship of the flow within the sewer is not directly considered. The routing of the sewer flow is done by shifting the inflow hydrograph by t_f and no consideration is given to the unsteady and nonuniform nature of the sewer flow. Shifting of hydrographs approximately accounts for the sewer flow translation time but offers no wave attenuation. However, the computational procedure through interpolation introduces numerical attenuation.

11.4 MINIMUM COST DESIGN OF STORM SEWER SYSTEMS

In the minimum-cost design of a storm sewer system the trade-off between pipe cost and excavation cost is considered. To convey a specified quantity of runoff, using a steeper pipe slope, the required pipe size is smaller, hence a lower pipe cost. However, the cost and amount of excavation is larger. A methodology is demonstrated in this section that can be used to determine the least-cost combination of sizes and slopes of the sewers and the depths of the manholes for a sewer network to collect and drain the stormwater runoff from an urban drainage basin. Because sewer slope depends on the end elevations of the sewer, the design variables are the diameters and the upstream and downstream crown elevations of the sewers, and the depths of the manholes. The methodology considers a network layout connecting manholes at various points within the drainage basin. The design inflows into these manholes are predetermined. The principal tasks in the development and formulation of an optimization model for the design of storm sewer systems are twofold.

a. Representation of the set of manholes in a form suitable for digital manipulation.

b. Selecting optimization techniques for the overall model which are flexible enough to handle design constraints and assumptions, various forms of cost functions, risk models, and hydraulic or hydrologic models, and to incorporate all design information.

Dynamic programming (DP) can be applied to the least-cost design of storm sewer systems (Mays and Yen, 1975; Tang et al., 1975; Mays and Wenzel, 1976; Mays et al., 1976; and Yen et al., 1976). The DP terms stage, state, decision, return, and transformation are defined below in terms of their counterparts in storm sewer systems.

Stage. Imaginary lines called **isonodal lines** are used to divide the sewer network into stages. These lines are defined such that they pass through manholes which are separated from the system outlet by the same number of pipe sections (links). An arbitrary stage i includes the pipes connecting upstream manholes on line i to downstream manholes on line $i+1$. For a system with $i = 1, 2, \ldots, I$ isonodal lines and having $I - 1$ stages, the manholes on any line i are connected to the system outlet by $I - i$ pipe sections. The isonodal lines are essentially constructed by starting at the outlet and proceeding upstream and are numbered reversely by starting with the upstream and proceeding downstream, as is shown in Fig. 11.4.1 for the example sewer system.

States. The states at each stage i are the crown elevations of the pipes. The input states $\mathbf{S}_{m_i, m_{i+1}}$ for a pipe connecting manholes m_i and m_{i+1} on isonodal lines i and $i+1$, respectively, are the crown elevations at the upstream end (manhole m_i). Similarly, the output states $\tilde{S}_{m_i, m_{i+1}}$ for this same pipe are the crown elevations at the downstream end (manhole m_{i+1}). The upstream manholes at an arbitrary stage i are numbered $m_i = 1, 2, \ldots M_i$ for M_i manholes on line i, and the downstream manholes are numbered $m_{i+1} = 1, 2, \ldots, M_{i+1}$ for M_{i+1} manholes on line $i+1$.

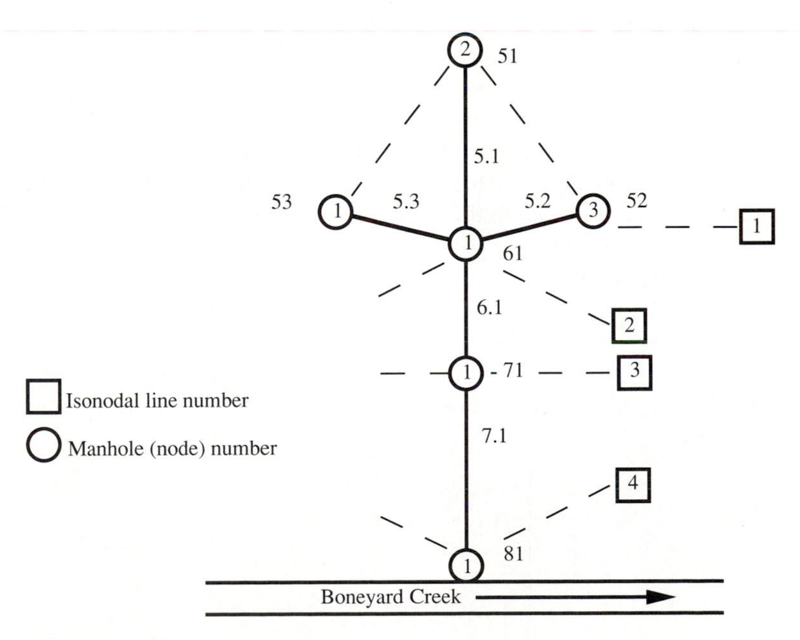

FIGURE 11.4.1
Goodwin Avenue storm sewer system showing isonodal lines.

Decisions. The decisions are the drops in crown elevations of the pipes across the stage; that is, the decision $\mathbf{D}_{m_i,m_{i+1}}$ represents the vector of drops from the upstream manhole m_i to the downstream manhole, m_{i+1}. There is a set of decisions for each pipe across the stage. Each drop in elevation represents a slope, so that by using Manning's equation, assuming full pipe flow, the pipe diameter can be determined for the design flow rate.

Return. The return at each stage is the cost of installing each pipe and the respective upstream manholes of these pipes; that is, the total return for each stage i is $r_i = \sum_{m_i}^{M_i} r_{m_i,m_{i+1}}(\mathbf{S}_{m_i,m_{i+1}}, \mathbf{D}_{m_i,m_{i+1}})$. Each manhole on an upstream isonodal line must be drained to a manhole on the downstream isonodal line, resulting in a total of M_i pipes connecting across stage i.

Transformation. The transformation function defines the manner in which the input states (or upstream crown elevations) $\mathbf{S}_{m_i,m_{i+1}}$ is transformed into output states (or downstream crown elevations) $\tilde{\mathbf{S}}_{m_i,m_{i+1}}$ through the decision variables (or drops in elevations) $\mathbf{D}_{m_i,m_{i+1}}$.

$$\tilde{\mathbf{S}}_{m_i,m_{i+1}} = \mathbf{S}_{m_i,m_{i+1}} - \mathbf{D}_{m_i,m_{i+1}} \tag{11.4.1}$$

Discrete differential dynamic programming described in Section 4.2 can be used as the optimization technique. This iterative technique is one in which the recursive equation of dynamic programming is used to search for an improved or lower-cost storm sewer system within the state space, defined by a corridor or set of possible crown elevations at the upstream and downstream end of each pipe. The first step in defining a corridor is to assume a sequence of states or crown elevations called the **trial trajectory**, $\bar{S}_{m_i,m_{i+1}}, i = 1, 2, \ldots I$. The initial and final states of the trial trajectory (i.e., the upstream crown elevation at stage 1 and the downstream crown elevation at stage I for a serial or nonbranching system) are specified, and the slope of all the sewers is assumed to be the same. Accordingly, the trial trajectory and the crown elevations and corresponding decisions for each stage of the system can be computed. When branches are considered, the same slope is used to compute the trial trajectories of the main and the branches.

Next, several other states in the neighborhood of the trial trajectory are introduced. These neighborhood states for each stage form a band around the trial trajectory called the **corridor** (Fig. 11.4.2). At every stage i, an increment, ΔS_i, is used in the state space S to form neighborhood states. Thus, the corridor for a storm sewer system, using five states or lattice points at the upstream and downstream ends of a sewer, is defined by

$$\mathbf{S}_{m_i,m_{i+1}} = \begin{bmatrix} \bar{S}_{m_i,m_{i+1}} + 2\Delta S_i \\ \bar{S}_{m_i,m_{i+1}} + \Delta S_i \\ \bar{S}_{m_i,m_{i+1}} \\ \bar{S}_{m_i,m_{i+1}} - \Delta S_i \\ \bar{S}_{m_i,m_{i+1}} - 2\Delta S_i \end{bmatrix} \tag{11.4.2}$$

where $\bar{S}_{m_i,m_{i+1}}$ defines the trial trajectory. The corridor must fall within the admissible domain of the state space. The upper boundary of the domain can be defined by the

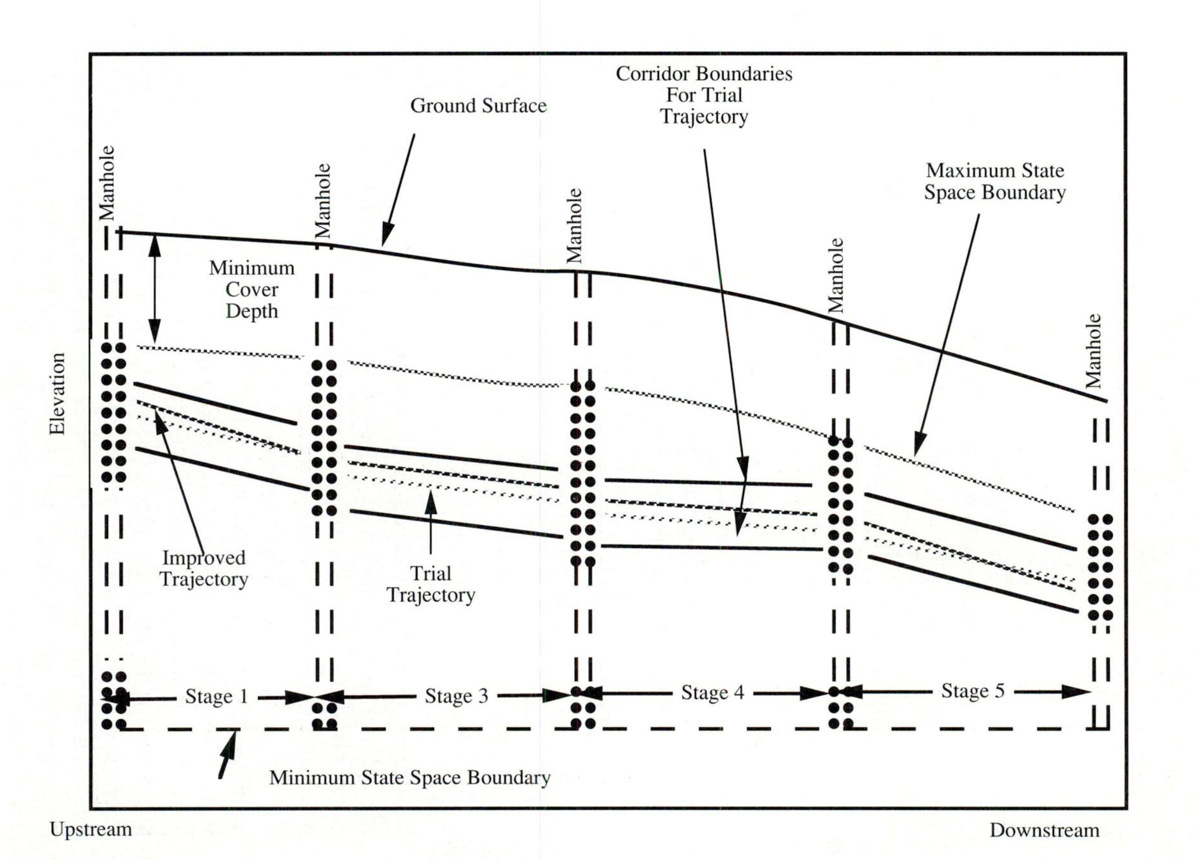

FIGURE 11.4.2
Example of stage-corridor representation for nonserial optimization approach.

required minimum soil coverage of the sewers or other more restrictive constraints. The lower boundary is defined by the elevation of the lowest sewer in the system or preferably by other more restrictive constraints such as that imposed by minimum sewer slopes.

The recursive equation for each pipe at each stage is given as

$$f_i(\tilde{\mathbf{S}}_{m_i,m_{i+1}}) = \min_{\mathbf{D}_{m_i,m_{i+1}}} [r_{m_i,m_{i+1}}(\mathbf{S}_{m_i,m_{i+1}}, \mathbf{D}_{m_i,m_{i+1}}) + f_{i-1}(\tilde{\mathbf{S}}_{m_i,m_{i+1}})] \qquad i = 1, \ldots, I-1 \tag{11.4.3}$$

where $f_i(\tilde{\mathbf{S}}_{m_i,m_{i+1}})$ represents the minimum cost of the system that is connected to manhole m_{i+1} through upstream manhole m_i. This recursive equation is for only one of the M_i pipes across stage i. The recursive equation for all the pipes in the state is

$$\sum_{\mathbf{T}_{m_i,m_{i+1}}} f_i(\tilde{\mathbf{S}}_{m_i,m_{i+1}}) = \sum_{\mathbf{T}_{m_i,m_{i+1}}} \min_{\mathbf{D}_{m_i,m_{i+1}}} [r_{m_i,m_{i+1}}(\mathbf{S}_{m_i,m_{i+1}}, \mathbf{D}_{m_i,m_{i+1}}) + f_{i-1}(\tilde{\mathbf{S}}_{m_i,m_{i+1}})] \qquad i = 1, \ldots, I-1 \tag{11.4.4}$$

where $\mathbf{T}_{m_i,m_{i+1}}$ represents the known vector of connections of the M_i pipes for the given layout from isonodal lines i to $i+1$. This vector represents the connection of manholes between isonodal lines i and $i+1$; that is, $T_{m_i,m_{i+1}} = 0$ implies that upstream manhole m_i does not connect to downstream manhole m_{i+1} and a '1' implies the connection of these two manholes. This vector of connections describes the system layout.

The conventional discrete DP approach is applied to the states within the corridor, by using the above recursive relationship to search for the least-cost trajectory among the introduced states. The least-cost trial trajectory is then adopted as the improved trajectory to form a new corridor. This process of corridor formation, optimization with respect to the states within the corridor and trace-back to obtain an improved trajectory, is called an **iteration.** A schematic representation of a trial trajectory, the corridor and the improved trajectory of the first DDDP iteration for a serial storm sewer system with five stages is shown in Fig. 11.4.2.

The DDDP computations for this model proceed from upstream to downstream in the sewer network. Each pipe in the particular stage being optimized is considered separately. The least cost drop associated with each downstream crown elevation (output state) of the pipe is determined. This minimum cost drop is selected by considering each of the upstream crown elevations (input states) for each output state representing drops for the particular pipe. The input state is selected which represents the minimum total cost of obtaining the design to the output state. This minimum total cost includes the minimum cost of the pipes and manholes required upstream of the input state for the pipe, in addition to the cost of the pipe being designed. This procedure is repeated for each pipe connection at each stage before proceeding to the next downstream stage.

When the computations for the last stage of the sewer network have been performed, a trace-back is performed through the system from downstream to upstream, stage by stage, to recover the minimum-cost system. The upstream and downstream crown elevations for each pipe of the minimum-cost design represent the least-cost trajectory to establish the improved corridor for the next iteration. This procedure is repeated beyond some iteration which produces a cost and set of states such that further iterations only produce a reduction in cost less than a specified tolerance. At this time, the state increment or distance ΔS between crown elevations at the respective manholes is reduced, resulting in a smaller corridor around the last improved trajectory. The iterations continue, reducing ΔS accordingly, until no further improvement can be made within a tolerance equal to some allowable error in cost or to a specified minimum ΔS.

Corridor size. The criterion used to determine when the magnitude of ΔS should be reduced is based on the relative change of the minimum cost for two successive iterations f_n, that is,

$$|f_n - f_{n-1}|/f_{n-1} \le E_r \tag{11.4.5}$$

When the ratio is equal to or smaller than a specified value E_r, say 1 percent, the state increment ΔS is reduced to one-half or any other desired fraction of its previous value, and then iterations are resumed.

It should be mentioned that for any iteration, the values of ΔS can vary for difference stages. Appropriate selection of the initial values of ΔS for different stages can improve the efficiency of DDDP. Also, the reduction rate of ΔS can be varied during iterations to improve computational efficiency and accuracy. Finally, to save computer time and to achieve high accuracy, a variable E_r with a smaller value for later iterations can be used.

Several assumptions and constraints are made in the development of optimization models which are in accordance with engineering practice. These include the assumptions of gravity flow, minimum and maximum permissible flow velocities, minimum cover, nondecreasing pipe sizes in the downstream direction, and a set of cost equations for pipes and manholes.

Example 11.4.1. Determine the minimum-cost sewers for pipes 5.1, 6.1, and 7.1 for the Goodwin Avenue problem (Example 11.2.1). Use dynamic programming with a state space increment of 1 ft. Table 11.4.1 lists the ground elevations, sewer lengths and peak design inflows. A minimum cover depth of 3.5 ft is required. At a manhole the crown elevation of the downstream pipe draining the manhole must be equal to or less than the crown elevation of the upstream pipe draining into the manhole. The unit cost (\$/ft) of available commercial pipe sizes are

Diameter (in)	12	15	18	21	24	27	30	36	42	48	54
Unit cost (\$/ft)	3.40	4.45	5.90	7.40	9.20	11.05	14.20	19.05	25.00	30.85	39.45

The unit cost for excavation is 6.00 $\$/yd^3$ and the manhole cost is \$100.00/ft depth. A 5 ft excavation width is assumed for all pipes.

TABLE 11.4.1
Information for Goodwin Avenue Example 11.4.1

	Manhole number		Ground elevation at upstream manhole	Sewer length	Peak inflow
Sewer	**Upstream**	**Downstream**	**(ft)**	**(ft)**	**(cfs)**
5.1	51	61	721.2	230	36.8
6.1	61	71	718.1	161	42.0
7.1	71	81	715.4	251	47.0
	81 (outlet)		715.1		

Solution. The state space is established using three states (crown elevations) and a state space increment of $\Delta S = 1.0$ ft, the elevation of the trail trajectory is $\bar{S}_1 = 721.2 - 3.5 - 1.0 = 716.7$ ft which is the ground surface elevation minus the minimum cover depth minus the state space increment. In a similar manner $\bar{S}_2 = 713.6$ ft, $\bar{S}_3 = 710.9$ ft, $\bar{S}_4 = 710.6$ ft, for manholes 61, 71, and 81, respectively. The state-space for manhole 5.1 is

$$S_1 = \begin{bmatrix} \bar{S}_1 + \Delta S \\ \bar{S}_1 \\ \bar{S}_1 - \Delta S \end{bmatrix} = \begin{bmatrix} 716.1 + 1 \\ 716.7 \\ 716.7 - 1 \end{bmatrix} = \begin{bmatrix} 717.7 \\ 716.7 \\ 715.7 \end{bmatrix}$$

The discretized state space (crown elevations) for each stage, defined by manholes 51, 61, 71, and 81, are shown in Fig. 11.4.3.

To start the dynamic programming computations, the state space for stage 1 (between manholes 51 and 61) is considered. For a fixed downstream crown elevation, the DP recursive Eq. (11.4.3) is applied for each upstream crown elevation in Stage 1. As an example, considering the downstream crown elevation 714.6 ft, the sewer pipe connecting upstream crown elevation 717.7 ft and downstream crown elevation 714.6 ft has a slope of $(717.7 - 714.6)/230 = 0.01348$ ft/ft. The required diameter for a design discharge of 36.8 cfs and slope of 0.01348 ft/ft is computed using Manning formula as

$$d = \left(2.16 \frac{n}{\sqrt{S_0}} Q_p\right)^{3/8} = \left(2.16 \frac{0.014}{\sqrt{0.01348}} (36.8)\right)^{3/8}$$

$$= (9.58)^{0.375} = 2.333 \text{ ft } = 28 \text{ in}$$

A 30 inch commercial pipe size is used. The average depth of excavation is $3.5 + 2.5 = 6.0$ ft and assume an average width of excavation of 5 ft then the volume of excavation is $(6.0)(5)(230) = 6900 \text{ ft}^3 = 255.5 \text{ yd}^3$. The costs for the pipe, the excavation, and the upstream manhole are

Pipe cost	$= 14.20$ (\$/ft) $\times$ 230 ft	= \$3,266.00
Excavation Cost	$= 6.0$ (\$/yd^3) $\times$ 255.6 yd^3	= \$1,533.60
Manhole Cost	$= 100.$ (\$/ft) $\times$ 6 ft.	= \$ 600.00
Return	$= r_{51,61,1,1}(\quad)$	= \$5,399.60

The above computations are shown on the first line of Table 11.4.2. The procedure is repeated to connect the current downstream crown elevation, that is, 714.6 ft, to the remaining upstream crown elevation, that is, 716.7 ft and 715.7 ft. The results are shown on the second and third lines of Table 11.4.2.

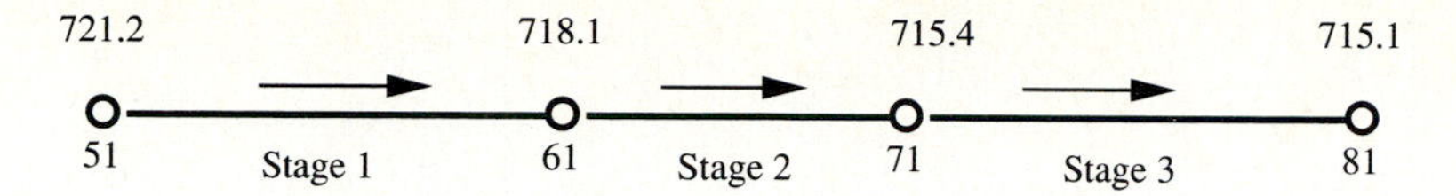

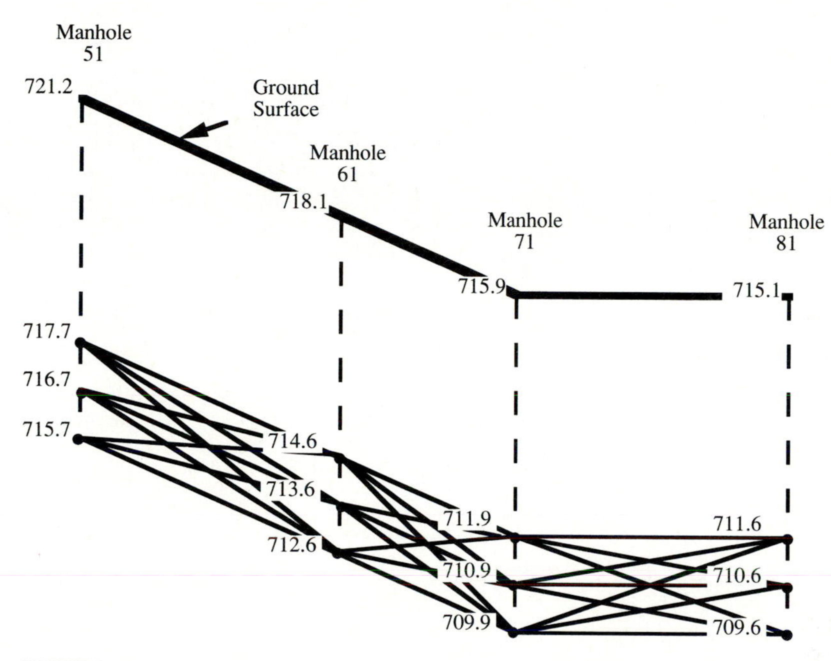

FIGURE 11.4.3.
State space for Example 11.4.1.

The DP recursive Eq. (11.4.3) is applied to the downstream crown elevation 714.6 ft at manhole 61

$$
\begin{aligned}
f_1(\tilde{S}_{51,61,1}) &= \min[r_{51,61,1,1}() + 0,\ r_{51,61,2,1}() + 0,\ r_{51,61,3,1}() + 0] \\
&= \min(\$5399.60,\ \$6920.38,\ \$7148.16) \\
&= \$5399.60
\end{aligned}
$$

The optimum (minimum-cost) pipe connected to crown elevation 714.6 ft (state $j = 1$ at manhole 61) is a 30 inch pipe.

Once the least cost connection for downstream crown elevation 714.6 ft is identified, the same procedure is repeated for downstream crown elevation 713.6 ft ($j = 2$) and for crown elevation 712.6 ft ($j = 3$) at manhole 61. Table 11.4.2 shows the DP computations for the storm sewer design. Columns (1) through (12) are self explanatory. The return which is the sum of the costs for the pipe, excavation, and upstream manhole is given in Column (13). Column (14) lists the costs of the upstream portion of the sewer system connected to the upstream state for the pipe being considered. Column (15) lists

TABLE 11.4.2

Dynamic programming computation for storm sewer design (Example 11.4.1)

(1) Stage	(2) Up-stream man-hole	(3) Down-stream man-hole	(4) Up-stream state k	(5) Crown (ft)	(6) Down-stream State j	(7) Crown (ft)	(8) Slope ft/ft	(9) Dia-meter (in)	(10) Pipe cost ($)	(11) Exca-vation cost ($)	(12) Up-stream cost ($)	(13) Return $r(\ )$ ($)	+	(14) Up-stream $f(\ )$ ($)	=	(15) Total $f(\ )$ ($)	(16) Min $f(\ )$	(17) Optimum states Up-stream	(18) Optimum states Down-stream
1	51	61	1	717.7	1	714.6	0.01348	30	3266.00	1533.60	600.00	5399.60	+	0	=	5399.60			
			2	716.7	1	714.6	0.00910	36	4381.50	958.20	750.00	6089.70	+	0	=	6089.70	$5399.60	1	1
			3	715.7	1	714.6	0.00478	36	4381.50	1916.66	850.00	7148.16	+	0	=	7148.16			
			1	717.7	2	713.6	0.01783	27	2541.50	1597.22	575.00	4713.72	+	0	=	4713.72			
			2	716.7	2	713.6	0.01348	30	3266.00	1788.88	700.00	5754.88	+	0	=	5754.88	$4713.72	1	2
			3	715.7	2	713.6	0.00910	36	4381.50	2044.44	850.00	7275.94	+	0	=	7275.94			
			1	717.7	3	712.6	0.02217	27	2541.50	1725.00	575.00	4841.50	+	0	=	4841.50			
			2	716.7	3	712.6	0.01783	27	2541.50	1852.78	675.00	5069.28	+	0	=	5069.28	$4841.50	1	3
			3	715.7	3	712.6	0.00135	30	3266.00	2044.44	800.00	6110.44	+	0	=	6110.44			
2	61	71	1	714.6	1	711.9	0.01677	30	2286.20	1073.34	600.00	3959.54	+	5399.60	=	9359.14			
			2	713.6	1	711.9	0.01056	36	3067.05	1252.22	750.00	5069.27	+	4713.72	=	9782.99	$9359.14	1	1
			3	712.6	1	711.9	0.00435	42	4025.00	1431.12	900.00	6356.12	+	4713.72	=	11069.84			
			1	714.6	2	710.9	0.02298	27	1779.05	1118.06	575.00	3472.11	+	5399.60	=	8871.71			
			2	713.6	2	710.9	0.01677	30	2286.20	1252.22	700.00	4238.42	+	4713.72	=	8952.14	$8871.71	1	2
			3	712.6	2	710.9	0.01056	36	3067.05	1431.12	850.00	5348.17	+	4713.72	=	10061.89			
			1	714.6	3	709.9	0.02919	27	1779.05	1207.50	575.00	3561.55	+	5399.60	=	8961.15			
			2	713.6	3	709.9	0.02298	27	1779.05	1296.94	675.00	3750.99	+	4713.72	=	8464.71	$8464.71	2	3
			3	712.6	3	709.9	0.01677	30	2286.20	1431.10	800.00	4517.30	+	4713.72	=	9231.02			
3	71	81	1	711.9	1	711.6	0.00120	54	9901.95	2231.12	800.00	12933.07	+	9359.14	=	22292.21			
			2	710.9	1	711.6	Infeasible	—	10^{20}	10^{20}	10^{20}	3×10^{20}	+	8871.71	=	3×10^{20}	$22292.21	1	1
			3	709.9	1	711.6	Infeasible	—	10^{20}	10^{20}	10^{20}	3×10^{20}	+	8464.71	=	3×10^{20}			
			1	711.9	2	710.6	0.00518	42	6275.00	2091.66	700.00	9066.66	+	9359.14	=	18425.80			
			2	710.9	2	710.6	0.00120	54	9901.95	510.00	900.00	13311.95	+	8871.71	=	22183.66	$18425.80	1	2
			3	709.9	2	710.6	Infeasible	—	10^{20}	10^{20}	10^{20}	3×10^{20}	+	8464.71	=	3×10^{20}			
			1	711.9	3	709.6	0.00916	36	4781.55	2091.66	650.00	7523.21	+	9359.14	=	16882.35			
			2	710.9	3	709.6	0.00518	42	6275.00	2370.56	800.00	9445.56	+	8871.71	=	18317.27	$16882.35	1	3
			3	709.9	3	709.6	0.00120	54	9901.45	2788.88	1000.00	13690.33	+	8464.71	=	22155.04			

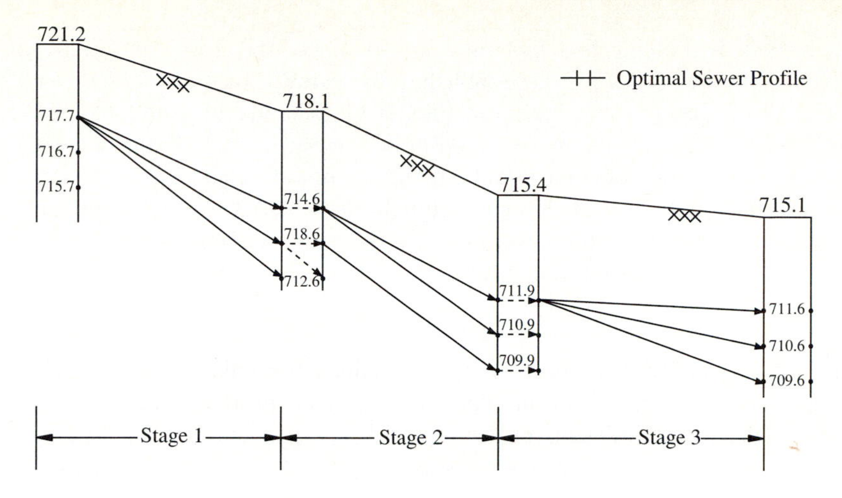

FIGURE 11.4.4
Optimal connections for dynamic programming computations (Example 11.4.1).

the total cost and Column (16) lists the minimum cost for the particular downstream state. As an example in stage 1, the minimum-cost pipe connecting to downstream state $j = 1$ is the 30 inch pipe at a cost of \$5399.60. Columns (17) and (18) list the upstream state (k) and the downstream state (j), respectively.

Once calculations have been performed for the first upstream stage 1, the computations proceed to the next downstream pipe (stage 2). From the statement of the problem, at a manhole, the crown elevation of the downstream pipe draining the manhole must be equal to or less than the crown elevation of the pipe draining into the manhole. For upstream state $k = 1$ in stage 2 the only possible crown elevation is 714.6 ft. The cost associated with an infeasible connection, that is, downstream crown elevation is higher than the upstream crown elevation, is assumed to be infinite. For upstream state $k = 2$ (crown elevation = 713.6 ft) in stage 2, the possible crown elevations for upstream pipes connecting to the manhole are 714.6 ft ($j = 1$) and 713.6 ft ($j = 2$) in stage 1. This is shown in Fig. 11.4.4. Note that the cheapest upstream pipe is chosen from \$5399.60 and \$6089.70. For upstream state $k = 3$ (crown elevation 712.6 ft) in stage 2 the cheapest upstream pipe in stage 1 is connected to $j = 2$ (crown elevation 713.6 ft) in stage 1. These minimum cost possibilities are shown by the dashed lines in the manholes in Figure 11.4.4.

The DP computations for stages 2 and 3 are shown in Table 11.4.2. Note that the cheapest cost system (\$16,882.35) has a downstream crown elevation of 709.6 ft. A traceback can be performed as shown in Table 11.4.2 with the optimal trajectory shown in Figure 11.4.4.

11.5 RELIABILITY ANALYSIS OF STORM SEWERS

11.5.1 Reliability Computation

Design of urban drainage facilities are inevitably subject to many uncertainties, and the design of storm sewer systems is no exception. The uncertainties involved in sewer design include both hydrologic and hydraulic uncertainties. By using probability theory, risk-safety factor curves can be established for various locations (cities) for the purpose of sewer design. Methods for computing the reliability of hydraulic structures using load-resistance inference are described in Section 5.4. In the context of storm sewers, the loading is the magnitude of surface runoff generated by a storm whereas the resistance is the sewer capacity. Reliability in storm sewer design can be defined as the probability that the sewer capacity Q_C, can accommodate the storm runoff, Q_L (Tang et al., 1975; and Yen, 1975).

$$\alpha = P(Q_L \leq Q_C) \tag{11.5.1}$$

Q_L and Q_C are commonly determined using the rational formula and Manning's equation, respectively. Values of mean capacity ($\bar{Q}_C$) and mean loading ($\bar{Q}_L$) as well as the corresponding coefficients of variation ($\Omega_{Q_C}, \Omega_{Q_L}$) can be estimated by performing a first-order analysis of uncertainties (Section 5.3.2) on the rational formula and Manning's equation.

Using Manning's equation, $\bar{Q}_C$ is estimated using

$$\bar{Q}_C = \frac{0.463}{\bar{n}} \bar{S}_0^{1/2} \bar{d}^{8/3} \tag{11.5.2}$$

in which n is the Manning roughness factor, S_0 is the sewer slope (in ft/ft); and d is the sewer diameter (ft). The coefficient of variation Ω_{Q_C} is computed by

$$\Omega_{Q_C}^2 = \Omega_n^2 + \frac{1}{4}\Omega_{S_0}^2 + \frac{64}{9}\Omega_d^2 \tag{11.5.3}$$

in which Ω_n, Ω_{S_0}, and Ω_d are the coefficients of variation for n, S_0 and d, respectively.

Using the rational formula, $\bar{Q}_L$ is evaluated by

$$\bar{Q}_L = \bar{C}\bar{i}\bar{A} \tag{11.5.4}$$

in which $\bar{Q}_L$ is in cfs, $\bar{C}$ is the mean runoff coefficient, $\bar{A}$ is the mean basin area in acres, and $\bar{i}$ is the mean rainfall intensity in in/hr. The coefficient of variation of surface runoff Ω_{Q_L} is computed by

$$\Omega_{Q_L}^2 = \Omega_C^2 + \Omega_i^2 + \Omega_A^2 \tag{11.5.5}$$

in which Ω_C, Ω_i, and Ω_A are the coefficients of variation of C, i, and A, respectively.

In case that the watershed consists of several subcatchments with different runoff coefficients, a weighted runoff coefficient C in the rational formula is computed from

$$C = \sum_j C_j \beta_j \tag{11.5.6}$$

in which $\beta_j = a_j/A$ where A is the total area of the drainage basin and a_j is the subarea having a runoff coefficient C_j.

By applying a first-order analysis to Eq. (11.5.6), the mean and coefficient of variation are, respectively,

$$\bar{C} = \sum_j \bar{C}_j \bar{\beta}_j \tag{11.5.7}$$

and

$$\Omega_C = \frac{1}{\bar{C}}\left[\sum_j \bar{C}_j^2 \bar{\beta}_j^2 \Omega_{C_j}^2\right]^{1/2} \tag{11.5.8}$$

Although all β_j are somewhat dependent because they should add up to unity, statistical independence among all C_j and β_j are assumed here for simplicity. The effect of dependence among β_j will diminish as j becomes larger.

Example 11.5.1. Determine the risk (probability) that the surface runoff (loading) exceeds the capacity of a storm sewer pipe. Loading is determined using the rational formula and the sewer capacity is determined using Manning's equation for full-pipe flow. Assume that both the loading and capacity are normally distributed. The mean and coefficient of variation of the parameters are:

Parameter	Mean	Coefficient of variation
C	0.825	0.071
i	7.2in/hr	0.177
A	12 acres	0.05
n	0.015	0.0553
d	5 ft.	0.010
S_0	0.001	0.068

Solution. The mean runoff (loading) is, from Eq. (11.5.4),

$$\bar{Q}_L = \bar{C}\bar{i}\bar{A} = (0.825)(7.2)(12)$$

$$= 71.28 \text{ cfs}$$

The coefficient of variation of the surface runoff is given by Eq. (11.5.5)

$$\Omega_{Q_L} = \left[\Omega_C^2 + \Omega_i^2 + \Omega_A^2\right]^{1/2}$$

$$= \left[(0.071)^2 + (0.177)^2 + (0.05)^2\right]^{1/2}$$

$$= 0.1972$$

and the standard deviation of the loading is

$$\sigma_{Q_L} = \bar{Q}_L \Omega_{Q_L}$$

$$= (71.3)(0.197)$$

$$= 14.04 \text{ cfs}$$

The mean capacity of the storm sewer pipe is given by Eq. (11.5.2)

$$\bar{Q}_C = \frac{0.463}{\bar{n}} \bar{S}_0^{1/2} \bar{d}^{8/3}$$

$$= \frac{0.463}{0.015}(0.001)^{1/2}(5)^{8/3}$$

$$= 71.35 \text{ cfs}$$

The coefficient of variation of the sewer capacity is given by Eq. (11.5.3)

$$\Omega_{Q_C} = \left[\Omega_n^2 + \frac{1}{4}\Omega_{S_0}^2 + \frac{64}{9}\Omega_d^2\right]^{1/2}$$

$$= \left[(0.0553)^2 + \frac{1}{4}(0.068)^2 + \frac{64}{9}(0.01)^2\right]^{1/2}$$

$$= 0.0702$$

The standard deviation of the capacity is

$$\sigma_{Q_c} = \bar{Q}_C \Omega_{Q_C}$$

$$= (71.4)(0.0702)$$

$$= 5.01 \text{ cfs}$$

Since both surface runoff loading (Q_L) and sewer capacity (Q_C) are normal random variables, the reliability of the storm sewer can be calculated by using the safety-margin method. Using Eq. (5.4.6) the mean safety margin is

$$\mu_{SM} = \bar{Q}_C - \bar{Q}_L$$

$$= 71.35 - 71.28$$

$$= 0.07 \text{ cfs}$$

The standard deviation of the safety margin using Eq. (5.4.7) is

$$\sigma_{SM} = \left[\sigma_{Q_C}^2 + \sigma_{Q_L}^2\right]^{1/2}$$

$$= \left[(5.01)^2 + (14.04)^2\right]^{1/2}$$

$$= 14.9 \text{ cfs}$$

The reliability α is then

$$\alpha = \Phi\left(\frac{\mu_{SM}}{\sigma_{SM}}\right) = \Phi\left(\frac{0.07}{14.9}\right) = \Phi(0.0047)$$

$$= 0.5$$

and the corresponding risk $(1 - \alpha)$ is 50 percent that the surface runoff exceeds the capacity of a storm sewer pipe.

Example 11.5.2. Using the data presented in Example 11.5.1, determine the contribution of each parameter (C, i, A) to the variability of loading and (n, d, S_0) to the variability of the capacity. Also determine the contribution of each parameter to the total variability.

Solution. The coefficient of variation of the loading was computed in Example 11.5.1 to be $\Omega_{Q_C} = 0.1972$. Using Eq. 11.5.5 the percent of the variability due to the runoff coefficient is the squared coefficient of variation of the runoff coefficient divided by the squared coefficient of variation of the loading, that is,

$$\frac{\Omega_C^2}{\Omega_{Q_L}^2} = \frac{(0.071)^2}{(0.197)^2} = 0.129 \text{ or } 12.9\%$$

Similarly

$$\frac{\Omega_i^2}{\Omega_{Q_L}^2} = \frac{(0.177)^2}{(0.197)^2} = 0.807 \text{ or } 80.7\%$$

and

$$\frac{\Omega_A^2}{\Omega_{Q_L}^2} = \frac{(0.05)^2}{(0.197)^2} = 0.064 \text{ or } 6.4\%$$

These three percentages add up to 100 percent. The contributions of n, d, and S_0 to the capacity are summarized in Column (4) of Table 11.5.1.

The contribution of each parameter to the total variability is determined using

$$\frac{\eta\Omega_x^2}{(\Omega_{Q_L}^2 + \Omega_{Q_C}^2)}$$

where all η are the coefficients in Eq. (11.5.3) and (11.5.5). As an example the contribution of the slope S_0 is

$$\frac{\left(\frac{1}{4}\right)(0.068)^2}{[(0.197)^2 + (0.0705)^2]} = 0.0265 \text{ or } 2.65\%$$

Column (5) in Table 11.5.1 lists the contribution of each parameter to the total variability.

TABLE 11.5.1
First-order analysis of uncertainty for a storm sewer pipe for Example 11.5.2

(1) Parameter x	(2) Mean $\overline{x}$	(3) Coefficient of variation Ω_x	(4) $\eta\frac{\Omega_x^2}{\Omega_{Q_L}^2}$ %	(5) $\frac{\eta\Omega_x^2}{\left(\Omega_{Q_L}^2 + \Omega_{Q_C}^2\right)}$ %
C	0.825	0.071	12.9	11.43
i	7.2 in/hr	0.177	80.7	71.53
A	12. acres	0.050	6.4	5.71
Q_L	71.3 cfs	0.197	$\sum$ = 100.0	$\sum$ = 88.7
			$\eta\frac{\Omega_x^2}{\Omega_{Q_C}^2}$ %	$\frac{\eta\Omega_x^2}{\left(\Omega_{Q_L}^2 + \Omega_{Q_C}^2\right)}$ %
n	0.015	0.0553	62.1	6.99
d	5.0 ft	0.010	14.1	1.62
S_0	0.001 ft/ft	0.068	23.5	2.65
Q_C	71.4 cfs	0.0705	$\sum$ = 100.0	$\sum$ = 11.3

11.5.2 Risk-Safety Factor Relationship

Section 11.2 describes the common design practice for storm sewers of determining the loading using the rational method for a specified return period, then computing a diameter for Q_L. A safety factor is inherently built into the choice of the return period. Alternatively, the loading could be multiplied by a safety factor (SF) to determine the capacity that is, $Q_C = SF \times Q_L$. Section 11.5.1 discussed various kinds of uncertainty associated with Q_C and Q_L. Using the reliability analysis, a probability of failure $P(Q_L > Q_C)$ can be calculated for selected return periods and safety factors. The corresponding risks and safety factors for each return period can be plotted to derive the risk-safety factor relationship for each return period. Such an example is shown in Fig. 11.5.1 for a 100-year return period for the Derby Catchment in England.

The procedure to determine risk-safety factor relationships for storm sewers is as follows:

1. Select the return period (T).
2. Select a rainfall-runoff model, for example, the rational model, and perform uncertainty analysis to compute $\bar{Q}_L$ and Ω_{Q_L}.
3. Select a pipe diameter and compute $\bar{Q}_C$ and Ω_{Q_C}.

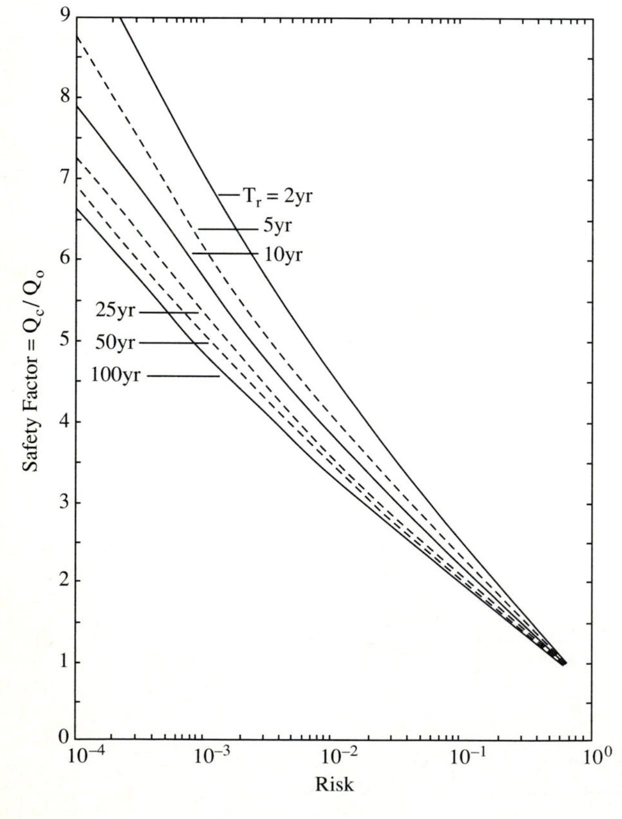

FIGURE 11.5.1
Risk-safety factor curves for Derby Catchment (Yen, 1975).

4. Compute the risk and safety factor.
5. Return to 3 for as many diameters as needed.
6. Return to 2 for as many rainfall durations as needed.
7. Return to 1 for each return period considered.

11.6 STORMWATER DETENTION

11.6.1 Urbanization Effects and Stormwater Detention

During the past considerable attention has been placed upon the effects of urbanization on the flood potential of small urban watersheds. The effects of urbanization result in increased total-runoff volumes and peak-flow rates as depicted in Fig. 11.6.1. In general, the major changes in flow rates in urban watersheds are caused by:

a. The volume of water available for runoff increases because of the fabrication of increased impervious covers such as parking lots, streets, and roofs that reduce the amount of infiltration.

b. Changes in hydraulic efficiency associated with artificial channels, curbing, gutters, and storm drainage collection systems increase the magnitude of flood peaks and reduce the time of concentration.

STORMWATER DETENTION. Because the effects of urbanization increase peak-runoff rates and volume, efforts have been made in urban areas to institute management methods to offset these effects. A very effective method of reducing peak runoff rates and minimizing the detrimental impact of urbanization is to use detention. **Detention** refers to holding runoff for a short period of time and then releasing it to the natural water course where it returns to the hydrologic cycle. This is basically capturing stormwater runoff for later controlled release. A stormwater detention basin can range from simply backing up water behind a highway or road culvert to a reservoir with sophisticated control devices.

Retention involves holding water contained in the storage facility for a considerable length of time. The water may never be discharged to a natural watercourse, but instead consumed by plants, evaporation, or infiltration into the ground. Detention facilities generally do not significantly reduce the total volume of surface runoff, but simply reduce peak-flow rates by redistributing the runoff over time as shown in Fig. 11.6.1. However, there are exceptions; for example, the reduced surface runoff from detention basins on granular soils.

The **on-site detention** of stormwater refers to the storage of precipitation runoff excess followed by gradual release of the stored runoff during and after the peak of the runoff. In some applications, the runoff may first be conducted short distances by collector sewers located on or adjacent to the site of the detention facility. On-site detention can be differentiated from downstream detention by its proximity to the upper end of a basin and the use of small detention facilities as opposed to large dams

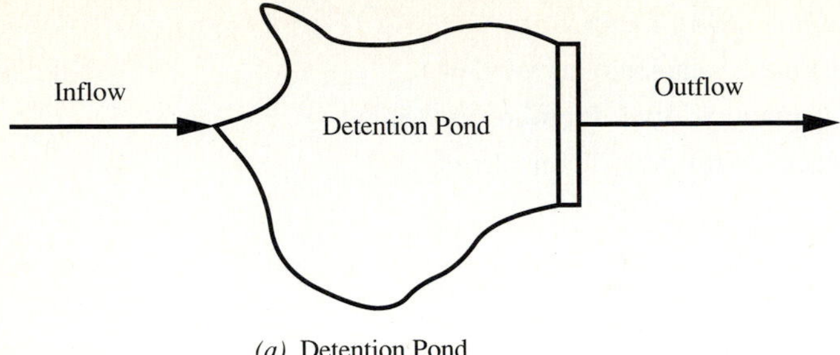

(a) Detention Pond

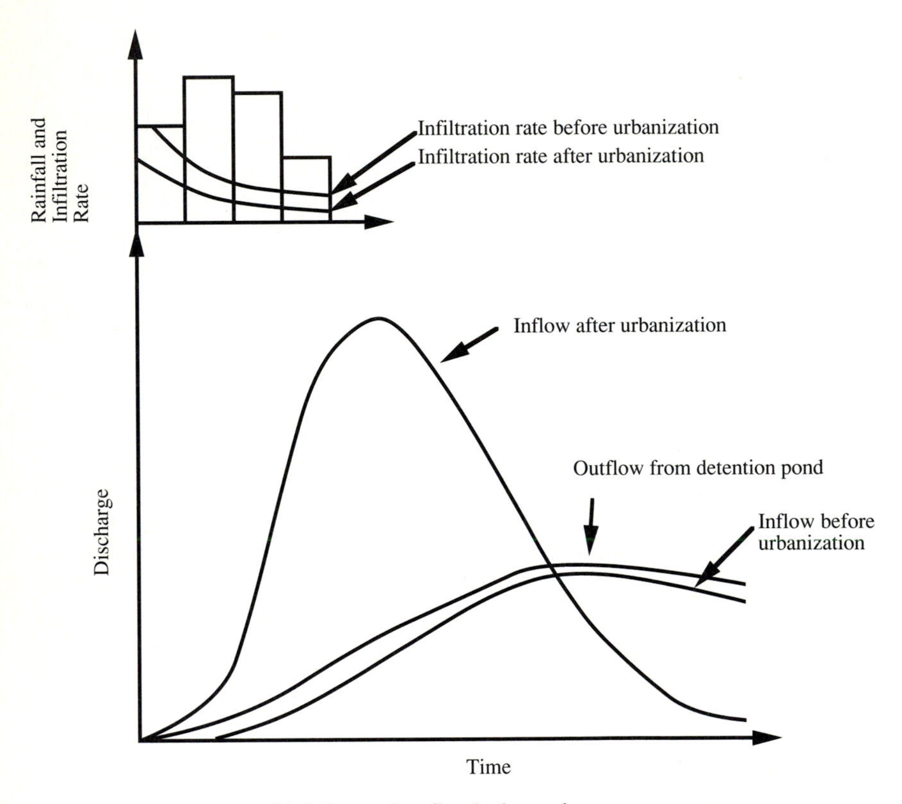

(b) Inflow and outflow hydrographs

FIGURE 11.6.1
Hydrologic effect of urbanization on storm runoff and function of detention ponds.

normally associated with downstream detention. Detention storage can be provided at one or a combination of locations: (1) on, or near the sites where the precipitation occurs; (2) in storm sewer systems; or (3) in downstream impoundments.

Detention storage is extremely important in areas having appreciable topographic relief for the purpose of attenuating peak-flow rates and the high kinetic energy of surface runoff. Such flow attenuation can reduce soil erosion and the amounts of contaminants of various kinds that are assimilated and/or transported by urban runoff from land, pavements, and other surfaces. Several methods exist for the detention of stormwater runoff including underground storage, basins and ponds on ground surfaces, parking lot storage, and rooftop detention.

11.6.2 Selection of Detention Pond Size—Modified Rational Method

The modified rational method can be used to determine the preliminary design, which is the detention pond volume requirements for contributing drainage areas of 30 acres or less. For larger contributing areas a more detailed rainfall-runoff analysis with a detention basin flow routing procedure should be used. The modified rational method is an extension of the rational method to develop hydrographs for storage design, rather than only peak discharges for storm sewer design. The shape of hydrographs produced by the modified rational method is either a triangular or trapezoidal shape constructed by setting the duration of the rising and recession limbs equal to the time of concentration t_c and computing the peak discharge assuming various durations. Figure 11.6.2*a* illustrates modified rational method hydrographs.

An allowable discharge, Q_A, from a proposed detention basin can be the requirement that the peak discharge from the pond be equal to the peak of the runoff hydrograph for predeveloped conditions. The required detention storage, V_s, for each rainfall duration can be approximated as the cumulative volume of inflow minus the outflow as shown in Fig. 11.6.2*b*.

The assumptions of the modified rational method include:

1. the same assumptions as the rational method;
2. the period of rainfall intensity averaging is equal to the duration of the storm;
3. because the outflow hydrograph is either triangular or trapezoidal, then the effective contributing drainage area increases linearly with respect to time.

An equation for the **critical storm duration**, that is, the storm duration that provides the largest storage volume can be determined for small watersheds based upon the modified rational method. Consider a rainfall intensity-duration equation of the general form

$$i = \frac{a}{(t_D + b)^c} \tag{11.6.1}$$

where i is the average rainfall intensity (in/hr) for the specific duration and return period; t_D is storm duration in minutes; and a, b, and c are coefficients for a specific

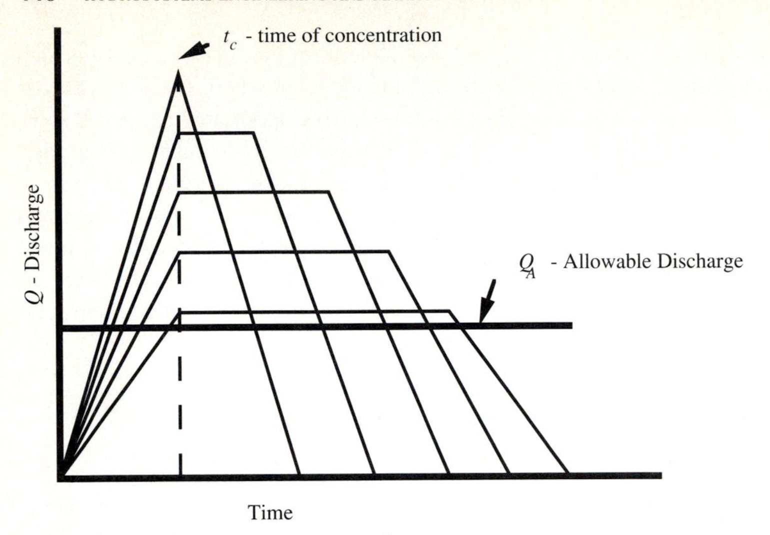

(a) Hydrographs for different durations

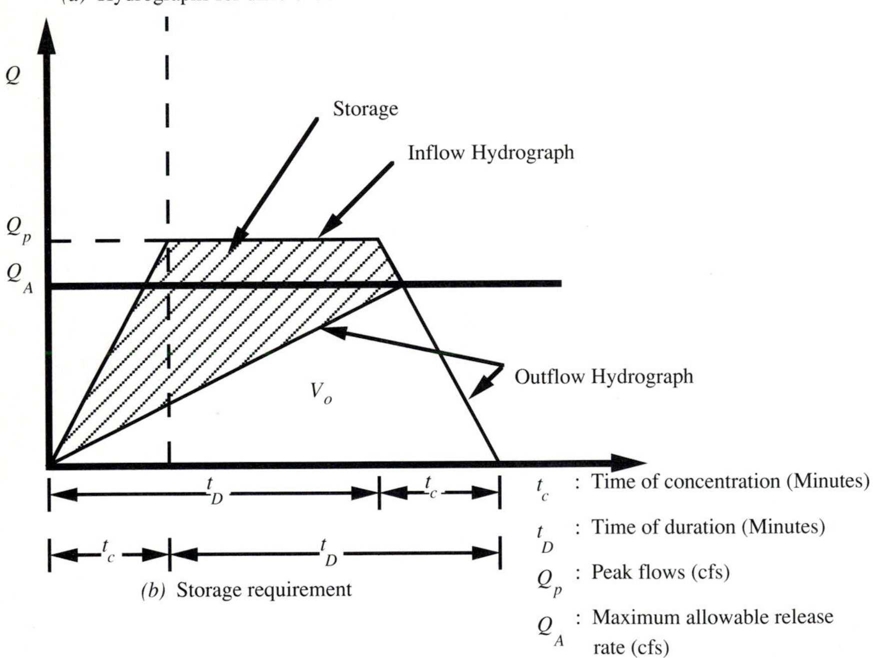

(b) Storage requirement

FIGURE 11.6.2
Modified rational method hydrographs.

return period and location. Consider the trapezoidal shaped inflow hydrograph and outflow hydrograph shown in Fig. 11.6.2.

Using the rational formula the peak discharge can be expressed in terms of the storm duration

$$Q_p = C_p i A$$
$$= C_p \left[\frac{a}{(t_D + b)^c}\right] A \tag{11.6.2}$$

The inflow hydrograph volume V_i in ft^3, is expressed as

$$V_i = 60(0.5)Q_p\,[(t_D - t_c) + (t_D + t_c)] \tag{11.6.3}$$

where t_c is the time of concentration for proposed conditions. The outflow hydrograph volume V_0 in ft^3, is expressed as

$$V_0 = 60(0.5)Q_A\,(t_D + t_c) \tag{11.6.4}$$

where Q_A is the allowable peak flow release in ft^3. The storage volume V_s in ft^3 is computed using the above expressions for V_i and V_0

$$V_s = V_i - V_0$$
$$= 60(0.5)Q_p\Big[\big(t_D - t_c\big) + \big(t_D + t_c\big)\Big] - 60(0.5)Q_A\,(t_D + t_c)$$
$$= 60Q_p t_D - 30Q_A\,(t_D + t_c) \tag{11.6.5}$$

The duration for the maximum detention is determined by differentiating Eq. (11.6.5) with respect to t_D and setting the derivative equal to zero:

$$\frac{dV_s}{dt_D} = 0 = 60t_D\frac{dQ_D}{dt_D} + 60Q_p - 30Q_A$$
$$= 60t_D C_p A\frac{di}{dt_D} + 60C_p i A - 30Q_A \tag{11.6.6}$$

where

$$\frac{di}{dt_D} = \frac{d}{dt_D}\left[\frac{a}{(t_D + b)^c}\right] = \frac{-ac}{(t_D + b)^{c+1}} \tag{11.6.7}$$

so

$$\frac{dV_s}{dt_D} = 0 = 60C_p A(-ac)\frac{t_D}{(t_D + b)^{c+1}} + 60C_p\left[\frac{a}{(t_D + b)^c}\right] - 30Q_A \tag{11.6.8}$$

Simplifying results in

$$\frac{t_D(1 - c) + b}{(t_D + b)^{c+1}} - \frac{Q_A}{2C_p A} = 0 \tag{11.6.9}$$

t_D in Eq. (11.6.9) can be solved by using Newton's iteration technique where the iterative equation is

$$t_{D_{i+1}} = t_{D_i} - \frac{F(t_{D_i})}{F'(t_{D_i})} \tag{11.6.10}$$

where

$$F(t_D) = \frac{[t_D(1-c)+b]}{(t_D+b)^{c+1}} - \frac{Q_A}{2C_pA} = 0 \tag{11.6.11}$$

and

$$F'(t_{D_i}) = \frac{d\left[F(t_{D_i})\right]}{dt_D} = -\frac{[t_D(1-c)+b](c+1)}{(t_D+b)^{c+2}} + \frac{(1-c)}{(t_D+b)^{c+1}} \tag{11.6.12}$$

Example 11.6.1. Determine the critical duration t_D for a 15.24-acre watershed with a developed runoff coefficient of $C_p = 0.85$. The allowable discharge is the predevelopment discharge of $Q_A = 32.17$ cfs. The time of concentration for proposed conditions is 21.2 min. The applicable rainfall-intensity-duration relationship is

$$i = \frac{97.86}{(t_D + 16.4)^{0.76}}$$

Solution. The critical storm duration is found from solving Eq. (11.6.9) by use of Newton's method with an initial guess of the duration as 30 min. The procedure begins by using Eq. (11.6.11)

$$F\left(t_{D_{i=1}}\right) = \frac{[30(1-0.76)+16.4]}{(30+16.4)^{0.76}} - \frac{32.17}{2(0.85)(15.24)} = 0.0148$$

and Eq. (11.6.12)

$$F'\left(t_{D_{i=1}}\right) = -\frac{[30(1-0.76)+16.4](0.76+1)}{(30+16.44)^{0.76+2}} + \frac{(1-0.76)}{(30+16.4)^{0.76+1}} = -0.00076$$

Applying Eq. (11.6.10)

$$t_{D_{i=2}} = 30 - \frac{0.0148}{-0.00076} = 49.42 \text{ min}$$

The procedure continues using $t_{D_{i=2}} = 49.42$ min in the next iteration. The results are presented below in Table 11.6.1.

The procedure actually converges to a duration of 71.6772 min (or 71.68 min) after five iterations using a convergence criteria of

$$\frac{F\left(t_{D_i}\right)}{F'\left(t_{D_i}\right)} < 0.5$$

TABLE 11.6.1
Application of Newton's Method (Example 11.6.1)

Iteration	t_{D_i}	$F(t_{D_i})$	$F'(t_{D_i})$	$t_{D_{i=1}}$
1	30.0000	0.0148436	−0.0007643	49.4203
2	49.4203	0.0051300	−0.0003251	65.1982
3	65.1982	0.0011547	−0.0001949	71.1222
4	71.1222	0.0000911	−0.0001653	71.6732
5	71.6732	0.0000006	−0.0001629	71.6772

Example 11.6.2. Determine the maximum detention storage for the watershed in Example 11.6.1.

Solution. The peak discharge for the duration of $t_D = 71.68$ min is

$$Q_P = C_P A\left[\frac{a}{(t_D + b)^c}\right]$$

$$= 0.85(15.24)\left[\frac{97.86}{(71.68 + 16.4)^{0.76}}\right]$$

$$= 42.16 \text{ cfs}$$

Using Eq. (11.6.5) the maximum detention storage is

$$V_s = 60Q_p t_D - 30Q_A(t_D + t_c)$$

$$= 60(42.16)(71.68) - 30(32.17)(71.68 + 21.2)$$

$$= 91674 \text{ ft}^3$$

11.6.3 Hydrograph Design Method

There are several major design determinations involved in the engineering design of stormwater detention facilities. These are: (1) the selection of a design rainfall event; (2) the volume of storage needed; (3) the maximum permitted release rate; (4) pollution control requirements and opportunities; and (5) identification of practical detention methods and techniques for the specific project.

A simple design procedure for detention basins is now outlined that is useful in practice.

1. Determine the watershed characteristics and location of the detention basin.
2. Determine the design inflow hydrograph to the detention basin using a rainfall runoff model.
3. Determine the detention storage-discharge relationship.
   a. Determine the storage-elevation relationship.
   b. Determine the discharge-elevation relationship for the discharge structure (culvert, spillway, etc.).
   c. Using the above relationships develop the storage-discharge relationship.
4. Perform the computations described in Section 10.3.1 to route the inflow hydrograph through the detention basin using hydrologic routing.
5. Once the routing computations are completed the reduced peak can be checked to see that the reduction is adequate and also check the delay of the peak outflow.
6. This procedure using steps 3(b) through 5 can be repeated for various discharge structures.

11.7 MINIMUM COST DESIGN OF REGIONAL STORMWATER DETENTION SYSTEMS

Comprehensive stormwater management plans consider methods to help mitigate the hydrologic effects of urbanization. On-site detention in many cases has questionable benefits and the unplanned placement of multiple detention ponds could aggravate potential flooding. A better approach is to develop an integrated regional stormwater program to select detention sites and regulate outflows based on a watershed wide or regional analysis. Such an approach considers larger regional facilities as opposed to on-site detention. This can allow for significant improvement in the efficiency of the control of stormwater runoff resulting from future developments.

Major questions concerning the feasibility of regional detention storage are the size (volume or amount of acreage) and the location of the detention basins. Conventional design methods can be used in the framework of trial-and-error methods to determine the minimum (optimal) cost, size, and location of detention basins within a detention system for an urban watershed. Conventional design methods do not account for the cost interactions of the various components of stormwater runoff control systems in an optimal framework to find minimum cost design.

A regional detention basin system can be divided into stages through the use of **iso-drainage** lines as shown in Fig. 11.7.1 to match the requirements of a DP procedure (Mays and Bedient, 1982; Bennett and Mays, 1985; and Taur et al., 1987). The iso-drainage lines are analogous to the isonodal line for storm sewers. The stages consist of existing and candidate detention facility locations. State variables are discretized at each individual detention facility location into states which represent the set of all the possible detention basin sizes and the allowable downstream peak discharges from the basin. Decisions are made by the DP algorithm at each stage which transforms the current state of the system into a state associated with the next stage. The decision variable represents the choice of state which is a combination of detention basin size, allowable peak discharge, and design of the outlet structure.

The recursive equation used in the model is

$$f_n(S_{n,i,j,k}) = \min \left[r_n(S_{n,i,j,k}, D_{n,i,j,k}) + f_{n-1}(S_{n-1,i,l,m})\right] \tag{11.7.1}$$

in which $r_n(S_{n,i,j,k}, D_{n,i,j,k})$ represents the cost of the detention basin i on iso-drainage line n of size denoted by subscript k and the allowable downstream peak flow associated with subscript j. $f_{n-1}(S_{n-1,i,l,m})$ represents the minimum cost of the detention system that is connected to detention basin $(n-1, i)$ of size m and allowable peak associated with index l. The LHS of Eq. (11.7.1), $f_n(S_{n,i,j,k})$, is the minimum cost of the upstream detention system that is connected to the detention basin (n, i) with the basin size k and allowable peak discharge j.

Equation (11.7.1) considers only one immediate upstream detention facility which drains to the downstream detention basin. There may be one or more upstream facilities on the previous stage that could drain to a downstream basin. The recursive equation for a branched system is

$$f_n(S_{n,i,j,k}) = \text{Min} \left[r_n(S_{n,i,j,k}, D_{n,i,j,k}) + \sum_{i}^{P} f_{n-1}(S_{n-1,i,l,m})\right] \tag{11.7.2}$$

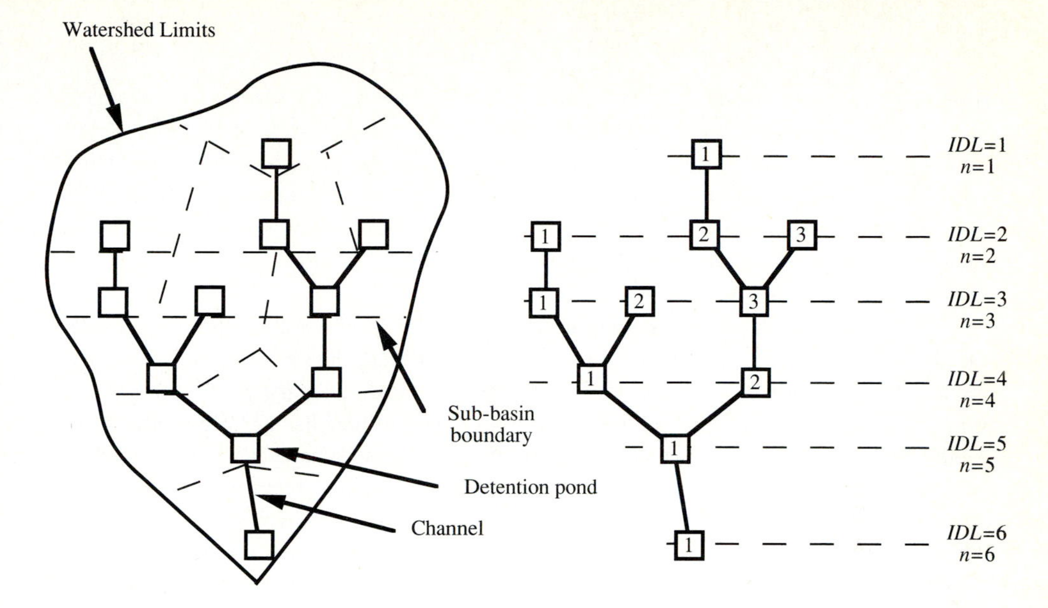

FIGURE 11.7.1
Application of iso-drainage line to the detention basin systems.

in which P is the number of upstream detention facilities which can drain to the downstream detention basin i on stage n.

Objective function. The objective of the model is to minimize the total land and construction costs for an entire detention basin network while satisfying peak flow and land availability constraints. The equation, which relates the land cost and construction cost for detention structures, is represented as

$$r_n(S_{n,i,j,k}, D_{n,i,j,k}) = f(L_w) + f(C_L, A_D) + f(V) \tag{11.7.3}$$

in which L_w is the weir length (ft) for the outflow structure, C_L is the land cost (\$/acre), A_D is the land area (acre) of the detention basin and V represents the total storage size (acre-foot) of the basin.

Two system constraints are used to match the realistic conditions for the detention basin system. The first constraint is the maximum water surface elevation for each basin, so that the maximum volume $V_{\max}$, of the basin cannot be exceeded,

$$\int_o^t (Q_{\text{in}} - Q_{\text{out}})dt \leq V_{\max} \tag{11.7.4}$$

The second constraint assures that the peak discharge, $Q_{p,n,i}$ from detention basin (n, i) is less than or equal to the maximum allowable discharge, $Q_{\max}$, in the receiving channel

$$Q_{p,n,i} \leq Q_{\max} \tag{11.7.5}$$

in which $Q_{\max}$ is selected based on floodplain considerations in the receiving channel.

Algorithm. The computational procedure begins at the upstream end (stage 1) of the watershed and moves downstream stage by stage. The procedure starts by finding the optimal combination of states for the candidate basin in the initial stage. Then recursive Eqs. (11.7.1) and (11.7.2) are applied to relate all downstream stages until the optimal policy for the last stage is found.

The DP algorithm at any detention facility (n, i) will pick the least cost upstream facility (stage $n-1$) from all the possible state combinations of sizes and peak discharges, as shown in Fig. 11.7.2. Accomplishing the search for the least cost facility, the downstream state will be identified and indexed. When the DP algorithm completes the computations in a stage, the routing procedure will be used for the channels associated with the optimum facility. The outflow hydrograph from the optimum upstream facility is routed through the selected receiving channel to a downstream state. This is then combined with the lateral inflow hydrograph at the next downstream facility. The final hydrograph will be routed by a detention basin routing procedure through the chosen detention basin design of state (j, k) at stage n. The two system constraints are utilized here to check that the outflow hydrograph from the detention basin design does not exceed the maximum allowable discharge of the channel and the maximum volume of the basin. If exceedance occurs, the basin storage volume

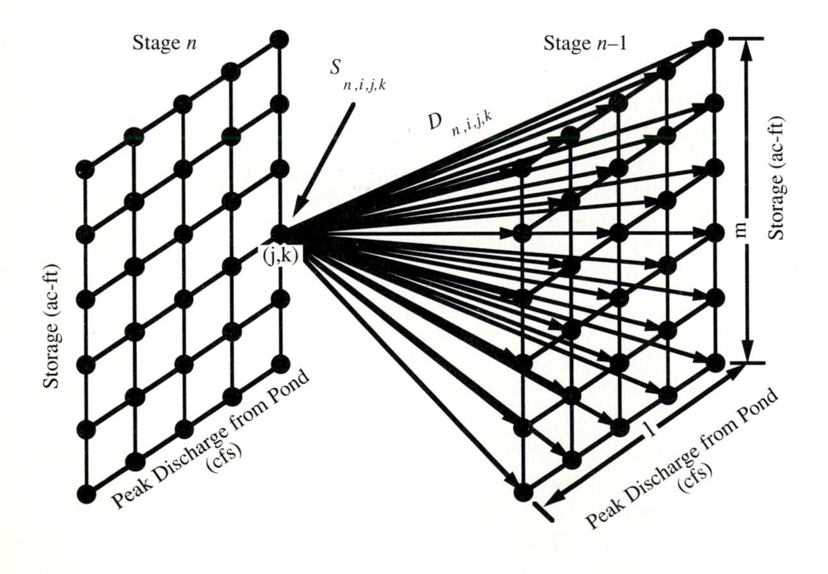

FIGURE 11.7.2
State variables for DP model for regional detention basin design. Each state variable (j, k) on pond (n, i) has $l^* m$ choices to select the least-cost design.

will be increased by a predefined amount to reduce the outflow hydrograph. However, the total volume of the basin design cannot be larger than the available storage at the location of the facility. Otherwise, the decision will be abandoned, a new state will be chosen and the computation will be resumed at this stage. The system constraints are used to guarantee a final solution compatible with the amount of land available at each candidate site and the flow capacity of the receiving channels.

Once the computation has been completed through the most down-stream detention location, a traceback procedure is performed to retrieve the least cost detention facility system for the watershed. The final step is to route the stormwater through the optimal system. The basin and channel routing techniques use the generated detention characteristics from the optimization procedure and route the respective inflows through the proposed system. This procedure cannot guarantee global optimality and should be thought of as a heuristic approach.

REFERENCES

Bennett, M. S. and L. W. Mays: "Optimal Design of Detention and Drainage Channel Systems," *Journal of the Water Resources Planning and Management Division,* ASCE, vol. 111, no. 1, pp. 99–112, January 1985.

Chow, V. T., D. R. Maidment, and L. W. Mays: *Applied Hydrology*, McGraw-Hill, Inc., New York, 1988.

Donohue, J. R., R. H. McCuen, and T. R. Bondelid: "Comparison of Detention Basin Planning and Design Models," *J. Water Res., Planning and Management Div.*, Am. Soc. Civ. Eng., vol. 107, no. WR2, pp. 385–400, October 1981.

Huber, W. C., J. P. Heaney, M. A. Medina, W. A. Peltz, H. Sheikhj, and G. F. Smith: *Storm Water Management Model User's Manual*, version II, Environmental Protection Technology Series, EPA-670/2-75-017, Municipal Environmental Research Laboratory, USEPA, March 1975.

Mays, L. W. and P. B. Bedient: "Model for Optimal Size and Location of Detention Basins," *Journal of the Water Resources Planning and Management Division*, ASCE, vol. 108, no. WR3, p. 220–285, October 1982.

Mays, L. W. and H. G. Wenzel Jr.: "Optimal Design of Multilevel Branching Sewer Systems," *Water Resources Research*, AGU, vol. 12, no. 5, pp. 913–917, October 1976.

Mays, L. W., H. G. Wenzel Jr., and J. C. Liebman: "Model for Layout and Design of Sewer Systems," *Journal of the Water Resources Planning and Management Division*, ASCE, vol. 102, no. WR2, pp. 385–405, November 1976.

Mays, L. W. and B. C. Yen: "Optimal Cost Design of Branched Sewer Systems," *Water Resources Research*, vol. 11, no. 1, pp. 37–47, February 1975.

Tang, W. H., L. W. Mays, and B. C. Yen: "Optimal Risk-Based Design of Storm Sewer Networks," *Journal of Environmental Engineering Division*, ASCE, vol. 101, no. EE3, pp. 381–398, June 1975.

Taur, C. K., G. Toth, G. E. Oswald, and L. W. Mays: "Austin Detention Basin Optimization Model," *Journal of Hydraulic Engineering*, ASCE, vol. 113, no. 7, pp. 860–878, July 1987.

Yen, B. C.: "Risk Based Design of Storm Sewers," Report no. INT 141, Hydraulics Research Station, Wallingford, England, July 1975.

U.S. Army Corps of Engineers: Hydrologic Engineering Center, *Introduction and Application of Kinematic Ware Routing Techniques Using HEC-1*, Davis, Calif. May 1979.

Yen, B. C., ed.: *Storm Sewer System Design*, Department of Civil Engineering, University of Illinois at Urbana-Champaign, 1978.

Yen, B. C., H. G. Wenzel, Jr., L. W. Mays, and W. H. Tang: "Advanced Methodologies for Design of Storm Sewer Systems," Research Report 112, Water Resources Center, University of Illinois at Urbana-Champaign, August 1976.

PROBLEMS

11.2.1 Design sewers 5.1, 6.1, and 7.1 of the Goodwin Avenue, Urbana, Illinois drainage basin (Example 11.2.1) shown in Figure 11.2.1 using the hydrograph time lag method.

11.4.1 Complete one more iteration of the DDDP procedure in Example 11.4.1 storm server design.

11.5.1 Consider an urbanized drainage basin consisting of 40 percent roofs, 20 percent asphalt streets, and 40 percent driveways and sidewalks. The prediction error for the percentages is assumed to be 0.10 and the ranges of runoff coefficients are listed in the following table. Determine the coefficient of variation of the runoff coefficient, assuming a uniform distribution for the runoff coefficient.

	Driveways and sidewalks	**Roofs**	**Streets**
β_j	0.40	0.40	0.20
Range of C_j	0.75-0.85	0.75-0.95	0.70-0.95
$\bar{C}_j$	0.800	0.850	0.825
Ω_{C_j}	0.036	0.068	0.087

11.5.2 Determine the risk (probability) of the surface runoff loading exceeding the sewer capacity using a 66-inch pipe for the problem posed in Example 11.5.1. Compare the risk value obtained to that in Example 11.5.1 with a 60 inch pipe. Assuming that the coefficient of variation of all the parameters remain the same as in Example 11.5.1 and both runoff (loading) and sewer capacity are independent normal random variables.

11.5.3 Determine the risk of the loading exceeding the capacity of the 66-inch storm sewer pipe in Problem 11.5.2 assuming that the capacity and loading are independent and lognormal random variables. Data on all the parameters are given in Example 11.5.1.

11.5.4 Determine the coefficient of variation of the loading and the capacity for the following parameters. Assume a uniform distribution to define the uncertainty of each parameter.

Parameter	**Mode**	**Range**
C	0.75	0.70–0.80
i	7.0 in/hr	7.2–7.8 in/hr
A	12 acres	11.9–12.1 acres
n	0.015	0.0145–0.0155
d	5 ft	4.96–5.04 ft
S_0	0.001 ft/ft	0.0009–0.0011 ft/ft

11.5.5 Solve Problem 11.5.4 assuming the use of a triangular distribution to compute the uncertainty of each parameter.

11.5.6 Using the results of Problem 11.5.4 determine the risk of the loading exceeding the capacity of the sewer pipe. Assume that the safety margin is normally distributed.

11.5.7 Using the results of Problem 11.5.5 determine the risk of the loading exceeding the capacity of the sewer pipe. Assume that the safety margin is normally distributed.

11.5.8 Using the results of Problems 11.5.4 and 11.5.6 determine the risk of the loading exceeding the capacity of the sewer pipe. Assume that Q_L and Q_C are independent and log-normally distributed.

11.5.9 Using the results of Problem 11.5.5 determine the risk of the loading exceeding the capacity of the sewer pipe. Assume that Q_L and Q_C are independent and log-normally distributed.

11.5.10 Solve Example 11.5.1 using a mean drainage area of $A = 11$ acres with all other data remaining the same.

11.5.11 Determine the risk of the loading exceeding the capacity of a storm sewer, for a loading condition determined using the rational formula and the capacity determined using the Darcy-Weisbach equation. Assume a triangular distribution for describing the uncertainties of each parameter. Also assume that Q_L and Q_C are independent and log-normally distributed.

Parameter	Mode	Range
C	0.5	0.45–0.55
i	48.0 mm/hr	45.–51. mm/hr
A	87700 m^2	87600–87800 m^2
d	914 mm	905– 923 mm
f	0.0297	0.0290–0.0304
S_0	0.005	0.0045–0.0055

11.5.12 Solve Problem 11.5.11 assuming that the safety margin is normally distributed.

11.5.13 Determine the risk-safety factor curve for the the Example 11.5.3 problem for a return period of 2 years.

11.5.14 Determine the risk-safety factor curve for the Example 11.5.3 problem for a return period of 25 years.

11.5.15 Derive a general expression for the coefficient of variation of the rainfall intensity and apply it to Urbana, Illinois, for which the rainfall intensity can be expressed as

$$i = \frac{120T^{0.175}}{27 + t_D}$$

where i is in in/hr, T is in years, and t_D is in minutes.

11.6.1 Determine the critical duration, t_D, and maximum detention storage for a 31.39 acre fully developed watershed with a runoff coefficient of $C_p = 0.95$. The allowable discharge is the predevelopment discharge of $Q_A = 59.08$ cfs. The time of concentration for proposed conditions is 21.2 minutes. The applicable rainfall intensity duration relationship is

$$i = \frac{97.86}{(t_D + 16.4)^{0.76}}$$

Use an initial guess of $t_D = 75.8858$ minutes.

11.6.2 Solve Problem 11.6.1 using a runoff coefficient of 0.75.

11.6.3 Determine the critical duration, t_D, and maximum detention storage for a 4.25 acre watershed with a runoff coefficient of 0.95 for fully developed conditions. The allowable discharge is the predevelopment discharge of $Q_A = 9.2$ cfs. The time of concentration for developed conditions is 20 minutes. The applicable rainfall intensity-duration relationship is

$$i = \frac{129.03}{(t_D + 17.83)^{0.7625}}$$

11.6.4 Write a computer program using Newton's method to determine the critical duration, t_D. Input to the program would be the allowable discharge from the detention pond, the runoff coefficient for developed conditions, the time of concentration for developed conditions, and the coefficients a, b, and c for the following rainfall intensity duration relationship.

$$i = \frac{a}{(t_D + b)^c}$$

11.6.5 Use the computer program developed in Problem 11.6.4 to solve Problem 11.6.1.

11.6.6 Use the computer program developed in Problem 11.6.4 to solve Problem 11.6.3.

CHAPTER 12

FLOODPLAIN MANAGEMENT SYSTEMS

Flooding results from conditions of hydrology and topography in floodplains such that the flows are large enough that the channel banks overflow resulting in overbank flow that can extend over the floodplain. For large floods, the floodplain acts as a conveyance and as a temporary storage for flood flows. The main channel is usually a defined channel that can meander through the floodplain carrying low flows. The overbank flow is usually shallow as compared to the channel flow and also flows at a much slower velocity than the channel flow.

The objective of flood control is to reduce or alleviate the negative consequences of flooding. Alternative measures that modify the flood runoff are usually referred to as flood-control facilities and consist of engineering structures or modifications. Construction of flood-control facilities, referred to as **structural measures**, are designed to consider the flood characteristics including reservoirs, diversions, levees or dikes, and channel modifications. Flood-control measures that modify the damage susceptibility of floodplains are usually referred to as **nonstructural measures** and may require minor engineering works. Nonstructural measures are designed to modify the damage potential of permanent facilities and provide for reducing potential damage during a flood event. Nonstructural measures include flood proofing, flood warning, and land use controls. Structural measures generally require large sums of capital investment. **Floodplain management** considers the integrated view of all engineering, nonstructural, and administrative measures for managing (minimizing) losses due to flooding on a comprehensive scale.

12.1 FLOOD-CONTROL ALTERNATIVES

12.1.1 Structural Alternatives

Table 12.1.1 summarizes several damage reduction measures and the parametric relationships which are modified. The basic functional relationships required to assess the value of flood damage reduction alternatives are shown in Fig. 12.1.1. **Stage-damage relationships** define the flood severity in terms of damage cost for various stages. **Stage-discharge relationships**, also referred to as **rating curves**, are modified by various flood-control alternatives. Flood flow frequency relationships (described in Section 10.4) define the recurrence nature in terms of the flood magnitudes. Flood-control alternatives are designed to modify the flood characteristics by modifying one or more of the above relationships. The major types of flood-control structures are reservoirs, diversions, levees or dikes, and channel modifications. Each of these are discussed below defining the resulting changes in the basic relationships.

Flood-control reservoirs are used to store flood waters for subsequent release after the flood event, reducing the magnitude of the peak discharge. Reservoirs modify the flood-flow frequency curve, which is lowered because of the decrease of the peak discharge of a specific event. Figure 12.1.2 illustrates the effect of flood-control reservoirs. Long-term effects of reservoir storage modify the streamflow regime and can result in channel aggradation or degradation at downstream locations altering the rating curve.

TABLE 12.1.1
Effect of floodplain management measures (U.S. Army Corps of Engineers, 1988)

	Impacted relationship*				
	Stage-discharge	Stage-damage	Discharge-damage	Discharge-frequency	Damage-frequency
Reservoir†	NC	NC	NC	M‡	M
Levee or floodwall†	M	M	M	M‡	M
Channel modification†	M	NC	M	M	M
Diversion†	NC	NC	NC	M	M
Flood forecasting	NC	NC	NC	M	M
Flood proofing	NC	M	M	NC	M
Relocation	NC	M	M	NC	M
Flood warning	NC	M	M	NC	M
Land use control§	NC	M	M	M	M

*The following codes apply to the table above:
NC = No Change in parametric relationship
M = Modification to parametric relationship

†Long-term effects resulting from a change in stream regime induced by these measures could affect the basic stage-flow relationship and thus other derived relationships at some future date.

‡Elimination of significant amounts of flood plain storage can result in downstream effects on flow-frequency relationship.

§The impact indicated is that which would occur to a future condition in the absence of the measure.

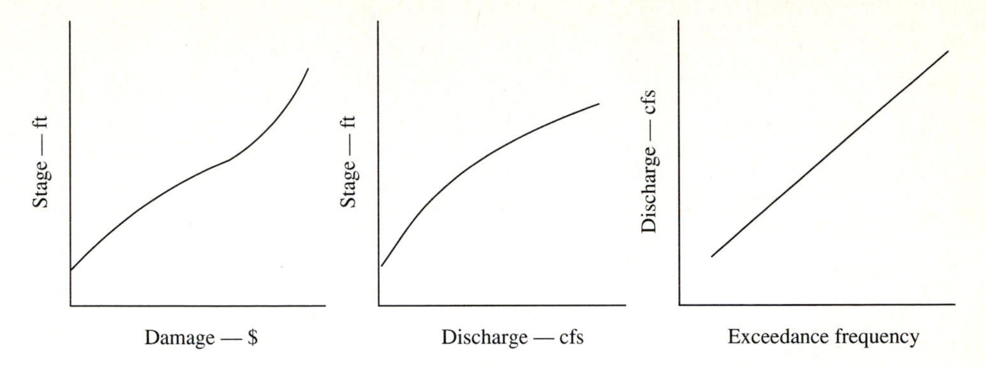

FIGURE 12.1.1
Flood assessment functional relationships. (Source: U.S. Army Corps of Engineers Hydrologic Engineering Center)

Diversion structures are used to reroute or bypass flood flows from damage centers in order to reduce the peak flows at the damage centers. Diversion structures are designed to modify (lower) the frequency curve so that the flow magnitude for a specific event is lowered at the damage center. Figure 12.1.3 illustrates the effect of diversions on the functional relationships. The stage-damage and stage-discharge relationships remain the same if there are no other induced effects. Long-term effects of diversions can cause aggradation or degradation at downstream locations and result in sediment depositions in bypass channels.

Levees or dikes are used to keep flood flows from floodplain areas where damage can occur. Levees essentially modify all three of the functional relationships. The effect of levees is to reduce the damage in protected areas from water surface stages within the stream or main channel. This effect essentially truncates the stage-damage relationship for all stages below the design elevation of the levee, as illustrated in Fig. 12.1.4. The effect of excluding flood flows from portions of the floodplain outside the levees constricts the flow to a smaller conveyance area resulting in increasing

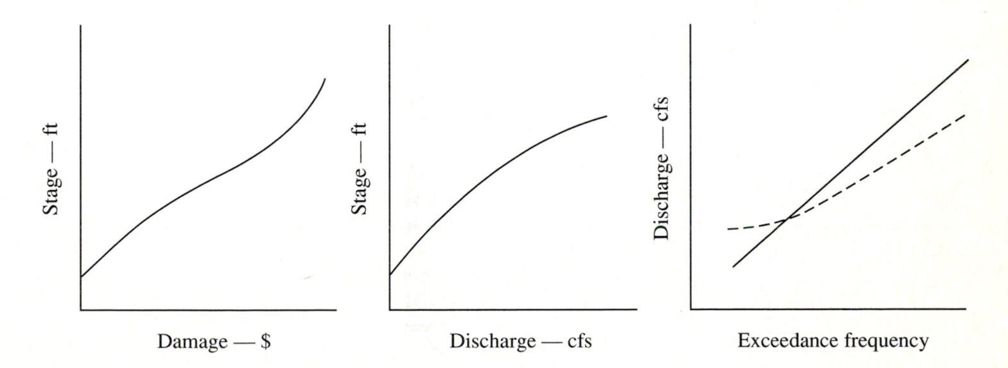

FIGURE 12.1.2
Effect of reservoir. (Source: U.S. Army Corps of Engineers Hydrologic Engineering Center)

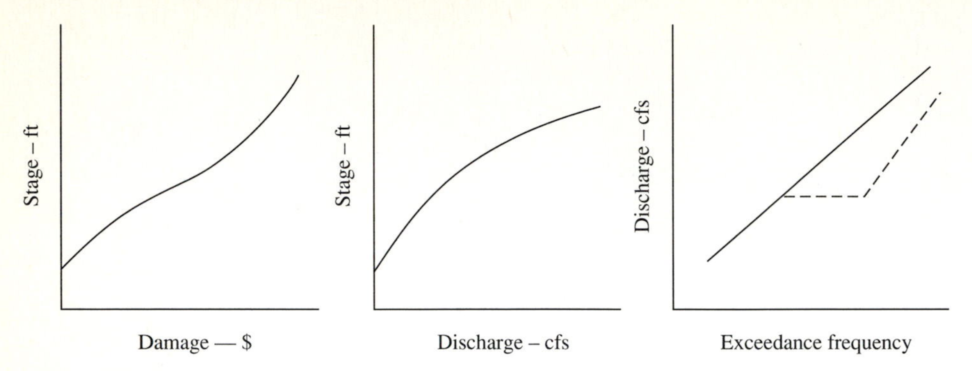

FIGURE 12.1.3
Effect of diversion. (Source: U.S. Army Corps of Engineers Hydrologic Engineering Center)

the stage for the various discharges. This upward shift of the stage-discharge relationship is shown in Fig. 12.1.4. Constricting the flow to within levees reduces the amount of natural storage of a flood wave causing an increase in peak discharges downstream. This effect increases the discharge for various exceedance frequencies shifting the frequency upwards, as shown in Fig. 12.1.4. Long-term effects of levees can cause aggradation or degradation of channels in downstream reaches. Even though levees are for the purpose of protecting property and lives, they also have the potential for major disasters when design discharges are exceeded and areas are inundated that are thought of as being safe.

Channel modifications are performed to improve the conveyance characteristics of a stream channel. The improved conveyance lowers the stages for various discharges having the effect of lowering the stage-discharge relationship, as illustrated in Fig. 12.1.5. The peak discharges for flood events are passed at lower stages decreasing the effect of natural valley storage during passage of a flood wave. This effect results in higher peak discharges downstream than would occur without the channel modifications causing an upward shift of the frequency curve, as illustrated

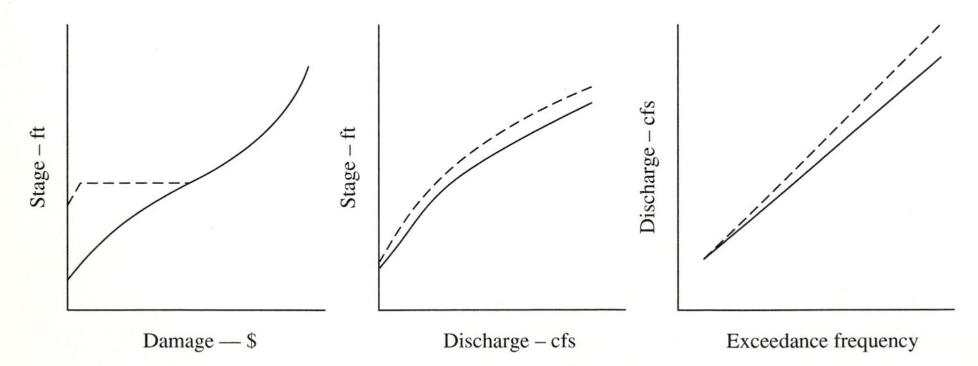

FIGURE 12.1.4
Effect of levee. (Source: U.S. Army Corps of Engineers Hydrologic Engineering Center)

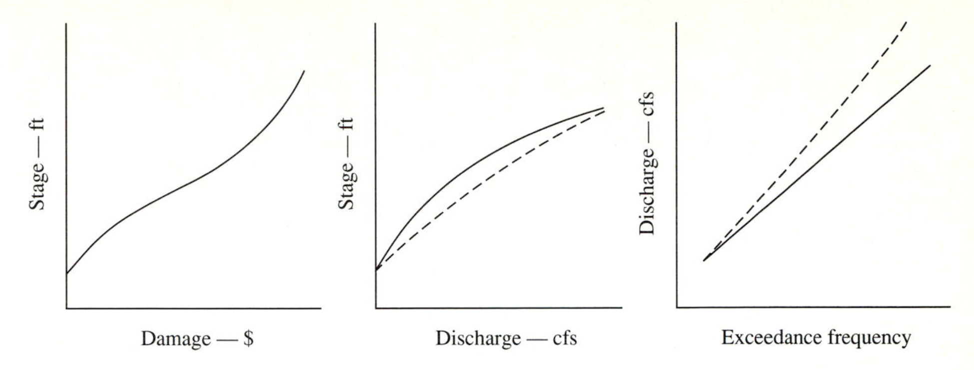

FIGURE 12.1.5
Effect of channel modification. (Source: U.S. Army Corps of Engineers Hydrologic Engineering Center)

in Fig. 12.1.5. Long-term effects of channel modification can cause aggradation and degradation of downstream channel reaches. Channel modifications are usually for local protection alternatives but can be integrated with other flood control alternatives to provide a more efficient flood-control system.

12.1.2 Nonstructural Measures

Nonstructural measures are used to modify the damage potential of permanent structures and facilities in order to decrease the susceptibility of flooding to reduce potential damages. Nonstructural measures include flood proofing, flood warning, and various types of land use control alternatives. These measures are characterized by their value in reducing future or potential unwise floodplain use. Of the nonstructural measures mentioned above, only flood proofing has the potential to modify present damage potential.

Flood proofing consists of a range of nonstructural measures designed to modify the damage potential of individual structures susceptible to flood damage. These measures include elevating structures, water proofing exterior walls, and rearrangement of structural working space. Flood proofing is most desirable on new facilities. Flood proofing changes only the stage-damage relation, as illustrated in Fig. 12.1.6, shifting the relationship upwards.

Flood warning provides lead time notice to floodplain occupants in order to reduce potential damage (see Section 13.1). The lead time provides the opportunity to elevate contents of structures, to perform minor proofing, and to remove property susceptible to flooding. The greatest value of flood warning is to reduce or eliminate the loss of life. Flood warning requires real-time flood forecasting and communication facilities to warn inhabitants of floodplains.

Land use controls refer to the many administrative and other actions in order to modify floodplain land use so that the uses are compatible with the potential flood hazard. These controls consist of zoning and other building ordinances, direct acquisition of land and property, building codes, flood insurance, and information programs by local, state, and federal agencies.

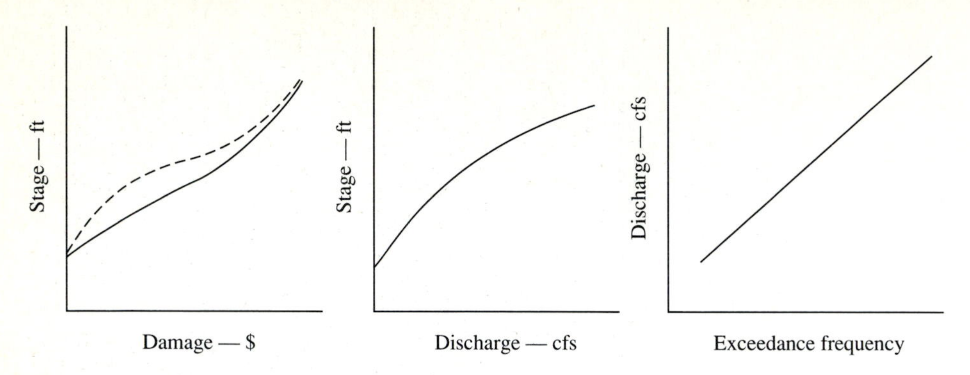

FIGURE 12.1.6
Effect of flood proofing. (Source: U.S. Army Corps of Engineers Hydrologic Engineering Center)

12.2 FLOOD DAMAGE ESTIMATION

12.2.1 Damage Relationships

Flood damages are usually reported as **direct damage** to property which is only one of five empirical categories of damages: direct damages, indirect damages, secondary damages, intangible damages, and uncertainty damages. **Indirect damages** result from lost business and services, cost of alleviating hardship, rerouting traffic and other related damages. **Secondary damages** result from adverse effects by those who depend on output from the damaged property or hindered services. **Intangible damages** include environmental quality, social well-being, and aesthetic values. **Uncertainty damages** result from the ever-present uncertainty of flooding.

Various techniques have been used to calculate direct damages. Grigg and Helweg (1975) used three categories of techniques: aggregate formulas; historical damage curves; and empirical depth-damage curves. One of the more familiar aggregate formulas is that suggested by James (1972)

$$C_D = K_D U M_S h A \tag{12.2.1}$$

where C_D is the flood damage cost for a particular flood event; K_D is the flood damage per foot of flood depth per dollar of market value of the structure; U is the fraction of floodplain in urban development; M_S is the market value of the structure inundated in dollars per developed acre; h is the average flood depth over the inundated area in feet; and A is the area flooded in acres. Eckstein (1958) presented the historical damage curve method in which the historical damages of floods are plotted against the flood stage.

The use of empirical depth-damage curves, which is the most common method, requires a property survey of the floodplain and either an individual or aggregated estimate of depth (stage) vs. damage curves for the structures, roads, crops, utilities, etc., that are in the floodplain. This stage-damage is then related to the relationship for the stage-discharge to derive the damage-discharge relationship which is then used along with the discharge-frequency relationship to derive the damage-frequency curve, as illustrated in Fig. 12.2.1.

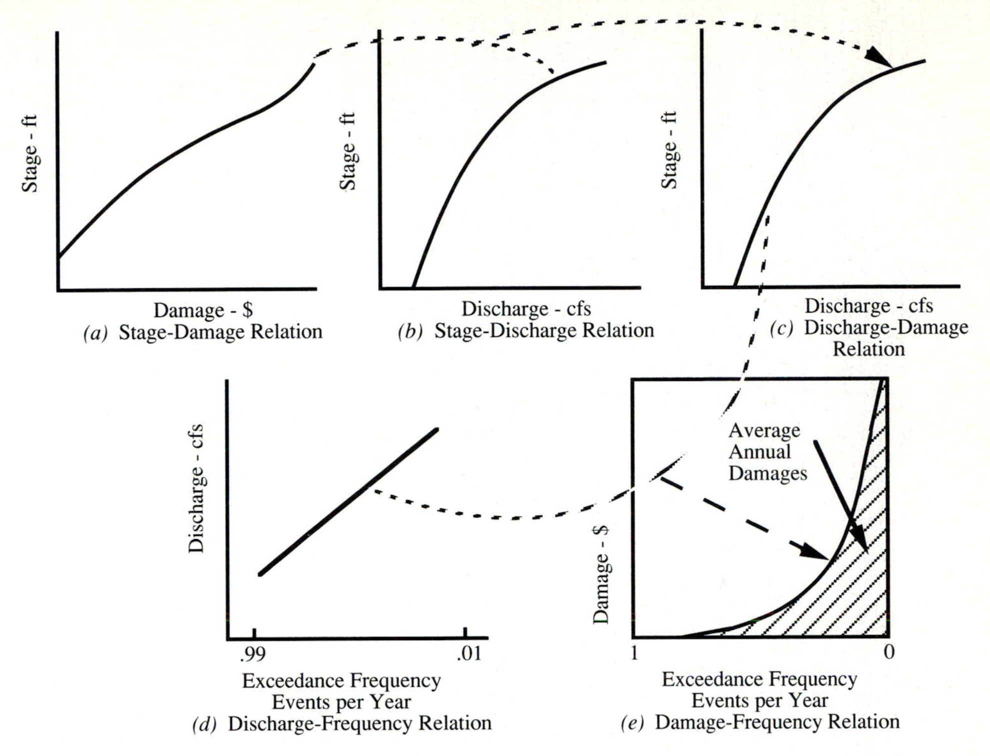

FIGURE 12.2.1
Computation of average annual damages. (Source: U.S. Army Corps of Engineers Hydrologic Engineering Center)

Various depth-damage tables and curves have been published by various agencies, for example, Grigg and Helweg (1975) presented depth-damage curves for various types of residential structures based upon data from various agencies. Corry et al. (1980) presented the depth-damage curves in Fig. 12.2.2. Table 12.2.1 shows that the percent damage to various crops as a function of the duration and depth of inundation.

12.2.2 Expected Damages

The annual expected damage cost $E(D)$ is the area under the damage frequency curve as shown in Figure 12.2.1e, which can be expressed as

$$E(D) = \int_{q_c}^{\infty} D(q_d) f(q_d) dq_d = \int_{q_c}^{\infty} D(q_d) dF(q_d) \tag{12.2.2}$$

where q_c is the threshold discharge beyond which damage would occur; $D(q_d)$ is the flood damage for various discharges, (q_d), which is the damage-discharge relationship; and $f(q_d)$ and $F(q_d)$ are the probability density functions (PDF) and the cumulative distribution function (CDF), respectively, of discharge, q_d. In practical applications,

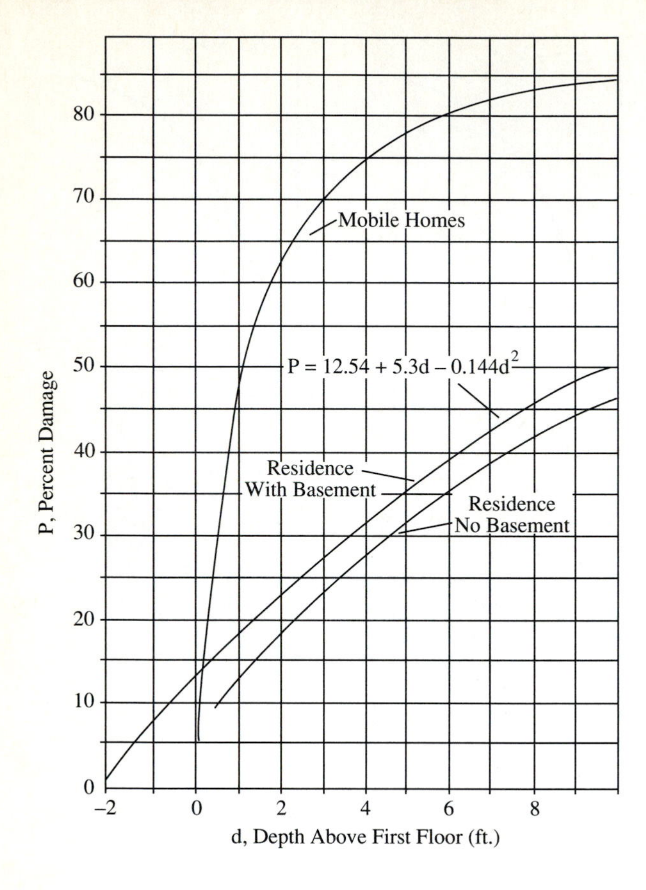

FIGURE 12.2.2
Percent damage, mixed residences (Corry et al., 1980).

the evaluation of E(D) by Eq. (12.2.2) is carried out using numerical integration because of the complexity of damage functions and probability distribution functions. Therefore, the shaded area in Figure 12.2.1e can be approximated, numerically, by the trapezoidal rule as an example,

$$E(D) = \sum_{j=1}^{n} \frac{\left[D(q_j) + D(q_{j+1})\right]}{2} \left[F(q_{j+1}) - F(q_j)\right], \text{ for } q_c = q_1 \leq q_2 \leq \ldots \leq q_n < \infty \tag{12.2.3}$$

in which q_j is the discretized discharge in the interval (q_c, ∞).

Example 12.2.1. A bridge is to be designed to cross the Leaf River at Hattiesburg, Mississippi. The bridge opening under consideration is 280 feet and the embankment elevation is 147 feet above the datum. At the bridge site, there is a stream flow gage that has been operated by the USGS from 1939 to 1981. Based on results of a flood frequency analysis, the peak discharges for the various return periods are shown in Columns (1) and (2) of Table 12.2.2. Columns (3) and (4) list the average water surface elevations with and without the bridges that were determined by a water surface profile analysis.

Because the highway crossing encroaches onto part of the natural flood way, additional upstream property damage due to backwater effect could occur. To assess

TABLE 12.2.1
Percent damage to crops (Corry et al., 1980)

	% Damage			
	Less than 24 hours inundation		More than 24 hours inundation	
Crop	0 to 2 ft	Over 2 ft	0 to 2 ft	Over 2 ft
Corn	54	88	75	100
Soybeans	92	100	100	100
Oats	67	97	81	100
Hay	60	82	70	97
Pasture	50	75	60	90
Winter wheat	57	87	72	100

TABLE 12.2.2
Average surface water elevations upstream of bridge site

		Surface water elevation	
(1) Discharge (cfs)	(2) Return period (yr)	(3) Without bridge (ft)	(4) With bridge (ft)
54,200	10	146.97	146.97
62,000	15	148.09	148.23
68,000	20	148.72	148.79
73,000	25	149.28	149.35
90,500	50	150.87	150.94
110,000	100	152.47	152.51
121,000	160	153.27	153.32
131,000	200	153.99	154.04
164,000	500	155.13	155.46

such potential property damage, residential buildings in the floodplain that are potentially affected by the bridge are identified. Table 12.2.3 lists the number of homes, their first floor elevations and the total incremental property values. Evaluate the property damage due to backwater effect assuming that all buildings have basements.

Solution. The computations of property damage with and without the bridge corresponding to different discharges are shown in Table 12.2.4. The upper half of the table contains the percentage of damage for properties of different first-floor elevations under various discharge conditions. The percentages of damage are obtained from Fig. 12.2.2. As an

TABLE 12.2.3
Summary of first-floor elevations and total incremental values for houses in floodplain upstream of bridge site

Representative 1st floor elevation (ft.)*	Number of houses	Approximate total incremental value ($1000)
148	25	1,500
149	49	2,900
150	66	3,960
151	65	3,900
152	55	3,300
153	22	1,320
154	10	600
155	3	180

*Rounded to the nearest foot.

example, for a discharge of 54,200 cfs, the average water suface elevation is 146.97 ft which is a depth of 146.97 – 148 = –1.03 ft. This represents 1 foot below the first floor elevation. From Fig. 12.2.2 for a residence with a basement, the damage is 7 percent. For a given discharge, the corresponding total damage can be computed by summing up the multiplication of property value and the percentage of damage for all houses on the floodplain.

Similarly, the percentages of damage and the total damage, when the bridge is present, can be estimated. These are given in the lower half of Table 12.2.4. In the last column, the incremental damages between the conditions for with and without the bridge are computed by subtracting the total damage, for a given discharge level, without a bridge from the total damage with a bridge. For example, when Q = 62,000 cfs, the incremental damage is obtained as $540,400 – $509,400 = $31,000.

The incremental damage, given in the last column of Table 12.2.4, and the corresponding discharge are regenerated in Columns (5) and (1), respectively, of Table 12.2.5, along with the return period. Table 12.2.5 is used to illustrate the computations of the annual expected damage, using Eq. (12.2.3), due to the backwater effect. Applying such an equation requires the assessment of the exceedance probabilities associated with each flood discharge listed in Column (1) of Table 12.2.5. Knowing the return period associated with the flood discharge, the annual probability of exceedance P in Column (3) can be calculated as $P = 1/T$ (see Section 10.4). Therefore, the annual probability of nonexceedance F in Column (4) can be calculated easily as $F = 1 - P$.

To assess the total annual expected damage, the averaged backwater damages for the two consecutive return periods are computed in Column (6). Next, the incremental probabilities between the two consecutive nonexceedance probabilites in Column (4) are calculated in Column (7). The incremental backwater damage in Column (8) can be obtained by multiplying Columns (6) and (7) together. The sum of incremental backwater damages in Column (8) yields the total annual expected cost of $2,784.

12.3 HEC FLOOD DAMAGE ANALYSIS PACKAGE

The U.S. Army Corps of Engineers Hydrologic Engineering Center Flood Damage Analysis (FDA) package is a system of hydrologic, hydraulic, and flood damage

TABLE 12.2.4
Damage calculations for Example 12.2.1

Discharge (cfs)	Average water surface elevation (ft)	% Damage (without bridge)								Total damage ($K)	
		$1500K 148′	2900K 149′	3960K 150′	3900K 151′	3300K 152′	1320K 153′	600K 154′	180K 155′		
54,200	146.97	7.00%								105.00	
62,000	148.09	13.00	8.00	2.00						509.40	
68,000	148.72	16.00	11.00	6.00						801.00	
73,000	149.28	19.50	14.00	9.00	3.00					1177.50	
90,500	150.87	26.50	22.00	12.50	12.00	7.00				2238.30	
110,000	152.47	33.00	29.00	25.00	19.00	15.00	10.00	4.50		3732.60	540.40 − 509.40
121,000	153.27	36.00	32.00	28.00	24.00	19.50	14.00	9.00	3.50	4414.20	= 31
131,000	153.99	39.00	35.00	31.00	28.00	23.00	18.00	13.00	7.50	5021.70	
164,000	155.13	42.50	39.50	35.50	31.50	27.50	23.50	18.50	13.00	5785.20	
Discharge (cfs)	**Average water surface elevation (ft)**	**% Damage (with bridge)**									**Incremental damage ($K)**
54,200	146.97	7.00%								105.00	0.00
62,000	148.23	13.37	8.37	2.37						540.40	31.00
68,000	148.79	16.21	11.21	6.21						819.00	18.00
73,000	149.35	19.70	14.20	9.20	3.20					1201.80	24.30
90,500	150.94	26.60	22.20	12.60	12.10	7.10	1.50			2278.00	39.70
110,000	152.51	33.10	29.10	25.20	19.10	15.10	10.20	4.60		3754.70	22.10
121,000	153.32	36.20	32.10	28.20	24.10	19.70	14.10	9.20	3.60	4440.60	26.40
131,000	154.04	39.20	35.10	31.20	28.10	32.20	18.10	13.20	7.60	5047.00	25.30
164,000	155.46	43.40	40.40	36.40	32.40	28.40	24.40	19.40	13.90	5938.40	153.20

TABLE 12.2.5
Computation of annual expected damage cost due to backwater (Example 12.2.1)

(1) Q (cfs)	(2) T (yrs)	(3) P	(4) F (1-P)	(5) Backwater damage ($K)	(6) Average backwater cost ($\bar{D}$)	(7) Incremental probability (ΔF)	(8) $\bar{D}\Delta F$
54,200	10	0.1000	0.9000	0	$ 15.5K	0.0333	$0.516K
62,000	15	0.0667	0.9333	31	24.5K	0.0167	0.409K
68,000	20	0.0500	0.9500	18	21.2K	0.0100	0.212K
73,000	25	0.0400	0.9600	24.3	32.0K	0.0200	0.640K
90,500	50	0.0200	0.9800	39.7	30.9K	0.0100	0.309K
110,000	100	0.0100	0.9900	22.1	24.3K	0.00375	0.091K
121,000	160	0.00625	0.99375	26.4	25.9K	0.00125	0.032K
131,000	200	0.0050	0.9950	25.3	89.3K	0.0030	0.268K
164,000	500	0.0020	0.9980	153.2	153.2K	0.0020	0.306K
∞	∞	0	1.000	153.2			
						Total	2.784K$

programs linked by a data-storage system to automatically transfer data between the programs. The data-storage system is referred to as the Hydrologic Engineering Center's Data Storage System (HECDSS) (U.S. Army Corps of Engineers, 1983). The various computer programs are shown in Fig. 12.3.1 with a brief description of the programs in Table 12.3.1. The FDA package includes three hydrologic and hydraulic

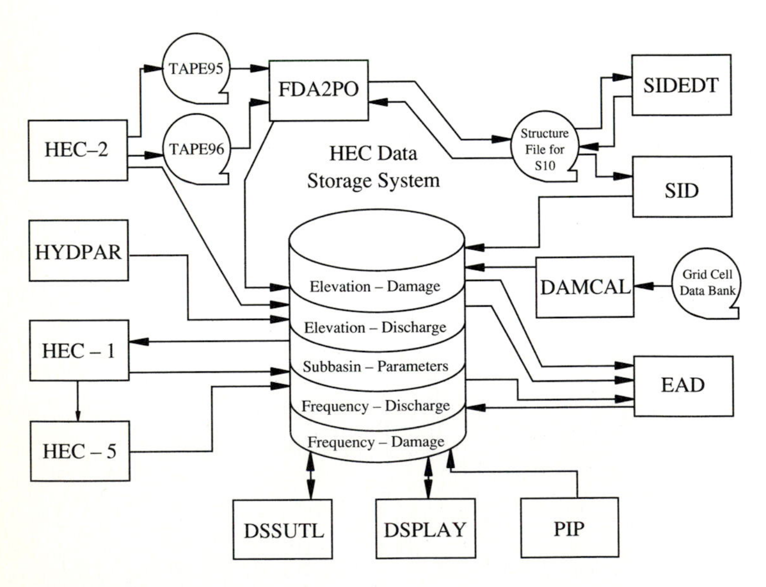

FIGURE 12.3.1
Flood damage analysis package (U.S. Army Corps of Engineers, 1983).

TABLE 12.3.1

Computer programs in the HEC flood damage analysis package (U.S. Army Corps of Engineers, 1983)

1. Hydrologic Analysis Computer Programs

- HEC-1 Flood Hydrograph Package; simulates rainfall-runoff, simple reservoirs and hydrologic channel routing; used to develop existing, without conditions, and modified conditions flow-frequency curves.
- HEC-2 Water Surface Profiles; computes steady-state, uniform flow profiles; used to develop elevation-flow rating curves.
- HEC-5 Simulation of Flood Control and Conservation Systems; simulates complex reservoir systems; used to develop existing, without and modified flow-frequency curves.

2. Flood Damage Analysis Computer Programs

- SID, Structure Inventory for Damage analysis; processes inventories of structures located in the flood plain; used to develop elevation-damage relationships.
- SIDEDT, Structure Inventory for Damage Analysis Edit Program edits structure inventory and damage function files used for the SID program.
- DAMCAL, Damage Reach State-Damage Calculation; performs same analysis as SID except based on a geographic (spatial) unit; used to develop elevation-damage relationships.
- EAD, Expected Annual Damage Computation; computes expected (or equivalent) annual damage and inundation reduction benefits; used to compare flood damage mitigation plans.
- FDA2PO, HEC-2 post-processor program; computes the reference flood elevation at structures and stores stage-flow rating curves in a HECDSS data file.

3. HEC-DSS (Data Management) Utility Programs

- PIP, Interactive Paired-Function Input Program; directly inputs paired function relationships to a DSS data file, for example, and elevation-damage relationship derived by hand from field data.
- DSSUTL, HEC-DSS Utility Program; provides the means of performing utility functions on data stored in the HEC-DSS data file, for example, cataloging, editing, and deleting data.
- DSPLAY, HEC-DSS Display Program; provides the means to tabulate and plot data stored in a HEC-DSS data file.

programs, five flood damage analysis programs, three data management programs, and a library of data management software.

HEC-1 can be used to develop a base condition discharge-exceedance frequency relation through a rainfall-runoff analysis by determining the discharge hydrograph for various frequency rainfalls. The peak discharge and frequency are used to construct the discharge-exceedance frequency curve. If a gaging station exists near the flood damage site, then the base condition discharge-frequency relation is determined through a frequency analysis.

Discharge-frequency curves for modified conditions are developed for each alternative plan considered using HEC-5 and/or HEC-1. HEC-1 is used to derive the modified condition frequency curves for plans such as ungated reservoirs. Peak discharges and associated hydrographs are completed for each frequency level. HEC-5 is used in a similar manner to simulate gated reservoirs. HEC-5 does not model the rainfall-runoff process whereas HEC-1 does and is used in conjunction with HEC-5. Both HEC-1 and HEC-5 are run in the "multiratio" and "multiplan" model. The multiratio uses specified ratios of the precipitation volumes or runoff hydrograph ordinates. Multiplan refers to various structural alternatives. The base and modified discharge-frequency relationships are stored in the HECDSS.

HEC-2 is used to compute steady-state, water surface profiles and in turn develop rating curves (elevation-discharge relationships) for base conditions. HEC-2 can then be used to develop modified stage-discharge relationships for channel modifications and levees. The rating curves are stored in the HECDSS either directly or through the utility program FDA2PO.

The SID (Structure Inventory Damage) and DAMCAL (Damage Calculation) programs compute the aggregated elevation-damage curves for each damage reach. DAMCAL is used along with spatial data analysis techniques. SID is used with structure inventories where there is no spatial analysis to perform nonstructural flood damage analysis. All damage reaches are analyzed at one time for one plan, so that separate runs are required for each damage reduction measure. The generated elevation-damage curves are stored in HECDSS.

The SID depth-damage curves and structure inventories are managed and manipulated by SIDEDT (SID Editing) program. FDA2PO is the link between the HEC-1, HEC-2 and HEC-5 programs and the flood damage analysis programs. Rating curves are stored in HECDSS for selected damage reach index locations using FDA2PO, which computes reference flood elevations at each structure in the SID inventory and at the index location.

The EAD (Expected Annual Damage) program merges all of the basic parametric relationships computed by HEC-1, HEC-2, HEC-5, SID, DAMCAL, and FDA2PO. EAD derives the frequency-damage curve from the frequency curves computed by HEC-1 and HEC-5, the rating curves computed by HEC-2, and the elevation-damage curves computed by SID or DAMCAL. EAD calculates the expected annual damage by integrating the frequency-damage curve; performs the equivalent annual damage computations procedure; and computes inundation reduction benefit for all plans and summarizes the expected annual damage by category, reach, and plan. The FDA package has been adapted to the microcomputer (U.S. Army Corps of Engineers, 1988).

12.4 OPTIMIZATION MODEL FOR PLANNING FLOOD CONTROL

This section describes an optimization approach for the planning and design of flood-control systems. Such a model can be used in preliminary selection of alternative flood-control configurations. An advantage of using an optimization model is that a complete set of alternatives are examined in a systematic manner, automatically searching for the optimal solution.

Consider a simple example of three potential flood-control reservoirs with gated spillways and two potential channel improvements to protect two damage centers as illustrated in Fig. 12.4.1. Annual cost of channel improvement $f_j(Q_j)$ is expressed as a function of the peak discharge Q_j at the damage center j. The annual cost of flood storage $f_i(ST_{\max_i})$ in the reservoir is expressed as a function of maximum storage capacity $ST_{\max_i}$ in reservoir i. The annual spillway cost $f_i(C_i)$ is expressed as a function of the spillway discharge capacity C_i. Trade-offs in costs exist between the reservoir capacity, the spillway structure, and the level of downstream channel improvement. As an example for a large flood-control reservoir the spillway capacity and the level of downstream channel improvement would be minimal. On the other hand, for a small flood-control reservoir the downstream spillway capacity would be greater and the level of downstream channel improvement would be greater.

The objective of the optimization model is to determine the minimum cost channel improvement (i.e., Q_j, which defines the capacity of the improved channel). The following model can be solved for various reservoir storage capacities, $ST_{\max}$; and spillway capacities. For the system shown in Fig. 12.4.1 the objective function is expressed as

$$\text{Min } Z = \sum_{j=1}^{2} f_j(Q_j) + \sum_{i=1}^{3} \left[f_i(ST_{\max_i}) + f_i(C_i) \right] \tag{12.4.1}$$

which is to minimize the sum of annual costs for specified $ST_{\max_i}$ and C_i. The constraints for this model include reservoir-storage constraints, reservoir-release constraints, reservoir mass-balance constraints, channel-routing constraints, maximum-flow constraints at protected-damage centers, and continuity at channel junctions. Each of these constraints are explained below.

The reservoir-storage constraints state that at the beginning of each time period t the actual reservoir storage $ST_{i,t}$ must be less than or equal to the maximum reservoir storage, $ST_{\max_i}$, that is,

$$ST_{i,t} \leq ST_{\max_i}, \qquad \text{for all } i, t \tag{12.4.2}$$

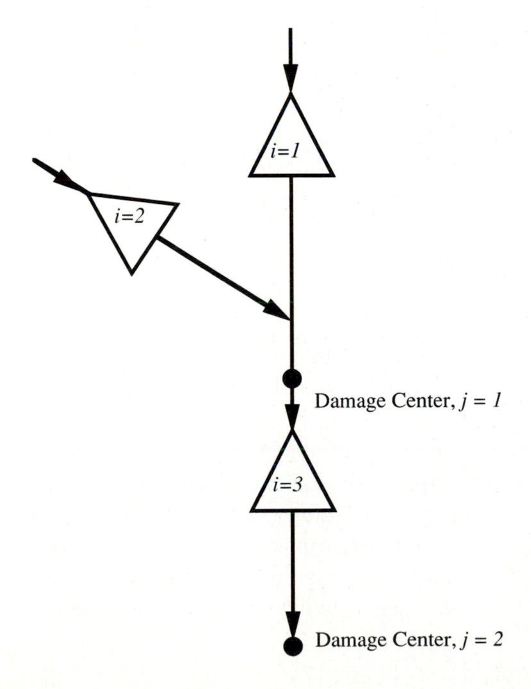

FIGURE 12.4.1
Hypothetical reservoir system consisting of three potential flood-control reservoirs.

where $ST_{i,t}$ is the storage in reservoir i at the beginning of time period t and $t = 1, ..., T$ where T is the number of time periods for a flood hydrograph.

Releases are limited by spillway capacities and storage in the reservoir. Expressing the release as a function of the storage $R_{i,t}(ST_{i,t})$ then the reservoir-release constraint can be expressed as

$$QO_{i,t} \leq R_{i,t}(ST_{i,t}) \leq C_i, \qquad \text{for all } i, t \tag{12.4.3}$$

where $QO_{i,t}$ is the controlled- or minimum-required outflow.

Reservoir mass-balance constraints are used to perform routing of a flood hydrograph through a reservoir. They are expressed in the form of a continuity equation where the average inflow volume minus the average outflow volume for a time period t must equal the change in reservoir storage for that time interval, that is,

$$(QI_{i,t+1} + QI_{i,t})\frac{\Delta t}{2} - (QO_{i,t+1} + QO_{i,t})\frac{\Delta t}{2} = ST_{i,t+1} - ST_{i,t} \tag{12.4.4}$$

where QI is the reservoir inflow.

Channel routing through a channel reach k is performed using constraints expressed in the form of the Muskingum routing expressed as

$$QO_{k,t+1} = C1_k QI_{k,t} + C2_k QI_{k,t+1} + C3_k QO_{k,t}, \qquad \text{for all } k, t \tag{12.4.5}$$

where $QO_{k,t+1}$ is the discharge from reach k at the end of time period $t, QI_{k,t}$ and $QI_{k,t+1}$ are the inflows to reach k at the beginning of and at the end of time period t; $C1_k, C2_k$, and $C3_k$ are the Muskingum routing coefficients for reach k (see Section 10.3.2).

Channel flows at protected areas must be less than or equal to the improved river channel for the design flood condition, expressed as

$$QO_{k,j,t} \leq Q_j, \qquad \text{for all } k, j, t \tag{12.4.6}$$

where $QO_{k,j,t}$ is the flood discharge in channel reach k at protected area j at the beginning of time period t.

A constraint is required to define the continuity at a river junction, that is,

$$QI_{k,t} = \sum_{1 \epsilon L} QO_{1,t} \tag{12.4.7}$$

where $QI_{k,t}$ is the inflow to the channel reach k downstream of the junction; $QO_{1,t}$ is the outflow from the reaches flowing into the junction; and L is the set of reaches flowing into the junction.

The decision variable Q_j defines the capacity of the improved channel at j which is required to contain the flood hydrograph. The above optimization problem, Eqs. (12.4.1)–(12.4.7), can be solved by linear programming if the cost functions are linear. If the cost functions are nonlinear and convex, a piece-wise linearization of the objective function can be utilized. A nonlinear programming procedure (such as GRG2, GAMS-MINOS or MINOS) can be used to solve the model without the piece-wise linearization.

12.5 OPTIMAL SELECTION OF FLOOD-CONTROL ALTERNATIVES

A procedure for flood-control planning is outlined in Fig. 12.5.1. Goals are typically to protect against the 100-year flood. Procedures can be used to determine the optimal return period design, which account for the cost trade-offs between increasing costs for larger (greater return period) projects and the decreasing damage costs for larger projects. Example 12.5.1 illustrates such computations.

Example 12.5.1. Determine the optimal design return period for a flood-control project with the damage costs and capital costs for the respective return periods listed in Columns (5) and (6) in Table 12.5.1.

Solution. The solution is presented in Table 12.5.1. Incremental expected damages, $\Delta E(D)$ in Column 4, are computed using the trapezoidal rule as explained in Eq. (12.2.3), that is, $\Delta E(D) = 1/2[D(T_j) + D(T_{j+1})][P(Q \leq Q_{T_{j+1}}) - P(Q \leq Q_{T_j})] = 1/2[0 - 40{,}000][1 - 0.5] = \$10{,}000$ for a return period of $T = 2$ years. The incremental expected damage cost for the other return periods $T = 5, 10, \ldots, 100$, and 200 years can be computed in the same fashion and the results are listed in Column (4).

Now consider the existing condition where there is no flood-control project in place. The annual expected damage cost for the existing condition then is equal to the sum of all incremental damage costs listed in Column (4), that is, $98,196/year. Since there is no flood-control project in place, the annual capital cost is zero (see Column (6)). The total annual cost for the existing condition is $98,196 as shown in Column (7).

When the flood-control project is designed for protecting floods with a return period of $T = 2$ years, the corresponding incremental expected damage cost of $10,000 then would be saved. Therefore, the corresponding annual expected flood damage is

TABLE 12.5.1
Calculation for optimal design return period

(1) Return period T (years)	(2) Annual exceedance probability	(3) Damage $D(T)$ ($)	(4) Incremental expected damage ($/year)	(5) Damage risk cost ($/year)	(6) Capital cost ($/year)	(7) Total cost ($/year)
1	1.0	0	—	98,196	0	98,196
2	0.500	40,000	10,000	88,196	6,000	94,196
5	0.200	120,000	24,000	64,196	28,000	92,196
10	0.100	280,000	20,000	44,196	46,000	90,196
15	0.067	354,000	10,566	33,630	50,000	83,630
20	0.050	426,000	6,500	27,130	54,000	81,130
25	0.040	500,000	4,630	22,500	58,000	80,500
50	0.020	600,000	11,000	11,500	80,000	91,500
100	0.010	800,000	7,000	4,500	120,000	124,500
200	0.005	1,000,000	4,500	0	160,000	160,000

The sum of the incremental expected damages is $98,196/year.

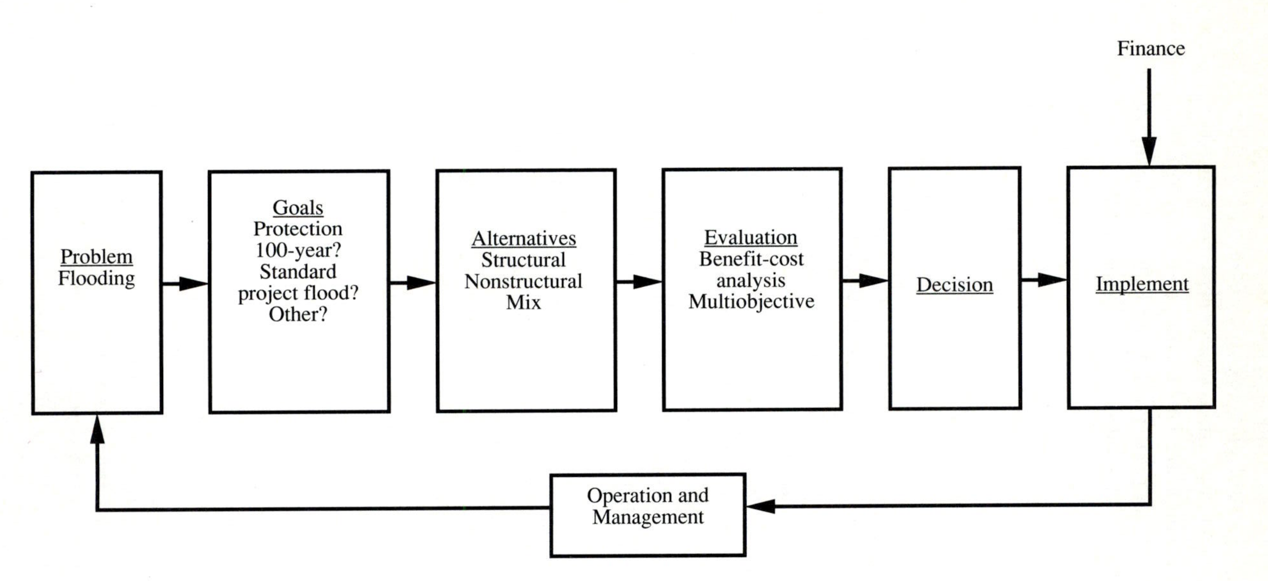

FIGURE 12.5.1
Flood-control planning procedure (Grigg, 1985).

\$98,196 − \$10,000 = \$88,196 (see Column (5)). In a similar fashion, the annual expected flood damage for T = 5 years is \$88,196 − \$24,000 = \$64,196, and so on. As can be seen, more savings in flood damage, which corresponds to less damage risk cost, can be obtained as the design return period increases. Of course, savings from flood damage cannot be obtained without paying the price in the form of capital cost, listed in Column (6). For example, \$6,000 must be invested for protection of 2-year floods. The higher the level of protection in terms of return period the more expensive the design. The thrust of this Example is to show the trade-off between savings in flood damage and the associated capital cost. The optimal return period for this example is 25 years which corresponds to the lowest total annual cost of \$80,500 (see Column (7)).

12.6 RISK-BASED DESIGN

Conventional risk-based design procedures for hydraulic structures consider only the inherent hydrologic uncertainty. The probability of failure of the hydraulic structure is generally evaluated by means of frequency analysis as described in Section 10.4. Other aspects of hydrologic uncertainties are seldom included. Methodologies have been developed to integrate various aspects of hydrologic uncertainties into risk-based design of hydraulic structures (Tung and Mays, 1982).

Risk-based design approaches integrate the procedures of uncertainty analysis and reliability analysis in design. Such approaches consider the economic trade-offs between project costs and expected damage costs through the risk relationships. The risk-based design procedure can be incorporated into an optimization framework to determine the optimal risk-based design. Therefore, in an optimal risk-based design, the expected annual damage is taken into account in the objective function, in addition to the installation cost. The problem is to determine the optimal structural sizes/capacities associated with the least total expected annual cost (TEAC). Mathematically, the optimal risk-based design can be expressed as

$$\underset{\mathbf{x}}{\text{Min}}\ \text{TEAC} = FC(\mathbf{x}) \cdot CRF + E(D|\mathbf{x}) \qquad (12.6.1a)$$

$$\text{subject to design specifications, } g(\mathbf{x}) = 0 \qquad (12.6.1b)$$

where FC is the total installation cost (first cost) of the structure, CRF is the capital recovery factor for conversion of cost to an annual basis, $\mathbf{x}$ is a vector of decision variables relating to structural sizes and/or capacities, and $E(D|\mathbf{x})$ is the expected annual damage cost due to structural failure. A diagram illustrating the cost computations for optimal risk-based design is shown in Fig. 12.6.1.

In general, quantification of the first cost is straightforward. The thrust of risk-based design is the assessment of the annual expected damage cost. Depending on the types of uncertainty to be considered, assessment of $E(D|\mathbf{x})$, as described in Section 12.2, varies. Current practice in risk-based water-resources engineering design considers only the inherent hydrologic uncertainty due to the randomness of the hydrologic process. The mathematical representation of the annual expected damage under this condition can be computed using Eq. (12.2.2).

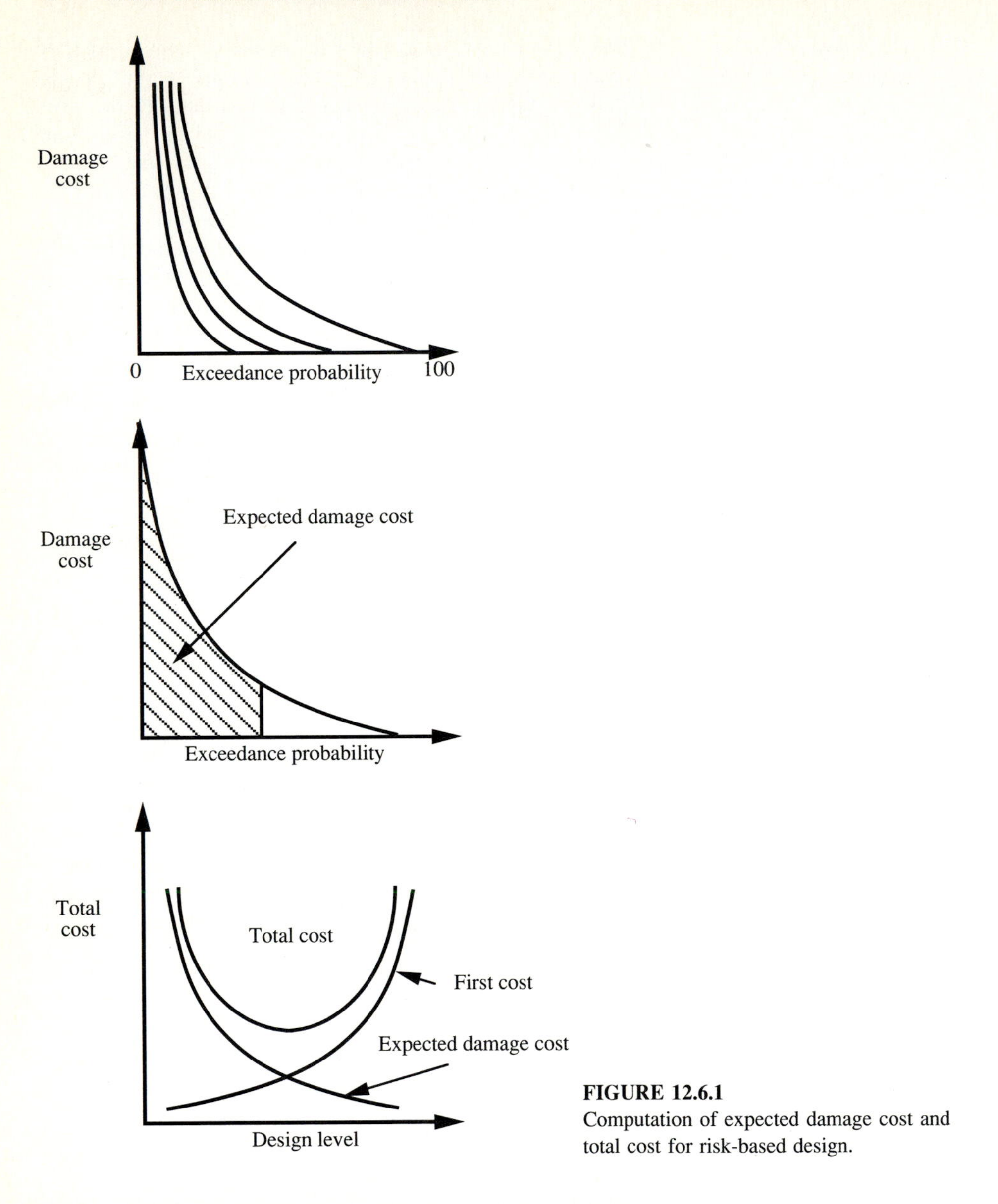

FIGURE 12.6.1
Computation of expected damage cost and total cost for risk-based design.

12.7 RISK-BASED DESIGN OF HIGHWAY DRAINAGE STRUCTURES

12.7.1 Design of Roadway Crossing Structures

Hydraulic structures for roadway crossings consist of bridges, pipe culverts, and box culverts. Important considerations in the hydraulic design of roadway crossing structures include the flood hazard associated with the proposed highway crossing and the impact of the roadway crossing on human life, property, and stream stability. Hydraulic design of highway crossings specifically involves the selection of the location of the crossing, the determination of the embankment height, and the size of the structural opening. Conventional design procedures are based on: (1) the selection of flood-peak discharge, known as the design flow, for a certain predetermined return period depending on the type of the road; and (2) the size of structural opening and embankment height which will allow this selected flood flow to pass. Due to many uncertainties involved in the hydraulic design of roadway crossing structures, the conventional design procedures usually assign an additional embankment height as a safety margin to protect the structures. Thus, existing conventional design procedures fail to explicitly account for the cost interaction between project components and the expected damages. Also, they do not provide a means to systematically account for uncertainties in the design. The following subsections describe the optimal risk-based hydraulic design of highway bridges which considers a balance between the project cost and potential flood damages or economic losses. The procedure and analysis are similar for designing culverts.

COST COMPONENTS IN RISK-BASED DESIGN OF HIGHWAY BRIDGES. To perform risk-based hydraulic design of highway bridges, several types of flood-related costs, in addition to construction and operation/maintenance costs, must be included. The quantification and assessment of flood-related damage costs requires significantly additional effort in collecting and analyzing data. This section briefly describes some relevant cost components to be considered in risk-based hydraulic design of highway bridges.

CONSTRUCTION COSTS (C_1). This usually is the largest single-cost item of the project. Various cost estimation procedures are used by highway agencies to estimate this cost. Construction costs generally include initial embankment, pavement, and structures. In general, operation and maintenance costs are included. Any other costs associated with bridge accessories such as spur dikes and riprap for scouring protection, should also be included. The construction cost largely depends on the physical layout of structures which usually increase with bridge opening and embankment height. A qualitative representation of construction cost in relation to the bridge opening and embankment height is shown in Fig. 12.7.1.

FLOOD DAMAGE COST (C_2). This is the cost of property damage caused by flooding which would not be affected by the project construction. To assess this type of cost, it is necessary to gather the information about the stage-discharge relationships, land use pattern, and property evaluation in the floodplain. Flood damage costs are evaluated under natural conditions, that is, no bridge construction. For a given location and with the assumption of steady land use conditions, this flood damage cost depends on the magnitude of the flood flow as shown in Fig. 12.7.2.

BACKWATER DAMAGE COST (C_3). This is the difference between the flood damage cost with the bridge of a certain layout in place, and the flood damage cost, C_2, under natural conditions. In other words, it is the added flood damage cost due to the backwater effect caused by the presence of a bridge and embankment. Their

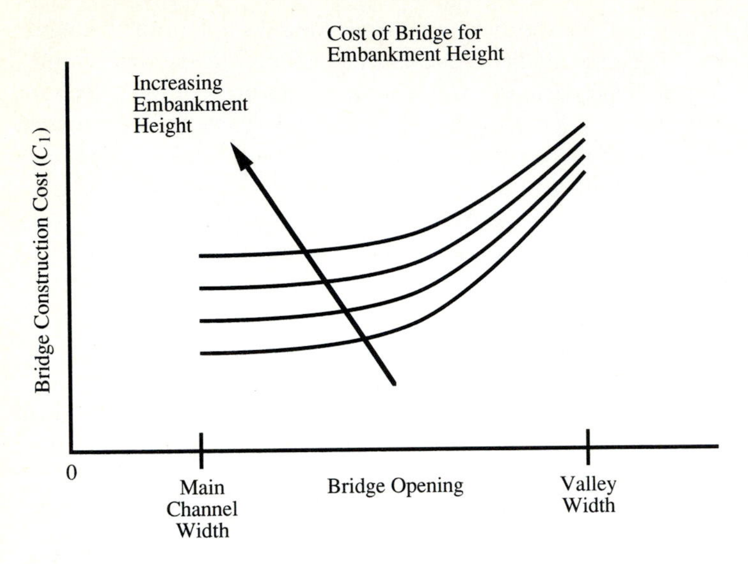

FIGURE 12.7.1
Variation of construction costs with bridge opening and embankment height (after Schneider and Wilson, 1980).

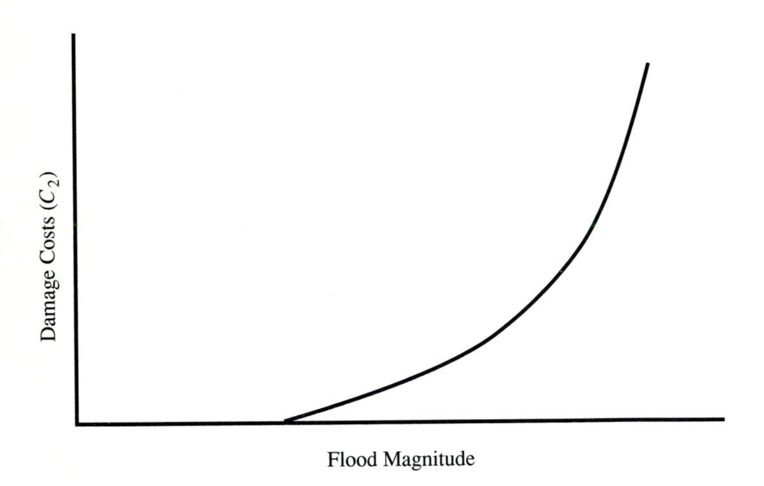

FIGURE 12.7.2
Schematic diagram of damage cost corresponds to flood magnitude before project construction (after Schneider and Wilson, 1982).

presence usually raises the surface water elevation upstream and inundates more land under natural conditions. In order to assess this backwater damage cost, backwater computations are needed which, in turn, require hydraulic characteristics and geometry of the floodplain. Backwater damage usually increases with embankment height for a fixed bridge opening; however, it decreases with an increase in bridge opening, as shown in Fig. 12.7.3.

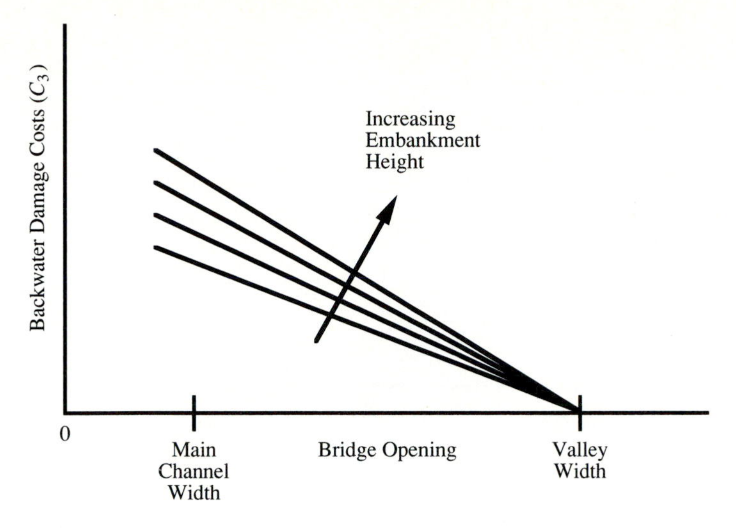

FIGURE 12.7.3
Variation of backwater damage with bridge opening and embankment height (after Schneider and Wilson, 1986).

TRAFFIC-RELATED COST (C_4). This is an additional cost of driving a vehicle on a primary detour opposite the usual route when the road is out of service due to overtopping of a bridge structure. There are basically three types of traffic-related costs (Corry et al., 1980), namely, (1) increased running cost due to a detour, (2) lost time of vehicle occupants, and (3) increased accidents on the detour. Information required for the assessment of traffic-related costs include, but are not limited to, average daily traffic, traffic mixture, vehicle running cost, lengths of normal and detour routes, period of inundation and its frequency, fatality rate, unit cost of injuries and property damage, etc. It is not difficult to imagine that some of the cost items involved in calculating traffic-related damage cannot be obtained with accuracy. Since traffic-related damages are mainly caused by traffic interruption due to inundation of bridge decks, this cost usually decreases with an increase in embankment height and bridge opening, as shown in Fig. 12.7.4. More detailed descriptions are given by Schneider and Wilson (1980) and Corry et al. (1980).

EMBANKMENT AND PAVEMENT REPAIR COST (C_5). When floodflows overtop a bridge and embankment over a sustained period of time, they cause erosion of embankment material as well as damaging the pavement. The assessment of embankment repair cost requires an understanding of erosion mechanics to quantify the extent of erosion. The cost of embankment and pavement repair depends on such variables as the total volume of the embankment and the area of pavement damaged, the rates of embankment and pavement repair, the unit costs of embankment and pavement repair, and the mobilization cost. In general, the amount of embankment erosion and pavement damage can depend on the duration and depth of overflow. Heuristically,

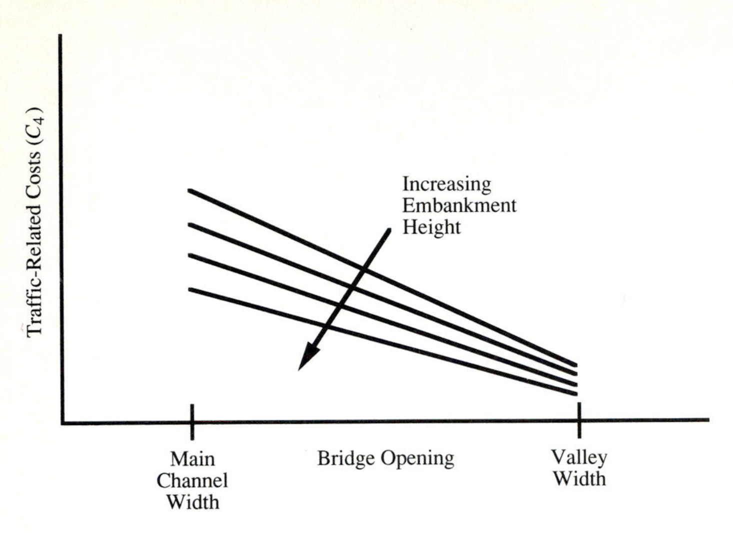

FIGURE 12.7.4
Variation of traffic-related costs with bridge opening and embankment height (after Schneider and Wilson, 1980).

for a given flood hydrograph C_5 decreases with an increase in embankment height and bridge opening, as shown in Fig. 12.7.5.

12.7.2 Expected Damages and Objective Function

Considering the above cost components in risk-based hydraulic design of highway bridges, the backwater damage (C_3), the traffic-related cost (C_4), and the embankment and pavement repair cost (C_5), each depends on the flood magnitude, bridge opening, construction method, material, and embankment height. Therefore, for a given flood magnitude q, width of bridge opening W, and embankment height H, the resulting total damage $D(q, W, H)$, can be expressed as

$$D(q, W, H) = C_2 + C_3 + C_4 + C_5 \tag{12.7.1}$$

The total expected annual cost of damage $E(D)$ for a given W and H can be expressed as

$$E(D|W, H) = \int_0^{\infty} D(q, W, H) f(q) dq \tag{12.7.2}$$

in which $f(q)$ is the probability density function (PDF) of the discharge q.

The decision variables involved in the hydraulic design of highway bridges are the embankment height, H, and the length of the bridge opening, W. The objective of the optimization is to seek the optimal embankment height and bridge opening which minimizes the total expected annual cost (TEAC) which is the sum of the annual bridge construction cost and expected annual damage cost. The objective function in terms of expected annual cost is expressed as

$$\text{Minimize TEAC} = C_1(W, H) \cdot CRF + E(D|W, H) \tag{12.7.3}$$

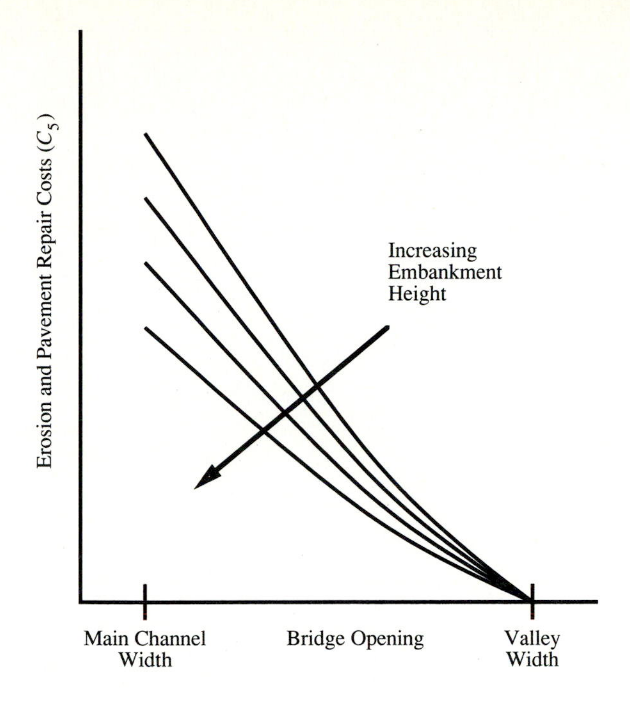

FIGURE 12.7.5
Variation of erosion damage potential with bridge opening and embankment height (after Schneider and Wilson, 1980).

where CRF is the capital recovery factor (see Chapter 2) which converts a lump sum of the construction cost to an annual basis for a specified interest rate and service period of the bridge.

Example 12.7.1. Table 12.7.1 summarizes the economic losses for a bridge design which is 280 feet long with a 147 feet embankment elevation. Determine the expected annual damages and the total expected annual costs when the annual construction cost is \$272,300.

Solution. The solution is presented in Table 12.7.1. Cols. (1), (2), (3), and (8) present the frequency, corresponding discharge, exceedance probabilities and total losses. Column (9) presents the average losses between each total loss. Column (10) presents the incremental probabilities. The incremental expected damages in Column (11) are determined by multiplying the average losses times the incremental probabilities.

The total expected annual damages for the specified bridge configuration is

$$\begin{aligned} E(D) &= \$1{,}324 + \$2{,}169 + \$2{,}624 + \$11{,}923 + \$8{,}941 + \$3{,}556 + \$1{,}209 \\ &\quad + \$3{,}156 + \$2{,}261 \\ &= \$37{,}163 \end{aligned}$$

The TEAC is the sum of the annual construction cost and the expected annual damage cost, by Eq. (12.2.2). For an annual construction cost of \$272,300, the TEAC is

$$\text{TEAC} = \$272{,}300 + \$37{,}163 = \$309{,}463$$

TABLE 12.7.1
Summary of economic losses

(1)	(2)	(3)	(4)	(5)	(6)	(7)	(8)	(9)	(10)	(11)
				Indirect & secondary damage ($1000)						
T (yrs.)	Q_T (cfs)	P	$F-$ $(1-P)$	Traffic	Structural*	Back-water	Total	Average cost $(\bar{D})$	Incremental probability (ΔF)	$\bar{D} \cdot \Delta F$
10	51,300	0.1000	0.9000	0	0	0	0			
15	59,900	0.0667	0.9333	48.6	0	30.9	79.5	39.75	0.0333	1.324
20	66,500	0.0500	0.9500	130.9	31.4	18.0	180.3	129.90	0.0167	2.169
25	71,900	0.0400	0.9600	249.4	68.1	24.3	341.8	262.40	0.0100	2.624
50	90,600	0.0200	0.9800	600.0	210.7	39.8	850.5	596.15	0.0200	11.923
100	112,600	0.0100	0.9900	644.8	270.8	22.1	937.7	894.10	0.0100	8.941
160	129,600	0.00625	0.99375	663.5	270.8	26.4	960.7	949.20	0.00375	3.556
200	138,400	0.0050	0.9950	677.3	270.8	25.2	973.3	967.00	0.00125	1.209
500	179,400	0.0020	0.9980	706.6	270.8	153.2	1130.6	1051.95	0.0030	3.156
	∞	0	1.0000	706.6	270.8	153.2	1130.6	1130.60	0.0020	2.261
									Total	$37.163K

*Embankment and pavement repairs.

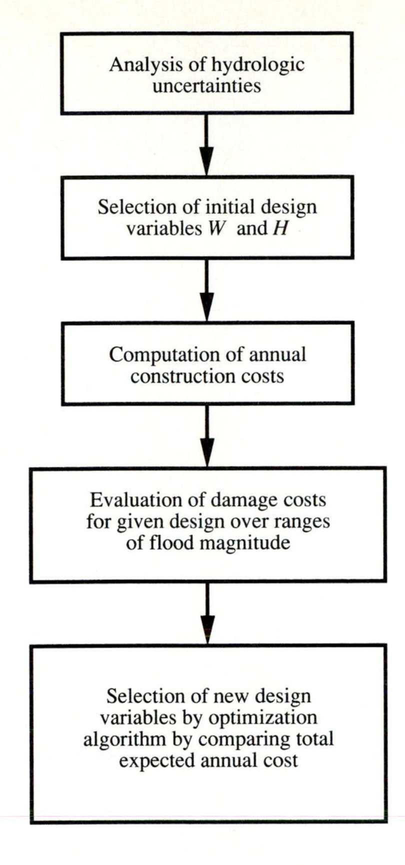

FIGURE 12.7.6
Flowchart of optimal risk-based hydraulic design procedure.

OPTIMIZATION TECHNIQUE. Because there are only two decision variables, H and W, a two-dimensional golden-section (Section 4.2) method can be used to determine the optimal values of H and W. The procedure for the optimal risk-based hydraulic design of highway bridges is shown in Fig. 12.7.6. The constraints in the optimal risk-based design model are the maximum and minimum embankment heights and lengths of the bridge opening, which depend on the topography of the bridge site, the foundation conditions, and the technical aspects of construction. Procedures for the risk-based hydraulic design of highway bridges have been reported by Schneider and Wilson (1980) and Tung and Mays (1982).

REFERENCES

Corry, M. L., J. S. Jones, and D. L. Thompson: "The Design of Encroachments of Floodplains Using Risk Analysis," Hydraulic Engineering Circular No. 17, U.S. Department of Transportation, Federal Highway Administration, Washington, D.C., July 1980.

Eckstein, O.: *Water Resources Development; The Economics of Project Evaluation*, Harvard University Press, Cambridge, 1958.

Grigg, N. S. and O. J. Helweg: "State-of-the-Art of Estimating Flood Damage in Urban Areas," *Water Resources Bulletin*, vol. 11, no. 2, pp. 379–390, April 1975.

Grigg, N. S.: *Water Resources Planning*, McGraw-Hill, Inc., 1985.

James, L. D.: "Role of Economics in Planning Floodplain Land Use," *Journal of the Hydraulics Division*, ASCE, vol. 98, no. HY6, pp. 981–992, 1972.

Schneider, V. R. and K. V. Wilson: *Hydraulic Design of Bridges with Risk Analysis*, Report, FHWA-TS-80-226, Department of Transportation, Federal Highway Administration, Washington, D.C., March 1980.

Tung, Y. K. and L. W. Mays: "Optimal Risk-Based Hydraulic Design of Bridges," *Journal of Water Resources Planning and Management Division*, ASCE, vol. 108, no. WR2, pp. 191–203,1982.

U.S. Army Corps of Engineers, Hydrologic Engineering Center, HEC-2, *Water Surface Profiles, User's Manual*, Davis, Calif. 1990.

U.S. Army Corps of Engineers, Hydrologic Engineering Center, HECDSS *User's Guide and Utility Program Manual*, Davis, Calif., 1983.

U.S. Army Corps of Engineers, Hydrologic Engineering Center, *Flood Damage Analysis Package Users Manual*, Davis, Calif., April, 1988.

PROBLEMS

12.1.1 A city in an alluvial valley is subject to flooding. As a matter of good fortune, no serious floods have taken place during the past 100 years and, therefore, no flood-control measures of any consequence have been taken. However, a serious flood threat was developed in the last year; people realize the danger they are exposed to and a flood investigation is under way.

From the hydrologic flood frequency analysis of past streamflow records and hydrometric surveys, the streamflow frequency curve, rating curve, and damage curve are derived as shown in Figs. 12.P.1, 12.P.2, and 12.P.3, respectively.

Three flood-control alternatives are considered to protect the city against the 100-year flood and they are (1) construction of a diking system with a capacity of 15,000 cfs throughout the city, (2) designing an upstream permanent diversion that will give protection up to a natural flow of 15,000 cfs, and (3) channel modification by increasing conveyance capacity of the river up to 15,000 cfs. Note that when discharge exceeds the capacity of a diking system, it completely fails. Develop the damage-frequency curve for the existing conditions and for the alternatives.

12.1.2 Referring to Problem 12.1.1, a flood detention facility that is located upstream of the city is added in the consideration. The flood detention facility is to have a conduit with maximum capacity of 12,000 cfs, which is the natural flow capacity in the river flowing through the city. Assume that all flow rates less than 12,000 cfs will pass through the conduit without being detained behind the detention pond. The shape of the flood hydrograph flowing into the reservoir is shown in Fig. 12.P.4 and the storage-elevation relationship at the reservoir site is shown in Fig. 12.P.5.

(*a*) Determine the height of spillway crest of the reservoir above the river bed which could accommodate a 15,000 cfs peak discharge.

(*b*) Develop the damage-frequency curve for the flood-control reservoir. The spillway capacity can be calculated as $Q_s = 400H^{1.5}$ in which Q_s = spillway capacity (in cfs) and H = head of water above the spillway crest in ft. (You need to perform reservoir routing to assess the peak outflow for assessing the downstream damage.)

12.2.1 Based on the information given in Table 12.7.1 about the flow discharges, traffic interruption cost, structural damage and property damage due to backwater effect, construct the damage-discharge curves for each indirect cost item. Assume that the annual maximum flood flows at the bridge site follow a log-Pearson type III distribution with $\overline{\ln Q} = 9.90$, $s_{\ln Q} = 0.75$ and $G_{\ln Q} = 0.30$, estimate the annual expected damage cost due to the presence of a bridge structure.

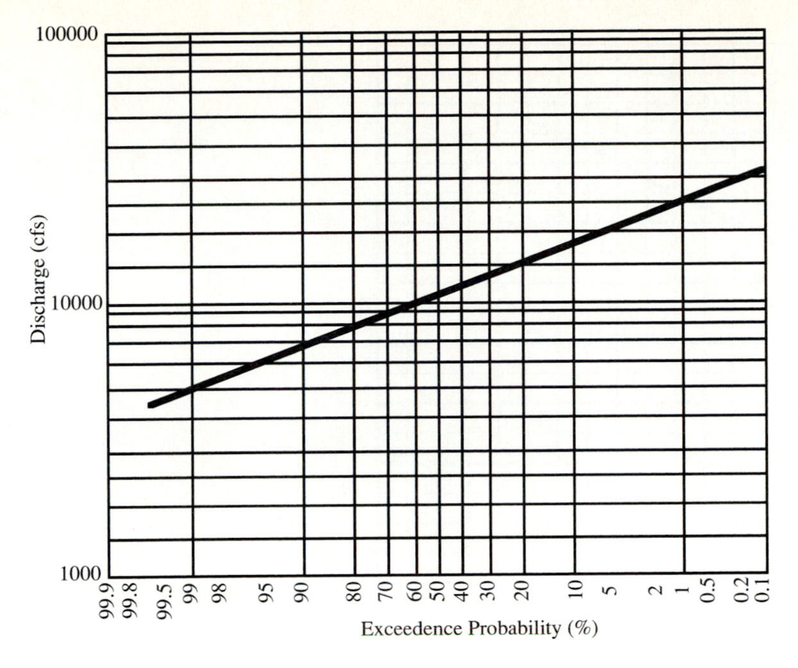

FIGURE 12.P.1
Frequency curve.

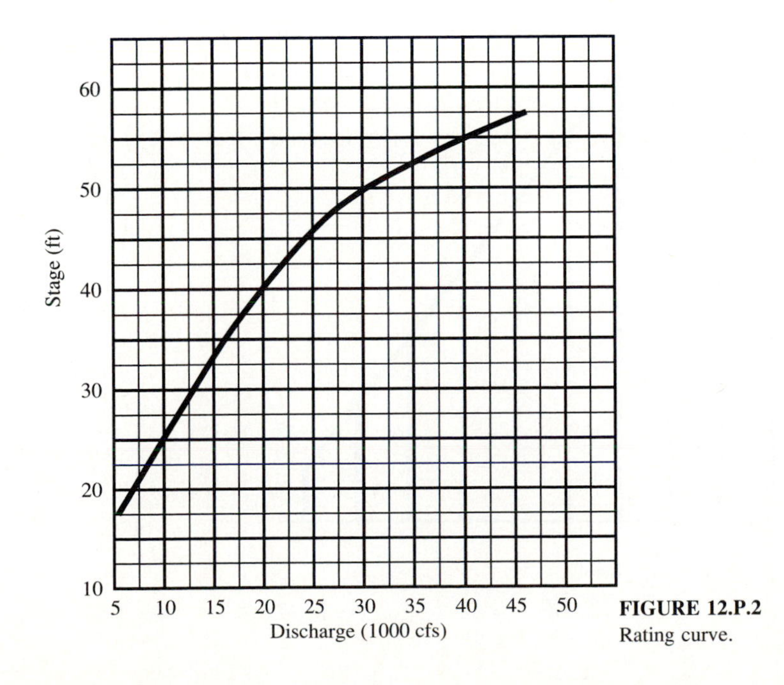

FIGURE 12.P.2
Rating curve.

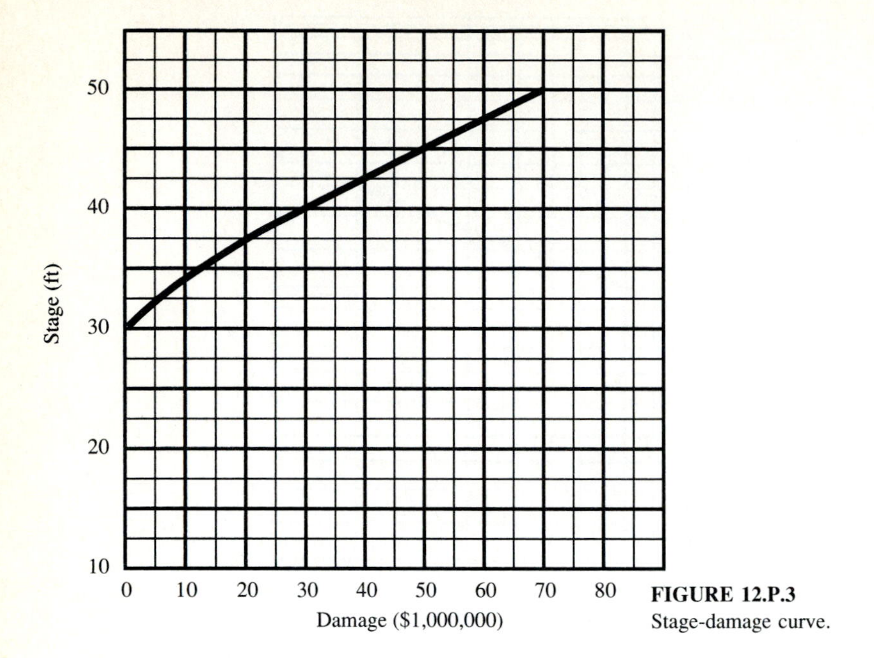

FIGURE 12.P.3
Stage-damage curve.

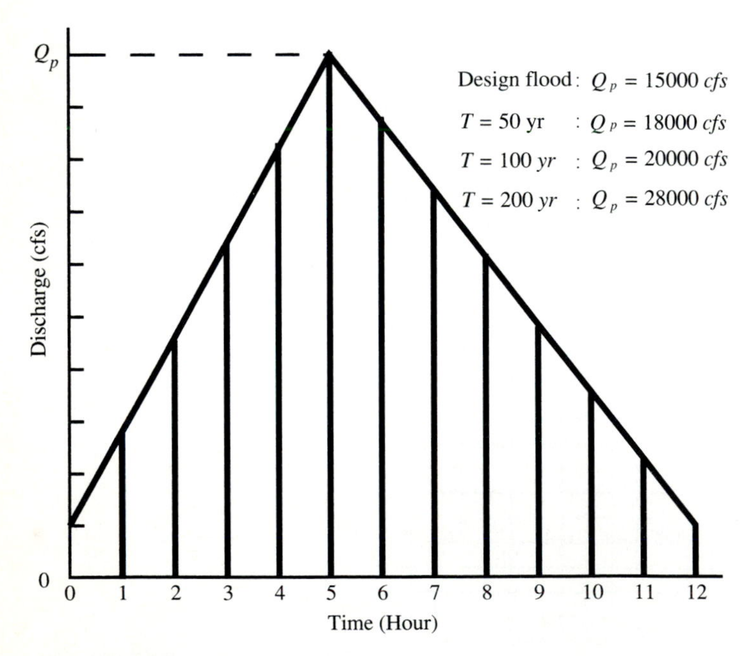

FIGURE 12.P.4
Flood hydrograph.

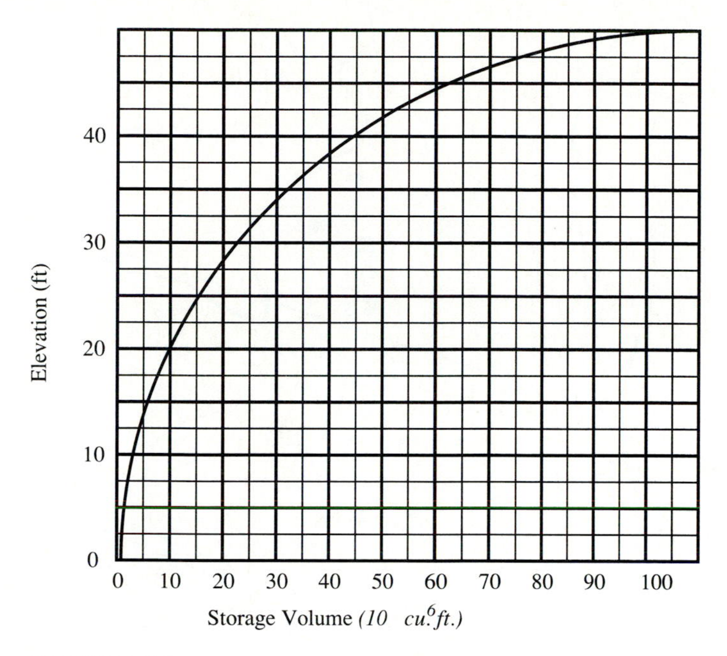

FIGURE 12.P.5
Elevation-Storage Volume Curve

12.2.2 Repeat Problem 12.2.1 to assess the annual expected damage by assuming that the annual maximum flows follow a log-normal distribution.

12.2.3 Repeat Problem 12.2.1 to assess the annual expected damage by assuming that the annual maximum flows follow a Pearson type III distribution. The mean, standard deviation, and skew coefficient are 26,400 cfs, 22,940 cfs, and 3.5, respectively.

12.2.4 Determine the backwater damage costs for the floodplain crossing presented in Fig. 12.P.6 using the depth-damage relationship in Fig. 12.2.2. All residences have basements; the first floor elevation, building value, and contents value are presented in Table 12.P.1 below. Results of backwater analysis for various discharges for the three cross-sections A, B and C, are presented in Table 12.P.2.

12.5.1 From the damage-frequency curves developed in Problems 12.1.1 and 12.1.2 for the four flood-control alternatives, rank their merits on the basis of expected flood damage reduction.

12.6.1 Referring to Problem 12.1.1, determine the return period for the diking system that maximizes the annual expected benefit.

12.6.2 Referring to Problem 12.1.1, determine the return period corresponding to the upstream diversion capacity with the maximum expected annual benefit.

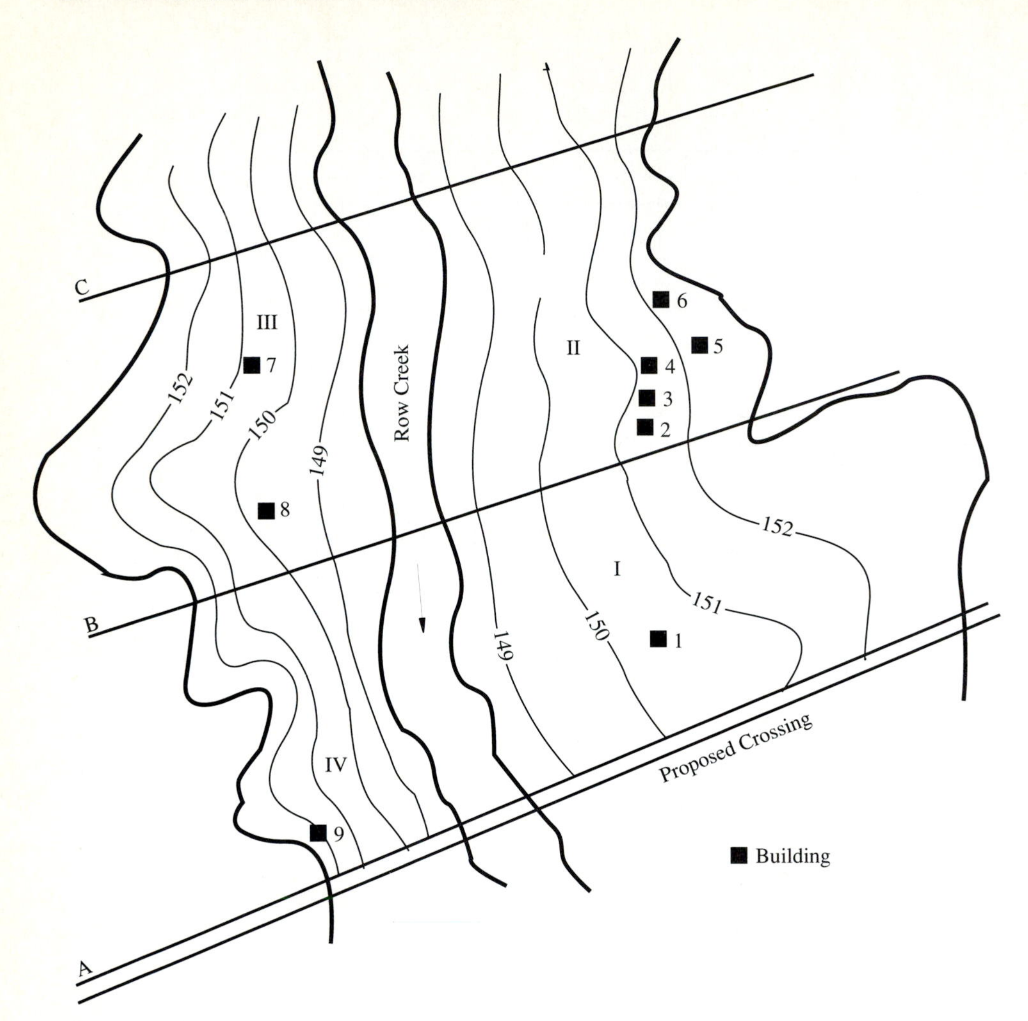

FIGURE 12.P.6
Row Creek crossing (Corry et al., 1980).

12.6.3 Referring to Problem 12.1.1, determine the return period associated with the flow capacity that maximizes the expected annual net benefit in the channel modification alternative.

12.6.4 Referring to Problem 12.1.2, determine the return period associated with the reservoir storage capacity that maximizes the expected annual net benefit.

12.6.5 Based on the results obtained from Problems 12.6.1–12.6.4, state your decision on which flood-control alternative should be pursued.

12.7.1 Design a 100 ft long circular culvert to be located under a two-lane highway. The equivalent average daily traffic is 3000 vehicles per day. The discount rate is 7.128 per-

TABLE 12.P.1
Residence information

Cross-sections	Symbol	Representative 1st floor elevation* (ft)	Value Building ($)	Contents ($)	Total ($)
$A-B$	1	151	60,000	30,000	90,000
$A-B$	9	152	15,000	3,750	18,750
$B-C$	8	150	20,000	150,000	170,000
$B-C$	3	151	90,000	45,000	
$B-C$	4	151	8,000	2,000	
$B-C$	7	151	80,000	40,000	265,000
$B-C$	6	152	2,000	2,000	
$B-C$	2	152	10,000	2,500	
$B-C$	5	152	80,000	40,000	136,500

*Rounded to nearest foot

TABLE 12.P.2
Results of backwater analysis

Discharge cfs	Return period (yr)	Cross-sections	Average elevation (ft) Natural	Backwater
20,000	10	$A-B$	150.9	151.1
25,000	25	$A-B$	151.7	152.4
30,000	50	$A-B$	152.0	153.5
35,000	100	$A-B$	152.6	154.0
40,000	200	$A-B$	152.9	154.4
20,000	10	$B-C$	152.4	152.7
25,000	25	$B-C$	153.0	153.5
30,000	50	$B-C$	153.4	153.9
35,000	100	$B-C$	153.7	154.4
40,000	200	$B-C$	154.2	154.9

cent and the useful life of the structure is 35 years. At the culvert site, a flood frequency analysis yields the following information.

Return Period (yrs)	5	10	20	40	80	160
Discharge (cfs)	100	150	170	190	200	230

Four alternative design diameters of 48, 54, 60, and 66 inches, are considered; each with an embankment elevation of 316 ft. Based on the hydraulic and economic analyses, the economic losses due to traffic interruption, backwater damage, and damage to embankment/pavement have been assessed and the results are given in the following table. The maximum economic loss is $928. Determine the design associated with the least total annual expected cost.

	Economic losses ($)					
Design alternatives	**Return period (yrs.)**					
(diameter, embankment elevation)	**5**	**10**	**20**	**40**	**80**	**160**
(48″, 316′)	$ 0	150	375	490	650	928
(54″, 316′)		0	105	275	460	710
(60″, 316′)				0	159	510
(66″, 316′)					0	248

The capital cost and annual maintenance cost for each alternative design are:

Culvert diameter	Capital cost	Annual maintenance cost
48″	$4090	$25
54″	5340	20
60″	6600	15
66”	8320	10

Determine the design associated with the least total annual expected cost.

12.7.2 Consider a reach of river, 3 miles long, that passes through a low-lying area. A levee embankment is to be constructed around it. The schematic diagram of the river cross-section, with levee geometry, is shown in Fig. 12.P.7. The flow carrying capacity of the levee system can be estimated approximately by the following Manning's formula

$$Q_c = \left[\frac{1.49}{n_c} A_c^{5/3} P_c^{-2/3} + \frac{2.98}{n_b} A_b^{5/3} P_b^{-2/3}\right] S_f^{1/2}$$

in which n_c and n_b are Manning's roughness factors for the main channel and flood plain, respectively. Other quantities in the above equation are defined in Fig. 12.P.7. The flow carrying capacity of a levee depends also on the embankment height and width of encroachment. The relevant information regarding the channel cross-section is given below.

	Parameter	Value
Channel width	B	230 feet
Channel depth	D	15 feet
Longitudinal channel slope	S_c	0.0005 ft/ft
Channel roughness	n_c	0.0500
Bank roughness	n_b	0.0750
Transverse bank slope	t	0.0017 ft/ft
Longitudinal bank slope	S_b	0.0015 ft/ft

Consider that the shape of the levees has a slope of 3 (horizontal) : 1 (vertical) on the river side and a 4 (horizontal) : 1 (vertical) on the land side. The crown width

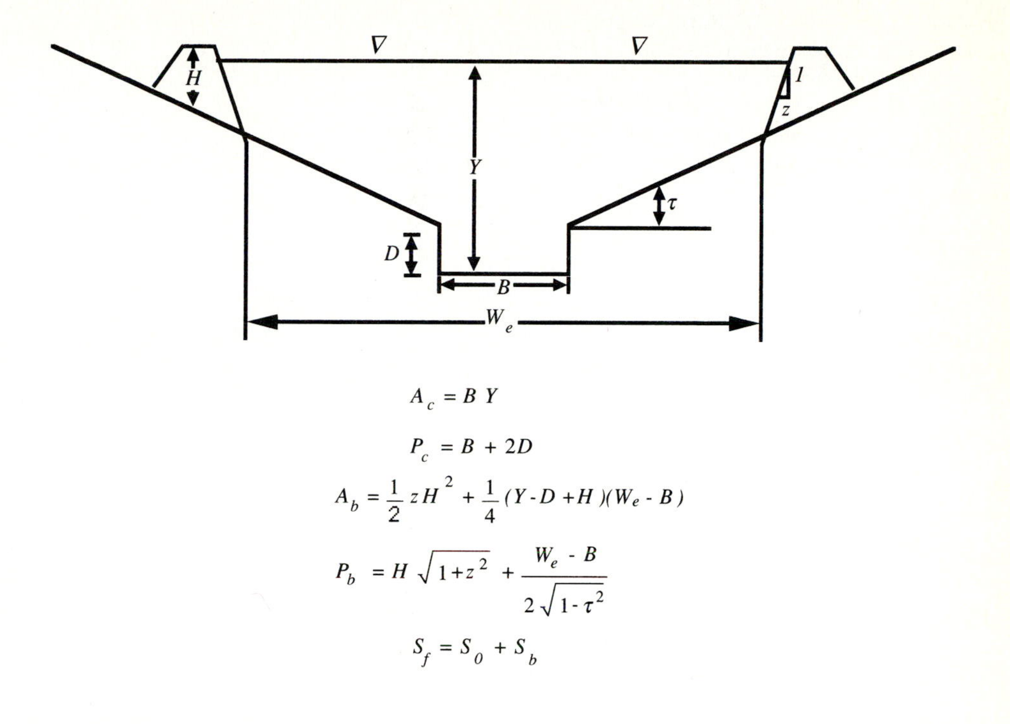

FIGURE 12.P.7
River Cross-section (Problem 12.7.2).

of the levee is 6 feet. Hence, the volumetric content of the levee is a function of levee height only and can be expressed in cubic yards per mile as $V_{LV} = 1173.3\,h + 683.3\,h^2$, in which h is the levee height in feet. The cost of levee construction is largely due to the cost of soil compaction. For simplicity, the cost of installing other accessories is not considered. Using \$0.86 per cubic yard as the unit cost of soil compaction for common road construction, the cost of levee construction per mile of given height, h, is $Z_1 = 2(0.1009\,h + 588.6\,h^2)$. The unit cost of land in the levee construction area is \$3000/acre. The expected service period for the levee is 100 years and the interest rate is 8 percent. A hypothetical relationship between the volume of water, the flood stage, and the associated flood damage costs are tabulated in Table 12.P.3. Based on the historical flood records for the study area the time base for flood hydrographs is assumed to be 50 hours. For simplicity, a triangular flood hydrograph (see Fig. 12.P.8) with a time base of 50 hours is used in the computations of flood damage associated with volume of water that exceeds the levee capacity. Hydrologic frequency analysis based on a 75-year record indicates that the annual maximum floods follow a log-Pearson type III distribution with mean 9.524, standard deviation 0.938, and skew coefficient 0.142. Consider four design alternatives: (embankment height, encroachment width) = (8′, 300′), (5′, 800′), (8′, 500′), (5′, 300′), determine the design with the least total annual expected cost.

TABLE 12.P.3
Hypothetical relationships between the volume of water, the flood stage, and the flood damage cost

Volume (AF)	Damage ($)	Average water depth in floodplain (ft)
0.0	0	3.0
290.0	5000.0	25.0
1250.0	20000.0	36.0
2360.0	100000.0	47.0
12390.0	400000.0	58.0
19860.0	600000.0	69.0
23500.0	1000000.0	70.0

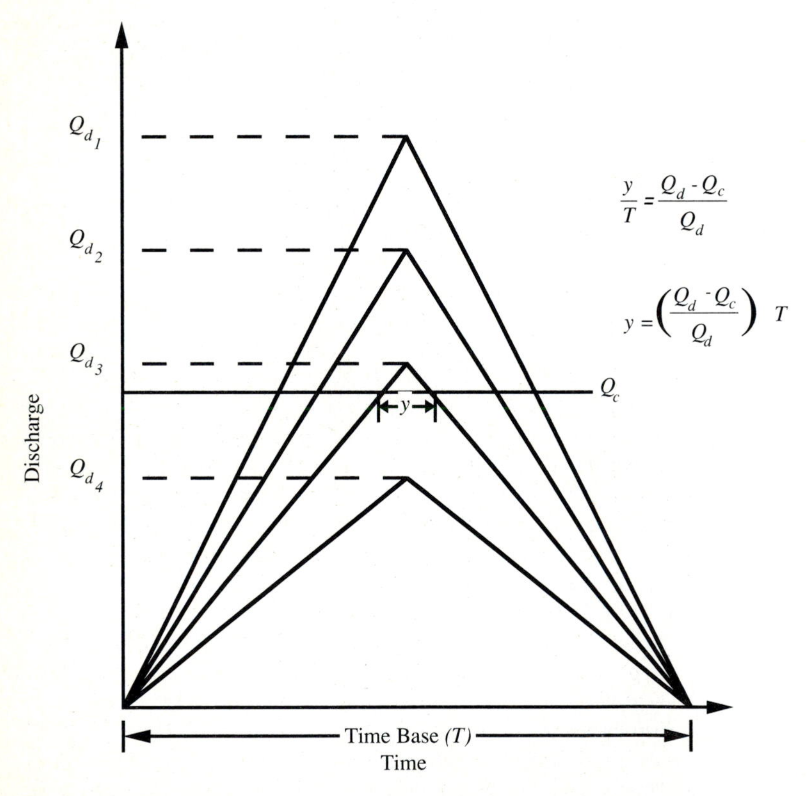

FIGURE 12.P.8
Hypothesized triangular inflow hydrograph (Problem 12.7.2).

CHAPTER 13

OPERATION OF SURFACE WATER SYSTEMS FOR FLOOD CONTROL

Flooding has caused the most prevalent and costly natural disasters in the United States and in many other countries. Floods cause impacts to society that go beyond cost and fatalities; including impacts such as family and community disruptions, dislocation, injuries and unemployment. The seriousness of flooding has been recognized for years and large amounts of money and effort have been expended to mitigate flood hazards. This chapter presents methodologies for the forecasting and optimal operation of flood control systems.

One of the most important aspects of minimizing the impacts of floods is the operation of flood control systems. In order to operate these systems, the forecasting of flood events is very important. **Hydrological forecasting** is the prior estimate of future states of hydrological phenomena in real time (Nemac, 1986). **Prediction** refers to the process of data computation for predicting hydrological values such as discharge for design. In this chapter interest is given to **flood forecasting** for the purposes of operation of flood control systems. The term **"forecasting"** refers to the determination of the discharges and water surface elevations at various points within a river system as a result of observed or predicted inflow flood hydrographs. There is an ongoing effort in real-time hydrological forecasting for both developed and developing countries throughout the world. Developed countries are concerned with the expansion and improvement of existing flood-forecasting systems for floodplain management and developing countries are concentrating on the establishment of basic data networks and forecasting capabilities for the prevention of fatalities and flood damage.

13.1 REAL-TIME FLOOD FORECASTING

13.1.1 Concepts

The three characteristics that classify hydrological forecasts are: (a) the forecast variable; (b) the purpose of the forecast; and (c) the forecast period (Nemac, 1986). In flood forecasting the **forecast variable** is water level for rivers, lakes, and reservoirs. The purpose of flood forecasting is to determine the water levels resulting from seasonal floods, flash floods, dam breaks, and storm surges on estuaries and coastal areas resulting from combined river and sea flooding. The **forecasting period** could range from a short-period to a long-period. For example the forecasting period in a small urban watershed could be in the order of just a few hours where as the forecasting period on the lower Mississippi may be in the order of several weeks.

One of the most important criteria in flood forecasting is the **lead time**, which is the interval of time between the issuing of a forecast (or warning) and the expected occurrence of the forecasted event. Both time and location are important in flood forecasting. As an example, a relatively short lead time for a short river reach may become a long lead time for locations much further downstream. To illustrate this, Fig. 13.1.1 shows three urban areas; A, B, and C; with a major rainfall in the upper portion of the watershed. There would be a short lead time for urban area A, with a longer time for urban area B and the longest lead time for urban area C. The longer lead time is obviously due to the travel time of a flood (floodwave) propagating down the river. The respective flood hydrographs at A, B, and C are shown in Fig. 13.1.2. From this simple example the lead time for urban area A is very short whereas for C it is fairly long. In fact the beginning of the flood hydrograph at C is approximately at the same time that the rainfall ends.

This example brings out another point that to make forecasts for urban area A precipitation forecasts are also required whereas for urban area C the precipitation will be known throughout the rainfall event when it is needed for the forecast. There may be several times during the flood event when forecasts will be made. As illustrated in Fig. 13.1.2 four forecasts are needed at A whereas additional ones will be required at C, past the time of the fourth one at A.

Flood forecasting includes the steps of (1) obtaining real-time precipitation and stream flow data through a microwave, radio, or satellite communications network, (2) inserting the data into rainfall-runoff and stream flow routing programs, and (3) forecasting flood flow rates and water levels for periods of from a few hours to a few days ahead, depending on the size of the watershed. Flood forecasts are used to provide warnings for people to evacuate areas threatened by floods, and to help

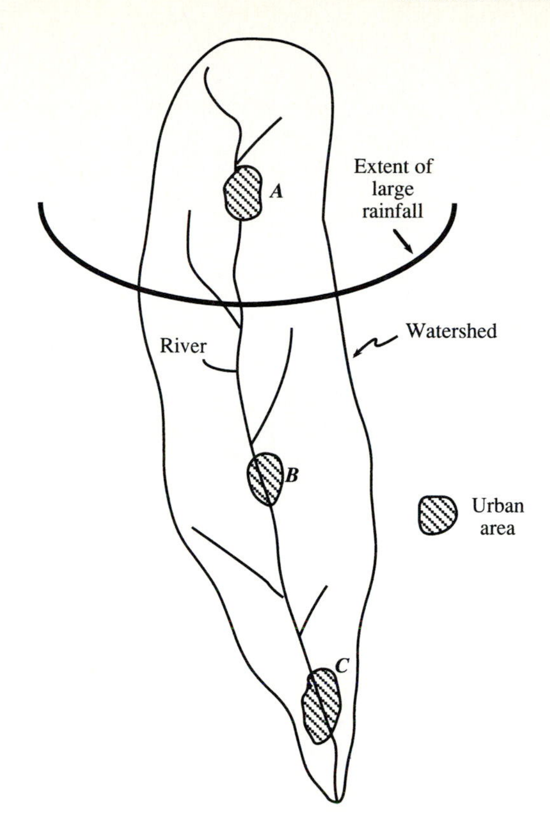

FIGURE 13.1.1
Effect of lead time.

water management personnel operate flood-control structures, such as gated spillways on reservoirs.

The **real-time reservoir operation** problem involves the operation of a reservoir system by making decisions on reservoir releases as information becomes available, with relatively short time intervals which may vary between several minutes to several hours. Real-time operation of multireservoir systems involves various hydrologic, hydraulic, operational, technical, and institutional considerations. In order to make operation decisions for flood control systems in real-time, the operations involved are the decisions on releases from the reservoir(s) in order to control flood waters. For efficient operation, a monitoring system is essential that provides the reservoir operator with the flows and water levels at various points in the river system including upstream extremities, tributaries and major creeks as well as reservoir levels, and precipitation data for the watersheds whose outputs (runoff from rainfall) are not gaged. Many river-reservoir systems in the U.S., such as the lower Colorado River in central Texas (Fig. 13.1.3), now have real-time data collection and transmission systems.

Flood forecasting in general, and real-time flood forecasting in particular, have always been an important problem in operational hydrology, especially when the operation of flood-control reservoirs is involved. The forecasting problem, as in most hydrological problems, can be viewed as a system with inputs and outputs. In the

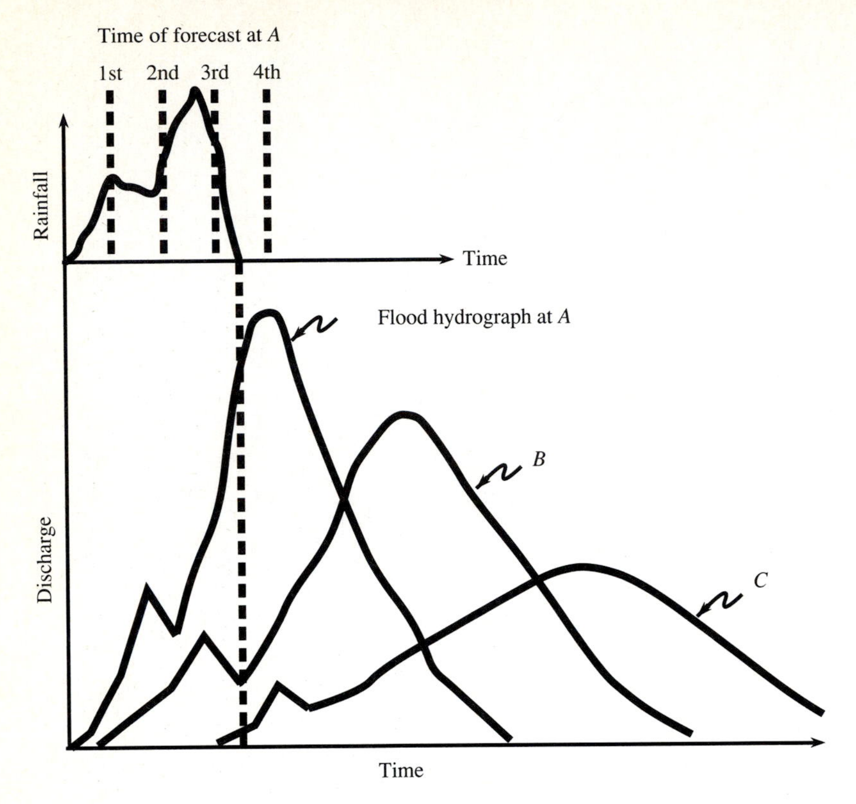

FIGURE 13.1.2
Flood hydrograph at downstream location in a watershed.

reservoir management problem, the system is the river system that includes a main river and its tributaries, catchments, and natural and constructed structures on the path of the flood waters. The system inputs are inflow hydrographs at the upstream ends of the river system, and runoff from the rainfall (and snowmelt, where applicable) in the intervening catchments. The system outputs are flow rates and/or water levels at control points of the river system.

13.1.2 Real-time Data Collection Systems for River-Lake Systems

Real-time data collection and transmission can be used for flood forecasting on large river-lake systems covering hundreds to thousands of square miles, as shown in Fig. 13.1.4 for the lower Colorado River in central Texas. The data collection system is called a **hydrometeorological data acquisition system** (EG&G Washington Analytical Services Center, Inc., 1981) and is used to provide information for a flood forecasting model. This information is of two types: (a) the water surface elevations at various locations throughout the river-lake system, and (b) rainfall from a rain gage network for the ungaged drainage areas around the lakes. The hydromet system

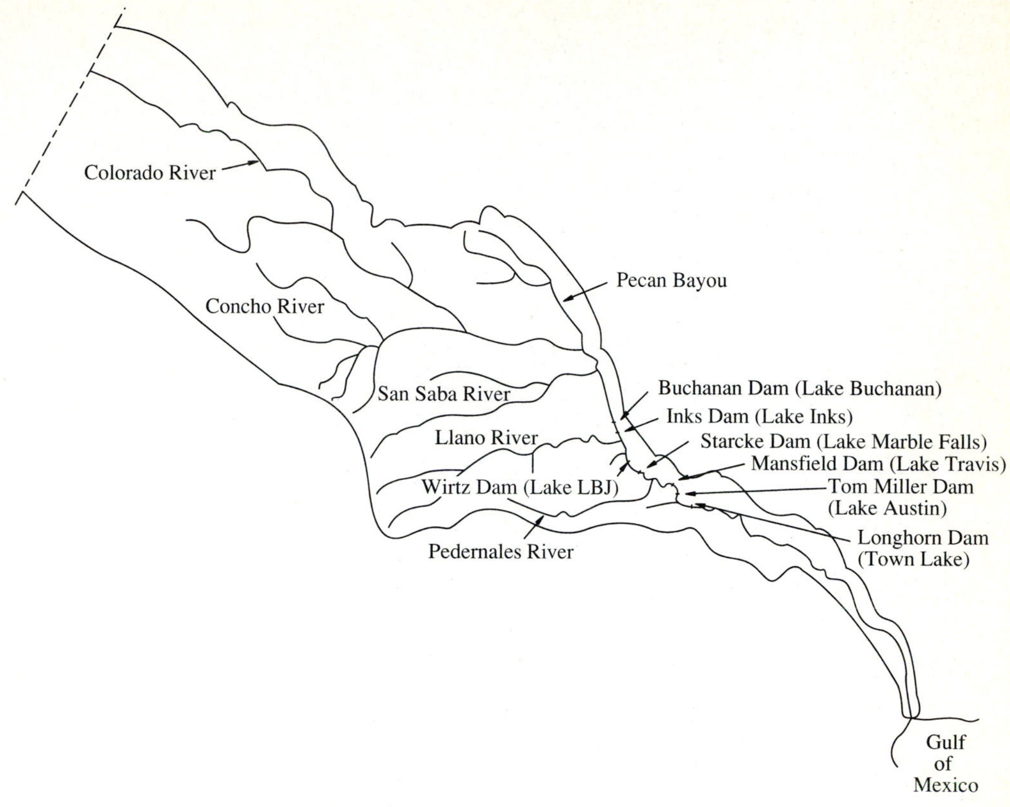

FIGURE 13.1.3
Lower Colorado River Basin (Source: Unver, Mays, and Lansey, 1987).

consists of: (a) remote terminal unit (RTU) hydrometeorological data acquisition stations installed at U.S. Geological Survey river gage sites; (b) microwave terminal unit (MTU) microwave to UHF radio interface units located at microwave repeater sites, which convert radio signals to microwave signals; and (c) a central control station located at the operations control center in Austin, Texas, which receives its information from the microwave repeating stations. The system is designed to automatically acquire river levels and meteorological data from each RTU; then telemeter this data on request to the central station via the UHF/microwave radio system; determine the flow rate at each site by using rating tables stored in the central system memory; format and output the data for each site; and maintain a historical file of data for each site which may be accessed by the local operator, a computer, or a remote dial-up telephone line terminal. The system also functions as a self-reporting flood alarm network.

13.1.3 Flood Early Warning System for Urban Areas

Because of the potential for severe flash flooding and consequent loss of life in many urban areas throughout the world, **flood early warning systems** have been constructed

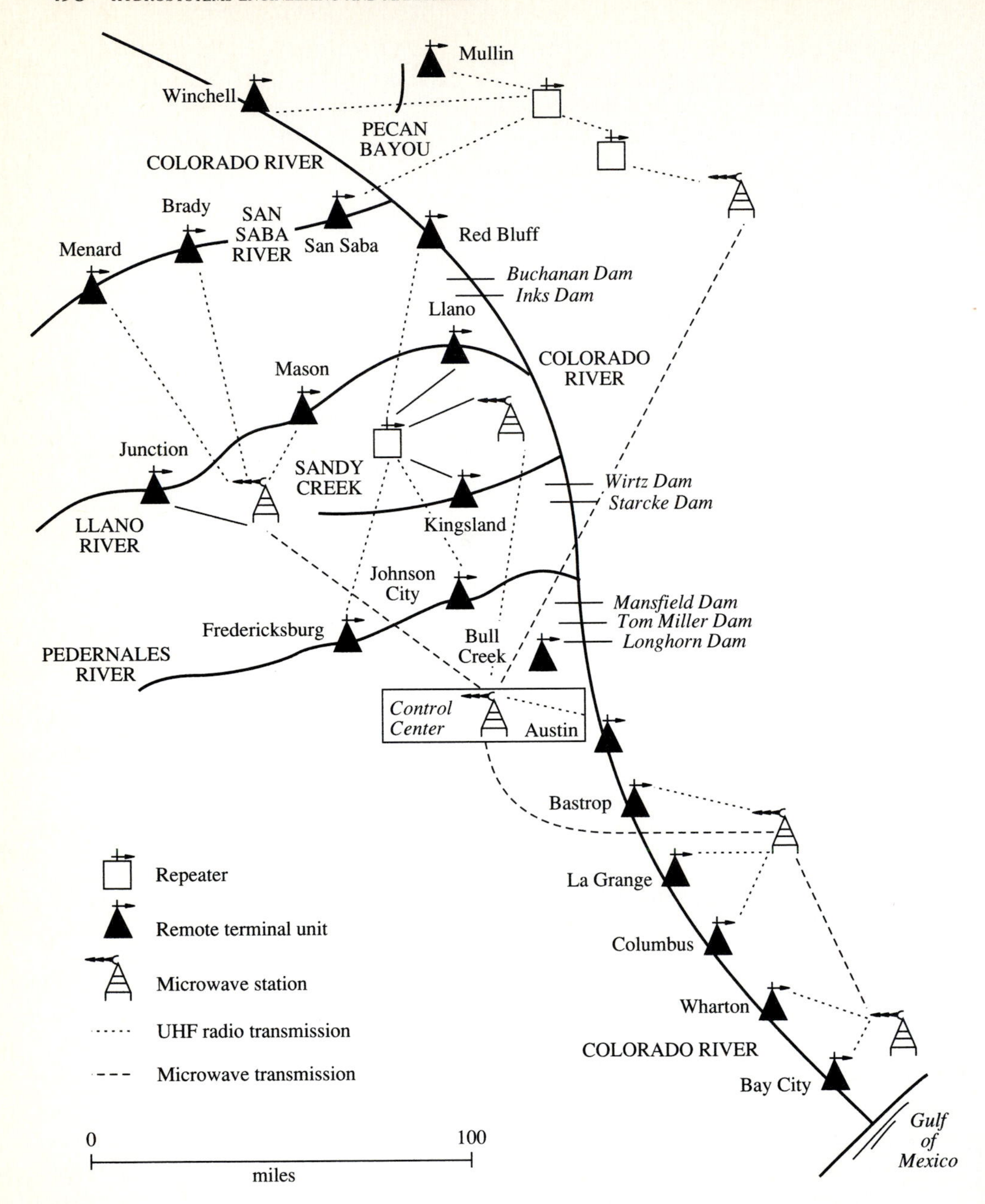

FIGURE 13.1.4
Real-time data transmission network on the lower Colorado River, Texas. Water level and rainfall data are automatically transmitted to the control center in Austin every 3 hours to guide releases from the dams. During floods data are updated every 15 minutes (Chow et al., 1988).

and implemented. Flood early warning systems (Fig. 13.1.5) are real-time event reporting systems that consist of remote gaging sites with radio repeater sites to transmit information to a base station. The overall system is used to collect, transport, and analyze data, and make a flood forecast in order to maximize the warning time to occupants in the flood plain. Such systems have been installed in Austin and Houston, Texas, and elsewhere (Sierra/Misco, Inc., 1986).

The remote stations (Fig. 13.1.6) each have a tipping bucket rain gage, which generates a digital input to a transmitter whenever 1 mm of rainfall drains through the funnel assembly. A transmission to the base station is made for each tip of the bucket. The rain gage is completely self-contained, consisting of a cylindrical stand pipe housing for the rain gage, antenna mount, battery, and electronics.

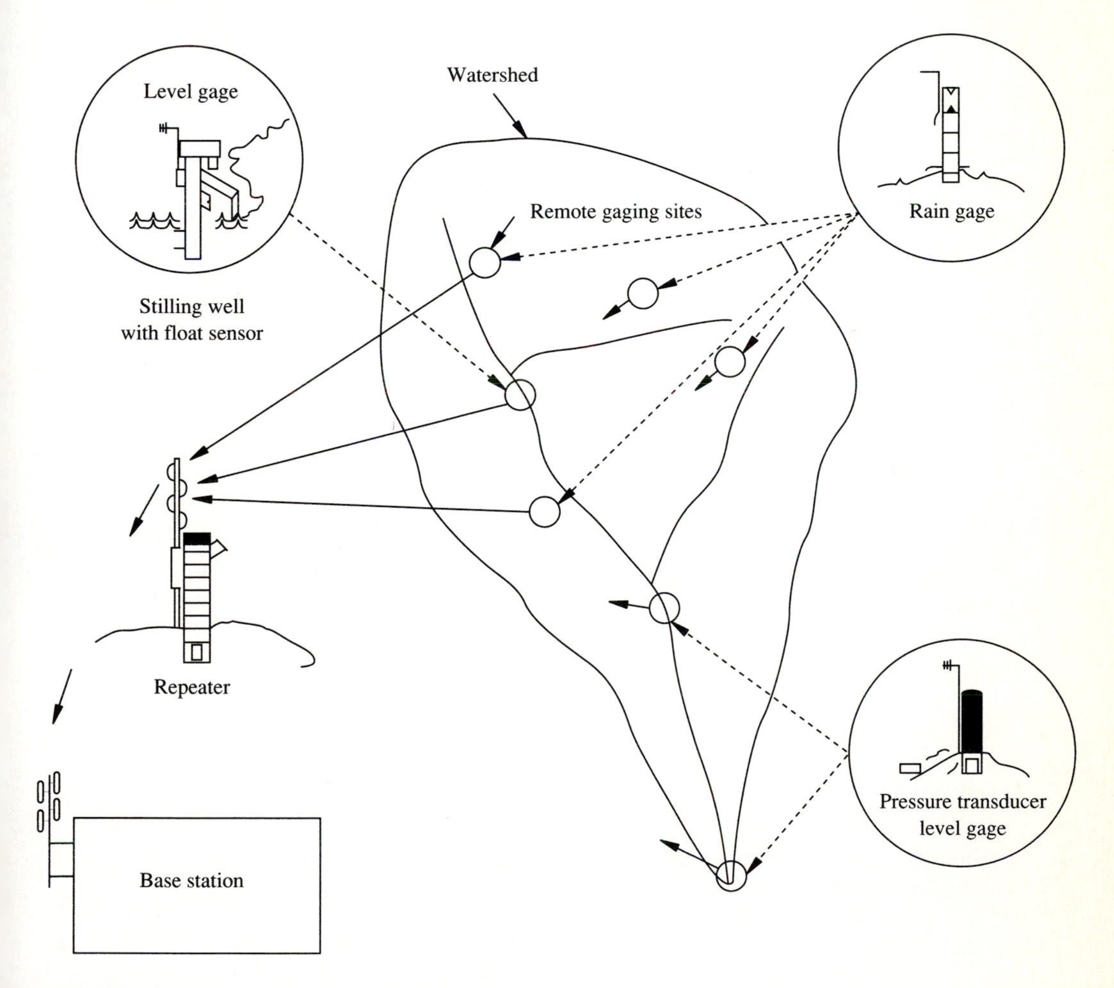

FIGURE 13.1.5
Example of a flood early warning system for urban areas (Chow et al., 1988).

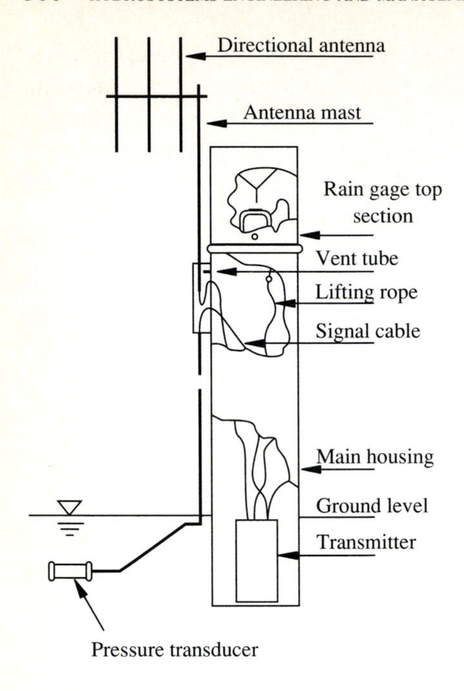

FIGURE 13.1.6
Remote station combining precipitation and stream gages. (Courtesy of Sierra/Misco, Inc., 1986. Used with permission.)

Some remote stations have both rainfall and streamflow gages. The remote stations can include a stilling well or a pressure transducer water level sensor. The pressure transducer measures changes of the water level above the pressure sensor's orifice. The electronic differential pressure transducer automatically compensates for temperature and barometric pressure changes with a one-percent accuracy over the measured range. Automatic repeater stations, located between the remote stations and the base station, receive data from the remote stations, check the data for validity, and transmit the data to the base station.

Incoming radio signals are transformed from radio analog form to digital format and are forwarded to the base station computer through a communications port. After data quality checks are made, the data are formatted and filed on either hard or floppy disk media. Once the data filing is complete, the information can be displayed or saved for analysis.

The base station has data management software which can handle up to 700 sensors with enough on-line storage to store three years of rainfall data. It can cover 12 separate river systems with up to 25 forecast points possible in each; each forecast point can receive inflow from up to 10 different sources. Different future rainfall scenarios can be input for each individual forecast point, and optional features can be added to control pumps, gates, remote alarms, and voice synthesized warnings (Sierra/Misco, Inc., 1986).

13.1.4 Flood-Forecasting Models

Flood-forecasting models include a variety of flood routing procedures ranging from models that are based upon rainfall-runoff models using hydrologic routing such as the U.S. Army Corps of Engineers HEC-1 computer program to the U.S. National Weather Service DWOPER computer program which is a distributed or hydraulic routing model (Chapter 10). Some programs such as the HEC-5 (U.S. Army Corps of Engineers, 1973, 1979) is based upon hydrologic routing and includes the ability to operate reservoirs. HEC-5 has been discussed in Section 7.4 and is discussed further in Section 13.2. Another model that considers continuous hydrologic simulation along with hydrologic routing models and the DWOPER program is the U.S. National Weather Service River Forecast System (NWSRFS), which is described in more detail below. There have been several theoretical developments of flood forecasting models reported in the literature. These include Kitanidis and Bras (1980*a*, *b*), Georgakakos and Bras (1982), Krzysztofowicz and Davis (1983*a*, *b*, *c*), Georgakakos (1986*a*, *b*), and Melching et al. (1987).

The National Weather Service River Forecast System (NWSRFS) Operational Forecast System (Brazil and Smith, 1989) generates streamflow forecasts using observed and forecasted precipitation and temperature data in hydrologic and hydraulic models that simulate snow accumulation and ablation, rainfall/runoff, watershed routing, and channel routing processes to produce streamflow. The NWSRFS software consists of three separate systems (Fig. 13.1.7): the calibration system; the operational forecast system, and the extended streamflow prediction (ESP) system. The calibration system is used to calibrate the various hydrologic/hydraulic models for a watershed. The operational forecast system performs all the tasks for operational river forecasting. The various operations are listed in Table 13.1.1. The ESP system is used to make forecasts for periods of weeks or months into the future (Day, 1985). Conceptual hydrologic and hydraulic models are used along with current watershed conditions, historical meteorological data, and forecasted meteorological data to make extended probabilistic forecasts for a number of streamflow variables.

13.2 RIVER-RESERVOIR OPERATION FOR FLOOD CONTROL

13.2.1 Reservoir Operation Models

Multireservoir operation can be characterized by the integrated operation of multiple facilities on river systems for multiple objectives. Flood control is one of the major purposes of many reservoirs in the U.S. Many reservoirs were built several years ago and operation policies were established. However, many of these reservoirs cannot be operated in the manner that they were initially intended to be operated. One of the major reasons is the uncontrolled urbanization into the floodplains of the rivers and reservoirs. Other reasons are due to inadequate spillways for passing floods, legal constraints, and reduced downstream conveyance capacities.

Many of the reservoir systems are characterized by conditions that result in significant backwater conditions due to gate operation, tributary flows, hurricane surge flows, and tidal conditions, flow constrictions in the rivers. These conditions cannot be described by the use of hydrologic routing methods, and as a result must be described by more accurate hydraulic routing models such as DWOPER, which is based upon a

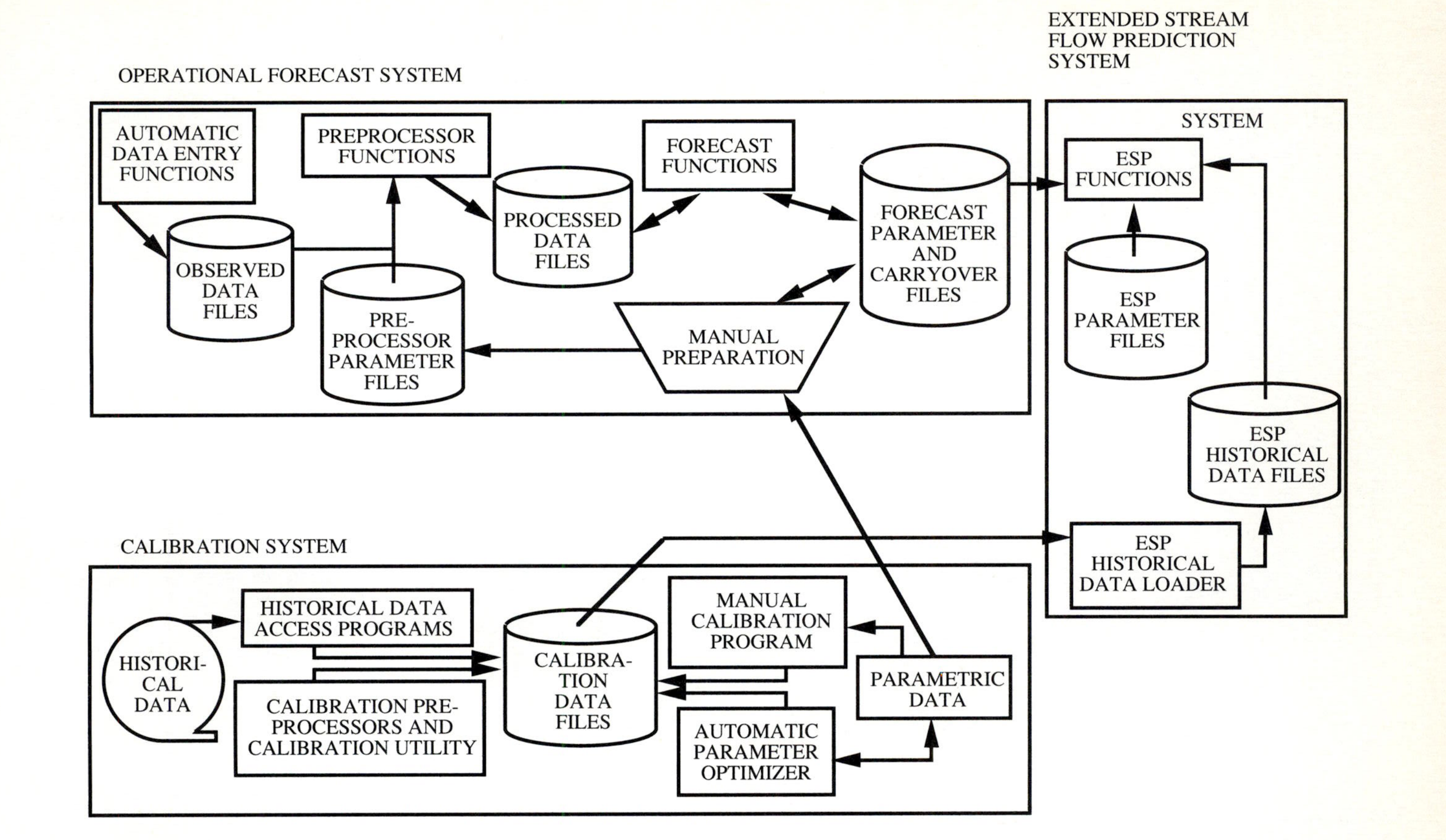

FIGURE 13.1.7
National Weather Service River Forecast System software overview (Day, 1985).

TABLE 13.1.1

Operations planned for forecast component of NWSRFS Version 5 (Day, 1985)

Hydrologic/hydraulic models (1)	Arithmetic computations (2)	Updating and verification procedures (3)	Displays (4)
API/MKC–Antecedent precipitation index rainfall-run-off model for the Missouri Basin and north central RFC's	ADD/SUB–Add or subtract time series	ADJUST-Q–Adjust simulated to observed discharge and blend into future	INSQPLOT–Plots instantaneous discharge time series
SAC-SMA–Sacramento soil moisture accounting model	CLEAR-TS–Clear time series	CHAT–Computer hydrograph adjustment technique	WY-PLOT–Water year mean-daily flow plot
UNIT-HG–Unit hydrograph operation	WEIGH-TS–Weight time series	SACFIL1–Estimation theory (Kalman filter) formulation of the SAC-SMA and UNIT-HG for lumped, non-snow headwater basins	SAC-PLOT–Sacramento type mean-daily flow plot
SNOW-17–HYDRO-17 snow accumulation and ablation model	CHANGE-T–Change time interval of a time series	STAT-OP–Statistical package for measuring NWSRFS effectiveness	PLOT-TS–General time series plotting utility
LAG/K–Lag and K routing	MEAN-Q–Computation of mean discharge for specified time interval		PLOT-TUL–Time-series plotting routine specifically designed for real-time operational forecasting
LAY-COEF–Layered coefficient routing			STAT-QME–Computes statistical summary of mean-daily discharge
MUSKROUT–Muskingum routing			
TATUM–Tatum routing			
DWOPER–Dynamic wave operational model			
CHANLOSS–Empirical chanel-loss/gain routine			
CHANLEAK–Conceptual channel-loss/gain routing			
STAGE-Q–Converts river stage to discharge or vice-versa			
RES-SNGL–Single reservoir control operation			

finite difference solution of the Saint Venant equations (see Sections 10.6 and 10.7). Also, flows through reservoirs having considerable length are not properly predicted by the simple hydrologic methods, particularly when the inflow hydrograph is a **flash flood**, that is, has a short time base.

There have been many reservoir operation models reported in the literature but only a few have been directed at reservoir operation under flooding conditions. Jamieson and Wilkinson (1972) developed a DP model for flood control with forecasted inflows being the inputs to the model. Windsor (1973) employed a recursive linear programming procedure for the operation of flood control systems, using the Muskingum method for channel routing and the mass balance equation for reservoir computations.

The U.S. Army Corps of Engineers (1973, 1979) developed HEC-5 and HEC-5C for reservoir operation for flood control, where the releases are selected by applying a fixed set of heuristic rules and priorities that are patterned after typical operation studies (see Section 7.4). These models are based upon hydrologic routing techniques and provide no optimal strategy for operation. One application of these models was to the Kanawha River Basin (U.S. Army Corps of Engineers, 1983) which contributes flow to the Ohio River at Pt. Pleasant, West Virginia. Figure 13.2.1 illustrates observed and forecasted hydrographs at Kanawha Falls for the March 1967 event. The vertical dashed line represents the time of the forecast.

The Tennessee Valley Authority (1974) developed an incremental dynamic programming and successive approximations technique for real-time operations with flood control and hydropower generation being the objectives. Can and Houck (1984) developed a goal programming model for the hourly operations of a multireservoir system and applied it to the Green River basin in Indiana. The model objective is defined by a hierarchy of goals, with the best policy being a predetermined rule curve.

Wasimi and Kitanidis (1983) developed an optimization model for the daily operations of a multireservoir system during floods which combines linear quadratic gaussian optimization and a state-space mathematical model for flow forecasting. Yazicigil (1980) developed an LP optimization model for the daily real-time operations of the Green River basin in Indiana, a system of four multipurpose reservoirs. The model inputs are deterministic. The objective of operation is to follow a set of target states, deviations from which are penalized. The channel routing is performed using a linear routing procedure similar to the Muskingum method, called multi-input linear routing. The reservoir calculations are based on mass-balance equations which take into account precipitation input.

13.2.2 Lower Colorado River Flood Forecasting System

The flood forecasting model for the Lower Colorado River—Highland Lakes system in Texas developed by Unver et al. (1987) was developed for a real-time framework to make decisions on reservoir operations during flooding. This model is an integrated computer program with components for flood routing, rainfall-runoff modeling, and graphical display, and is controlled by interactive software. Input to the model includes

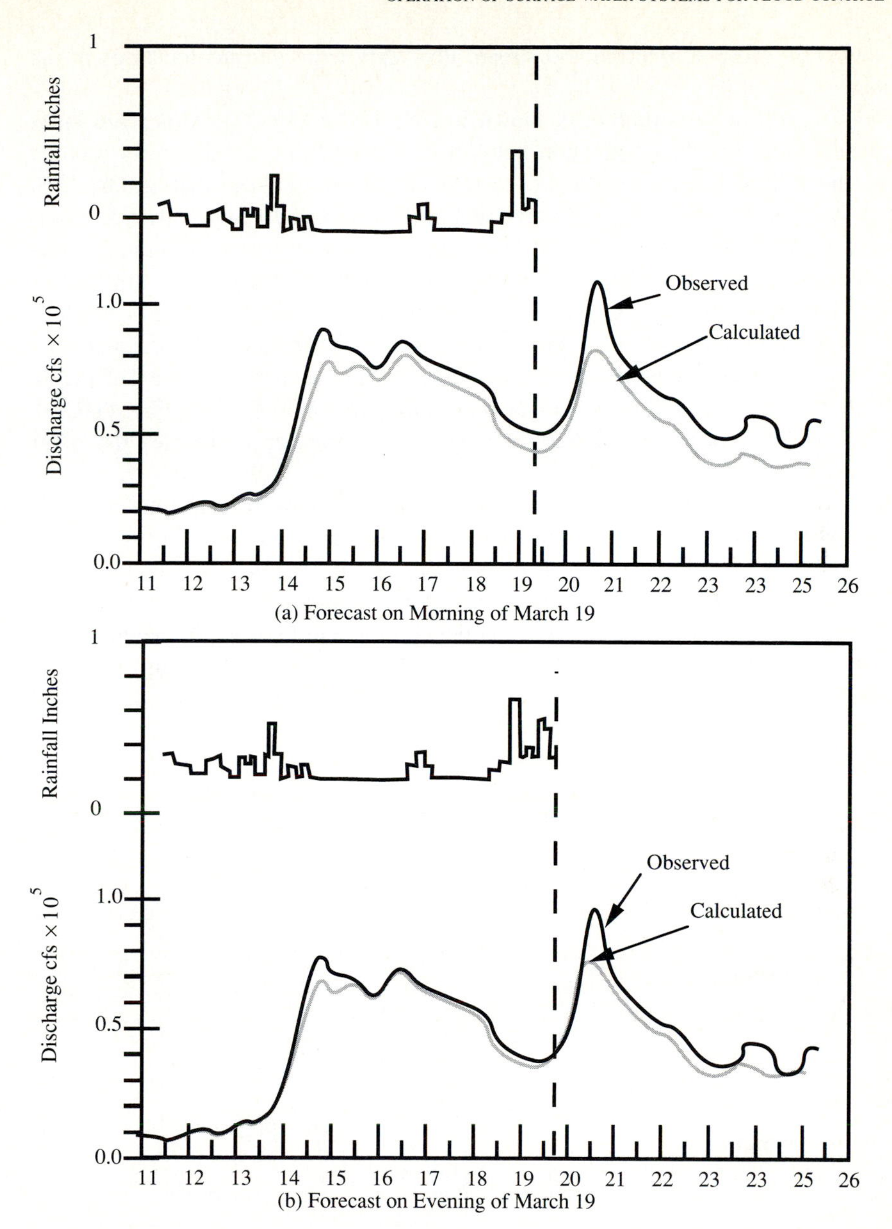

FIGURE 13.2.1
Observed and forecasted hydrographs at Kanawha Falls, resulting from a forecast of the March 1967 flood event. (U.S. Army Corps of Engineers, 1983)

automated real-time precipitation and stream flow data from various locations in the watershed.

The overall model structure is shown in Fig. 13.2.2. Real-time data are input to the model from the data collection network. The real-time flood control module includes the following submodules: (1) a DWOPER submodule, that is, the U.S. National Weather Service Dynamic Wave Operational model for unsteady flow routing; (2) a GATES submodule, which determines gate operation information (internal boundary conditions) for DWOPER, such as the gate discharge as a function of the head on the gate; (3) a RAINFALL-RUNOFF submodule which is a rainfall-runoff model based upon the unit hydrograph approach for the ungaged drainage area surrounding the lakes for which stream flow data is not available; (4) a DISPLAY submodule, which contains graphical display software; and (5) an OPERATIONS submodule which is the user-control software that interactively operates the other submodules and data files.

The input for this flood forecasting model includes both the real-time data and the physical description of system components that remain unchanged during a flood. The physical data include: (1) DWOPER data describing stream cross section information, roughness relationships, and so on; (2) characteristics of the reservoir spillway structures for GATES; and (3) drainage area description and hydrologic parameter estimates for RAINFALL-RUNOFF. The real-time data include: (1) stream-flow data from automated stations and headwater and tailwater elevations at each dam; (2) rain-

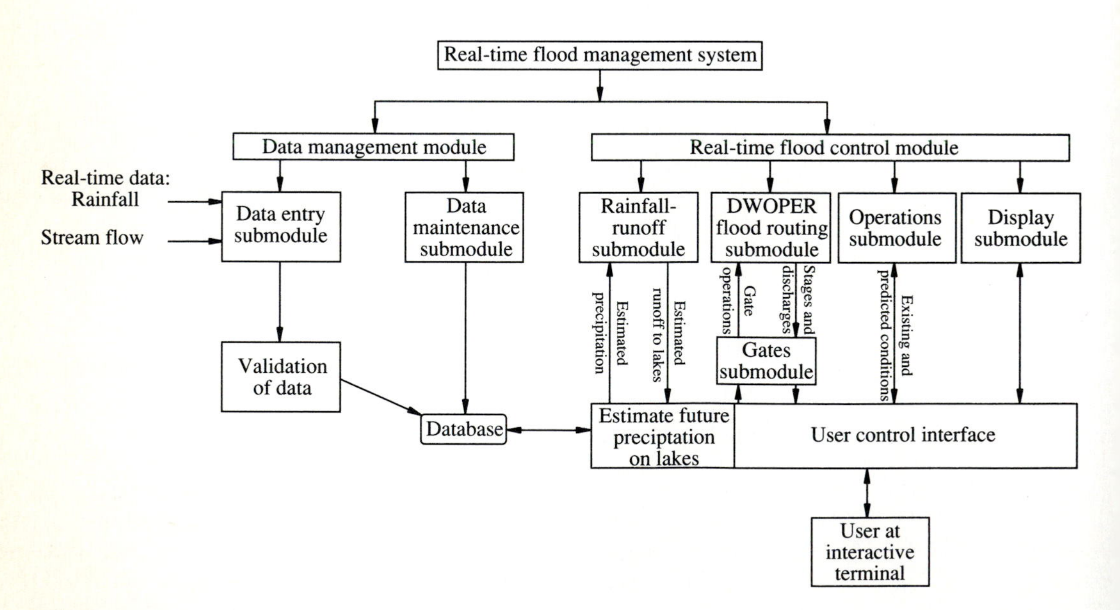

FIGURE 13.2.2
Structure of real-time flood management model. This model is used by the Lower Colorado River Authority to manage the river-lake system shown in Fig. 13.1.3. The real-time data collection system is shown in Fig. 13.1.4. (Source: Unver, Mays, and Lansey, 1987).

fall data from recording gages; (3) information as to which subsystem of lakes and reservoirs will be considered in the routing; and (4) reservoir operations.

13.3 OPTIMIZATION MODELS FOR DEVELOPING OPERATION POLICIES

Optimization models can be used to determine the flood control operation of the reservoir systems as shown in Fig. 13.3.1. The objective of such a model is to determine an operating policy (releases) that minimizes the flood damage for reaches A and B. Inputs to the model consist of the damage functions and weekly average historical flood flows representing the largest floods recorded at the gaging station (see Fig. 13.3.1). The minimum storage in reservoir 1 is the conservation pool storage and the maximum storage is the full flood pool storage.

A DP model is developed in which the decision variable is the release QA from reservoir 1 and the state variable is the storage ST in reservoir 1. The objective of the optimization is to minimize the total flood damages Z:

$$\text{Min } Z = \sum_{n=1}^{N} r_n(QA_n) \tag{13.3.1}$$

where $n = 1, \ldots N$ are the stages which are the weekly time periods in chronological order and $r_n(QA_n)$ is the return for the nth stage from the decision (release) QA from reservoir 1. Damages for both reaches A and B are functions of QA. Constraints on the operation include the continuity equation which is the transformation function

$$ST_{n+1} = ST_n + I_n - QA_n, \qquad n = 1, \ldots, N \tag{13.3.2}$$

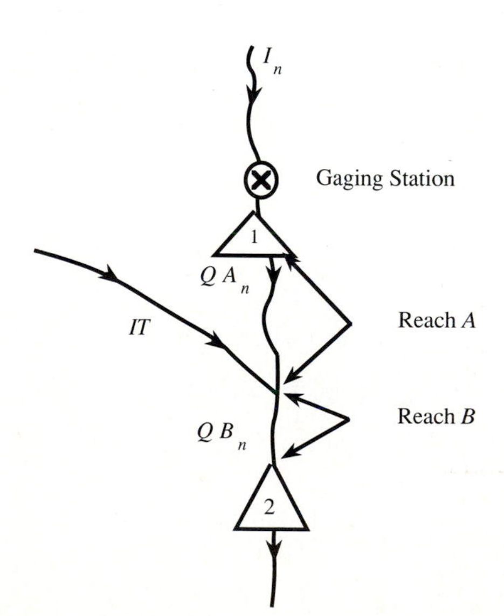

FIGURE 13.3.1
Reservoir system for Example 13.3.1.

where ST_n and ST_{n+1} are the beginning and end of the nth week storage in reservoir 1 and I_n is the inflow into reservoir 1 during the nth week. The bounds on the storage are

$$ST_{\min} \leq ST_n \leq ST_{\max} \qquad n = 1, \ldots, N \tag{13.3.3}$$

where $ST_{\min}$ is the minimum storage in the reservoir which is the conservation pool storage and $ST_{\max}$ is the maximum storage which is the full flood pool storage. The maximum release constraint for reservoir 1 is

$$QA_n \leq QA_{\max}, \qquad n = 1, \ldots, N \tag{13.3.4}$$

Flow in reach B is defined as

$$QB_n = QA_n + IT_n, \qquad n = 1, \ldots, N \tag{13.3.5}$$

where IT_n is the tributary inflow.

The DP recursive equation for a forward (chronological order) algorithm would be

$$f_n(ST_n) = \underset{QA_n}{\text{Min}} \left[r_n(QA_n) + f_{n-1}(ST_{n-1}) \right] \tag{13.3.6}$$

where $f_n(ST_n)$ is the cumulative return (damages) from the optimal policy for the system through the first n stages (time periods). The stage return is the sum of the damages for both reaches A and B,

$$r_n(QA_n) = D_A(QA_n) + D_B(QB_n = QA_n + IT_n) \tag{13.3.7}$$

where $D_A(\ )$ and $D_B(\ )$ are the damage functions.

Example 13.3.1. Develop an optimization model to determine the flood control operation of the reservoir system shown in Fig. 13.3.2. This system is similar to the system shown in Fig. 13.3.1, but also includes the damage reach C which is downstream of reservoir 2.

Solution. This problem must consider two state variables, the storages in reservoirs 1 and 2, ST1 and ST2, respectively; and two decision variables, the releases from reservoirs 1 and 2, QA and QC, respectively. The DP model can be stated as

$$\text{Min } Z = \sum_{n=1}^{N} r_n(QA_n, QC_n)$$

The stage return (damage for time period n) is

$$r_n(QA, QC) = D_A(QA_n) + D_B(QA_n + IT_n) + D_C(QC_n)$$

where $D_A(\ ), D_B(\ )$, and $D_C(\ )$ are the damage functions. The constraints on continuity (transformation function) are

$$ST1_{n+1} = ST1_n + I_n - QA_n$$

for reservoir 1 where ST1 refers to the storage in reservoir 1. Similarly the continuity equation for reservoir 2, assuming negligible lateral inflow in reaches A and B, is

$$ST2_{n+1} = ST2_n + QB_n - QC_n$$

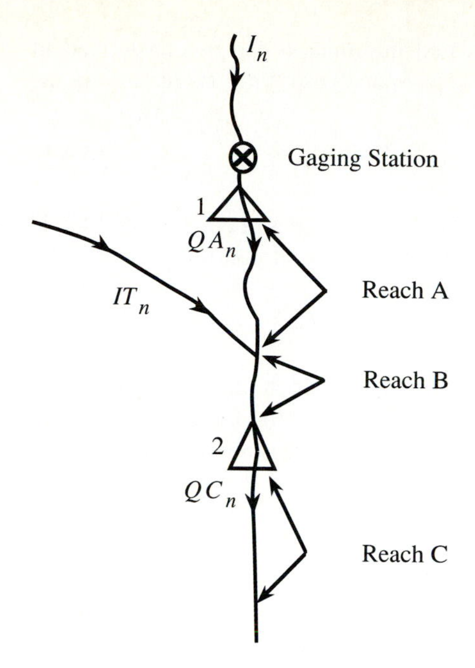

FIGURE 13.3.2
Reservoir system for Example 13.3.1.

or

$$ST2_{n+1} = ST2_n + QA_n + IT_n - QC_n$$

where $ST2$ is the storage in reservoir 2. The bounds on storage for each reservoir are

$$ST1_{\min} \leq ST1_n \leq ST1_{\max}$$

and

$$ST2_{\min} \leq ST2_n \leq ST2_{\max}$$

Maximum releases from the two reservoirs are expressed as

$$QA_n \leq QA_{\max}$$

and

$$QC_n \leq QC_{\max}$$

13.4 OPTIMIZATION MODELS FOR REAL-TIME OPERATION OF RESERVOIRS

13.4.1 Optimization Model Using Hydrologic Routing

Models that optimize real-time operations should be able to consider the trade-offs between long-term operational goals and short-term restrictions on operation. The restrictions on operation are due to physical constraints of the system such as outlet capacities and storage capacities. Uncertain information on future inflows, tributary flows and precipitation should also be considered in the trade-offs.

A mathematical model based on forecasted information can be constructed in which the objective function would be to minimize penalties on the operation, that is,

$$\text{Min } Z = \sum_{n=1}^{N} [Z_1(ST_n) + Z_2(Q_n)] \tag{13.4.1}$$

where $Z_1(ST_n)$ is a penalty function based on the expected storage and $Z_2(Q_n)$ is a penalty function based on expected downstream flow at a control station. The reservoir continuity equation is

$$ST_{n+1} - ST_n + R_n = I_n, \qquad n = 1, \ldots, N \tag{13.4.2}$$

where ST_n and ST_{n+1} are the beginning and ending reservoir storages for time period n; R_n is the reservoir release during time period n; and I_n is the forecasted reservoir inflow during time period n. The discharge, Q_n, at a downstream control station during time period n is

$$Q_n - C(R_n) = 0, \qquad n = 1, \ldots, N \tag{13.4.3}$$

where $C(R_n)$ is a function representing the hydrologic flood routing equation between the reservoir and the downstream control station. An example of a routing equation would be the Muskingum equation (see Section 10.3) which results in Eq. (13.4.3) being linear. Constraints on the maximum storage, $ST_{\max}$, and maximum release, $R_{\max}$, are

$$ST_n \leq ST_{\max} \qquad n = 1, \ldots, N \tag{13.4.4}$$

$$R_n \leq R_{\max} \qquad n = 1, \ldots, N \tag{13.4.5}$$

Several solutions of models similar to the above have been reported in the literature. Sigvaldason (1976) and Yazicigil (1980) solved the mathematical program at the beginning of each day of real-time operation. Solution of the model provides optimal releases and the optimal reservoir storages and the discharges at the control point for all days in the operating horizon. These results for each day are dependent, of course, on the forecasted reservoir inflows. Only the current day's releases are used in operating decisions. For the next day, the model is reconstructed using the actual beginning storage which may be different than the model results because the actual inflow would not be exactly the same as the forecasted inflow.

Example 13.4.1. Develop an LP model for the single reservoir system shown in Fig. 13.4.1. Use the storage penalty function shown in Fig. 13.4.2 to develop the model. No penalty on discharges at the control point will be considered. Storages ST_{α_1} and ST_{α_2} are storages lower than the target storage, ST_{TAR}, and ST_{β_1}, ST_{β_2}, and ST_{β_3} are the storages greater than ST_{TAR}.

Solution. The objective function is expressed as

$$\text{Min } Z = \sum_{n=1}^{N} \left[\alpha_1 ST_{\alpha_1}(n) + \alpha_2 ST_{\alpha_2}(n) + \beta_1 ST_{\beta_1}(n) + \beta_2 ST_{\beta_2}(n) + \beta_3 ST_{\beta_3}(n)\right]$$

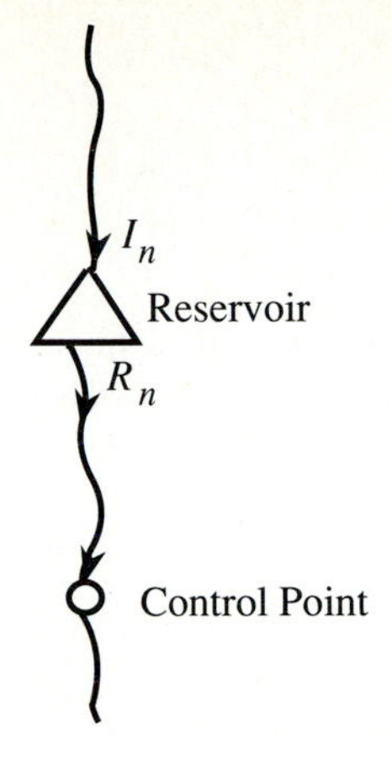

FIGURE 13.4.1
Reservoir with a downstream control point.

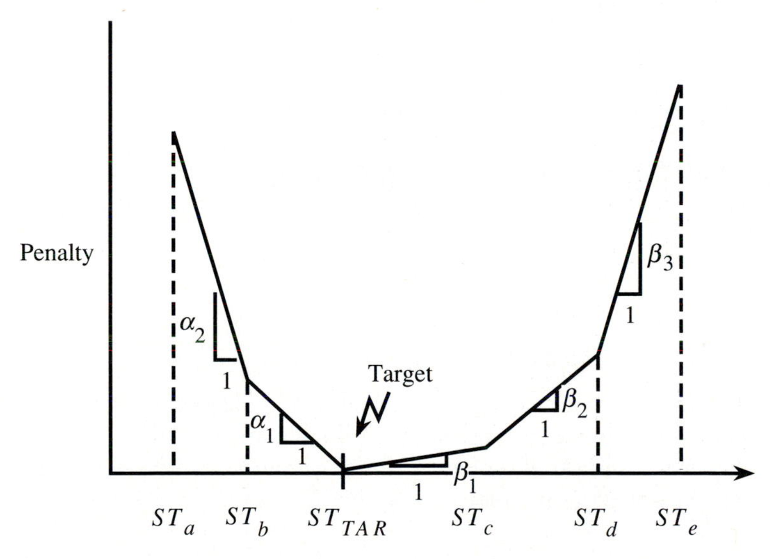

FIGURE 13.4.2
Storage penalty function for Example 13.5.1.

The storage constraints for the reservoir can be expressed as

$$ST_n + ST_{\alpha_1}(n) + ST_{\alpha_2}(n) - ST_{\beta_1}(n) - ST_{\beta_2}(n) - ST_{\beta_3}(n) = ST_{TAR}$$

to define storage. Bounds on storages are

$$ST_{\alpha_1}(n) \leq ST_{TAR} - ST_b$$

$$ST_{\alpha_2}(n) \leq ST_b - ST_a$$

$$ST_{\beta_1}(n) \leq ST_c - ST_{TAR}$$

$$ST_{\beta_2}(n) \leq ST_d - ST_c$$

$$ST_{\beta_3}(n) \leq ST_e - ST_d$$

The continuity equation for each time period is

$$ST_{n+1} + R_n = ST_n + I_n$$

The linear routing constraints to the downstream control point are

$$Q_n - f(R_n) = 0.$$

If discharge penalties were considered then this constraint becomes important to the objective function because the penalty term would also be included as expressed by Eq.(13.4.1). (See Problem 13.4.2). Maximum storage and reservoir release constraints are respectively

$$ST_n \leq ST_{max}$$

and

$$R_n \leq R_{max}$$

Nonnegativity constraints are

$$ST_n, R_n, Q_n \geq 0$$

$$ST_{\alpha_1}(n), ST_{\alpha_2}(n), ST_{\beta_1}(n), ST_{\beta_2}(n), ST_{\beta_3}(n) \geq 0$$

13.4.2 Optimization Model Using Hydraulic Routing

The optimization problem for the operation of multireservoir systems under flooding conditions can be stated as follows:

Objective.

$$\text{Minimize } Z = f(\mathbf{h}, \mathbf{Q}) \quad (13.4.6)$$

where **h** and **Q** are the vectors of water surface elevations and discharges, respectively. The objective is defined by minimizing: (a) the total flood damages; (b) deviations from target levels; (c) water surface elevations in the flood areas; or (d) spills from reservoirs or maximizing storage in reservoirs.

Constraints.

a. Hydraulic constraints are defined by the Saint-Venant equations for one-dimensional gradually varied unsteady flow (see Section 10.6) and other relationships such as upstream, downstream, and internal boundary conditions and initial conditions that describe the flow in the different components of a river-reservoir system,

$$\mathbf{g}(\mathbf{h}, \mathbf{Q}, \mathbf{r}) = 0 \quad (13.4.7)$$

where **h** is the matrix of water surface elevations; **Q** is the matrix of discharges; and **r** is the matrix of gate settings for spillway structures, all given in matrix form to consider the time and space dimensions of the problem.

b. Bounds on discharges defined by minimum and maximum allowable reservoir releases and flow rates at specified locations,

$$\underline{\mathbf{Q}} \leq \mathbf{Q} \leq \overline{\mathbf{Q}} \quad (13.4.8)$$

Bars above and below a variable denote the upper and lower bounds, respectively, for that variable.

c. Bounds on elevations defined by minimum and maximum allowable water surface elevations at specified locations (including reservoir levels),

$$\underline{\mathbf{h}} \leq \mathbf{h} \leq \overline{\mathbf{h}} \tag{13.4.9}$$

d. Physical and operational bounds on spillway gate operations,

$$\mathbf{0} \leq \underline{\mathbf{r}} \leq \mathbf{r} \leq \overline{\mathbf{r}} \leq \mathbf{1} \tag{13.4.10}$$

e. Other constraints such as operating rules, target storages, storage capacities, etc.

$$\mathbf{W}(\mathbf{r}) \leq \mathbf{0} \tag{13.4.11}$$

Unver and Mays (1990) have developed a model based upon an optimal control approach for the real-time operation of river-reservoir systems and applied it to Lake Travis of the Highland lake system on the Colorado River in Texas (see Fig. 13.1.3). The constraints of the model can be divided into two groups: the hydraulic constraints (Eq. 13.4.7) and the operational constraints (Eqs. 13.4.8–13.4.11). The hydraulic constraints are equality constraints consisting of the equations that describe the flow in the system. These are: (a) the Saint-Venant equations for all computational reaches except internal boundary reaches; (b) relationships to describe the upstream and downstream boundary conditions for the extremities; and (c) internal boundary conditions which describe flow that cannot be described by the Saint-Venant equations such as critical flow resulting from flow over a spillway or waterfall. The hydraulic constraints are solved implicitly by the simulation model, DWOPER (see Section 10.7), each time the optimizer needs these constraints evaluated, whereas the operational constraints are solved by the optimizer, such as GRG2 (Lasdon and Waren, 1983).

The operational constraints are basically greater-than or less-than type constraints that define the variable bounds, operational targets, structural limitations, and capacities. Options for the operator to set or limit the values of certain variables are also classified under this category. Bound constraints are used to impose operational or optimization-related requirements. Nonnegativity constraints on discharges are not used because discharges are allowed to take on negative values in order to be able to realistically represent the reverse flow phenomena (backwater effects) due to a rising lake or large tributary inflows into a lake or tidal condition. Nonnegativity of water surface elevations is always satisfied since the system hydraulics are solved implicitly by the simulation model, DWOPER. The lower limits on elevations and discharges can be used to indirectly impose water quality considerations, minimum required reservoir releases, and other policy requirements. The upper bounds on elevations and discharges can be used to set the maximum allowable levels (values beyond are either catastrophic or physically impossible) such as the overtopping elevations for major structures, spillway capacities, etc. Damaging elevations and/or discharges must be given to the model through the constraints, as the objective functions do not have any terms to control them.

The third model variable, gate opening, is allowed to vary between zero and one, which corresponds to zero and one hundred percent opening of the available total spillway gate areas, respectively. The bounds on gate settings are intended primarily to reflect the physical limitations on gate operations as well as to enable the operator to prescribe any portion(s) of the operation for any reservoir(s). Operational constraints other than bounds can be imposed for various purposes. The maximum allowable rates of change of gate openings, for instance, for a given reservoir, can be specified through this formulation, as a time-dependent constraint. This particular formulation may be very useful, especially for cases where sharp changes in gate operations, that is, sudden openings and closures, are not desirable or physically impossible. It is handled by setting an upper bound to the change in the percentage of gate opening from one time step to the next. This constraint can also be used to model another important aspect of gate operations for very short time intervals, that is, the gradual settings that have to be followed when opening or closing a gate. For this case, the gate cannot be opened (or closed) by more than a certain percentage during a given time interval. This can be expressed in mathematical terms as follows:

$$-r_c \leq r_i^{j+1} - r_i^j \leq r_o \qquad i \epsilon I_r \tag{13.4.15}$$

where r_c and r_o are the maximum allowable (or possible) percentages by which to open and close the gate and I_r is the set of gates that apply. This constraint can be used to model manually operated gates, for example, for all or a portion of the time intervals. The same constraint can be used, for example, to incorporate an operational rule that ties the operations of a reservoir to those of the upstream reservoir such as a multi-site constraint. Refer to the work by Unver et al. (1987) and Unver and Mays (1990) for more detail.

REFERENCES

Brazil, L. E., and G. F. Smith: "Development and application of hydrologic forecast systems," in *Computerized Decision Support Systems for Water Managers*, Labadie, J. W., L. E. Brazil, I. Corbu, and L. E. Johnson, eds., ASCE, 1989.

Can, E. K. and M. H. Houck: "Real-Time Reservoir Operations by Goal Programming," *Journal Water Resources Planning and Management*, ASCE, **110**(3), pp. 297–309, 1984.

Chow, V. T., D. R. Maidment, and L. W. Mays: *Applied Hydrology*, McGraw-Hill, Inc., New York, 1988.

Day, G. N.: "External Streamflow Forecasting Using NWSRFS," *Journal of Water Resources Planning and Management,* ASCE, vol. III, no. 2, pp. 157–170, February 1985.

EG&G Washington Analytical Services Center, Inc.: *Lower Colorado River Authority Software User's Manual*, Albuquerque, N. Mex., December 1981.

Georgakakos, K. P.: "A Generalized Stochastic Hydrometeorological Model for Flood and Flash-Flood Forecasting–1. Formulation," *Water Resources Research*, vol. 22, no. 13, pp. 2083–2095, December 1986a.

Georgakakos, K. P.: "A Generalized Stochastic Hydrometeorological Model for Flood and Flash-Flood Forecasting–2. Case Studies," *Water Resources Research*, vol. 22, no. 13, pp. 2096–2106, December 1986b.

Georgakakos, K. P. and R. L. Bras: "Real-Time, Statistically Linearized, Adaptive Flood Routing," *Water Resources Research*, vol. 18, no. 3, pp. 513–524, June 1982.

Jamieson, D. G. and J. C. Wilkinson: "River Dee Research Program, 3, A Short-Term Control Strategy for Multipurpose Reservoir Systems," *Water Resources Research*, vol. 8, pp. 911–920, 1972.

Kitanidis, P. K. and R. L. Bras: "Real-Time Forecasting With a Conceptual Hydrologic Model—1. Analysis of Uncertainty," *Water Resources Research*, vol. 16, no. 6, pp. 1025–1033, December 1980a.

Kitanidis, P. K. and R. L. Bras: "Real-Time Forecasting With a Conceptual Hydrologic Model—2. Applications and Results," *Water Resources Research*, vol. 16, no. 6, pp. 1034–1044, December 1980b.

Krzysztofowicz, R. and D. R. Davis: "A Methodology for Evaluation of Flood Forecast-Response Systems—1. Analysis and Concepts," *Water Resources Research*, vol. 19, no. 6, pp. 1423–1429, December 1983*a*.

Krzysztofowicz, R. and D. R. Davis: "A Methodology for Evaluation of Flood Forecast-Response Systems—2. Theory," *Water Resources Research*, vol. 19, no. 6, pp. 1431–1440, December 1983*b*.

Krzysztofowicz, R. and D. R. Davis: "A Methodology for Evaluation of Flood Forecast-Response Systems—3. Case Studies," *Water Resources Research*, vol. 19, no. 6, pp. 1441–1454, December 1983c.

Lasdon, L. S. and A. D. Waren: *GRG2 User's Guide*, Dept. of General Business, The University of Texas, Austin, 1983.

Melching, C. S., B. C. Yen, and H. G. Wenzel, Jr.: "Incorporation of Uncertainties in Real-Time Catchment Flood Forecasting," *Research Report 208*, Water Resources Center, University of Illinois at Urbana-Champaign, September 1987.

Nemac, J.: *Hydrological Forcasting*, Dr. Reidel Publishing Company, Boston, 1986.

Sierra/Misco, Inc.: Flood early warning system for City of Austin, Texas; Berkeley, Calif., 1986.

Sigvaldason, O. T.: "A Simulation Model for Operating a Multipurpose Multireservoir System," *Water Resources Research*, vol. 12, no. 2, 1976.

Tennessee Valley Authority: "Development of a Comprehensive TVA Water Resource Management Program," *Technical Report*, Div. of Water Cont. Plan., Tennessee Valley Authority, Knoxville, 1974.

Unver, O, L. W. Mays, and K. Lansey: "Real-Time Flood Management Model for the Highland Lakes System," *J. Water Res. Planning and Management Div.*, American Society of Civil Engineers., vol. 13, no. 5, pp. 620–638, 1987.

Unver, O. and L. W. Mays: "Model for Real-Time Optimal Flood Control Operation of a Reservoir System," *Water Resources Management*, vol. 4, Kluwer Academic Publishers, Netherlands, pp. 21–46, 1990.

U.S. Army Corps of Engineers, Hydrologic Engineering Center, "HEC-5C, A Simulation Model for System Formulation and Evaluation," Davis, Calif., 1973.

U.S. Army Corps of Engineers, Hydrologic Engineering Center, "HEC-5, Simulation of Flood Control and Conservation Systems," Davis, Calif., 1979.

U.S. Army Corps of Engineers, Hydrologic Engineering Center, "Real-Time Floodcasting and Reservoir Control for the Kanawha," Special Projects Memo. no. 83–10, Davis, Calif., December 1983.

Wasimi, S. A. and P. K. Kitanidis: "Real-Time Forecasting and Daily Operation of a Multireservoir System During Floods by Linear Quadratic Gaussian Control," *Water Resources Research*, vol. 19, pp. 1511–1522, 1983.

Windsor, J. S.: "Optimization Model for the Operation of Flood Control Systems," *Water Resources Research*, vol. 9, HY5, pp. 1219–1226, 1973.

Yazicigil, H.: "Optimal Operation of a Reservoir System Using Forecasts," *Ph.D. Dissertation*, Purdue University, West Lafayette, Indiana, 1980.

PROBLEMS

13.3.1 Develop a dynamic programming model for the flood control operation of the three reservoir system in Fig. 13.P.1. For this formulation consider only damages in reaches A, B, and C, assuming no damages in reach D. How many state variables are required for this problem?

13.3.2 Develop the DP model for the three reservoir system (Fig. 13.P.1) in Problem 13.3.1 and consider the damages in reaches A, B, C, and D.

13.4.1 Develop a linear programming model for the flood operation of the two reservoir system shown in Fig. 13.P.2. Use the penalty functions shown in Fig. 13.P.3 for the two control points.

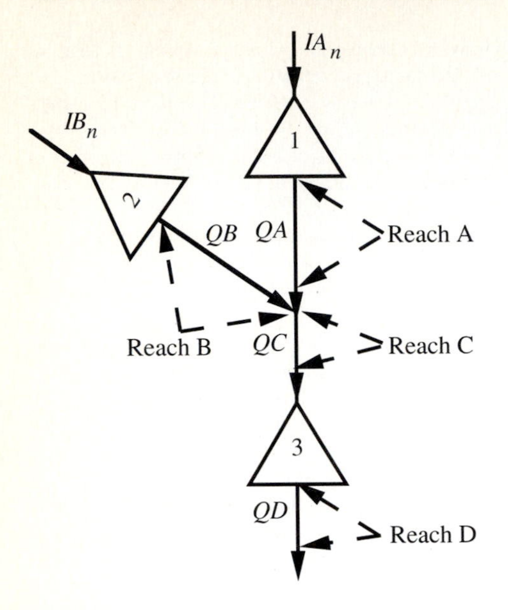

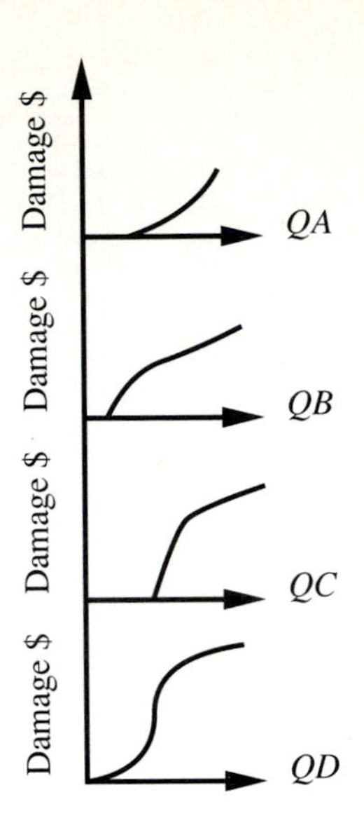

FIGURE 13.P.1
Reservoir system for Problems 13.3.1 and 13.3.2.

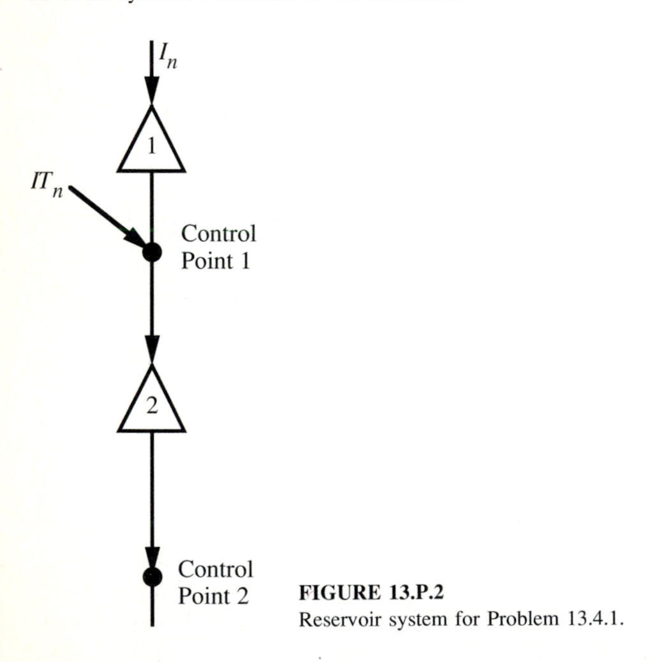

FIGURE 13.P.2
Reservoir system for Problem 13.4.1.

13.4.2 Develop the linear programming model for the single reservoir system shown in Fig. 13.4.1 (Example 13.4.1). Use the discharge penalty function in Fig. 13.P.3 to develop the model.

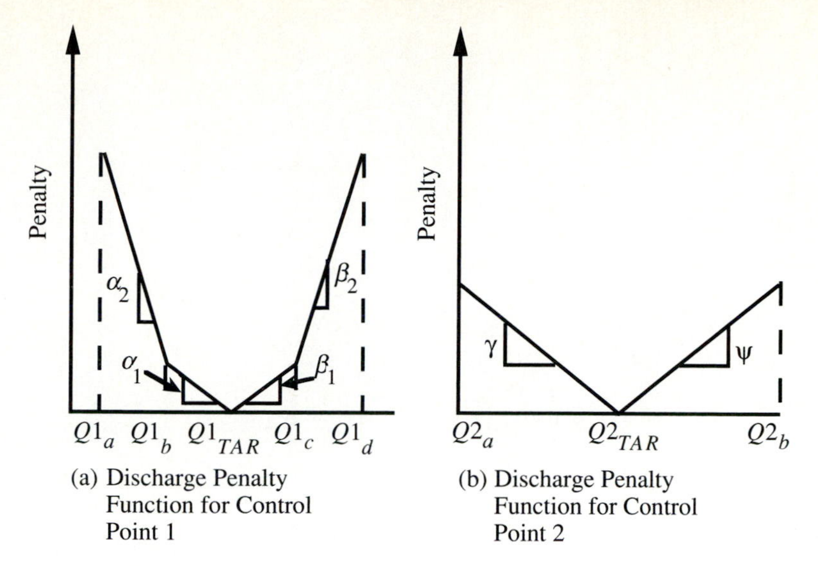

FIGURE 13.P.3
Discharge penalty function for Problems 13.4.1 and 13.4.2.

AUTHOR INDEX

INDEX